I0880719

Building Resilience Through Multi-Hazard Early Warning Systems

Jaime Santos-Reyes
https://orcid.org/0000-0002-3758-9862
Instituto Politecnico Nacional, Mexico

IGI Global
Scientific Publishing
Publishing Tomorrow's Research Today

Vice President of Editorial	Melissa Wagner
Director of Acquisitions	Mikaela Felty
Director of Book Development	Jocelynn Hessler
Production Manager	Mike Brehm
Cover Design	Jose Rosado

Published in the United States of America by
IGI Global Scientific Publishing
701 East Chocolate Avenue
Hershey, PA, 17033, USA
Tel: 717-533-8845 | Fax: 717-533-7115
Website: https://www.igi-global.com E-mail: cust@igi-global.com

Copyright © 2026 by IGI Global Scientific Publishing. All rights reserved. No part of this publication may be reproduced, stored or distributed in any form or by any means, electronic or mechanical, including photocopying, without written permission from the publisher. Use of this publication to train generative artificial intelligence (AI) technologies is expressly prohibited. The publisher reserves all rights to license its use for generative AI training and machine learning model development.

Product or company names used in this set are for identification purposes only. Inclusion of the names of the products or companies does not indicate a claim of ownership by IGI Global Scientific Publishing of the trademark or registered trademark.

Library of Congress Cataloging-in-Publication Data

Names: Santos-Reyes, Jaime (Lecturer at National Polytechnic Institute, Mexico) editor
Title: Building resilience through multi-hazard early warning systems / edited by: Jaime Santos-Reyes.
Description: Hershey PA : IGI Global Scientific Publishing, [2026] | Includes bibliographical references. | Summary: "This book provides a multidisciplinary exploration of early warning systems for multi-hazard risk management. It examines key components across diverse hazards, highlighting global best practices that enhance community preparedness, resilience, and policy development for effective disaster risk reduction"-- Provided by publisher.
Identifiers: LCCN 2025054511 (print) | LCCN 2025054512 (ebook) | ISBN 9798337373454 hardcover | ISBN 9798337373461 paperback | ISBN 9798337373478 ebook
Subjects: LCSH: Natural disaster warning systems | Natural disasters--Forecasting | Emergency management | Environmental economics | Resilience (Ecology)
Classification: LCC GB5030 .B85 2026 (print) | LCC GB5030 (ebook) | DDC 363.34/72--dc23/eng/20260323
LC record available at https://lccn.loc.gov/2025054511
LC ebook record available at https://lccn.loc.gov/2025054512

British Cataloguing in Publication Data
A Cataloguing in Publication record for this book is available from the British Library.

All work contributed to this book is new, previously-unpublished material.
The views expressed in this book are those of the authors, but not necessarily of the publisher.
This book contains information sourced from authentic and highly regarded references, with reasonable efforts made to ensure the reliability of the data and information presented. The authors, editors, and publisher believe the information in this book to be accurate and true as of the date of publication. Every effort has been made to trace and credit the copyright holders of all materials included. However, the authors, editors, and publisher cannot assume responsibility for the validity of all materials or the consequences of their use. Should any copyright material be found unacknowledged, please inform the publisher so that corrections may be made in future reprints.

Table of Contents

Detailed Table of Contents

AI technologies in mitigating the impacts of natural and human-induced disasters. It explores the historical evolution of AI applications in disaster management and also addresses ethical considerations. Moreover, the incorporation of AI in social media analytics and natural language processing for emergency communication highlights its role in improving information dissemination and community engagement during emergencies. The review highlights the use of AI in disaster management, ultimately building more resilient and inclusive communities worldwide. It also covers related terms and definitions for AI applications in disaster management.

The practice of forecasting the meteorological conditions at a precise location and forthcoming time has been a quest for ages. From archaic society to contemporary meteorological models scraping varied amounts of data, the pursuit of precise estimates is witnessing a continual rise. Intensifying challenges of climate change bolster the need for forecasting technologies to safeguard the economies and communities against the unforeseen impacts. With the conventional methods having their confines, the predictive models nowadays that integrate AI offer better scalability and interpretability. The article delves into the groundbreaking role of AI in weather predictions, as well as exploring its benefits, applications, future potential, and challenges.

The chapter explores the emergence of Cognitive-Adaptive Early Warning Systems (CA-EWS) that integrate Artificial Intelligence and cognitive science to enhance multi-hazard preparedness. Traditional Early Warning Systems emphasize data precision but often neglect how humans perceive and act on alerts. The proposed framework combines machine learning, behavioral analytics, and perception modeling to create adaptive, human-centered warning communication. By leveraging real-time feedback from social media sentiment, mobility data, and community responses, CA-EWS dynamically refines alert clarity, timing, and personalization. Case studies from Japan's AI-enhanced tsunami warning and COVID-19 pandemic systems reveal measurable improvements in public responsiveness, trust, and compliance. The study concludes that embedding cognitive intelligence within EWS transforms reactive systems into proactive, empathetic, and resilient disaster communication ecosystems.

Integrated multi-hazard early warning systems (MHEWS) are critical for reducing disaster risks amid increasing climate uncertainty and complex hazard interactions. Advances in technologies such as remote sensing, artificial intelligence, and sensor networks have improved hazard detection and forecasting, but system effectiveness also depends on governance, institutional coordination, and community response. This chapter examines MHEWS as socio-technical systems that integrate technological innovation with governance and behavioral dynamics. It analyzes the transition from single-hazard to integrated approaches and explores key implementation challenges, including institutional fragmentation, infrastructure gaps, data governance issues, and public trust. Drawing on global case studies, the chapter highlights practical lessons on system performance across different contexts. The findings emphasize that strengthening MHEWS requires a balanced approach combining technology, governance capacity, and inclusive communication to enhance resilience and support effective disaster preparedness.

Urban heat island (UHI) effects have intensified with rapid urbanisation, elevating temperatures by 5-12°C above rural areas and imposing significant health, environmental, and economic burdens on urban populations projected to reach 68% globally by 2050. Importance: Effective thermal management strategies are critical for sustainable urban development, with cooling energy costs rising 10-120% during hot months and productivity losses potentially reaching 2.6% of global GDP by 2100. While individual green and blue infrastructure interventions demonstrate cooling benefits, a comprehensive understanding of their synergistic integration remains limited. Critical knowledge gaps exist in optimal spatial configurations, implementation strategies, and governance frameworks for integrated Blue-Green systems, particularly regarding equity considerations and unintended consequences such as green gentrification.

South Africa's warming climate is amplifying multi hazard events that overwhelm critical infrastructure. This chapter assesses the performance of South Africa's multi-hazard early warning systems (MHEWS) and offers evidence-based measures to boost climate-resilient infrastructure. A PRISMA guided systematic review of peer reviewed studies and grey literature (2015-2025). Findings show fragmented coordination, inadequate multilingualism, and severe financing gaps. The chapter recommends: (a) embedding real time SAWS dashboards in municipal IDP and budgets; (b) allocating a "climate risk" line item; (c) adopting a three tier (national provincial municipal) coordination protocol; (d) deploying multilingual/voice based alert platforms; (e) creating a ring fenced MHEWS maintenance fund financed by a modest municipal levy; and (f) forming community "Flood Watch" committees. Implementing these actions will enhance the robustness, redundancy, timeliness, and resourcefulness of warnings, reduce disaster losses, and align practice with the DMA and the Climate Change Act.

Traditional seismic rehabilitation, reliant on static designs and isolated hazard treatment, proves inadequate against today's overlapping natural hazards and sustainability imperatives. This chapter introduces a paradigm integrating Multi-Hazard Early Warning Systems (MHEWS) with AI-enhanced Structural Health Monitoring (SHM) for dynamic, lifecycle-oriented resilience. By fusing MHEWS data streams with real-time sensor data from retrofitted structures, artificial intelligence serves as the critical nexus. Advanced machine learning models analyze this multi-modal data for real-time vulnerability assessment and predictive damage simulation under compound threats, enabling context-aware mitigation. This anticipatory framework transcends conventional code-based approaches, embedding proactive intelligence into infrastructure systems. Ultimately, we reposition seismic rehabilitation as an adaptive component within a broader hazard-response ecosystem—enhancing public safety, optimizing resources, and advancing sustainable development in an era of compounded environmental challenges.

Modern societies face disruptions such as climate change, pandemics, cyberattacks, geopolitical instability, economic fragility, and infrastructure failure. Early Warning Systems (EWS) must not be seen only as technical tools. They function as governance frameworks reliant on institutional capacity, ethical technology use, and leadership. This chapter examines the dual role of leadership in integrating disruption management and AI-enabled EWS in public-sector governance, drawing on legislation, global frameworks, and municipal reports. It contends that although technology recognizes risks, leadership dictates whether this awareness translates into resilience.Findings show that, despite South Africa's strong legal and policy frameworks, resilience outcomes vary. These differences relate to leadership quality, coordination, public trust, and governance of emerging technologies. The chapter shows that resilience depends not on technology alone, but on effective leadership and governance.

This chapter examines the critical role of education in enhancing the effectiveness of multi-hazard early warning systems (MHEWS) for disaster risk reduction. MHEWS integrate detection, assessment, and communication of diverse hazards, providing timely information to enable proactive community responses. However, technical, institutional, and social challenges often impede their full potential. Education emerges as a pivotal factor in bridging the gap between early warning dissemination and effective community action by fostering hazard awareness, shaping risk perceptions, and promoting preparedness behaviors. The chapter explores various educational types including formal, non-formal, and informal—and targets diverse audiences such as children, adults, and vulnerable populations. It discusses strategic approaches to curriculum development, training programs for key stakeholders, and public awareness campaigns coupled with community engagement.

Multi-hazard early warning systems (MHEWS) are critical instruments for reducing disaster-related losses in an era marked by intensifying seismic and hydrometeorological risks. Despite advances in sensor technologies and predictive analytics, the effectiveness of early warning systems depends largely on how warnings are communicated, interpreted, and acted upon by diverse populations. This chapter examines the transition from sensor-based detection to actionable public response, focusing on emerging technologies and human-centred dissemination strategies for earthquake and flash-flood warnings. Integrating perspectives from disaster risk reduction, communication theory, and socio-technical systems, the chapter highlights gaps between technological capability and social uptake. It proposes an integrative framework that aligns real-time sensing, decision support, and inclusive communication to strengthen resilience across vulnerable communities.

Droughts have considerable affectations to economic, environmental, and water shortages in urban environments. Traditionally, drought management has been reactive rather than proactive in mitigating the impact of drought hazard. The chapter presents a drought disaster management system (DDMS) where a drought early warning system (DEWS) plays a key role in decision-making to mitigate the impact of droughts for the case of Mexico City. The proactive approach is based on risk and vulnerability assessment. As part of the function of DEWS, an illustrative analysis of rainfall (RF) prediction has been conducted by employing fuzzy logic theory. The results produced an acceptable fuzzy model and have the potential not only to predict RF but also droughts. Furthermore, the results highlight the importance of DEWS in decision-making. Overall, to be able to mitigate the impact of droughts in Mexico City, it becomes necessary to have a coherent DDSM system. More generally, key decision-makers of the megalopolis should design and implement a multi-hazard early warning system.

Preface

INTRODUCTION

The accelerating complexity of the modern risk landscape demands a shift in how we understand, anticipate, and respond to hazards. No longer confined to singular events or isolated domains, today's threats span hydrometeorological extremes, pandemics, technological failures, and even extraterrestrial phenomena. These hazards increasingly interact, overlap, and compound—often with profound consequences for communities worldwide. Climate change, widely recognized as a critical driver, is intensifying the frequency, duration, and severity of extreme events, underscoring the urgency of strengthening preparedness and resilience.

At the heart of effective disaster risk reduction lies the concept of early warning systems (EWS). As defined by the United Nations Office for Disaster Risk Reduction, an early warning system encompasses "the set of capacities needed to generate and disseminate timely and meaningful warning information, enabling individuals, communities, and organizations at risk to prepare and act appropriately to reduce harm or loss." This definition highlights not only the technical dimensions of forecasting and detection but also the human dimensions of communication, trust, and response. A timely earthquake alert, for instance, can mean the difference between safety and catastrophe in seismically active regions.

This edited volume adopts an interdisciplinary and integrated perspective on multi-hazard early warning systems. It explores the four foundational components of EWS: risk knowledge; monitoring and warning; dissemination and communication; and response capability. By examining these elements across a diverse spectrum of hazards—including meteorological, hydrological, geophysical, environmental, biological, technological, and societal risks—the book provides a comprehensive understanding of both the challenges and opportunities inherent in designing effective systems.

Importantly, the chapters in this volume do not treat early warning systems as isolated technical solutions. Rather, they situate EWS within broader risk governance and disaster management frameworks, emphasizing their role as end-to-end (E2E) systems. Attention is given to both qualitative and quantitative approaches, as well as to public perception, education, and community engagement—factors that are often ant in determining whether warnings translate into meaningful action.

The contributions assembled here also highlight global examples where early warning systems have demonstrably enhanced preparedness, reduced risk, and strengthened resilience. These case studies offer valuable insights for policymakers and practitioners seeking to design systems that are not only technologically robust but also socially inclusive and operationally effective.

This book is intended for a wide and interdisciplinary audience: academics, researchers, and graduate students across fields such as engineering, environmental science, atmospheric studies, hydrology, seismology, volcanology, economics, and artificial intelligence. It will also serve as a practical resource for policymakers, urban planners, emergency responders, civil protection agencies, insurers, and humanitarian organizations. Ultimately, it speaks to all those engaged in the vital work of building resilient societies.

As editor, it is my hope that this volume contributes to advancing both the science and practice of early warning systems, and that it supports the development of policies and innovations capable of addressing the multi-hazard challenges of our time.

ORGANIZATION OF THE BOOK

As editor, I have sought to curate contributions that collectively illuminate both the breadth and depth of contemporary thinking on multi-hazard early warning systems (MHEWS). Each chapter offers a distinct yet interconnected perspective, reflecting the inherently interdisciplinary and evolving nature of this field.

Chapter 1 sets the foundation by presenting a comprehensive overview of advances in MHEWS, situating them within the increasingly complex global risk landscape shaped by climate change, urbanization, and technological interdependence. The chapter critically examines a wide spectrum of hazards and highlights the transformative role of emerging technologies such as artificial intelligence, the Internet of Things, and big data analytics. At the same time, it underscores governance challenges and the importance of community preparedness, drawing on global frameworks to illustrate best practices and persistent gaps.

Chapter 2 builds on this technological trajectory by offering an expansive exploration of artificial intelligence in disaster management. It traces the evolution of AI applications and demonstrates how these tools are reshaping prediction, response, and recovery processes. The discussion extends beyond technical capabilities to address ethical considerations and the growing importance of AI-driven communication tools, particularly in enhancing public engagement and inclusivity during emergencies.

Chapter 3 narrows the focus to weather forecasting, examining how AI-driven models are redefining predictive accuracy and scalability. By contrasting traditional forecasting methods with contemporary data-driven approaches, the chapter highlights both the promise and limitations of AI, emphasizing its critical role in addressing climate-induced uncertainties and safeguarding vulnerable economies and communities.

Chapter 4 introduces an important paradigm shift through cognitive-adaptive early warning systems, which integrate human perception into technological design. Moving beyond purely data-centric approaches, this chapter demonstrates how behavioral analytics and real-time feedback can significantly improve public responsiveness and trust. The case studies presented illustrate how embedding human cognition into warning systems can transform them into more empathetic and effective tools for disaster communication.

Chapter 5 further reinforces the need for integration by framing MHEWS as socio-technical systems that require a careful balance between technological innovation and governance capacity. It critically examines the transition from single-hazard to multi-hazard frameworks, identifying institutional fragmentation, data governance issues, and trust deficits as key barriers. The chapter provides practical insights drawn from global experiences, emphasizing that resilience depends as much on coordination and inclusivity as on technological advancement.

Chapter 6 broadens the discussion to urban resilience, focusing on the growing challenge of urban heat islands and the role of blue-green infrastructure in climate adaptation. While not traditionally framed within early warning systems, the chapter highlights the importance of anticipatory planning and integrated design strategies. It identifies critical knowledge gaps related to implementation, governance, and equity, thereby linking environmental management with broader resilience-building efforts.

Chapter 7 offers a regionally grounded analysis of MHEWS in South Africa, revealing the practical challenges of implementation in a context marked by climate stress and infrastructure vulnerability. Through a systematic review, the chapter identifies coordination failures, funding constraints, and communication barriers, while proposing actionable strategies to strengthen system performance, enhance inclusivity, and align policy with practice.

Chapter 8 presents a forward-looking integration of MHEWS with AI-enhanced structural health monitoring systems. By linking real-time hazard data with infrastructure performance, the chapter proposes a dynamic and predictive approach to resilience. This integration enables proactive decision-making and redefines traditional rehabilitation practices, embedding them within a broader, adaptive hazard-response ecosystem.

Chapter 9 shifts the focus to leadership and governance, arguing that the effectiveness of early warning systems ultimately depends on institutional capacity and decision-making processes. The chapter highlights the dual role of leadership in enabling or constraining resilience, demonstrating that even the most advanced technologies require strong governance frameworks, coordination, and public trust to be effective.

Chapter 10 underscores the central role of education in bridging the gap between warning dissemination and community response. It explores how formal, non-formal, and informal educational strategies can enhance risk awareness, shape behavior, and empower diverse populations to act effectively in the face of multi-hazard threats.

Chapter 11 further develops the human-centred perspective by examining how warnings are communicated and acted upon, particularly in the context of earthquakes and flash floods. The chapter highlights the disconnect between technological capability and social uptake, proposing integrative frameworks that align sensing technologies with inclusive communication strategies to ensure actionable outcomes.

Finally, Chapter 12 addresses drought as a slow-onset hazard, emphasizing the importance of early warning systems in proactive decision-making. Through a case study approach, it demonstrates how predictive models, including fuzzy logic, can support risk assessment and mitigation strategies. The chapter reinforces the need for integrated systems that move beyond reactive responses toward anticipatory governance.

Taken together, these chapters reflect the central theme of this volume: that effective multi-hazard early warning systems are not merely technological constructs, but complex, adaptive systems that integrate science, governance, and society.

CONCLUSION

In bringing this volume together, it has become increasingly clear that the future of multi-hazard early warning systems lies not in isolated advances, but in integration—across disciplines, technologies, institutions, and communities. The chapters presented here collectively demonstrate that while significant progress has been made in detection, forecasting, and communication, the true measure of effectiveness rests in whether these systems lead to timely, informed, and equitable action.

A recurring insight throughout this book is that early warning systems are as much social constructs as they are technical infrastructures. Advances in artificial intelligence, sensor networks, and predictive analytics are undoubtedly reshaping the field; however, without strong governance, inclusive communication, public trust, and informed leadership, these innovations risk falling short of their potential. The interplay between human behavior, institutional capacity, and technological capability emerges as the defining challenge—and opportunity—of modern early warning systems.

Equally important is the recognition that resilience must be proactive rather than reactive. Whether addressing rapid-onset hazards such as earthquakes and floods or slow-onset risks like drought and urban heat, the contributions in this volume emphasize anticipatory approaches grounded in risk knowledge, continuous monitoring, and adaptive response. The shift toward end-to-end systems—linking data to decision-making and action—represents a critical step forward in reducing disaster losses and strengthening societal resilience.

This volume also highlights the importance of context. Solutions must be adaptable to diverse geographic, socio-economic, and cultural settings. What works in one region may not translate directly to another without careful consideration of local governance structures, community needs, and resource constraints. As such, the path forward requires not only innovation but also collaboration—across sectors, borders, and knowledge systems.

Looking ahead, the integration of emerging technologies with human-centered design, ethical frameworks, and participatory governance will define the next generation of early warning systems. Continued investment in education, capacity-building, and policy alignment will be essential to ensure that these systems are accessible, trusted, and effective for all.

As editor, I hope this volume serves not only as a reference but also as a catalyst—for research, for policy development, and for practical implementation. The challenges posed by multi-hazard risks are complex and evolving, but so too are the tools and frameworks available to address them. By advancing a holistic understanding of early warning systems, this book contributes to the broader goal of building safer, more resilient, and more adaptive societies.

Chapter 1
Advances in Multi-Hazzard Early Warning Systems: Emerging Technologies, Challenges, and Future Perspectives

Khasim Saheb Shaik
http://orcid.org/0000-0002-6458-8494
Sreenidhi Institute of Science and Technology, India

Sreenivasarao Devavarapu
Sreenidhi Institute of Science and Technology, India

Surya Kiran Akku
http://orcid.org/0009-0007-5604-2946
Sreenidhi Institute of Science and Technology, India

ABSTRACT

Multi-hazard early warning systems (MHEWS) play a critical role in minimizing loss of life, property damage, and socioeconomic disruption caused by natural and anthropogenic hazards. The convergence of climate change, rapid urbanization, technological dependency, and environmental degradation has intensified the complexity and interconnection of hazards faced globally. This chapter presents a comprehensive analysis of MHEWS, examining approaches related to meteorological, hydrological, extraterrestrial, geohazard, environmental, biological, chemical, technological, and societal threats. It explores the integration of artificial intelligence (AI), the Internet of Things (IoT), and big data analytics in forecasting, detection, and communication

DOI: 10.4018/979-8-3373-7345-4.ch001

ight © 2026, IGI Global Scientific Publishing. Copying or distributing in print or electronic forms without written permission of IGI Global Scientific Publishing is prohibited. Use of this chapter to n generative artificial intelligence (AI) technologies is expressly prohibited. The publisher reserves all rights to license its use for generative AI training and machine learning model development.

systems. Furthermore, it highlights governance, community preparedness, and challenges in implementing effective multi-hazard systems. Case studies and examples from global frameworks such as the Sendai Framework for Disaster Risk Reduction are discussed to demonstrate best practices and lessons learned.

INTRODUCTION

Disasters triggered by natural and anthropogenic hazards are among the most significant threats to global security and sustainable development. In recent decades, climate change, unplanned urbanization, and technological dependency have increased the exposure and vulnerability of human populations to multi-hazard risks (Amaratunga et al., 2021), (Rokhideh et al., 2025). According to the United Nations Office for Disaster Risk Reduction (UNDRR), disasters have affected over 4 billion people in the past two decades, with economic losses exceeding USD 3 trillion.

The increasing frequency of compound and cascading disasters—where one hazard triggers or amplifies another—underscores the need for a robust and integrated approach to early warning. Multi-hazard early warning systems (MHEWS) are designed to detect, monitor, and communicate potential threats from multiple hazards simultaneously (Budimir et al., 2025), (Ayuba et al., 2025). They aim to ensure that at-risk communities receive timely, actionable, and understandable information to prevent or reduce disaster impacts.

Traditional early warning systems have often been hazard-specific, focusing on isolated threats such as floods, cyclones, or earthquakes. However, this fragmented approach fails to address the interconnected nature of contemporary hazards (Basher, 2023). For instance, an earthquake can trigger landslides, tsunamis, or dam failures, while tropical cyclones can result in floods, storm surges, and disease outbreaks. Therefore, MHEWS represent a paradigm shift toward comprehensive risk management and resilience-building across all sectors.

This chapter presents a detailed exploration of the conceptual foundations, components, and practical implementation of MHEWS. It discusses the integration of advanced technologies—AI, machine learning (ML), IoT, remote sensing, and big data analytics—into hazard prediction and monitoring frameworks (Razavi-Termeh et al., 2025). The chapter also addresses governance, social dimensions, communication mechanisms, and capacity-building initiatives crucial for ensuring system effectiveness and sustainability.

CONCEPTUAL FRAMEWORK OF MULTI-HAZARD EARLY WARNING SYSTEMS

Early warning systems (EWS) are essential tools for disaster risk reduction (DRR). According to the UNDRR, an effective EWS consists of four interconnected components: risk knowledge, monitoring and warning services, dissemination and communication, and response capability. MHEWS expand this model by integrating multiple hazards and their interdependencies within a unified framework.

A Multi-Hazard Early Warning System (MHEWS) is an integrated system that detects, monitors, and evaluates multiple hazard types and disseminates timely, actionable information to minimize harm to life, property, and the environment. The fundamental principles include comprehensiveness, integration, accessibility, and sustainability.

Multi-Hazard Interactions

Hazards rarely occur in isolation. Compound hazards refer to multiple events happening simultaneously. For example cyclone and heavy rainfall, whereas cascading hazards occur when one event triggers another. For example earthquake → tsunami → nuclear failure (Kelman et al., 2025), (Xu et al., 2026). Understanding these interconnections is essential for accurate risk modeling and for developing integrated early warning systems.

The Sendai Framework and Global Policy Context

The Sendai Framework for Disaster Risk Reduction (Rokhideh et al., 2025) emphasizes the importance of "end-to-end" and "people-centered" early warning systems. It calls for strengthening MHEWS as part of national disaster management policies. The framework's priority areas—understanding disaster risk, strengthening governance, investing in resilience, and enhancing preparedness—form the policy foundation for MHEWS development globally.

CLASSIFICATION OF MULTI-HAZARDS

Meteorological and Hydrological Hazards

Meteorological hazards are dangerous events caused by extreme weather and atmospheric conditions. These include phenomena such as cyclones, thunderstorms, heat waves, cold waves, heavy rainfall, lightning, hailstorms, and droughts. They occur due to changes in temperature, air pressure, humidity, and wind patterns, and can cause damage to life, property, agriculture, and the environment.

Hydrological hazards are disasters related to the movement, distribution, and accumulation of water on the Earth's surface. These include floods, flash floods, river overflow, storm surges, landslides triggered by heavy rain, and dam failures. They are often influenced by prolonged rainfall, snowmelt, or poor water management, and can lead to severe loss of life, infrastructure damage, and environmental degradation.

Meteorological and hydrological hazards constitute the most frequent and destructive category of disasters, including coastal and fluvial floods, flash floods, storm surges, cyclones, droughts, and heatwaves. The increasing intensity of these events is closely linked to global climate change (Nunes, 2025).

Floods are the most common natural disasters worldwide. Early warning relies on rainfall forecasting, river basin monitoring, and real-time hydrological models. Integration of remote sensing data and IoT-enabled sensors enhances flood detection accuracy and timeliness.

Droughts occur when there is very little rainfall over a long period, leading to water scarcity, crop failure, and food shortages. Drought monitoring involves meteorological indices such as the Standardized Precipitation Index and Palmer Drought Severity Index, combined with satellite-derived soil moisture and vegetation health data. Agricultural drought forecasting systems can integrate AI-based anomaly detection to predict crop yield impacts.

Cyclones are powerful storm systems with strong winds and heavy rain that form over warm oceans, causing flooding, coastal erosion, and destruction of infrastructure. Thunderstorms are short-duration weather events with heavy rain, thunder, lightning, and strong winds, often damaging buildings, trees, and power lines. Cyclone early warning involves atmospheric pressure analysis, sea-surface temperature monitoring, and numerical weather prediction (NWP) models. AI algorithms can further improve trajectory predictions, providing extended preparation time.

Heatwaves and cold waves represent extreme temperature conditions. Heat waves are prolonged periods of high temperatures that can cause heat strokes, dehydration, and increased mortality, particularly among vulnerable populations. Cold waves involve temperatures dropping significantly below normal levels, affecting human health, livestock, and agriculture. Heatwaves have become more frequent and intense due to global warming. Early warning systems integrate temperature forecasts with public health systems to reduce heat-related mortality.

Heavy rainfall occurs when large amounts of precipitation fall within a short duration, often leading to floods, landslides, and infrastructure damage. Lightning is a sudden electrical discharge during storms that can cause fatalities, fires, and damage to electrical systems. Hailstorms involve ice pellets falling from clouds, damaging crops, vehicles, and buildings.

Tornadoes are violently rotating columns of air extending from clouds to the ground, causing severe localized destruction. Fog reduces visibility and increases accident risks, while dust storms carry fine particles that affect human health and transportation systems.

Extraterrestrial Hazards

Extraterrestrial hazards are natural threats that occur outside the Earth's atmosphere and can affect the planet, its environment, and living organisms. These hazards arise from cosmic or space-based phenomena and are not influenced by human activities. Although they occur less frequently compared to terrestrial hazards, their impacts can be widespread and highly destructive.

Major extraterrestrial hazards include asteroid and meteorite impacts, comets, solar flares, solar storms, cosmic radiation, and gamma-ray bursts. Asteroid or meteorite impacts can cause massive explosions, shock waves, fires, tsunamis, and long-term climate changes. Solar flares and geomagnetic storms release large amounts of energy and charged particles that can disrupt satellite communications, navigation systems, power grids, and space missions. Cosmic radiation and high-energy particles pose serious risks to astronauts and may affect Earth's atmosphere and electronic systems.

Extraterrestrial hazards are monitored using space telescopes, satellites, and ground-based observatories to predict possible threats and reduce their impact. Understanding these hazards is important for disaster preparedness, space exploration safety, and protecting modern technological systems that depend on satellites and electrical networks.

Solar flares and coronal mass ejections (CMEs) can disrupt satellite operations, GPS, and power transmission. Space weather monitoring systems use magnetometer readings and space-based observatories to issue alerts. Geomagnetic storms occur when charged particles from the Sun interact with Earth's magnetic field, potentially damaging power grids, causing blackouts, and interfering with communication networks. Asteroid and Meteorite Impacts.

Asteroids are rocky objects that orbit the Sun, and large impacts can cause explosions, earthquakes, wildfires, and tsunamis. Such events may also release dust into the atmosphere, blocking sunlight and causing global cooling. Meteorites are fragments that survive atmospheric entry and strike the Earth's surface, causing localized to widespread damage. Global coordination through space agencies supports detection and trajectory analysis of near-Earth objects.

Comets are icy bodies that travel through space in elongated orbits. If a comet collides with Earth, it can cause effects similar to asteroid impacts, including explosions and climate disturbances. Though rare, comet impacts are considered highly dangerous due to their high speed and energy Comets, which are icy bodies traveling in elongated orbits, can cause effects similar to asteroid impacts if they collide with Earth. Solar flares emit intense radiation that can disrupt communication and navigation systems. Cosmic radiation consists of high-energy particles that pose risks to astronauts and electronic systems. Gamma-ray bursts are extremely powerful cosmic explosions that could damage the ozone layer and increase radiation levels if directed toward Earth.

Geo hazards

Geo hazards are natural hazards originating from geological processes within the Earth or on its surface. These hazards are caused by movements and changes in the Earth's crust, rocks, soil, and internal energy, leading to significant damage to life, infrastructure, and the environment.

Major geo hazards include earthquakes, volcanic eruptions, landslides, tsunamis, sinkholes, and ground subsidence. Earthquakes result from the sudden release of energy due to tectonic plate movements, causing ground shaking and structural collapse. Volcanic eruptions occur when magma rises to the surface, releasing lava, ash, and gases that can destroy settlements and affect climate. Landslides involve the downward movement of soil and rock, often triggered by rainfall or seismic activity. Tsunamis are large ocean waves generated mainly by undersea earthquakes or volcanic eruptions, leading to extensive coastal flooding.

Geo hazards are influenced by natural factors such as tectonic activity, slope stability, and groundwater conditions, as well as human activities like mining and deforestation. Proper geological studies, hazard mapping, early warning systems, and land-use planning are essential for reducing their impacts (Bompotas et al., 2022).

Earthquake early warning systems use seismic sensor networks to detect primary waves (P-waves) before destructive secondary waves (S-waves) arrive. Advanced systems can provide critical seconds of warning, enabling protective actions. Earthquakes can destroy infrastructure and may also trigger secondary hazards such as landslides and tsunamis.

Tsunamis are generated by undersea earthquakes, volcanic eruptions, or submarine landslides. Warning systems rely on seismic data, sea-level sensors, and deep-ocean pressure recorders for early detection. Integrated regional systems enhance timely alerts and evacuation planning.

Landslides involve the downward movement of rock, soil, or debris, often triggered by rainfall, earthquakes, or human activities. Monitoring techniques include geotechnical sensors, rainfall thresholds, and satellite-based observations. Machine learning models can improve landslide risk prediction using terrain and environmental data.

Volcanic eruptions release lava, ash, and gases, causing destruction and environmental impacts. Sinkholes form when underground rock dissolves or collapses, posing risks to infrastructure. Ground subsidence occurs due to underground material movement, often linked to human activities such as groundwater extraction.

Avalanches are rapid flows of snow, ice, and debris down slopes, triggered by natural factors such as snowfall and temperature variations or human activities. They can bury settlements and disrupt transportation networks.

Environmental Hazards

Environmental hazards stem from the degradation of natural ecosystems due to human activity, climate change, or natural processes. These include air pollution, soil erosion, deforestation, land degradation, and biodiversity loss. Environmental hazards often exacerbate the effects of other disasters, making them a critical component of multi-hazard early warning systems (MHEWS).

Environmental hazards are natural or human-induced conditions that pose risks to human health, ecosystems, and infrastructure. These hazards arise due to imbalances in the environment and may be caused by physical, chemical, biological, or social factors. They can occur suddenly or develop gradually, with impacts ranging from local to global scales.

Environmental hazards include natural events such as floods, droughts, cyclones, earthquakes, wildfires, and volcanic eruptions, as well as anthropogenic hazards like air, water, and soil pollution, deforestation, climate change, industrial accidents, and improper waste disposal. Natural hazards disrupt ecosystems, while human-induced hazards degrade environmental quality and threaten public health.

Air pollution releases harmful gases and particulate matter that cause respiratory diseases and contribute to climate change. Water pollution contaminates rivers, lakes, and groundwater, affecting drinking water and aquatic ecosystems. Soil pollution reduces agricultural productivity and introduces toxins into the food chain. Climate change, driven by greenhouse gas emissions, increases the frequency and intensity of extreme weather events and biodiversity loss.

Environmental hazards can be mitigated through sustainable development, environmental protection policies, pollution control technologies, and public awareness. Understanding these hazards is essential for protecting human health, conserving ecosystems, and ensuring sustainable development.

Ambient air pollution is a leading cause of global mortality. Early warning systems rely on continuous monitoring of pollutants such as PM2.5, PM10, SO_2, NO_X,

O_3, and VOCs. Integrated forecasting systems combine satellite-based observations and ground monitoring stations to provide alerts when pollution exceeds safe limits. AI-based predictive analytics enhance air quality modeling by correlating emissions with meteorological data (Kumar et al., 2024), (Zhao et al., 2021). Public-oriented communication platforms support dissemination of warnings and awareness.

Air pollution occurs due to emissions from vehicles, industries, and fossil fuel combustion, leading to respiratory diseases, reduced visibility, and climate change.

Land and soil degradation reduce agricultural productivity and increase vulnerability to floods and droughts. Early warning systems integrate satellite remote sensing, soil moisture indices, and vegetation monitoring. These systems support land-use planning, reforestation, and sustainable agriculture. Soil pollution, caused by chemicals and industrial waste, affects crop production and introduces toxins into the food chain.

Wildfires are increasing due to rising temperatures and prolonged dry conditions linked to climate change. Modern early warning systems use satellite-based thermal detection, meteorological data, and AI-based spread models. These systems improve preparedness and response strategies.

Sea-level rise and coastal erosion are driven by global warming and melting ice sheets. Monitoring systems use satellite altimetry, tide gauges, and GIS-based mapping for early warning and planning (Razavi-Termeh et al., 2025). These tools support coastal protection and adaptation strategies.

Water pollution occurs when untreated sewage, industrial waste, and agricultural chemicals contaminate water bodies. It affects drinking water, harms aquatic life, and spreads diseases. Monitoring and early warning systems help detect contamination and prevent health risks.

Deforestation leads to biodiversity loss, soil erosion, and disruption of ecological balance. Climate change, driven by greenhouse gas emissions, results in rising temperatures, extreme weather events, and sea-level rise. These environmental changes intensify disaster risks and impact ecosystems.

Industrial accidents and chemical leaks release toxic substances into the environment, causing long-term ecological and health impacts. Effective monitoring, regulation, and early warning systems are essential to minimize risks.

Improper disposal of solid waste and plastics leads to land and water pollution. Plastics harm wildlife, block drainage systems, and persist in the environment for long periods. Sustainable waste management practices and awareness are crucial for mitigation.

Natural disasters such as floods, droughts, cyclones, earthquakes, and wildfires significantly impact ecosystems and human settlements. Their effects are often intensified by environmental degradation and climate change.

Biological Hazards

Biological hazards encompass infectious diseases, pandemics, zoonotic outbreaks, and pest infestations. The COVID-19 pandemic has underscored the global importance of biological early warning systems integrating public health surveillance, genomic data, and real-time reporting. Biological hazards, also known as biohazards, are threats to human health and the environment caused by living organisms or substances produced by them. These hazards originate from biological sources such as bacteria, viruses, fungi, parasites, and toxins, leading to diseases, epidemics, and ecological damage. They may occur naturally or due to human activities, with impacts ranging from localized outbreaks to global pandemics. Examples of biological hazards include infectious diseases, pandemics, epidemics, food-borne illnesses, vector-borne diseases, and harmful biological toxins. Pathogens can spread through air, water, food, contact, or insect vectors, making them highly contagious and difficult to control. Factors such as poor sanitation, overcrowding, climate change, deforestation, and global travel increase outbreak risks.

Biological hazards affect not only humans but also animals, plants, and ecosystems, leading to food insecurity and economic losses. Measures such as vaccination, public health surveillance, sanitation, hygiene, quarantine, and awareness programs are essential for mitigation. Understanding biological hazards is crucial for public health protection and environmental sustainability.

Disease Surveillance and Early Warning

Biological early warning systems rely on data from hospitals, laboratories, and community health networks. AI and big data analytics enhance early detection of abnormal disease patterns. Advanced systems use digital data sources and real-time reporting to identify emerging threats and support rapid response.

Zoonotic and Vector-Borne Diseases

Zoonotic diseases are transmitted from animals to humans and include rabies, avian influenza, swine flu, and COVID-19. These diseases spread through direct contact, animal products, or vectors. Vector-borne diseases such as malaria, dengue, chikungunya, and Zika are influenced by climatic conditions and sanitation levels. Early warning models integrate environmental variables such as temperature, rainfall, and vector population dynamics. Global coordination mechanisms support rapid data sharing and response strategies.

Agricultural and Plant Disease Hazards

Agricultural hazards such as crop diseases and pest infestations threaten food security and livelihoods. AI-based monitoring systems using drone imagery and remote sensing enable early detection of crop anomalies. Global systems provide early warning for transboundary pests and agricultural risks.

TYPES OF BIOLOGICAL HAZARDS

Viral Diseases: Viral infections such as influenza, COVID-19, dengue, and HIV/AIDS spread rapidly through air, water, and vectors, causing widespread illness and strain on healthcare systems.

Bacterial Diseases: Diseases like cholera, tuberculosis, and typhoid are caused by bacteria and spread through contaminated water, air, or food, leading to severe health impacts.

Food-Borne Diseases: These occur due to contaminated or improperly handled food, causing illnesses such as food poisoning and hepatitis.

Fungal Infections: Fungal diseases affect skin, lungs, or internal organs, particularly in immunocompromised individuals.

Biological Toxins: Toxic substances produced by organisms, such as aflatoxins and botulinum toxin, can cause severe illness or death when ingested or inhaled.

CHEMICAL HAZARDS

Chemical hazards are potential threats to human health, safety, property, and the environment caused by exposure to harmful chemical substances. These hazards arise from the production, storage, transportation, use, or disposal of chemicals in industrial, agricultural, medical, and domestic settings. They may occur suddenly through accidents and spills or develop gradually due to long-term exposure to toxic substances. Chemical hazards also arise from industrial accidents, pollution, and contamination of food, water, or air. Historical disasters such as the Bhopal Gas Tragedy (1984) and Chernobyl (1986) illustrate how chemical and technological failures can escalate into large-scale disasters.

Chemical hazards include toxic chemicals, corrosive substances, flammable and explosive materials, reactive chemicals, and hazardous industrial wastes. Exposure occurs through inhalation, ingestion, or skin contact, causing acute effects such as poisoning, burns, explosions, and fires, as well as chronic effects like cancer, respiratory disorders, and organ damage. Industrialization, improper handling of chemicals, lack of safety measures, and environmental pollution significantly increase these risks. Major chemical hazard events include industrial gas leaks, chemical spills, pesticide exposure, and contamination of air, water, and soil. Preventive measures such as proper labelling, safe storage, protective equipment, strict safety regulations, emergency preparedness, and environmental monitoring are essential to reduce risks. Understanding chemical hazards is crucial for protecting human life, maintaining environmental quality, and ensuring safe industrial and agricultural practices.

Industrial and Environmental Contaminants

Early warning systems for chemical hazards involve sensor networks that detect toxic releases such as ammonia, chlorine, and volatile organic compounds. IoT-enabled systems can trigger automated alarms and guide emergency response and evacuation strategies, (Haffarian et al., 2024).

Food and Water Safety

Chemical contaminants in food and water pose serious health risks. AI-enhanced laboratory analysis systems monitor heavy metals, pesticide residues, and contamination levels. Advanced data-driven technologies improve traceability and early detection of contamination events.

Hazardous Waste and Pesticides

Improper handling of hazardous waste leads to soil and groundwater contamination. Early warning systems rely on real-time monitoring and predictive modeling of pollutant dispersion. Emerging technologies such as AI-powered sensors and drones help detect illegal waste disposal and environmental pollution.

Types of Chemical Hazards

Toxic Industrial Chemicals: Substances such as chlorine, ammonia, and sulphur dioxide can cause respiratory issues, burns, and fatalities when released accidentally.

Industrial Chemical Accidents: Equipment failure or human error can lead to leaks, explosions, or spills, causing environmental pollution and long-term health effects.

Pesticides and Agrochemicals: Improper use contaminates soil and water, harms ecosystems, and affects human health through the food chain.

Corrosive Chemicals: Acids and alkalis can damage tissues and infrastructure upon contact.

Flammable and Explosive Chemicals: Substances such as petrol, LPG, and solvents pose fire and explosion risks if mishandled.

Chemical Waste: Improper disposal contaminates land and water, leading to chronic health issues and environmental degradation.

Air, Water, and Soil Contamination Chemical hazards also arise from pollution of air, water, and soil by harmful substances such as heavy metals, industrial effluents, and toxic gases. These pollutants contribute to respiratory diseases, water-borne illnesses, and reduced agricultural productivity.

TECHNOLOGICAL HAZARDS

Technological hazards are risks and disasters that arise from the failure, malfunction, misuse, or breakdown of technological systems. These hazards are primarily human-induced and occur in industrial processes, infrastructure systems, energy production, transportation, and modern technological activities. They can cause severe damage to human life, property, the environment, and economic stability. Major technological hazards include industrial accidents, chemical and gas leaks, nuclear accidents, power plant failures, dam failures, transportation accidents, fires, explosions, and cyber-related disruptions. These hazards often result from human error, equipment failure, poor design, inadequate maintenance, lack of safety regulations, or natural disasters triggering technological failures. For example, industrial explosions and chemical leaks release toxic substances, while nuclear accidents expose populations to harmful radiation (Tiggeloven et al., 2025).

Technological hazards have both immediate and long-term impacts. Immediate effects include loss of life, injuries, fires, and environmental contamination, while long-term impacts involve chronic health issues, displacement, and ecological damage. With increasing dependence on technology, the risk and complexity of such hazards continue to grow. Risk reduction measures such as strict safety standards, regular maintenance, disaster preparedness, emergency response systems, and personnel training are essential to minimize technological hazards. Understanding these hazards is critical for ensuring public safety, sustainable industrial development, and effective disaster management.

Energy Storage and Radioactive Waste

Energy systems, including battery storage and nuclear facilities, pose risks of fire, explosions, and radiation leaks. Predictive maintenance systems using AI and

IoT sensors help identify equipment failures before disasters occur. Advanced safety frameworks emphasize redundancy and multi-layered protection systems for critical infrastructure.

Cyber-Physical and IoT-Related Risks

The increasing reliance on digital infrastructure introduces new technological hazards such as cyberattacks on power grids, water systems, and early warning platforms. Cybersecurity early warning systems use AI-driven threat intelligence and anomaly detection techniques to prevent disruptions. Integration of cyber resilience into MHEWS ensures system reliability and data integrity.

Infrastructure Failures

Failures in bridges, tunnels, dams, and mines can have cascading impacts on communities and economies. Structural Health Monitoring (SHM) systems use sensors, vibration analysis, and AI-based prediction models to detect weaknesses early. Integration with smart city platforms enhances disaster anticipation and response.

Types of Technological Hazards

Industrial Accidents: Caused by equipment failure or human error, leading to explosions, fires, and release of toxic substances.
Chemical and Gas Leaks: Leakage of hazardous substances during production or transport can cause poisoning and respiratory problems.
Nuclear Accidents: Failures in nuclear reactors release radioactive materials, leading to long-term environmental and health effects.
Power Plant Failures: Disrupt energy supply and essential services such as healthcare, communication, and transportation.
Dam Failures: Structural weaknesses or excessive pressure can cause catastrophic flooding and large-scale destruction.
Transportation Accidents: Crashes involving trains, aircraft, ships, or vehicles cause fatalities, injuries, and environmental pollution.
Fires and Explosions: Often caused by faulty systems or mishandling of flammable materials, resulting in severe damage to infrastructure and life.

SOCIETAL HAZARDS

Societal hazards encompass events arising from human behavior, social unrest, or systemic breakdowns, such as human stampedes, crowd crushes, transportation accidents, financial crises, and misinformation-driven panic. These hazards often accompany other emergencies, amplifying their impact and complicating response mechanisms. Societal hazards are threats that arise from social conditions, human behavior, and collective actions within a society. They are closely linked to social, economic, political, and cultural factors, affecting public safety, stability, and community functioning.

These hazards include riots, civil unrest, terrorism, wars, armed conflicts, stampedes, mass migration, crime, poverty, and social inequality. They often result from population growth, unemployment, poor governance, political instability, and unequal resource distribution. Unlike natural hazards, societal hazards are deeply rooted in human decisions and social systems. The impacts of societal hazards are severe and long-lasting, including loss of life, injuries, displacement, destruction of property, breakdown of law and order, economic decline, and psychological trauma. They can weaken social cohesion and trust within communities. Reducing societal hazards requires effective governance, social justice, inclusive development, education, conflict resolution, and community participation. Understanding societal hazards is essential for maintaining peace, social harmony, and sustainable development.

Human Stampedes and Mass Gatherings

Mass gatherings can lead to stampedes due to overcrowding and poor crowd management. Early warning systems use AI-powered video analytics, crowd density monitoring, and real-time tracking technologies to prevent such incidents.

Transportation and Industrial Accidents

Transportation hazards such as train derailments, aviation incidents, and maritime collisions require proactive monitoring of traffic flow, mechanical systems, and environmental conditions. AI-based predictive analytics and automation systems help reduce human error and enhance safety.

Financial and Cyber Hazards

Financial instability and cyber threats can trigger widespread societal crises. Early warning models use AI-driven risk assessment, sentiment analysis, and economic indicators to forecast potential disruptions. Integration of cybersecurity within MHEWS ensures resilience against digital threats.

Types of Societal Hazards

Riots and Civil Unrest: Violent protests due to political or economic grievances can damage infrastructure and disrupt public services.
Terrorism: Deliberate attacks create fear, cause casualties, and disrupt economic and social systems.
Wars and Armed Conflicts: Large-scale conflicts lead to destruction, displacement, and long-term instability.
Crime and Violence: Criminal activities threaten public safety and reduce trust in governance systems.
Mass Migration and Refugee Crises: Forced migration creates pressure on resources and infrastructure in affected regions.
Poverty and Social Inequality: Economic disparities increase vulnerability to hazards and reduce resilience.
Public Health Crises: Large-scale disease outbreaks can disrupt societies and economies, especially in densely populated areas.
Approaches to Early Warning Systems

SAFETY MANAGEMENT SYSTEMS

A Safety Management System (SMS) is a systematic and organized approach used to identify, assess, manage, and reduce risks associated with hazards. In the context of early warning systems, SMS ensures that natural, technological, and societal threats are effectively monitored, communicated, and mitigated through coordinated mechanisms. The primary goal of SMS is to prevent accidents and disasters, protect lives, reduce environmental and property damage, and enhance preparedness and response capabilities. It integrates policies, procedures, responsibilities, and resources to promote a culture of safety and risk awareness. SMS is widely applied in industries such as aviation, healthcare, maritime, and disaster management to ensure systematic hazard control and continuous improvement. Safety management involves systematic hazard identification, risk assessment, and mitigation. Integrated MHEWS adopt a risk-based approach emphasizing redundancy, interoperability, and stakeholder coordination.

Core Components of SMS

Hazard Identification: Systematic identification of potential hazards across natural, technological, and societal domains.
Risk Assessment and Analysis: Evaluation of hazard likelihood and impact to prioritize mitigation strategies.
Safety Policies and Procedures: Establishment of SOPs for prevention, monitoring, and response.
Roles and Responsibilities: Clear definition of stakeholder responsibilities in hazard monitoring and emergency response.
Monitoring and Early Warning Systems: Continuous monitoring using sensors, satellites, and data analytics to detect abnormal conditions.
Communication and Dissemination: Timely delivery of warnings through multi-channel communication systems.
Training and Capacity Building: Enhancing preparedness through drills, awareness, and skill development.
Incident Reporting and Investigation: Analysis of incidents to prevent recurrence and improve safety measures.
Continuous Improvement: Regular updates and audits based on evolving risks and lessons learned.

INTEGRATED MULTI-HAZARD EARLY WARNING SYSTEMS (IMHEWS)

Integrated MHEWS link diverse data sources—meteorological, geological, health, industrial, and social—into a unified analysis platform. A "system-of-systems" approach ensures seamless data integration and warning dissemination. An Integrated Multi-Hazard Early Warning System (IMHEWS) provides coordinated monitoring and alerts for multiple hazards simultaneously, improving resilience and preparedness. Unlike single-hazard systems, IMHEWS address cascading and compound risks.

Key Features of IMHEWS

Multi-Hazard Monitoring: Continuous monitoring of natural and human-induced hazards using advanced technologies.
Risk Assessment and Prioritization: Evaluation of hazard severity to allocate resources effectively.
Integration of Data and Technology: Combining satellite, sensor, and environmental data for accurate predictions.

Early Warning Communication: Multi-channel dissemination including SMS, media, and alert systems.
Community Participation: Engaging communities in preparedness and response activities.
Coordination Across Agencies: Collaboration among government, NGOs, and emergency services.
Cascading Hazard Awareness: Recognition of interconnected hazards and their compound effects.
Continuous Improvement: Incorporating technological advancements and lessons learned.

DISASTER MANAGEMENT SYSTEMS (DMS)

Disaster Management Systems provide structured frameworks for disaster preparedness, response, mitigation, and recovery. They integrate early warning systems, risk assessment, and coordinated response strategies.

Risk Assessment and Hazard Identification: Mapping and evaluation of vulnerable regions and hazards.
Preparedness Planning: Development of SOPs, evacuation plans, and emergency drills.
Mitigation Measures: Implementation of structural and non-structural risk reduction strategies.
Early Warning Systems: Monitoring and forecasting systems for timely alerts,.
Response Mechanisms: Emergency rescue, relief, and medical support operations.
Recovery and Rehabilitation: Rebuilding infrastructure and supporting affected populations.
Community Participation: Enhancing resilience through local involvement.
Coordination Across Agencies: Multi-stakeholder collaboration for effective disaster response.
Continuous Monitoring and Evaluation: Improvement through post-disaster analysis and feedback.

EMERGING TECHNOLOGIES IN MHEWS

Emerging technologies enhance hazard detection, monitoring, forecasting, and communication.

Artificial Intelligence (AI) and Machine Learning (ML)

AI and ML analyze large datasets from sensors, satellites, and social media to improve prediction accuracy and decision-making. Applications include flood prediction, cyclone tracking, earthquake early warning, and wildfire detection.

Internet of Things (IoT) and Sensor Networks

IoT integrates distributed sensors for real-time monitoring of environmental and industrial parameters. These systems enable rapid detection of floods, landslides, earthquakes, and industrial hazards (Sena, 2024).

Big Data and Cloud Computing

Big data analytics and cloud platforms enable large-scale data integration, real-time processing, and collaborative decision-making.

Remote Sensing and GIS

Satellite-based remote sensing and GIS provide spatial analysis, hazard mapping, and real-time monitoring capabilities. These technologies support flood prediction, cyclone tracking, landslide assessment, and wildfire monitoring.

Communication and Dissemination Mechanisms

Effective communication is central to early warning systems, ensuring that alerts are timely, accurate, and actionable. Multi-channel platforms such as SMS, radio, television, social media, and satellite systems improve reach and reliability. Social media and crowdsourcing enhance situational awareness but require mechanisms to control misinformation. Inclusive communication strategies ensure accessibility for diverse populations, improving response effectiveness.

Governance, Policy, and Institutional Frameworks

Governance frameworks define policies, responsibilities, and coordination mechanisms for effective MHEWS implementation. Strong institutional structures ensure accountability, integration, and sustainability of early warning systems (Zheng et al., 2025). Governance determines the effectiveness and sustainability of early warning systems. Multi-level governance—spanning national, regional, and local institutions—ensures coherent policy alignment and operational capacity.

Clear mandates, accountability structures, and legal frameworks define institutional responsibilities in warning generation, dissemination, and response.

NATIONAL AND REGIONAL FRAMEWORKS

Countries such as Japan, the United States, and India have established national early warning authorities that coordinate multi-hazard operations.

Japan Meteorological Agency (JMA) integrates seismic, meteorological, and volcanic monitoring.

India's National Disaster Management Authority (NDMA) supports MHEWS for cyclones, floods, and heatwaves.

The European Union's Copernicus Emergency Management Service (CEMS) provides geospatial intelligence across member states.

Regional cooperation—such as the ASEAN Coordinating Centre for Humanitarian Assistance and the African Centre of Meteorological Applications for Development—enhances cross-border data exchange and capacity building.

LEGAL AND ETHICAL CONSIDERATIONS

Effective MHEWS require enabling legislation mandating data sharing, interagency collaboration, and resource allocation. Ethical dimensions—privacy, data security, and consent—are increasingly relevant as AI and IoT collect sensitive information. Data governance policies must balance public safety with individual rights.

Financing and Sustainability

Long-term sustainability depends on financial investment in technology, training, and maintenance. Public–private partnerships (PPPs) play an increasing role in financing infrastructure and innovation for early warning systems. International organizations such as the World Bank, UNDP, and WMO provide technical and financial assistance to bridge capability gaps.

Community Preparedness and Capacity Building

Community preparedness and capacity building enable communities to anticipate, respond to, and recover from hazards effectively. These efforts ensure that early warnings translate into timely and effective action, reducing vulnerability and enhancing resilience. Communities often act as first responders during disasters; hence, strengthening human, social, and institutional capacities is critical.

The Human Dimension of Early Warning

Technological sophistication alone cannot guarantee effective disaster prevention. Human awareness, preparedness, and response capacity determine whether early warnings save lives. Community participation is therefore both a practical necessity and a moral imperative.

Community-Based Early Warning Systems (CBEWS)

CBEWS emphasize local knowledge, decentralized decision-making, and grass-roots participation.

Bangladesh's Cyclone Preparedness Programme (CPP) demonstrates large-scale volunteer engagement.

Nepal's community flood warning networks utilize local communication channels such as radio and mobile alerts.

These systems enhance resilience through local ownership and sustainability.

Education and Training

Disaster education at schools, workplaces, and community centers builds a culture of preparedness. Simulation drills, workshops, and curriculum integration improve risk literacy and response capacity. Universities contribute through research and professional training.

Role of NGOs and Civil Society

NGOs and community-based organizations act as intermediaries between authorities and citizens. Their involvement in awareness campaigns, resource mobilization, and psychosocial support enhances inclusivity and outreach of MHEWS.

Case Studies and Global Practices

Global case studies show that effective MHEWS depend on technology, governance, and community participation.

Japan's Earthquake and Tsunami Early Warning System

Japan's EEWS demonstrates technological excellence with dense sensor networks and automated alerts. Integration of seismic and oceanographic data enables rapid response and minimizes losses.

India's Integrated Multi-Hazard Early Warning System (IMHEWS)

The Indian Meteorological Department (IMD) integrates meteorological, hydrological, and oceanographic data. Real-time alerts via the Common Alerting Protocol highlight effective inter-agency coordination under NDMA.

Copernicus Emergency Management Service (CEMS)

CEMS provides satellite-based geospatial intelligence for disaster monitoring and risk assessment. It exemplifies transnational cooperation and open data sharing.

African Regional Systems

Regional MHEWS supported by organizations such as WMO enhance climate forecasting and disaster preparedness.

Bangladesh Cyclone Preparedness Programme (CPP)

CPP is a globally recognized community-based early warning system that significantly reduces cyclone-related mortality.

Indonesia's Multi-Hazard Early Warning System

Indonesia uses integrated sensor networks and communication channels to provide rapid warnings for multiple hazards.

United States Wildfire Early Warning System

Advanced technologies such as GIS, remote sensing, and satellite imagery support wildfire detection and evacuation planning.

COVID-19 as a Multi-Hazard Event

The COVID-19 pandemic demonstrated the interaction of biological, social, and technological hazards. Countries with digital surveillance and communication systems managed risks more effectively.

Challenges and Limitations

Technological Gaps and Data Fragmentation, Fragmented systems hinder interoperability and data integration.

Accessibility and Equity Issues, Marginalized populations face barriers in accessing early warnings. Inclusive policies are essential.

False Alarms and Public Trust, Frequent false alarms can reduce public confidence.

Governance and Coordination Barriers, Institutional fragmentation can delay decision-making.

Financial and Capacity Constraints, developing countries face funding and infrastructure challenges.

FUTURE DIRECTIONS AND OPPORTUNITIES

Next-Generation Systems, Integration of AI, IoT, and advanced communication technologies will enhance predictive capabilities.

Climate Change Integration, Linking MHEWS with climate adaptation improves long-term resilience.

Citizen Science, Crowdsourced data enhances community participation and system effectiveness.

International Cooperation, Global initiatives such as Early Warnings for All promote universal access to warning systems.

Ethical AI, Transparency, accountability, and privacy protection are essential in AI-driven systems.

CONCLUSION

Multi-hazard early warning systems are central to global resilience against increasing disaster risks (Budimir et al., 2025). Their success depends on the integration of technology, governance, and community participation. Investment in both technological infrastructure and human capacity is essential to ensure inclusive and effective early warning systems.

REFERENCES

Amaratunga, D., Haigh, R., & Dias, N. (Eds.). (2021). *Multi-hazard early warning and disaster risks.* Springer Nature. DOI: 10.1007/978-3-030-73003-1

Ayuba, P., Hassan, A. B., Kuti, I. A., Morenikeji, G., Mohammed, I., Ambafi, J. G., & Ohize, H. O. (2025). Multi-hazard identification and early warning systems for emergency management: A review. *Confluence University Journal of Science and Technology.*

Basher, R. (2023). Challenges in multi-hazard early warning systems: Last-mile connectivity. *International Journal of Disaster Risk Reduction*, •••, 92.

Bompotas, A., Anagnostopoulos, C., Kalogeras, A., Kalogeras, G., Mylonas, G., Stefanidis, K., Alexakos, C., & Dandoulaki, M. (2022). A civil protection early warning system to improve the resilience of Adriatic-Ionian territories to natural and man-made risk. *arXiv.* https://arxiv.org/abs/2207.13941 DOI: 10.1109/ETFA52439.2022.9921697

Budimir, M., Šakić Trogrlić, R., Almeida, C., Arestegui, M., Chuquisengo Vásquez, O., Cisneros, A., Cuba Iriarte, M., Dia, A., Lizon, L., Madueño, G., Ndiaye, A., Ordoñez Caldas, M., Rahman, T., Rana Tharu, B., Sall, A., Uprety, D., Anderson, C., & McQuistan, C. (2025). Opportunities and challenges for people-centered multi-hazard early warning systems: Perspectives from the Global South. *iScience*, *28*(5), 112353. Advance online publication. DOI: 10.1016/j.isci.2025.112353 PMID: 40292326

Haffarian, S., Alidoost, F., Lagap, U., Chandramouli, P., Dzigan, Y., Grootes, M., Jalayer, F., & Kelman, I. (2024). Digital twins for early warning systems: Intricacies and solutions. DOI: 10.5194/egusphere-egu24-15738

Kelman, I., & Fearnley, C. J. (2025). From multi-hazard early warning systems (MHEWS) to all-vulnerability warning systems (AVWS). *iScience*, *28*(7), 112977. Advance online publication. DOI: 10.1016/j.isci.2025.112977 PMID: 40687808

Kumar, A., & Singh, R. (2024). Explainable AI for flood and air quality early warning systems. *Environmental Modelling & Software*, •••, 172.

Nunes, A. R. (2025). Health and multihazard early warning systems in the media. *Perspectives in Public Health*, *145*(6), 343–352. DOI: 10.1177/17579139251342156 PMID: 40838788

Razavi-Termeh, S. V., Sadeghi-Niaraki, A., Ali, F., Pradhan, B., & Choi, S. M. (2025). Optimizing ensemble learning for satellite-based multi-hazard monitoring and susceptibility assessment of landslides, land subsidence, floods, and wildfires. *Scientific Reports, 15*(1), 30968. DOI: 10.1038/s41598-025-15381-2 PMID: 40846725

Rokhideh, M., Fearnley, C. J., & Budimir, M. (2025). Multi-hazard early warning systems in the Sendai framework for disaster risk reduction: Achievements, gaps, and future directions. *International Journal of Disaster Risk Science, 16*(1), 103–116. DOI: 10.1007/s13753-025-00622-9

Sena, M. (2024). IoT-based early warning systems for natural disaster management. *International Journal of Innovation Studies.*

Tiggeloven, T., Pfeiffer, S., Matanó, A., van den Homberg, M., Thalheimer, L., Reichstein, M., & Torresan, S. (2025). The role of artificial intelligence for early warning systems: Status, applicability, guardrails, and ways forward. *iScience, 28*(11), 113689. Advance online publication. DOI: 10.1016/j.isci.2025.113689 PMID: 41280665

Xu, H., Lu, Y., Chen, Y., & Fan, Y. (2026). A broadly applicable warning response index: Cross-hazard validation for enhanced early warning communication. *Risk Analysis, 46*(2), e70189. Advance online publication. DOI: 10.1111/risa.70189 PMID: 41657267

Zhao, L., & Chen, X. (2021). Artificial intelligence applications in disaster prediction and response. *International Journal of Disaster Risk Science, 12*(3), 421–437.

Zheng, B., & Chu, V. W. (2025). Multi-hazard early warning systems for agriculture with featural-temporal explanations. *arXiv*. https://arxiv.org/abs/2507.22962

Chapter 2
Encyclopedia of Transformative Trends: AI Applications in Disaster Management

Ravinder Singh
http://orcid.org/0009-0007-7161-6709
SR University, India

Geetha Manoharan
http://orcid.org/0000-0002-8644-8871
SR University, India

Samrath Singh
http://orcid.org/0009-0009-3720-1432
Vellore Institute of Technology, Chennai, India

ABSTRACT

Artificial intelligence (AI) has emerged as a powerful tool for disaster management, offering real-time insights, situational awareness, and enhanced decision-making capabilities. In the past decade, AI has emerged as a powerful tool for predicting natural disasters. In this review, we explore the intersections of AI, disaster management, and societal resilience, highlighting the transformational potential of AI technologies in mitigating the impacts of natural and human-induced disasters. It explores the historical evolution of AI applications in disaster management and also addresses ethical considerations. Moreover, the incorporation of AI in social media analytics and natural language processing for emergency communication highlights its role in improving information dissemination and community engagement during

DOI: 10.4018/979-8-3373-7345-4.ch002

ght © 2026, IGI Global Scientific Publishing. Copying or distributing in print or electronic forms without written permission of IGI Global Scientific Publishing is prohibited. Use of this chapter to generative artificial intelligence (AI) technologies is expressly prohibited. The publisher reserves all rights to license its use for generative AI training and machine learning model development.

emergencies. The review highlights the use of AI in disaster management, ultimately building more resilient and inclusive communities worldwide. It also covers related terms and definitions for AI applications in disaster management.

1. INTRODUCTION

Disasters, whether natural phenomena like earthquakes, hurricanes, or floods, or man-made calamities such as industrial accidents or terrorist attacks, invariably leave behind a harrowing aftermath of devastation and sorrow. These events represent profound disruptions to the normal functioning of communities, often surpassing their intrinsic capacities to effectively cope with the ensuing challenges using local resources alone. Throughout history, humanity has grappled with the formidable task of mitigating and recovering from the impacts of disasters, a struggle that spans centuries and civilizations.

In recent decades, there has been a notable transformation in the approach to disaster management, marked by significant strides in research, policy development, and technological innovation. A concerted effort has been underway to enhance preparedness, response, mitigation, and recovery mechanisms, driven by a recognition of the escalating risks posed by a changing climate, urbanization, and global interconnectedness. This concerted effort has led to a paradigm shift in how societies conceptualize and address disasters, emphasizing proactive risk reduction strategies and adaptive resilience-building initiatives.

Central to this transformation is the rapid evolution of Artificial Intelligence (AI), which has permeated virtually every facet of human existence. From revolutionizing industries and transforming economies to fundamentally altering societal dynamics, the impact of AI is profound and pervasive. Within the realm of disaster management, AI holds immense potential to augment traditional approaches and revolutionize crisis response strategies. As AI's capabilities continue to expand and mature, so do the prospects for mitigating the human toll and socioeconomic disruptions caused by disasters. By harnessing the power of machine learning(ML) algorithms, data analytics, and advanced computing capabilities, AI offers unprecedented opportunities in all stages of disaster management.

The integration of AI into disaster management practices represents a convergence of cutting-edge technology and urgent humanitarian imperatives. As the capabilities of AI continue to expand and mature, so too do the prospects for mitigating the human toll and socioeconomic disruptions wrought by disasters. However, alongside the promise of technological innovation, there are ethical, legal, and societal considerations that demand careful attention and deliberation. Ensuring the responsible and equitable deployment of AI-enabled solutions requires a nuanced understanding of the complex interactions between technology, governance, and human behavior.

In this dynamic landscape, the evolution of AI in disaster management represents not only a technological revolution but also a transformative shift in how we perceive and respond to catastrophic events. By exploring the intersections of AI, disaster management, and societal resilience, we are poised to unlock new insights, strategies, and opportunities for building more resilient, adaptive, and inclusive communities in an increasingly uncertain world.

2. HISTORICAL CONTEXT

2.1 Evolution of Disaster Management.

When we examine the historical context of AI applications in disaster management, it becomes clear that humanity's response to disasters has evolved significantly over the centuries. From ancient civilizations' rudimentary strategies for coping with natural calamities to the modern era's sophisticated approaches informed by data-driven insights, the trajectory of disaster management reflects a continuous quest for resilience and adaptability. Disasters, as defined by Britannica, encompass any natural or human-generated calamitous event that produces substantial loss of human life or destruction of the natural environment, private property, or public infrastructure. A disaster may be relatively sudden, or it may unfold over a longer period of time. They can be classified as natural disasters or human-caused disasters. Other events, like the displacement of large populations due to war, can also fall under the term. However, UNDRR has defined disasters as '*a serious disruption of the functioning of a community or a society at any scale due to hazardous events interacting with conditions of exposure, vulnerability, and capacity, leading to one or more of the following: human, material, economic, and environmental losses and impacts.*' (United Nations Office for Disaster Risk Reduction, n.d.).

2.2 Historical Context of the Evolution of Disaster Management.

Ancient Greek mythological texts and some of the most significant religious works in Western culture, which view catastrophic events as acts of God, contain allusions to natural disasters (Sargiacomo et al., 2021). The dynamic field of disaster management underwent a significant transformation in the 20th century, marked by the reorientation of paradigms and strategic initiatives aimed at safeguarding lives and mitigating the destructive impacts of natural disasters. This period witnessed a notable transition towards centralized measures for citizen protection, sparking the inception of civil defense mechanisms (Coppola, 2015; Hecker & Domres, 2018; Singh et al., 2024). A pivotal moment in the history of disaster management transpired in 1987, as the United Nations General Assembly proclaimed the 1990s "The International Decade for Natural Disaster Reduction (IDNDR)" (United Nations, 1999). In subsequent years, the international community convened in Kobe in 2005 to establish the Hyogo Framework for Action (HFA). The Sendai Framework for Disaster Risk Reduction 2015–2030 (SFDRR) is a seminal global policy framework within the United Nations' post-2015 agenda, highlighting the importance of mental and physical health, resilience, and well-being in disaster risk reduction (DRR) strategies. All United Nations member states ratified the 2030 Agenda for Sustainable Development in 2015, outlining 17 Sustainable Development Goals (SDGs) (Singh et al., 2024).

2.3 Disaster Risk Management Approach.

Recently, there has been a noticeable shift in the approach to disasters, transitioning from mere response to proactive risk mitigation. This shift has led to the emergence of Disaster Risk Management (DRM), where measures aimed at reducing risks are seen as investments rather than mere expenses (Wisner et al., 2004). DRM incorporates diverse strategies for risk reduction, recognizing that effective disaster management is part of a larger, holistic system. It prioritizes managing underlying risks rather than solely reacting to disasters. While the Hyogo Framework outlined aspects like risk identification, monitoring, and assessment, the Sendai document notably emphasizes understanding and mitigating disaster risk (Aitsi-Selmi et al., 2016).

2.4 The Origin of AI and the Era of Unprecedented Growth.

2.4.1 Historical Development of AI.

Alan Turing and Gordon Welchman developed the Bombe Machine in 1941 to break Germany's ENIGMA code. ENIGMA was a sophisticated encoding device. The US Army's Ballistic Research Laboratory later developed ENIAC, the first electronic general-purpose computer (Kaynak, 2021). John McCarthy first introduced the term "AI" during the 1956 Dartmouth Summer Research Project on Artificial Intelligence (McCarthy et al., 2006). The Perceptron's debut in 1960 marked a milestone in AI, with the Mark 1 Perceptron demonstrating gender recognition in photos, a marvel at the time (Hay et al., 1960). Further advancements included expert systems like Feigenbaum's molecular structure determination, reflecting

AI's promising future alongside rapid progress in computing hardware and software development (Oke, 2008).

2.4.2 Periods of Decline: AI Winters.

AI winters denote periods of reduced funding due to disillusionment (Jiang et al., 2022). Two major AI winters occurred, the first in the 1970s, sparked by overconfidence in early machine translation results (Jones, 1994). Criticisms from reports like ALPAC and books like “Perceptrons” led to funding cuts (Jiang et al., 2022). Expert systems briefly revived interest until the late 1980s, when funding shifted elsewhere due to overhype and technical limitations. Common factors included biased evaluations, media influence, and underdeveloped tools (McCorduck & Cfe, 2004). Moreover, researchers lacked awareness of the risks and underestimated the time needed for foundational research. The cycles of substantial investments and divestments were the primary catalysts for the onset of AI winters. During the late 1980s, there was another downturn in AI advancement, marked by reduced funding and a redirection of attention towards emerging technologies like general-purpose computers (Jiang et al., 2022).

2.4.3 Driving Forces for the Resurgence of AI.

A multitude of pivotal elements fuel the resurgence of AI, each contributing to the advancement of this field. These factors encompass various aspects, spanning from technological innovations to societal perceptions. Breakthroughs in ML stand as a cornerstone, with sophisticated algorithms and techniques since the late 1980s. These advancements have laid the groundwork for contemporary ML methodologies and frameworks, often found in modern ML textbooks (Mohri et al., 2018). The exponential growth of digital data has provided AI systems with an abundant source of information for training and refinement (Zhang et al., 2018; Givigi & Jardine, 2018). The availability of large-scale datasets, coupled with advancements in data processing techniques, has significantly enhanced the effectiveness and accuracy of AI models (Yaseen et al., 2015). The rapid increase in computing power has empowered AI systems to tackle increasingly complex tasks (Yin & Kaynak, 2015). Advancements in hardware architecture, such as the development of specialized processors and high-performance computing clusters, have facilitated the efficient processing of vast amounts of data, enabling AI algorithms to achieve remarkable feats. The public demonstration of AI systems surpassing human capabilities in various domains has played a crucial role in reshaping perceptions and instilling confidence in AI technology. Instances where AI outperformed top human players in contests and competitions have garnered widespread attention and showcased the potential of AI to revolutionize diverse fields (Moravčík et al., 2017; Johnson et al., 2013).

Figure 1 shows the timeline for AI development milestones and AI winter periods, highlighting cycles of progress and stagnation in the field's evolution.

Figure 1. The timeline of AI

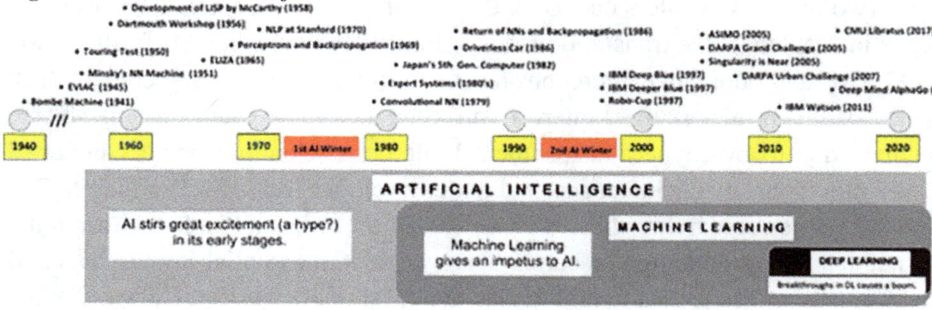

3. CURRENT TRENDS IN AI FOR DISASTER MANAGEMENT

3.1 Leveraging Technology for Disaster Prediction and Management.

Amidst the rising tide of natural disasters, there's an urgent call to leverage cutting-edge technology. As the world contends with the growing frequency and severity of these events, there's a pressing need for advanced tools to predict and manage disasters effectively. The gravity of this issue is highlighted by the swift rise in disasters causing irreversible damage to our environment and all living creatures, spurred by rapid urbanization (Singh, 2023). The convergence of natural and human-induced threats presents complex challenges, demanding a comprehensive strategy for disaster management. AI emerges as a potent ally in lessening the impact of catastrophes on our communities.

3.2 Disaster Management Cycle.

The disaster management cycle is crucial for minimizing the impact of disasters. It enables proactive measures such as risk assessment and mitigation, ensures readiness through preparedness efforts, facilitates swift and effective response when disasters strike, and supports long-term recovery and reconstruction, ultimately saving lives and safeguarding communities against future calamities. It has been described as a cyclical process aimed at reducing the impact of, responding to, and recovering from disasters. Goals include reduced loss, prompt assistance, and rapid and effective recovery (Murphy, 2023). UNDRR has explained disaster management cycle as the organization, planning, and application of measures preparing for, responding to, and recovering from disasters. It elaborates that disaster management may not completely avert or eliminate the threats; it focuses on creating and implementing preparedness and other plans to decrease the impact of disasters and "build back better" (United Nations Office for Disaster Risk Reduction, n.d.). According to the National Disaster Management Act 2005 of India, Disaster Management involves a continuous and integrated process of planning, organizing, coordinating, and implementing measures that are necessary or expedient for: prevention of danger, mitigation or reduction of risk of any disaster, capacity building, preparedness to deal with any disaster, response to any threatening disaster situation or disaster, assessing the severity or magnitude of effects of any disaster, evacuation, rescue and relief, rehabilitation, and reconstruction" (Disaster Management Division, n.d.). Figure 2 illustrates the Disaster Management Cycle, a comprehensive framework guiding disaster preparedness, response, recovery, and mitigation efforts.

Figure 2. Disaster management cycle

3.3 Advancing AI's Role in Revolutionizing Disaster Management.

AI's advancements will expand its role in disaster management, encompassing risk modeling, decision support, and humanitarian aid. Collaboration among academia, industry, and government can revolutionize disaster preparedness and response, potentially saving lives globally. Sun et al. (2020) outlined 26 AI methods across 17 application areas, illustrating AI's diverse potential in disaster management (Figure 3).

Figure 3. AI methods and application areas in disaster management (Sun et al., 2020)

3.4 AI Methodologies in Disaster Management.

Exploring the utilization of AI in disaster management requires categorizing AI methodologies into six distinct groups: supervised models, unsupervised models, deep learning (DL), reinforcement learning, deep reinforcement learning, and optimization. This classification aids in understanding the diverse approaches and their applications in enhancing disaster management strategies (Sun et al., 2020).

3.4.1 Supervised Models: Enhancing Human-Guided Algorithms.

Supervised models encompass algorithms trained on existing data with human guidance. These models are commonly employed for tasks such as information extraction, object recognition in computer vision, pattern recognition, and speech recognition, among others (Russell & Norvig, 2016).

3.4.2 Unsupervised Models: Identifying Anomalies and Reducing Data Dimensionality.

Unsupervised models excel at identifying anomalies and reducing data dimensionality, finding extensive applications in clustering and data aggregation tasks. Clustering algorithms, for instance, facilitate pattern recognition by segmenting unlabeled data into multiple groups based on specific similarity features (Maulik & Bandyopadhyay, 2002).

3.4.3 Deep Learning: Acquiring Abstraction in Computational Models.

DL enables computational models consisting of multiple layers of processing to acquire representations of data characterized by various levels of abstraction (LeCun et al., 2015). For instance, recursive neural networks (RvNN) and recurrent neural networks (RNN) have found successful applications in natural language processing (NLP) tasks (Socher et al., 2011; Graves et al., 2013).

3.4.4 Applications of Convolutional Neural Networks (CNN).

Additionally, convolutional neural networks (CNN) are well-suited for tasks like image recognition (Simonyan & Zisserman, 2014), computer vision (Krizhevsky et al., 2017), and speech processing (Dahl et al., 2011).

3.4.5 Real-Time Data Collection and Analysis through AI: Revolutionizing Monitoring, Simulation, and Disaster Prediction.

Collecting and analyzing data in real-time is made possible by the swift computational power of AI. This facilitates the instantaneous gathering and scrutiny of data from various sources, such as Internet of Things sensors, satellite images, and weather updates. These capabilities prove invaluable in monitoring tasks, leveraging sensors, satellite imagery, and even drones for comprehensive surveillance (Sun et al., 2020). A significant advantage of AI is its ability to efficiently perform simulations in a limited time frame, such as climate and geological models, scenario simulation, social media analysis, and forecasting. Incorporating AI into warning systems has revolutionized the prediction and understanding of disasters. This encompasses disseminating alerts, developing applications, and using digital platforms to deliver crucial information, readiness guidance, and real-time crisis updates. Moreover, it involves community-based alerts that tailor messages according to the distinct features of local areas, as well as coordinating relief efforts. Utilizing digital platforms, citizens and emergency management professionals can be trained on optimal approaches to disaster preparedness and response strategies (Zafar et al., 2019).

3.4.6 AI Driven Models for Earthquake Prediction.

Many existing prediction techniques for earthquakes result in high false alarm rates, leading to inaccurate predictions and contributing to the catastrophic consequences. AI-based methods offer a new potential for improving prediction accuracy, outperforming traditional techniques (Al Banna et al., 2020). Cataldi et al. (2024) conducted a study demonstrating that AI can effectively retrieve and corroborate data, providing accurate insights. The study emphasized the significance of using increasingly modern tools and utilizing multiple data sources for AI verification. Various researchers have contributed to the development of prediction models aimed at understanding the different characteristics of earthquakes. Bofeng and Yue (2005) introduced ESEP3.0, an explanatory system that tailors earthquake prediction information to individuals with varying levels of knowledge. Zhong and Zhang (2010) presented a mathematical model focused on forecasting the causes of reservoir-induced earthquakes. Mirrashid et al. (2016) investigated the potential of ANFIS in predicting the seismic moment of forthcoming earthquakes. Additionally, Rahmat et al. (2018) conducted a comparative analysis of predictive models to forecast the b-value in the Andaman-Nicobar region. Researchers have explored advanced techniques in earthquake prediction. Konstantaras et al. (2007) introduced an NFM utilizing an adaptive filter to detect electric earthquake precursors, showcasing a novel approach to early detection. Astuti et al. (2014) employed SVD and SVM to predict earthquake characteristics such as location, magnitude, and time, incorporating kernel functions to handle nonlinearity effectively. Furthermore, Li and Kang (2009) developed a hybrid algorithm combining KNN and polynomial regression (PR) to forecast earthquake aftershocks, demonstrating innovative methods for predicting seismic events and their aftermath. Several researchers have investigated synthetic and regional earthquake prediction strategies. Ikram and Qamar (2015) devised an expert system capable of forecasting earthquakes at least twelve hours before their occurrence, enhancing preparedness measures. Jiang et al. (2009) undertook the challenge of synthetically predicting the most significant annual earthquake magnitude in China, contributing to regional risk assessment efforts. Additionally, Hajikhodaverdikhan et al. (2018) introduced an SVR model, optimized by a particle filter, to predict the mean magnitude and number of earthquakes expected in Iran for the upcoming month, offering valuable insights for regional disaster management planning. Florido et al. (2015) developed localized earthquake prediction models, specifically tailoring a clustering-based approach for earthquake prediction in Chile.

This model aims to provide localized insights into seismic activity, offering valuable information for disaster preparedness and mitigation efforts in the region. A study by Ghani et al. (2022) shows that the artificial neural network (ANN) model is better at predicting the future than other models. This means that it can be used to figure out how likely it is that fine-grained soil deposits will melt. The study also shows liquefaction hazard mapping for the problematic area using the ANN model, highlighting its huge potential to improve risk prediction and preparedness for earthquakes.

3.4.7 Flood Forecasting with AI Models.

Due to their flexibility and interpretability, flood forecasting applications widely employ AI models like support vector machines (SVMs), random forests, decision trees, and gradient boosting machines (GBMs). These models utilize historical data to discern predictive relationships among meteorological and hydrological factors and flood occurrences. For instance, SVMs excel at categorizing flood events based on input variables and can effectively handle nonlinear relationships and complex data structures (Tariq, 2024). "DHARA", a flood-support mobile application, integrates a long short-term memory (LSTM) AI application with convolutional neural networks (CNN) (Kumar & Sud, 2020). This application was developed to predict the likelihood of flooding, enabling the implementation of early warning preparedness and restoration techniques prior to an incident. A rainfall prediction model in the proposed research generated projected data by evaluating environmental characteristics associated with rainfall (Aswin et al., 2018).

3.4.8 Fire Hazard Assessment in Forest Fires.

An AI algorithm analyzes various data points like temperature, humidity, wind speed, and precipitation to discern patterns and correlations, facilitating precise forecasts of future fire hazards. Its adaptability to changing meteorological conditions through continuous updates renders it a dynamic tool for swift and accurate evaluations of wildfire risk (Pettorru et al., 2023). Moreover, the AI model examines visual data from high-resolution satellite imagery using ML techniques, enhancing its capability to swiftly detect and locate wildfires (Lin, 2022). By leveraging vegetation indicators, the AI model assesses fuel loads in different areas, identifying regions with arid or distressed vegetation that could heighten wildfire risk. This data augmentation bolsters the model's proactive capabilities, enabling it to accurately target high-risk locations (Pang et al., 2022). Adding geospatial data like topography and land cover information also helps the AI model understand how spaces are connected. This makes risk assessments more accurate and helps find places where fires can spread quickly (Milanović et al., 2023). Incorporating air quality monitoring data into the AI model involves real-time monitoring of air quality indicators like PM2.5, carbon monoxide, and ozone levels. This enables the model to accurately gauge smoke dispersion and assess the impact of wildfires on public health (Khan et al., 2022; Shafik, 2024).

3.4.9 AI and Drone Integration for Disaster Management.

The strategic deployment and utilization of AI and drones to bolster disaster management in the wake of natural calamities will serve as a significant force multiplier for affected nations. As natural disasters increase in both frequency and severity globally, there's an escalating necessity to leverage advanced technologies for DRM endeavors (Singh et al., 2024). Large datasets and ML techniques power AI algorithms that make predictive analytics, early warning systems, and decision support tools possible. This improves situational awareness and the coordination of responses. Meanwhile, drones equipped with sophisticated sensors and imaging capabilities enable the execution of tasks inaccessible to humans during disasters (Vermiglio et al., 2022). Recently, there's been a surge in integrating innovative smart devices like drones into the Internet of Things (IoT) network. Drones offer new ways to deliver value-added IoT services, like monitoring, surveillance, and on-demand delivery. They also address the limitations of fixed IoT infrastructure. Due to their significant potential, they are anticipated to become an essential component of urban environments (Singh et al., 2024; Labib et al., 2021). The synergy between AI and drones revolutionizes disaster management. AI swiftly analyzes data for early warnings and resource distribution. Drones offer rapid aerial reconnaissance, aiding damage assessment and supply delivery to remote areas. Together, they heighten situational awareness, streamline decision-making, and foster inter-agency coordination, optimizing disaster response. This integration saves lives and minimizes the impact of disasters globally (Singh et al., 2024). Drones are indispensable tools in disaster management, offering rapid mapping, damage assessment, and search and rescue capabilities across various scenarios like floods, earthquakes, and wildfires (Daud et al., 2022; Singh et al., 2024). They outperform traditional methods by quickly providing high-quality imagery, aiding erosion and flood assessments, and facilitating remote site visits. Drones play a crucial role in hastening rescue operations, all the while prioritizing the safety of responders and efficiently covering vast and rugged terrains within shorter timeframes. Drones have the capability to deliver essential aid and supplies such as food, water, and medical resources to disaster-stricken areas that are challenging to reach through conventional means (Nihad & Nadjat, 2023). Drones can play a vital role in setting up temporary communication networks, particularly in disaster-affected or remote areas where traditional infrastructure may be compromised or lacking (Tanzi et al., 2016; Bushnaq et al., 2022). Drones are valuable tools for disaster managers, offering continuous real-time data for immediate decision-making support. They can assess the extent of damage and request additional assistance while still airborne, saving crucial time in disaster response (Restas, 2017). Overall, drones significantly bolster disaster response efficiency and effectiveness.

3.4.10 AI in Emergency Decision Making for Natural Disasters.

Emergency decision-making (EDM) is crucial in the context of natural disasters, acting as a cornerstone for rapid responses, life-saving interventions, and damage mitigation. Quick decisions made during natural disasters are essential for orchestrating swift evacuations, efficient resource allocation, and seamless coordination among response agencies (Singh & Manoharan, 2024). During the response stage of natural disasters, we can utilize AI for various tasks such as event mapping, damage assessment, early warning and alert systems, real-time disaster prediction and detection, training systems, and disaster evacuations (Sun et al., 2020). AI offers unparalleled advantages in speed, scale, and real-time data processing, enabling it to swiftly assimilate and analyze vast amounts of relevant information. This positions AI as a crucial tool for promptly recognizing potential threats and serving as a highly dependable early warning system. Furthermore, AI excels at recommending suitable actions for incident response. Researchers have integrated computer technology, GIS, and agents into emergency decision support systems to enhance aid efficiency. Developed countries prioritize GIS for natural disaster decision-making. Intelligent computer entities are used in agent-based emergency decision support systems. Other related systems include case-based reasoning (CBR) and the combination of ML and NLP (Zhou et al., 2018).

4. AI APPLICATIONS AND CASE STUDIES IN DISASTERS

4.1. AI Applications across Disasters.

4.1.1.

Artificial intelligence is widely applied across both pre-disaster monitoring and post-disaster response for a range of natural hazards. Its applications include the detection of meteorological disasters, flood monitoring, volcanic eruption observation, dynamic visualization of lava flows, and geological hazard assessment and early warning. These domains employ specialized AI systems such as intelligent earthquake, volcano, and flood monitoring platforms (Lu et al., 2022).

4.1.2.

Flood management systems typically integrate Internet of Things (IoT) sensing and control devices, cloud-based service centers, client interfaces, and communication networks to enable real-time monitoring, data analysis, dispatch and control, information dissemination, and decision-support services. In the context of geological hazards, machine learning techniques have been extensively used to develop models for landslide identification, susceptibility analysis, hazard evaluation, and risk assessment (Qi et al., 2021; Yang et al., 2022; Yu et al., 2022).

4.1.3.

In addition, speech recognition has emerged as a prominent AI technology in recent years. Researchers have designed intelligent speech-based robotic systems and tested them in flood prevention and disaster mitigation scenarios, demonstrating their ability to rapidly deliver critical information related to flood preparedness and response (Sun et al., 2020).

4.2. Case studies of AI in Disasters

4.2.1.

NOAA uses artificial intelligence to significantly enhance hurricane forecasting accuracy and timeliness. By integrating machine learning models with data from satellites, weather stations, and ocean sensors, NOAA improves predictions of storm formation, track, and intensity. AI also strengthens ensemble forecasting by reducing uncertainty across multiple models and enhances the prediction of rapid hurricane intensification, enabling earlier warnings. These advancements support better evacuation planning, optimized resource allocation, and improved emergency preparedness. As a result, AI-driven forecasting helps save lives, supports informed decision-making, and reduces the economic impact of hurricanes through proactive response and preparedness measures (Clement, 2024).

4.2.2.

AI-driven disaster prediction and management systems have strengthened resilience across multiple regions worldwide. In Japan, AI-powered earthquake early warning systems analyze seismic waveforms in real time to detect foreshocks and issue rapid alerts, reducing casualties and infrastructure damage. In the United States, AI-enhanced wildfire prediction systems such as FireCAST use satellite imagery and meteorological data to forecast fire spread and optimize firefighting resource deployment. In India, AI-based flood forecasting generates high-resolution inundation maps and early warnings, enabling timely evacuations. In Africa, AI models predict locust swarm movements, supporting targeted interventions that protect agriculture and food security (Rehan, 2022).

4.2.3.

A study by Sufi & Alsulami proposes an AI-driven disaster intelligence framework using GPT-based news analytics, geospatial clustering, and temporal forecasting to improve disaster preparedness. Analyzing 1.25 million news articles from 444 sources, the model identified 17,884 disaster events across 185 countries. The USA, India, and Australia emerged as the most disaster-prone regions, while storms, floods, and wildfires were the most frequent disasters. The framework offers valuable insights for policymaking, resource allocation, and climate-adaptive disaster management strategies. (Sufi & Alsulami, 2025).

4.2.4.

In the aftermath of sudden disasters such as earthquakes, hurricanes, and floods, social media platforms are quickly filled with images, videos, and text shared by affected residents, emergency responders, journalists, and bystanders. Recent research has applied advanced natural language processing and computer vision techniques to large volumes of disaster-related posts to detect infrastructure damage and location-specific needs. Geo-tagged content enables effective spatial mapping of impacts; however, uneven participation across communities limits comprehensive coverage of all affected areas. (Agbaje et al., 2024).

5. THE SOCIETAL AND ETHICAL IMPLICATIONS OF AI IN DISASTER MANAGEMENT

5.1. Ethical Challenges in AI Integration for Disaster Management.

The integration of AI in disaster management presents significant societal and ethical implications. AI can enhance predictive capabilities, optimize resource allocation, and streamline emergency responses, potentially saving countless lives. However, ethical concerns arise regarding data privacy, algorithmic biases, and equitable access to AI technologies. Ensuring that AI-driven decisions are transparent and inclusive is crucial to avoid exacerbating existing inequalities. Additionally, the reliance on AI systems necessitates robust safeguards against technological failures. Balancing the benefits of AI with ethical considerations is essential to foster trust and ensure the fair and effective use of AI in disaster management. The ethical dimensions can be described under transparency, accountability, and equitable deployment. Navigating the considerations of transparency, accountability, and equity necessitates a delicate balance, maximizing the positive impact of AI while upholding ethical imperatives (Singh & Manoharan, 2024).

5.2. Challenges in the Development and Deployment of AI.

The development of AI is based on big data, and the first ethical challenge is to design and develop an AI with an ethical face projecting concern for human dignity while addressing core issues such as data privacy, discrimination, human and social welfare, and the sustainability of the use of technology (West, 2018; Aalabaf-Sabaghi, 2020). An AI system must also be designed to resist unethical human interventions, given the significant risk of malicious actors and vested interest groups compromising data inputs and outputs, thereby threatening the system's stability and security. Additionally, AI systems have a unique vulnerability: even minor manipulations of inputs can lead to drastically different decisions (Shaw, 2019). This necessitates robust safeguards to ensure the integrity and reliability of AI processes. The use of satellite imagery combined with citizen reporting raises privacy concerns, necessitating the implementation of ethical and transparent data procedures (Feizizadeh et al., 2023).

5.3. Promoting Ethical AI Practices.

Ethical frameworks must ensure responsible AI deployment in disaster management. Transparent data collection, informed consent, and stringent privacy safeguards are essential to protecting individuals' rights. Continuous monitoring of AI algorithms is crucial to detect and mitigate biases. AI should complement, not replace, human decision-making, preserving human autonomy and dignity. Collaborative approaches involving local communities are vital for ethical AI implementation in disaster response. Transparent data collection protocols, informed consent mechanisms, and strict privacy safeguards are essential to protecting individuals' rights. Additionally, continuous monitoring and auditing of AI algorithms are crucial to detecting and mitigating biases (Singh et al., 2024). By clarifying the decision-making procedures, explainable AI improves the ethical aspects of AI implementation and establishes a basis for more knowledgeable and cooperative decision-making in wildfire prevention and response (Sevinç, 2023).

6. RELATED TERMS AND DEFINITIONS

6.1. Explainable Artificial Intelligence (XAI).

In the last decade, with the availability of large datasets and increased computing power, ML systems have achieved (super)human performance in tasks like image recognition, speech analysis, and strategic game planning. However, many state-of-the-art models lack transparency and interpretability, which is a significant drawback in fields like healthcare and finance, where understanding the rationale behind a model's decision is crucial for trust. Consequently, XAI has become a key area of interest in the research community (Došilović et al., 2018). XAI can be defined as *a suite of machine learning techniques that enables human users to understand, appropriately trust, and effectively manage the emerging generation of artificially intelligent partners* (Gunning et al., 2019). The related terms for XAI are explainability, interpretability, transparency, fairness, robustness, satisfaction, stability, and responsibility (Ali et al., 2023).

6.2. Decision Support Systems in Crisis Management.

A Decision Support System (DSS) is an information system that aids decision-making by generating and evaluating multiple alternatives to determine the best solution for various issues (Jung et al., 2020). By integrating new and diverse data sources, a DSS enhances emergency managers' decision-making in crisis situations (Fertier et al., 2020). A DSS should be capable of prioritizing specific teams for incidents by considering the importance of each team in emergencies, the sequence of tasks to be performed, and all possible orders (Nunes et al., 2020). AI can be used for event mapping, damage assessment, early warning and alert systems, real-time disaster prediction and detection, training systems, and disaster evacuations (Sun et al., 2020). In disaster response, AI offers unparalleled advantages in speed, scale, and real-time data processing. Additionally, AI excels at recommending suitable actions for incident response by analyzing vast amounts of real-time data to identify patterns and trends, allowing it to predict potential outcomes and suggest the most effective strategies for mitigating damage and managing resources (Singh & Manoharan 2024).

6.3. AI-powered Early Warning Systems.

Early warning systems (EWSs) aim to efficiently distribute disaster-related information to vulnerable communities before or during an event, enabling proactive measures to minimize loss. Rapid generation of vast amounts of data is a result of the rise of the Internet of Things and advanced sensor devices. To effectively monitor and manage disasters, EWSs must capture, store, and analyze this data. AI can help EWSs extract early warning signals from this data, enabling proactive disaster mitigation, preparedness, response, and recovery planning, and timely dissemination of alerts to stakeholders (Lamsal & Vijay Kumar, 2020). AI offers an alternative to traditional hydrologic and hydraulic models and can also be used for prediction and forecasting purposes. For instance, Google, through its Social Good initiative, utilizes AI to accurately forecast floods, integrating this data into the regional FEWS system (Perera et al., 2020). AI stands out as a crucial research technique that numerous researchers have analyzed and demonstrated to be highly effective in enhancing the performance of fire hazard detection in smart cities (Zhang et al., 2021). As part of the UrbanFlood FP7 project, the Flood Early Warning System (EWS) monitors sensor networks installed in flood defenses such as dikes, dams, and embankments. It detects abnormalities in sensor signals, assesses the probability of dike failure, and simulates potential scenarios of dike breaches and flood propagation. This information, along with simulation results, is integrated into an interactive decision support system. This system aids dike managers and city authorities in making informed decisions during emergencies and routine dike quality assessments.

6.4. AI enabled Evacuation Route Planning.

AI represents an innovative approach to emergency planning. It utilizes DL to analyze information or data, predicting evacuation routes, allocating emergency resources efficiently, and estimating disaster locations (Zhang et al., 2022). AI-driven emergency evacuation route planning is a powerful tool for ensuring population safety during emergencies. By leveraging advanced algorithms and ML techniques, these systems can analyze real-time data to determine the safest and most efficient evacuation routes for any situation. The objective is to swiftly devise evacuation routes for a community, ensuring the prompt evacuation of victims to safe zones. Addressing the evacuation route planning challenge involves identifying suitable routes and allocating a designated number of victims to each route (Liu et al., 2022).

6.5.Social Media Analytics for Disaster Response.

Social media analytics has emerged as a powerful tool for disaster management, offering real-time insights, situational awareness, and enhanced decision-making capabilities. Ethical considerations and privacy concerns should retain their prominence in social media analytics research. Delving deeper into privacy-preserving algorithms, methods for obtaining informed consent and anonymizing data, and setting up ethical guidelines can help to safeguard the responsible and ethical utilization of social media data (Karimiziarani, 2023). Social media analytics play a significant role in natural disaster management. Despite the abundance of metadata fields in social media data, four dimensions space, time, content, and network—have received special attention for extracting valuable information to enhance situational awareness and bolster disaster response efforts (Wang & Ye, 2018).

6.6. Resource Allocation with AI.

AI technologies provide data-driven insights, enhance predictive capabilities, optimize resource allocation, and ultimately strengthen the resilience of communities facing increased risks (Singh & Manoharan, 2024). Effective resource allocation is critical when resources are scarce. These systems can leverage optimization algorithms, such as linear programming and reinforcement learning, to generate optimal resource allocation strategies. AI-driven resource allocation systems assist emergency responders in making data-driven decisions, minimizing resource wastage, and ensuring that critical resources are directed to the most pressing needs. Additionally, these systems facilitate the coordination of resource sharing and distribution across different facilities and regions, enabling a more efficient and equitable response to crises (Fernando, 2023). The successful application of machine learning in disaster resource allocation hinges on several key factors: access to relevant data, robust modeling, and close collaboration between data scientists, domain experts, and disaster response agencies. Furthermore, distributing resources equitably to those in need during critical times requires ethical considerations, fairness, and transparency (Pachar et al., 2023).

6.7. Geospatial Intelligence and AI for Disaster Mapping.

Geospatial analysis plays a crucial role in disaster management across all phases, from mitigation to recovery. Given the essential need for spatial data, numerous projects utilize AI methods for spatial data analysis in disaster management (Ivić, 2019). Modern technologies such as AI, combined with geospatial data in a GIS environment, offer significant advancements in DRM (Haigh, 2017). Geospatial data is essential for detecting critical areas during floods, analyzing safer zones during earthquakes, and identifying suitable locations for shelters for homeless people. ML, a powerful AI tool in GIS, has been extensively discussed by researchers (Huang & Jensen, 1997; Allen et al., 2016; Mojaddadi et al., 2017; Chen et al., 2018) for its applications in geospatial risk assessments and susceptibility mapping. An automatic disaster detection system for landslide and flood detection analyzes satellite images using convolutional neural networks (Amit & Aoki, 2017). A framework for change detection on satellite images, employing CNNs, identifies the areas most affected by a disaster and aids in coordinating relief efforts (Doshi et al., 2018). In their study, Cooner et al. (2016) evaluated different algorithms for detecting earthquake damage from the 2010 Port-au-Prince earthquake in Haiti, using the Disaster Impact Index to quantify the impact on each area. Cooner et al. (2016) proposed a methodology for post-earthquake damage mapping in the city of Palu, Indonesia, using artificial neural networks (back propagation algorithm), support vector machines (radial basis function classifiers), and the decorrelation method. Post-earthquake maps are essential for risk assessment, evacuation planning, and mitigation (Syfia et al., 2019).

6.8. Mass Casualty Management with AI.

Disasters often result in mass casualty incidents, necessitating complex decisions at casualty collection points. Effective triage and coordinated evacuation are essential to prevent hospital overcrowding and additional casualties. Rapid transportation guidance based on triage categories, available ambulances, resources, and hospital capabilities is crucial (Nistor et al., 2020). Research has developed deep reinforcement learning models to redeploy ambulances, minimizing transportation time and increasing casualty survival rates (Ji et al., 2019). A novel algorithm, based on Twitter data from Hurricane Harvey in 2017, improves response to victims' requests during disasters (Yang et al., 2020). MobiRescue, a human mobility-based rescue team dispatching system, uses reinforcement learning to maximize rescue responses, minimize driving time, and reduce the number of rescue units needed (Yan et al., 2020). Traffic jams or blocked roads, especially in urban areas or during severe disasters, can delay ambulances in emergency situations. Traffic state detection is crucial for the MCI response. In Dhaka, a dynamic control system was tested, which measures traffic density using real-time video and image processing. Two cameras detect and count cars entering and leaving a lane using the adaptive learning-based Mixture of Gaussian (MoG) method. This information regulates traffic lights at intersections to improve flow (Chowdhury et al., 2018).

6.9. Humanitarian Robotics.

Robots in disaster-stricken areas are more efficient and effective when they use AI algorithms (Manoharan et al., 2024). According to Partigöç, NS (2022), this field of study plays a significant role in providing timely and relevant information and providing essential services to improve response. Robotic systems can inspect damaged structures, assess damage, and speed up resource delivery. Tadokoro (2020) designed robotic disaster rescue systems to mitigate disaster damage through search and rescue operations and recovery activities. Rescue robots assist in locating, assessing, stabilizing, and extricating victims in hard-to-reach areas. Small unmanned ground vehicles (SUGVs) can navigate tight or dangerous spaces inaccessible to humans and dogs (Murphy et al., 2016), using thermal imaging and computer vision algorithms for survivor detection. Large unmanned ground vehicles (LUGVs) serve as mobile base stations, transporting injured victims and supporting small UGVs. Equipped with powerful manipulators, LUGVs can perform heavy tasks like debris removal and excavation. They also provide an adaptable emergency communication backbone and various sensing capabilities, enhancing situational awareness for field operators and navigating rough terrain (De Cubber et al., 2013). Unmanned underwater robots expedite rescue efforts and support underwater research in flood- or tsunami-affected areas. In construction, robotics improve productivity, address labor shortages, and reduce safety risks, especially during disaster recovery (Zhang et al., 2019).

6.10. Natural Language Processing for Emergency Communication.

NLP originated in the 1950s at the intersection of AI and linguistics (Nadkarni et al., 2011). It encompasses methods for rendering human language comprehensible to computers (Eisenstein, 2019). By combining linguistic, statistical, and AI techniques, NLP can decipher the meaning of text or even generate responses resembling human language (Fanni et al., 2023). In times of natural disasters, overloaded emergency communication systems prompt individuals to seek assistance through social media. A study utilized ML and AI to automatically detect, identify, and categorize tweets relevant to first responders (Powers et al., 2023). Advanced research utilizes state-of-the-art ML, DL, and NLP techniques, employing both supervised and unsupervised learning methods. These approaches leverage social media datasets to extract real-time content related to emergency events, facilitating rapid response in critical situations (Shahbazi & Byun, 2021). A research initiative endeavors to create a comprehensive Information Technology (IT) solution for emergency response and disaster management, with social media data integration as its central component. Emergency response authorities employed text analysis methods to efficiently sift through vast amounts of automatically collected information to bolster their relief efforts. This endeavor utilized cutting-edge NLP, ML, and DL techniques (Khatoon et al., 2021).

7. CONCLUSION

In reviewing the historical evolution and current trends in AI applications for disaster management, this chapter delves into the transformative potential of AI technologies in mitigating the impacts of natural and human-induced disasters. The historical context illustrates the progressive integration of AI into disaster management, marking significant advancements in predictive analytics, risk assessment, and emergency response.

The emergence and resurgence of AI, despite periods of decline, highlight its pivotal role in enhancing disaster management strategies. AI's capabilities significantly support the disaster management cycle, a fundamental framework for preparedness, response, recovery, and mitigation. AI methodologies, including supervised and unsupervised models, DL, and reinforcement learning, offer diverse applications ranging from real-time data analysis and disaster prediction to resource allocation and emergency decision-making. AI's integration with other advanced technologies, such as drones and geospatial intelligence, further revolutionizes disaster management by providing rapid, accurate, and actionable insights. These technologies enhance situational awareness, facilitate efficient resource distribution, and support effective decision-making during crises. AI-driven innovations, such as earthquake prediction models, flood forecasting systems, fire hazard assessments, and humanitarian robotics, demonstrate AI's broad applicability in addressing various disaster scenarios. Moreover, the incorporation of AI in social media analytics and NLP for emergency communication highlights its role in improving information dissemination and community engagement during emergencies.

In conclusion, the chapter reveals that the synergy between AI and disaster management holds immense promise for the future. By continuing to advance AI methodologies and fostering collaboration among academia, industry, and government, we can develop more resilient communities and enhance our global capacity to predict, manage, and recover from disasters. AI stands as a crucial ally in safeguarding lives and minimizing the devastating impacts of disasters on our environment and society. Moreover, collaboration among diverse sectors ensures that AI tools are continuously refined and adapted to meet the evolving challenges posed by natural and human-made disasters. Such partnerships enable the sharing of critical insights, technological innovations, and best practices, fostering a holistic approach to disaster preparedness and response. As we harness the full potential of AI, we pave the way for a future where communities worldwide are better equipped to withstand and recover from the impacts of disasters, ensuring a safer and more resilient world for all.

REFERENCES

Aalabaf-Sabaghi, M. (2020). An Introduction to Data, Everything You Need to Know About AI: Big Data and Data Science.

Agbaje, T. H., Abomaye-Nimenibo, N., Ezeh, C. J., Bello, A., & Olorunnishola, A. (2024). Building Damage Assessment in Aftermath of Disaster Events by Leveraging Geoai (Geospatial Artificial Intelligence). *World Journal of Advanced Research and Reviews*, *23*(1), 667–687. DOI: 10.30574/wjarr.2024.23.1.2000

Aitsi-Selmi, A., Murray, V., Wannous, C., Dickinson, C., Johnston, D., Kawasaki, A., Stevance, A.-S., & Yeung, T. (2016). Reflections on a science and technology agenda for 21st century disaster risk reduction: Based on the scientific content of the 2016 UNISDR science and technology conference on the implementation of the Sendai framework for disaster risk reduction 2015–2030. *International Journal of Disaster Risk Science*, *7*(1), 1–29. DOI: 10.1007/s13753-016-0081-x

Al Banna, M. H., Taher, K. A., Kaiser, M. S., Mahmud, M., Rahman, M. S., Hosen, A. S., & Cho, G. H. (2020). Application of artificial intelligence in predicting earthquakes: State-of-the-art and future challenges. *IEEE Access : Practical Innovations, Open Solutions*, *8*, 192880–192923. DOI: 10.1109/ACCESS.2020.3029859

Ali, S., Abuhmed, T., El-Sappagh, S., Muhammad, K., Alonso-Moral, J. M., Confalonieri, R., Guidotti, R., Del Ser, J., Díaz-Rodríguez, N., & Herrera, F. (2023). Explainable Artificial Intelligence (XAI): What we know and what is left to attain Trustworthy Artificial Intelligence. *Information Fusion*, *99*, 101805. DOI: 10.1016/j.inffus.2023.101805

Allen, C., Tsou, M. H., Aslam, A., Nagel, A., & Gawron, J. M. (2016). Applying GIS and learning methods to Twitter data for multiscale surveillance of influenza. *PLoS One*, *11*(7), e0157734. DOI: 10.1371/journal.pone.0157734 PMID: 27455108

Amit, S. N. K. B., & Aoki, Y. (2017, September). Disaster detection from aerial imagery with convolutional neural network. In 2017 international electronics symposium on knowledge creation and intelligent computing (IES-KCIC) (pp. 239-245). IEEE.

Astuti, W., Akmeliawati, R., Sediono, W., & Salami, M. J. E. (2014). Hybrid technique using singular value decomposition (SVD) and support vector machine (SVM) approach for earthquake prediction. *IEEE Journal of Selected Topics in Applied Earth Observations and Remote Sensing*, *7*(5), 1719–1728. DOI: 10.1109/JSTARS.2014.2321972

Aswin, S., Geetha, P., & Vinayakumar, R. (2018, April). Deep learning models for the prediction of rainfall. In 2018 International Conference on Communication and Signal Processing (ICCSP) (pp. 0657-0661). IEEE. DOI: 10.1109/ICCSP.2018.8523829

Bofeng, Z., & Yue, L. (2005, November). Customized explanation in expert system for earthquake prediction. In *17th IEEE International Conference on Tools with Artificial Intelligence (ICTAI'05)* (pp. 5-pp). IEEE. DOI: 10.1109/ICTAI.2005.54

Bushnaq, O. M., Mishra, D., Natalizio, E., & Akyildiz, I. F. (2022). Unmanned aerial vehicles (UAVs) for disaster management. In *Nanotechnology-Based Smart Remote Sensing Networks for Disaster Prevention* (pp. 159–188). Elsevier., DOI: 10.1016/B978-0-323-91166-5.00013-6

Cataldi, D., Cataldi, G., & Straser, V. (2024). Earthquakes and AI: The Proposal of a New Research Model. New Concepts in Global Tectonics Journal, 12(1).

Chen, W., Peng, J., Hong, H., Shahabi, H., Pradhan, B., Liu, J., Zhu, A.-X., Pei, X., & Duan, Z. (2018). Landslide susceptibility modelling using GIS-based machine learning techniques for Chongren County, Jiangxi Province, China. *The Science of the Total Environment*, *626*, 1121–1135. DOI: 10.1016/j.scitotenv.2018.01.124 PMID: 29898519

Chowdhury, M. F., Biplob, M. R. A., & Uddin, J. (2018, June). Real time traffic density measurement using computer vision and dynamic traffic control. In 2018 Joint 7th International Conference on Informatics, Electronics & Vision (ICIEV) and 2018 2nd International Conference on Imaging, Vision & Pattern Recognition (icIVPR) (pp. 353-356). IEEE. DOI: 10.1109/ICIEV.2018.8641039

Clement, M. (2024). Case Studies: Successful Use of AI in Past Natural Disasters.

Cooner, A. J., Shao, Y., & Campbell, J. B. (2016). Detection of urban damage using remote sensing and machine learning algorithms: Revisiting the 2010 Haiti earthquake. *Remote Sensing (Basel)*, *8*(10), 868. DOI: 10.3390/rs8100868

Coppola, D. (2015). *Introduction to International Disaster Management. Butterworth-Heinemann*. Elsevier.

Dahl, G. E., Yu, D., Deng, L., & Acero, A. (2011). Context-dependent pre-trained deep neural networks for large-vocabulary speech recognition. *IEEE Transactions on Audio, Speech, and Language Processing*, *20*(1), 30–42. DOI: 10.1109/TASL.2011.2134090

Daud, S. M. S. M., Yusof, M. Y. P. M., Heo, C. C., Khoo, L. S., Singh, M. K. C., Mahmood, M. S., & Nawawi, H. (2022). Applications of drone in disaster management: A scoping review. *Science & Justice*, *62*(1), 30–42. DOI: 10.1016/j.scijus.2021.11.002 PMID: 35033326

De Cubber, G., Serrano, D., Berns, K., Chintamani, K., Sabino, R., Ourevitch, S., . . . Baudoin, Y. (2013, October). Search and rescue robots developed by the european icarus project. In 7th Int. Workshop on Robotics for Risky Environments. Citeseer.

Disaster Management Division. (n.d.). The Disaster Management Act, 2005. Ministry of Home Affairs. Government of India. https://ndmindia.mha.gov.in/images/The%20Disaster%20Management%20Act,%202005.pdf

Doshi, J., Basu, S., & Pang, G. (2018). From satellite imagery to disaster insights. arXiv preprint arXiv:1812.07033.

Došilović, F. K., Brčić, M., & Hlupić, N. (2018, May). Explainable artificial intelligence: A survey. In 2018 41st International convention on information and communication technology, electronics and microelectronics (MIPRO) (pp. 0210-0215). IEEE. DOI: 10.23919/MIPRO.2018.8400040

Eisenstein, J. (2019). *Introduction to natural language processing*. MIT press.

Fanni, S. C., Febi, M., Aghakhanyan, G., & Neri, E. (2023). Natural language processing. In *Introduction to Artificial Intelligence* (pp. 87–99). Springer International Publishing. DOI: 10.1007/978-3-031-25928-9_5

Feizizadeh, B., Omarzadeh, D., Mohammadnejad, V., Khallaghi, H., Sharifi, A., & Karkarg, B. G. (2023). An integrated approach of artificial intelligence and geoinformation techniques applied to forest fire risk modeling in Gachsaran, Iran. *Journal of Environmental Planning and Management*, *66*(6), 1369–1391. DOI: 10.1080/09640568.2022.2027747

Fernando, Y. S. (2023). TRANSFORMING EMERGENCY RESPONSE AND DISASTER MANAGEMENT WITH AI: DEVELOPING INTELLIGENT SYSTEMS FOR TRIAGE, RESOURCE ALLOCATION, AND SITUATIONAL AWARENESS IN HEALTHCARE CRISES. *International Journal of Applied Machine Learning and Computational Intelligence*, *13*(12), 11–20.

Fertier, A., Barthe-Delanoë, A. M., Montarnal, A., Truptil, S., & Bénaben, F. (2020). A new emergency decision support system: The automatic interpretation and contextualisation of events to model a crisis situation in real-time. *Decision Support Systems*, *133*, 113260. DOI: 10.1016/j.dss.2020.113260

Florido, E., Martínez-Álvarez, F., Morales-Esteban, A., Reyes, J., & Aznarte-Mellado, J. L. (2015). Detecting precursory patterns to enhance earthquake prediction in Chile. *Computers & Geosciences, 76*, 112–120. DOI: 10.1016/j.cageo.2014.12.002

Ghani, S., Kumari, S., Jaiswal, S., & Sawant, V. A. (2022). Comparative and parametric study of AI-based models for risk assessment against soil liquefaction for high-intensity earthquakes. *Arabian Journal of Geosciences, 15*(14), 1262. DOI: 10.1007/s12517-022-10534-3

Givigi, S., & Jardine, P. T. (2018). Machine learning for data-driven control of robots. *IEEE Potentials, 37*(4), 35–39. DOI: 10.1109/MPOT.2018.2824398

Graves, A., Mohamed, A. R., & Hinton, G. (2013, May). Speech recognition with deep recurrent neural networks. In 2013 IEEE international conference on acoustics, speech and signal processing (pp. 6645-6649). Ieee.

Gunning, D., Stefik, M., Choi, J., Miller, T., Stumpf, S., & Yang, G. Z. (2019). XAI—Explainable artificial intelligence. *Science Robotics, 4*(37), eaay7120. DOI: 10.1126/scirobotics.aay7120 PMID: 33137719

Haigh, R. (2017). *Disaster Management Lifecycle*. Learning Package.

Hajikhodaverdikhan, P., Nazari, M., Mohsenizadeh, M., Shamshirband, S., & Chau, K. W. (2018). Earthquake prediction with meteorological data by particle filter-based support vector regression. *Engineering Applications of Computational Fluid Mechanics, 12*(1), 679–688. DOI: 10.1080/19942060.2018.1512010

Hay, J. C., Martin, F. C., & Wightman, C. W. (1960, January). The mark-1 perceptron-design and performance. *Proceedings of the Institute of Radio Engineers, 48*(3), 398–399.

Hecker, N., & Domres, B. D. (2018). The German emergency and disaster medicine and management system—history and present. *Chinese Journal of Traumatology = Zhonghua Chuang Shang Za Zhi, 21*(2), 64–72. DOI: 10.1016/j.cjtee.2017.09.003 PMID: 29622286

Huang, X., & Jensen, J. R. (1997). A machine-learning approach to automated knowledge-base building for remote sensing image analysis with GIS data. *Photogrammetric Engineering and Remote Sensing, 63*(10), 1185–1193.

Ikram, A., & Qamar, U. (2015). Developing an expert system based on association rules and predicate logic for earthquake prediction. *Knowledge-Based Systems, 75*, 87–103. DOI: 10.1016/j.knosys.2014.11.024

Ivić, M. (2019). Artificial intelligence and geospatial analysis in disaster management. *The International Archives of the Photogrammetry, Remote Sensing and Spatial Information Sciences*, *42*(W8), 161–166. DOI: 10.5194/isprs-archives-XLII-3-W8-161-2019

Ji, S., Zheng, Y., Wang, Z., & Li, T. (2019). A deep reinforcement learning-enabled dynamic redeployment system for mobile ambulances. *Proceedings of the ACM on Interactive, Mobile, Wearable and Ubiquitous Technologies*, *3*(1), 1–20. DOI: 10.1145/3314402 PMID: 40144218

Jiang, C., Wei, X., Cui, X., & You, D. (2009). Application of support vector machine to synthetic earthquake prediction. *Earth Science*, *22*(3), 315–320. DOI: 10.1007/s11589-009-0315-8

Jiang, Y., Li, X., Luo, H., Yin, S., & Kaynak, O. (2022). Quo vadis artificial intelligence? *Discover Artificial Intelligence*, *2*(1), 4. DOI: 10.1007/s44163-022-00022-8

Johnson, N., Zhao, G., Hunsader, E., Qi, H., Johnson, N., Meng, J., & Tivnan, B. (2013). Abrupt rise of new machine ecology beyond human response time. *Scientific Reports*, *3*(1), 2627. DOI: 10.1038/srep02627 PMID: 24022120

Jones, K. S. (1994). Natural language processing: a historical review. Current issues in computational linguistics: in honour of Don Walker, 3-16.

Jung, D., Tran Tuan, V., Quoc Tran, D., Park, M., & Park, S. (2020). Conceptual framework of an intelligent decision support system for smart city disaster management. *Applied Sciences (Basel, Switzerland)*, *10*(2), 666. DOI: 10.3390/app10020666

Karimiziarani, M. (2023). Social Media Analytics in Disaster Response: A Comprehensive Review. arXiv preprint arXiv:2307.04046.

Kaynak, O. (2021). The Golden Age of Artificial Intelligence: Inaugural Editorial. *Discover Artificial Intelligence*, *1*(1), 1–7. DOI: 10.1007/s44163-021-00009-x

Khan, A., Hassan, B., Khan, S., Ahmed, R., & Abuassba, A. (2022). DeepFire: A Novel Dataset and Deep Transfer Learning Benchmark for Forest Fire Detection. *Mobile Information Systems*, *2022*, 1–14. DOI: 10.1155/2022/5358359

Khatoon, S., Alshamari, M. A., Asif, A., Hasan, M. M., Abdou, S., Elsayed, K. M., & Rashwan, M. (2021). Development of social media analytics system for emergency event detection and crisismanagement. *Computers, Materials & Continua*, *68*(3). Advance online publication. DOI: 10.32604/cmc.2021.017371

Konstantaras, A., Varley, M. R., Vallianatos, F., Makris, J. P., Collins, G., & Holifield, P. (2007). Detection of weak seismo-electric signals upon the recordings of the electrotelluric field by means of neuro-fuzzy technology. *IEEE Geoscience and Remote Sensing Letters*, *4*(1), 161–165. DOI: 10.1109/LGRS.2006.887068

Krizhevsky, A., Sutskever, I., & Hinton, G. E. (2017). ImageNet classification with deep convolutional neural networks. *Communications of the ACM*, *60*(6), 84–90. DOI: 10.1145/3065386

Kumar, T. V., & Sud, K. (Eds.). (2020). *AI and robotics in disaster studies*. Palgrave Macmillan. DOI: 10.1007/978-981-15-4291-6

Labib, N. S., Brust, M. R., Danoy, G., & Bouvry, P. (2021). The rise of drones in internet of things: A survey on the evolution, prospects and challenges of unmanned aerial vehicles. *IEEE Access : Practical Innovations, Open Solutions*, *9*, 115466–115487. DOI: 10.1109/ACCESS.2021.3104963

Lamsal, Rabindra, and TV Vijay Kumar. (2020). "Artificial intelligence and early warning systems." AI and Robotics in Disaster Studies: 13-32.

LeCun, Y., Bengio, Y., & Hinton, G. (2015). Deep learning. nature, 521(7553), 436-444.

Li, A., & Kang, L. (2009, December). Knn-based modeling and its application in aftershock prediction. In *2009 International Asia Symposium on Intelligent Interaction and Affective Computing* (pp. 83-86). IEEE. DOI: 10.1109/ASIA.2009.21

Lin, T. Z. (2022). Suppressing Forest Fires in Global Climate Change Through Artificial Intelligence: A Case Study on British Columbia. Proceedings – 2022 International Conference on Big Data, Information and Computer Network, BDICN 2022. IEEE. DOI: 10.1109/BDICN55575.2022.00086

Liu, L., Jin, H., Liu, Y., & Zhang, X. (2022). Intelligent Evacuation Route Planning Algorithm Based on Maximum Flow. *International Journal of Environmental Research and Public Health*, *19*(13), 7865. DOI: 10.3390/ijerph19137865 PMID: 35805524

Lu, S., Christie, G. A., Nguyen, T. T., Freeman, J. D., & Hsu, E. B. (2022). Applications of artificial intelligence and machine learning in disasters and public health emergencies. *Disaster Medicine and Public Health Preparedness*, *16*(4), 1674–1681. DOI: 10.1017/dmp.2021.125 PMID: 34134815

Manoharan, G., Razak, A., Rao, B. S., Singh, R., Ashtikar, S. P., & Nivedha, M. (2024). Navigating the Crescendo of Challenges in Harnessing Artificial Intelligence for Disaster Management. In Predicting Natural Disasters With AI and Machine Learning (pp. 64-94). IGI Global. DOI: 10.4018/979-8-3693-2280-2.ch003

Maulik, U., & Bandyopadhyay, S. (2002). Performance evaluation of some clustering algorithms and validity indices. *IEEE Transactions on Pattern Analysis and Machine Intelligence*, *24*(12), 1650–1654. DOI: 10.1109/TPAMI.2002.1114856

McCarthy, J., Minsky, M. L., Rochester, N., & Shannon, C. E. (2006). A proposal for the dartmouth summer research project on artificial intelligence, august 31, 1955. *AI Magazine*, *27*(4), 12–12.

McCorduck, P., & Cfe, C. (2004). *Machines who think: A personal inquiry into the history and prospects of artificial intelligence*. AK Peters/CRC Press. DOI: 10.1201/9780429258985

Milanović, S., Kaczmarowski, J., Ciesielski, M., Trailović, Z., Mielcarek, M., Szczygieł, R., Kwiatkowski, M., Bałazy, R., Zasada, M., & Milanović, S. D. (2023). Modeling and Mapping of Forest Fire Occurrence in the Lower Silesian Voivodeship of Poland Based on Machine Learning Methods. *Forests*, *14*(1), 46. DOI: 10.3390/f14010046

Mirrashid, M., Givehchi, M., Miri, M., & Madandoust, R. (2016). Performance investigation of neuro-fuzzy system for earthquake prediction.

Mohri, M., Rostamizadeh, A., & Talwalkar, A. (2018). *Foundations of machine learning*. MIT press.

Mojaddadi, H., Pradhan, B., Nampak, H., Ahmad, N., & Ghazali, A. H. B. (2017). Ensemble machine-learning-based geospatial approach for flood risk assessment using multi-sensor remote-sensing data and GIS. *Geomatics, Natural Hazards & Risk*, *8*(2), 1080–1102. DOI: 10.1080/19475705.2017.1294113

Moravčík, M., Schmid, M., Burch, N., Lisý, V., Morrill, D., Bard, N., Davis, T., Waugh, K., Johanson, M., & Bowling, M. (2017). Deepstack: Expert-level artificial intelligence in heads-up no-limit poker. *Science*, *356*(6337), 508–513. DOI: 10.1126/science.aam6960 PMID: 28254783

Murphy, L. (2023). Disaster Relief. In The Role of Child Life Specialists in Community Settings (pp. 191-214). IGI Global.

Murphy, R. R., Tadokoro, S., & Kleiner, A. (2016). Disaster robotics. In *Springer handbook of robotics* (pp. 1577–1604). Springer. DOI: 10.1007/978-3-319-32552-1_60

Nadkarni, P. M., Ohno-Machado, L., & Chapman, W. W. (2011). Natural language processing: An introduction. *Journal of the American Medical Informatics Association : JAMIA*, *18*(5), 544–551. DOI: 10.1136/amiajnl-2011-000464 PMID: 21846786

Nihad, B., & Nadjat, A. (2023). Drones Use in Disaster Management [Doctoral dissertation, university center of abdalhafid boussouf-MILA].

Nistor, M. S., Son, P. T., Pickl, S. W., Gaindric, C., & Cojocaru, S. (2020). A concise review of AI-based solutions for mass casualty management. *CEUR Workshop Proceedings*, *2805*, 222–232.

Nunes, I. L., Calhamonas, G., Marques, M., & Teodoro, M. F. (2020). Building a decision support system to handle teams in disaster situations-a preliminary approach. In Hybrid Intelligent Systems: 18th International Conference on Hybrid Intelligent Systems (HIS 2018) Held in Porto, Portugal, December 13-15, 2018 18 (pp. 551-559). Springer International Publishing. DOI: 10.1007/978-3-030-14347-3_54

Oke, S. A. (2008). A literature review on artificial intelligence. International journal of information and management sciences, 19(4), 535-570.

Pachar, S., Dudeja, D., Batra, N., Tomar, V., Bhimavarapu, J. P., & Singh, A. K. (2023). Recent Developments in Machine Learning Predictive Analytics for Disaster Resource Allocation. *Engineering Proceedings*, *59*(1), 19.

Pang, Y., Li, Y., Feng, Z., Feng, Z., Zhao, Z., Chen, S., & Zhang, H. (2022). Forest Fire Occurrence Prediction in China Based on Machine Learning Methods. *Remote Sensing (Basel)*, *14*(21), 5546. DOI: 10.3390/rs14215546

Partigöç, N. S. (2022). *The Role of Artificial Intelligence in Disaster Risk Management*. Bilişim Teknolojileri Dergisi.

Perera, D., Seidou, O., Agnihotri, J., Mehmood, H., & Rasmy, M. (2020). Challenges and technical advances in flood early warning systems (FEWSs). Flood impact mitigation and resilience enhancement.

Pettorru, G., Fadda, M., Girau, R., Sole, M., Anedda, M., & Giusto, D. (2023). Using Artificial Intelligence and IoT Solution for Forest Fire Prevention. *2023 International Conference on Computing, Networking and Communications, ICNC 2023*. IEEE. DOI: 10.1109/ICNC57223.2023.10074289

Powers, C. J., Devaraj, A., Ashqeen, K., Dontula, A., Joshi, A., Shenoy, J., & Murthy, D. (2023). Using artificial intelligence to identify emergency messages on social media during a natural disaster: A deep learning approach. *International Journal of Information Management Data Insights*, *3*(1), 100164. DOI: 10.1016/j.jjimei.2023.100164

Qi, W., Xu, C., & Xu, X. (2021). AutoGluon: A revolutionary framework for landslide hazard analysis. *Natural Hazards Review*, *1*(3), 103–108. DOI: 10.1016/j.nhres.2021.07.002

Rahmat, B., Afiadi, F., & Joelianto, E. (2018, December). Earthquake prediction system using neuro-fuzzy and extreme learning machine. In *International Conference on Science and Technology (ICST 2018)* (pp. 452-458). Atlantis Press. DOI: 10.2991/icst-18.2018.95

Rehan, H. (2022). Enhancing Disaster Response Systems: Predicting and Mitigating the Impact of Natural Disasters Using AI. *Journal of Artificial Intelligence Research*, *2*(1), 501.

Restas, A. (2017). Disaster management supported by unmanned aerial systems (UAS) focusing especially on natural disasters. Zeszyty Naukowe SGSP/Szkoła Główna Służby Pożarniczej, 61.

Russell, S. J., & Norvig, P. (2016). *Artificial intelligence: a modern approach*. Pearson.

Sargiacomo, M., Servalli, S., Potito, S., D'Andreamatteo, A., & Gitto, A. (2021). Accounting for natural disasters from a historical perspective: A literature review and research agenda. *Accounting History*, *26*(2), 179–204. DOI: 10.1177/10323732211003173

Sevinç, V. (2023). Mapping the forest fire risk zones using artificial intelligence with risk factors data. *Environmental Science and Pollution Research International*, *30*(2), 4721–4732. DOI: 10.1007/s11356-022-22515-w PMID: 35974271

Shafik, W. (2024). Artificial Intelligence Models to Prevent Forest Fires. In AI and IoT for Proactive Disaster Management (pp. 78-106). IGI Global. DOI: 10.4018/979-8-3693-3896-4.ch005

Shahbazi, Z., & Byun, Y. C. (2021). Blockchain-based event detection and trust verification using natural language processing and machine learning. *IEEE Access : Practical Innovations, Open Solutions*, *10*, 5790–5800. DOI: 10.1109/ACCESS.2021.3139586

Shaw, J. (2019, January–February). Harvard Magazine. Retrieved May 18, 2024, from Harvard Magazine: https://www.harvardmagazine.com/2019/01/artificial-intelligence-limitations#

Simonyan, K., & Zisserman, A. (2014). Very deep convolutional networks for large-scale image recognition. arXiv preprint arXiv:1409.1556.

Singh, C. R., & Manoharan, G. (2024). Strengthening Resilience: AI and Machine Learning in Emergency Decision-Making for Natural Disasters. In Internet of Things and AI for Natural Disaster Management and Prediction (pp. 249-278). IGI Global.

Singh, R. (2023). Turkey Earthquake 2023: Reducing Risks in the Indian Context. MP-IDSA. https://idsa.in/idsacomments/turkey-earthquake-2023-rsingh-120423

Singh, R., Manoharan, G., & Singh, S. (2024). Strategic Deployment of AI and Drones Enhancing Disaster Management in Natural Disasters. In Ouaissa, M., Ouaissa, M., Boulouard, Z., Iwendi, C., & Krichen, M. (Eds.), *AI and IoT for Proactive Disaster Management* (pp. 172–203). IGI Global., DOI: 10.4018/979-8-3693-3896-4.ch009

Socher, R., Lin, C. C., Manning, C., & Ng, A. Y. (2011). Parsing natural scenes and natural language with recursive neural networks. In *Proceedings of the 28th international conference on machine learning (ICML-11)* (pp. 129-136).

Sufi, F., & Alsulami, M. (2025). AI-Driven Global Disaster Intelligence from News Media. *Mathematics*, *13*(7), 1083. DOI: 10.3390/math13071083

Sun, W., Bocchini, P., & Davison, B. D. (2020). Applications of artificial intelligence for disaster management. *Natural Hazards*, *103*(3), 2631–2689. DOI: 10.1007/s11069-020-04124-3

Tadokoro, S. (2020). Disaster Response Robot. In Baillieul, J., & Samad, T. (Eds.), *Encyclopedia of Systems and Control*. Springer., DOI: 10.1007/978-1-4471-5102-9_181-2

Tanzi, T. J., Chandra, M., Isnard, J., Camara, D., Sebastien, O., & Harivelo, F. (2016). Towards" drone-borne" disaster management: future application scenarios. In XXIII ISPRS Congress, Commission VIII (Volume III-8) (Vol. 3, pp. 181-189). Copernicus GmbH.

Tariq, M. U. (2024). AI and IoT in Flood Forecasting and Mitigation: A Comprehensive Approach. In AI and IoT for Proactive Disaster Management (pp. 26-60). IGI Global. DOI: 10.4018/979-8-3693-3896-4.ch003

United Nations. (1999). Activities of the International Decade for Natural Disaster Reduction; Report of the General Secretary. UN. https://www.un.org/esa/documents/ecosoc/docs/1999/e1999-80.htm

United Nations Office for Disaster Risk Reduction. (n,d,). Disaster. https://www.undrr.org/terminology/disaster

Vermiglio, C., Noto, G., Rodríguez Bolívar, M. P., & Zarone, V. (2022). Disaster management and emerging technologies: A performance-based perspective. *Meditari Accountancy Research, 30*(4), 1093–1117. DOI: 10.1108/MEDAR-02-2021-1206

Wang, Z., & Ye, X. (2018). Social media analytics for natural disaster management. *International Journal of Geographical Information Science, 32*(1), 49–72. DOI: 10.1080/13658816.2017.1367003

West, D. M. (2018). *The future of work: Robots, AI, and automation*. Brookings Institution Press. DOI: 10.5040/9780815751878

Wisner, B., Blaikie, P., Cannon, T., & Davis, I. (2004). *At risk: Natural hazards, people's vulnerability, and disasters* (2nd ed.). Routledge.

Yan, L., Mahmud, S., Shen, H., Foutz, N. Z., & Anton, J. (2020, November). MobiRescue: Reinforcement learning based rescue team dispatching in a flooding disaster. In 2020 IEEE 40th International Conference on Distributed Computing Systems (ICDCS) (pp. 111-121). IEEE.

Yang, Z., Nguyen, L., Zhu, J., Pan, Z., Li, J., & Jin, F. (2020, December). Coordinating disaster emergency response with heuristic reinforcement learning. In *2020 IEEE/ACM International Conference on Advances in Social Networks Analysis and Mining (ASONAM)* (pp. 565-572). IEEE. DOI: 10.1109/ASONAM49781.2020.9381416

Yang, Z., Xu, C., & Li, L. (2022). Landslide detection based on ResU-Net with transformer and CBAM embedded: Two examples with geologically different environments. *Remote Sensing (Basel), 14*(12), 2885. DOI: 10.3390/rs14122885

Yaseen, Z. M., El-Shafie, A., Jaafar, O., Afan, H. A., & Sayl, K. N. (2015). Artificial intelligence based models for stream-flow forecasting: 2000–2015. *Journal of Hydrology (Amsterdam), 530*, 829–844. DOI: 10.1016/j.jhydrol.2015.10.038

Yin, S., & Kaynak, O. (2015). Big data for modern industry: Challenges and trends [point of view]. *Proceedings of the IEEE, 103*(2), 143–146. DOI: 10.1109/JPROC.2015.2388958

Yu, B., Xu, C., Chen, F., Wang, N., & Wang, L. (2022). HADeenNet: A hierarchical-attention multi-scale deconvolution network for landslide detection. *International Journal of Applied Earth Observation and Geoinformation, 111*, 102853. DOI: 10.1016/j.jag.2022.102853

Zafar, U., Shah, M. A., Wahid, A., Akhunzada, A., & Arif, S. (2019). Exploring IoT applications for disaster management: identifying key factors and proposing future directions. Recent trends and advances in wireless and IoT-enabled networks, 291-309.

Zhang, D., Han, X., & Deng, C. (2018). Review on the research and practice of deep learning and reinforcement learning in smart grids. *CSEE Journal of Power and Energy Systems*, *4*(3), 362–370. DOI: 10.17775/CSEEJPES.2018.00520

Zhang, H., Dolan, C., Jing, S. M., Uyimleshi, J., & Dodd, P. (2019). Bounce forward: Economic recovery in post-disaster Fukushima. *Sustainability (Basel)*, *11*(23), 6736. DOI: 10.3390/su11236736

Zhang, Y., Geng, P., Sivaparthipan, C. B., & Muthu, B. A. (2021). Big data and artificial intelligence based early risk warning system of fire hazard for smart cities. *Sustainable Energy Technologies and Assessments*, *45*, 100986. DOI: 10.1016/j.seta.2020.100986

Zhong, M., & Zhang, Q. (2010). Prediction of reservoir-induced earthquake based on fuzzy theory. *Proc. ISNNS*, 101-104.

Zhou, L., Wu, X., Xu, Z., & Fujita, H. (2018). Emergency decision making for natural disasters: An overview. *International Journal of Disaster Risk Reduction*, *27*, 567–576. DOI: 10.1016/j.ijdrr.2017.09.037

Chapter 3
Analysis of AI Models in Weather Forecasts:
A Leap Towards a Robust Future

Anusha Thakur
http://orcid.org/0000-0001-8761-2250
Independent Researcher, India

ABSTRACT

The practice of forecasting the meteorological conditions at a precise location and forthcoming time has been a quest for ages. From archaic society to contemporary meteorological models scraping varied amounts of data, the pursuit of precise estimates is witnessing a continual rise. Intensifying challenges of climate change bolster the need for forecasting technologies to safeguard the economies and communities against the unforeseen impacts. With the conventional methods having their confines, the predictive models nowadays that integrate AI offer better scalability and interpretability. The article delves into the groundbreaking role of AI in weather predictions, as well as exploring its benefits, applications, future potential, and challenges.

INTRODUCTION

Escalating weather conditions pose unparalleled challenges to ecological integrity, biodiversity losses, economic security, and societal stability (Seneviratne, 2021). The conditions, ranging from heatwaves to droughts to floods to severe storms, wield profound impacts on the environment and human livelihoods. The archetype

DOI: 10.4018/979-8-3373-7345-4.ch003

ght © 2026, IGI Global Scientific Publishing. Copying or distributing in print or electronic forms without written permission of IGI Global Scientific Publishing is prohibited. Use of this chapter to generative artificial intelligence (AI) technologies is expressly prohibited. The publisher reserves all rights to license its use for generative AI training and machine learning model development.

shift has begun to unfold, not driven by the physics-based simulations, but by the state-of-the-art technologies such as Artificial Intelligence.

AI, primarily transformer architectures and deep learning, is revolutionizing weather forecasting in ways beyond imagination. Unlike the conventional models that address the complicated equations for atmospheric flow, these pioneering models apprehend anomalies and patterns via high-dimensional data correlations and probabilistic reasoning (Leah, 2025). The convergence of data science and meteorology is not just a progressive development but a vital change in the way weather is being comprehended and anticipated. The integration caters to the major confines of conventional Numerical Weather Prediction (NWP) models (Amarel, 2025).

The role of AI goes beyond mere technological developments and poses to be a necessity in the continuing pursuit to comprehend and foretell the caprices of nature. The transformation is not just restricted to the exclusive meteorological centres or research labs, but with the declining cloud computing costs and the models becoming open-source, the AI-based estimate models are witnessing an increase in accessibility, especially in developing regions. From regulating the predictive models to offering customized weather updates, the technology is expected to augment the precision of the day-to-day estimates and play a crucial role in the preparedness and response to the weather conditions.

The AI-based models help create estimates that are dependable and faster than those generated with conventional methods. With significant transformation of the traditional models, factors such as improved grid reliability, better decision-making processes, and augmented disaster response are gaining traction (DeepMind, 2025). Artificial intelligence is significantly revolutionizing the segment of weather forecasting by providing unparalleled insights and accuracy. Leveraging the capabilities of ensemble forecasting, data assimilation, Machine Learning (ML), and other advanced techniques, the technology showcases effective readiness and responsiveness to weather events.

The article discusses real-use cases of companies offering models and platforms to augment the field of meteorological conditions. Despite the promising aspects of AI in weather conditions, challenges do exist. According to the meteorologists, at times, it becomes difficult to understand how AI makes its decisions and also to trust its results, posing a challenge in its implementation (Jones, 2025). The article further explores the different factors hindering the implementation of AI in weather prediction models. Addressing the important matters of climatic variability, training complexities, and data quality will take time and ongoing effort; it is expected to augment the weather predictability patterns in a precise manner.

Research Questions

RQ 1: Need for Artificial Intelligence in Weather Forecasting in the changing climatic conditions?
RQ 2: How do AI-based models augment Numerical Weather Prediction?
RQ 3: How the key market players respond to the changing weather conditions, and the way they incorporate AI in their offerings?
RQ 4: What are the implementation barriers of artificial intelligence in weather prediction?
Key Findings: Weather forecasting and applications are witnessing an explosive growth with the integration of artificial intelligence. The latest approaches illustrate the vast potential to boost economic prosperity, health, and public safety. The progress in today's epoch is being facilitated by the advances in Numerical Weather Prediction (NWP), the availability of more datasets, and the growth in computing power, which gathers Earth's physical conditions to be monitored more regularly. Despite the hurdles to overcome, the potential of AI models to offer insightful projections across varied climate states paves the way for new avenues for climate adaptation strategies and research.

LITERATURE REVIEW

Artificial Intelligence is redefining the boundaries of possibilities in climate forecasting. The incorporation of AI in weather prediction has led to substantial developments, augmenting both local and global forecasting methodologies. Precision in forecasting methods is pivotal for mitigating the impact of extreme weather events and optimizing resource allocation (Kumar, 2025). From the inception of the conventional numerical weather prediction models to the latest hyper-local predictions and real-time analysis, the technology has completely revolutionized weather forecast methodologies.

The weather prediction methods are anticipated to witness a radical change for better long-term estimations. The latest forecasting technologies are by bounds and leaps, cheaper, faster, and easier to produce with less computational energy, compared to conventional systems.

With the continuous technological evolution, AI plays a pivotal role in augmenting the dependability and precision of the weather estimates, thereby safeguarding individuals and communities. The emergence of these tools does not necessarily indicate a transition from conventional to modern forecasting methods. The AI-based weather forecasting system operates in conjunction with conventional methods and has been extensively tested in various real-world scenarios. The technology is no longer a substitute but has been developed in a way to support health, economic prosperity, and public safety (Bassetti, 2024). Even with challenges, the ongoing interdisciplinary collaboration and technological advancements pave the way for a future wherein the AI-driven forecasts protect the property, save lives, and assure sustainable development.

NEED FOR AI IN WEATHER FORECASTING

From monitoring the natural instances to developing early weather predictions to the latest medium-range predictions, the need for weather forecasting has been a crucial tool in advancing societal conditions. According to the collaboration of studies between Microsoft Research and the University of Washington illustrates how, artificial intelligence can contemplate the past climatic patterns to predict future events in a faster and precise manner, unlike current technologies (AI, 2021). AI-based models surpass conventional prediction systems for weather forecasts over varying lead times from seasons to hours. This can be attributed to the fact that the weather provides precisely what AI relies on: data, and a lot of it. The shifting trends also depend upon the development of sophisticated AI algorithms, the availability of datasets, and advancements in computational capabilities.

The incorporation of cutting-edge technology mechanisms enables multi-hazard decision support systems to become more accessible and capable of providing actionable and clear predictions to avoid disaster losses. The data assimilation techniques, deep learning, and machine learning models form a fundamental part of modern meteorology. The meteorological advancements focus on AI algorithms that analyze huge datasets of social media feeds, radar data, satellite imagery, and weather observations to recognize the relationships and patterns that may otherwise not be easily measurable by conventional models.

Localized and Personalized Forecasts

Artificial Intelligence enables hyper-local weather predictions, offering customized forecasts to specified locations. Unlike the conventional predictive models operating on bigger regional grids, AI helps incorporate the local data and downscale the outputs to enhance the microclimate projections. The hyper-local predictions driven by AI poses advantageous to sectors such as city management, renewable energy, and agriculture, wherein understanding block-by-block weather plays an important role. With the availability of granular details, the AI-based models enable the communities to better prepare and respond at their respective doorsteps (Morrill, 2024). In addition to this, harnessing artificial intelligence also enables the weather services to offer curated information, thereby helping individuals to make better and relevant decisions based on the forecasts.

Streamlining Atmospheric Modelling

The conventional NWPemphasizes solving the complicated physics equations on supercomputers, which can otherwise be very time-consuming and necessitate the usage of a lot of computational resources. With the growing number of events, ranging from heatwaves to droughts to floods to severe storms, the need for AI is witnessing a significant surge. Understanding, characterising, and modeling the extreme events acts as a key to advancing the mitigation and adaptation strategies (Valls, 2025). The technology helps process and evaluate large datasets to identify complex patterns, create advanced models, and subsequently, forecast severe weather events. The optimization techniques focus on rapid data analysis and complicated pattern recognition.

Augmented Forecast Accuracy

Integration of artificial intelligence is expected to bolster the precision of weather predictions by learning the complicated patterns that are otherwise absent in the case of conventional models. The researchers nowadays are using the data-driven AI hybrid systems, which amalgamate the conventional atmospheric models with AI; the forecasts need to be combined transparently. AI algorithms help analyze the meteorological data with unparalleled precision and speed, leading to more reliable weather predictions (Team G., 2025). The power of AI helps develop cutting-edge forecast models for evaluating the multifarious impacts of climate change.

The techniques, such as predictive analytics and machine learning algorithms in climate modeling, offer a strong framework for comprehending and predicting the complicated dynamics related to global climate change. Such models empower precise and quick projections of future historical eventualities, scenario testing, climate scenarios, their adaptation and magnitude.

AI-BASED MODELS AUGMENT NUMERICAL WEATHER PREDICTION

The advancements of artificial intelligence, lack of access to big data, and inherent computational limitations of NWP models, and lack of accessibility to big data, bolster the need for AI in terms of intricacy and resolution (Waqas, 2025). With the NWP models facing the complexities of fine-scale details and being computationally intensive, integration of artificial intelligence helps cater to the same. The integration of AI with the NWP models is anticipated to lead to much enhanced projections. The integration of AI can be illustrated as:

This includes:

i. *Hybrid Approaches:* The technology emphasizes on creating models that amalgamate physical simulations with data-driven insights. A blend of machine learning outputs with physics-based predictions is expected to augment the precision and speed of weather forecasting.
ii. *Error Correction:* AI-based models help find systematic errors in the Numerical Weather Prediction outputs, and subsequently make the adjustments.

Further, it can be elaborated as:

Data Pre-Processing

The data pre-processing techniques poses to be vital for the weather predictions. The raw weather data gathered from different sources undergo stringent pre-processing before the effective interpretation of the AI models. The data cleaning process comprises of removing the outliers, errors, and inconsistencies. The normalization processes ensure the presence of data at a uniform scale. Time series analysis methods allow the identification of trends and patterns in the weather data. Optimizing the data for the AI models also includes opting for the most appropriate variables and filtering the dataset for augmented forecasts.

Real-Time Data Assimilation

Artificial Intelligence also plays a significant role in real-time data assimilation by enabling the development of advanced algorithms that can adapt to diverse data sources, including non-traditional sources such as community-sourced data from smartphones, which further helps address skepticism in observations. With the effective analysis of inputs from radar systems, weather stations, and satellites, the AI-based algorithms help update predictions in a precise and regular manner. The blended outputs generate the estimates of the current atmospheric conditions. Such abilities play a pivotal role in high-impact weather events, such as tornadoes and hurricanes, wherein appropriate forecasts help protect lives. Despite the developments in Data Learning (DL) and Machine Learning, the computational costs related to these models necessitate prolonged processing times and extensive hardware resources, thereby posing a challenge. Further, the amalgamation of the AI algorithms in the NWP models also emphasize on the changes in the workflows and architectures.

Quantification of Uncertainty

Unpredictability in weather forecasts arises from the intrinsically chaotic behavior of atmospheric systems. Such uncertainties in the conventional methods are catered to with probabilistic predictions with ensembles of deterministic models, which require significant resources and are also computationally expensive. The efforts are limited to point-valued and deterministic forecasting, making it difficult to compute predictive uncertainties, which are vital for optimal decision-making in various applications (Bülte, 2025). Unlike the conventional models, artificial intelligence-based models enable augmented forecast generation at a small fraction of the computational costs. The AI-based advancements provide an alternative to the prevailing operational weather predictive paradigm, wherein the algorithmic resources kept by the statistical imitator are expected to be assigned to regularly issuing the forecasts or augmenting the resolution of the physics-based models (Li, 2024).

Data Post-Processing Techniques

The technique focuses on the applicability of the statistical methods to raw the outputs to augment the usability, reliability, and accuracy. The process helps in correcting the bias, refining the probabilistic predictions, augmenting precision, and blending the model outputs. A few of the post-processing techniques include Transformer-based models, Ensemble Learning, and Quantile Mapping (QM) (arxiv, n.d.), (Andrade, 2024).

HOW DOES ARTIFICIAL INTELLIGENCE IN WEATHER FORECASTING WORK?

The process of AI in weather forecasting includes a few steps to alter raw data into precise forecasts. This can be elaborated as:

Figure 1. How does artificial intelligence in weather forecasting work? (Climavision, 2025)

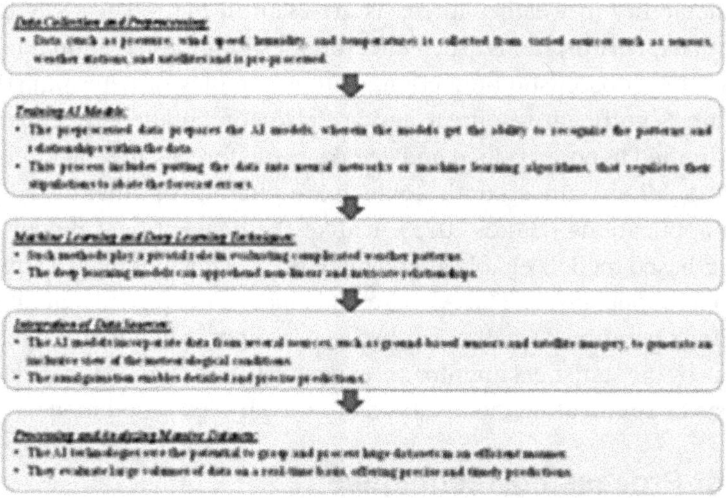

REAL USE-CASES OF ARTIFICIAL INTELLIGENCE IN WEATHER FORECASTING

With the escalating severity and frequency of weather events, such as tornadoes and hurricanes, it is more crucial than ever to enhance climate research and forecasts through innovative technologies.

1. MICROSOFT - AURORA

Aurora, a large-scale, state-of-the-art AI foundation model, is engaged in forecasting atmospheric processes and weather patterns. Microsoft announced the introduction of this groundbreaking new artificial intelligence model that promises to enhance the accuracy of environmental and weather forecasts.

An Advancement in the Prediction Technology:

- The model has the potential to forecast atmospheric changes in a faster way and generate high-resolution predictions for predicting complicated weather phenomena (such as air quality fluctuations and typhoons) in comparison to traditional methodologies.

- The model's ground-breaking architecture has been proficient in over a million hours of diversified climate and weather simulations, augmenting it to outshine in a wider spectrum of forecasting tasks while attaining a striking spatial resolution of 0.1° - nearly 11 km at the equator. This extent of detailedness augments the precision of operational estimates and provides an anticipated computational speed advantage of approximately 5,000 times over traditional numerical weather-forecast systems (Microsoft, n.d.).
- The company avows that Aurora can create a 10-day global prediction in a minute, a task that characteristically needs hours on high-performance systems. The model's exceptionality in utilizing machine learning to evaluate and detect patterns differentiates it from normal weather models, thereby enabling users to provide estimates with a resolution of nearly 6.2 miles (10 kilometres) and customized insights for global and domestic contexts (Singh, 2025).
- Aurora comprises an adjustable 3D Swin Transformer with Perceiver-based decoders and encoders, augmenting it to operate and project a variety of atmospheric variables across the pressure and space levels. By pre-training on the collection of varied data and regulation of precise tasks, the model has the potential to capture elaborate structure and patterns in the atmosphere, enabling it to excel even with the restricted training data when it is being adjusted for any particular task (Bruinsma, 2024).

Applications of Aurora

Aurora's predictions in a wide range of applications include: Predicting typhoons, ocean wave conditions, air quality, maritime, and public health safety. The model is further broadening its applicability by integrating with sectors such as public policy, disaster management, and agriculture. The features of scalability and adaptability make Aurora a favourable model for optimizing crisis management and resource allocation, enhancing weather-associated interferences, as well as bolstering strategic planning and decision-making processes.

The development of Aurora marks a leap onward in using machine learning for environmental monitoring. Hence, with actionable and timely intelligence, the model helps transform the above-mentioned areas in an unprecedented manner.

Challenges of Aurora

Despite the overarching benefits of AI-driven predictions, Aurora, too, has a few sets of challenges. Aurora's dependence on a wider range of datasets generates issues wherein the gathered data poses erratic or sparse data, leading to forecasting errors. In addition to this, the "black box" nature of AI systems often results in trust and transparency problems among the users, which poses a question in terms of the inferences drawn (Ferguson, 2025).

2. NVIDIA - EARTH-2

Nvidia announced the launch of Earth-2, a platform for unfolding AI-augmented, intensified, high-resolution weather and climatic solutions with responsive visualization. The platform has the potential to make better forecasts and estimate the challenges of extreme events (Geman, 2025). It is a digital twin cloud platform that amalgamates the power of computer graphics, physical simulations, and AI, - enabling visualization and simulation of climate and weather forecasts at a worldwide scale with unparalleled speed and accuracy. The Earth-2 platform offers an array of pioneering AI weather models, microservices, and tools for simulating and visualizing the world.

The NVIDIA Omniverse Blueprint

NVIDIA Omniverse Blueprint is a ground-breaking digital-twin platform adopted for more precise, energy-efficient, and high-resolution weather forecasts and disaster readiness. The platform allows independent software retailers to develop and position the AI-augmented visualization and simulation pipelines for the weather and climate domains at a global level with unparalleled speed and accuracy. The blueprint is a reference architecture that bolsters the development of climate tech applications for climate and weather analysis, and risk mitigation for the progress of digital twins of climate and weather. It further showcases (NVIDIA, n.d.):

i. How can the NVIDIA Omniverse platform be utilized to envisage geo-spatial data at scale?
ii. How are the AI weather models positioned through the NIMs to generate intensified weather predictions?
iii. How does a merged data architecture offer an adapter framework to link the open data repositories and partner platforms?

Innovative Approaches for Climate Tech - Generative AI Models and Their Impact

In the domain of climate technology, the Earth – 2 platform leverages Generative AI to steer in an era of accuracy in climate and weather forecasting. The revolutionary impact of the Generative AI models, emphasizing their efficacy and applicability in climate tech, is illustrated as:

- Corrdiff, a Generative AI model, is a set of software and services for climate and weather research. The model uses GenAI to ameliorate weather forecasts, offering high-resolution estimates that augment climate resilience and disaster response. This pioneering model, engaging diffusion modelling methods for super-resolution imaging, is gaining prominence in decision-making processes by integrating pivotal metrics at a speed and scale previously unachievable. Corrdiff minimizes the resolution of weather prediction models from nearly 25 km to 2 km, attaining this at 3,000 times the energy efficiency and 1,000 times the speed of traditional models (McClure, 2024).
- FourCastNet: FourCastNet, an AI model proposed by NVIDIA, augments delivering early warnings (such as nearly three weeks in advance), that play a pivotal role in readying and alleviating the impacts of severe weather conditions, if any (Szabo, 2024). The model approaches global weather modelling by applying a geometric and scalable ML approach to prospective ensemble projections. The approach is intended to precisely model the spatially associated likelihood nature of the issue, leading to a realistic and stable spectra dynamic across manifold scales.

FourCastNet provides prediction precision that exceeds leading traditional ensemble models and competes with the finest diffusion-based systems, whilst producing predictions 8 to 60 times quicker than these approaches. In comparison to other machine learning methods, the model illustrates exceptional possible calibration and withholds accurate spectra, even at protracted lead times of nearly 60 days. Such advancements ascertained via a righteously convolutional neural network architecture, particularly customized for spherical geometry (Bonev, 2025).

The newest version of FourCastNet, founded on the Spherical Fourier Neural Operator, quickly delivers months-long predictions and consents for the generation of groups of simulations or large ensembles with insignificant differences in initial conditions, for extreme weather risk predictions well in advance (Petty, 2023).

Applications in Climate Tech

The latest cloud Application Programming Interfaces (API) by Earth-2 enables users to virtually generate high-resolution, interactive, and AI-powered weather simulations, wherein the startup companies such as Tomorrow.io, ClimaSens, and North.io benefit from the platform's potential to build climatic AI developments. This can be illustrated as:

- Tomorrow.io: Tomorrow.io, based in Boston, provides weather-related and actionable insights to individuals, businesses, and countries by putting innovative machine learning and AI models to an exclusive global dataset collected from sensors, radars, and satellites. The climate adaptation and weather intelligence platforms offer precise and high-resolution predictions across long-term and short-term time zones. The company progresses by using Earth 2 to study the probable impacts on its group of satellites on the global model predictions. By operating OSSEs or observing-system simulation experiments in amalgamation with the Earth-2 AI prediction models, the company recognizes the optimum structure of satellites and other instruments to augment the weather-forecasting conditions (Walker, 2024).
- ClimaSens: ClimaSens, founded in 2017, is headquartered in Melbourne, Australia. The company is the provider of climate risk intelligence to augment communities, government, and businesses to comprehend and mitigate the risks associated with the changing meteorological conditions. The company's AI-powered high-resolution platform integrates the pioneering real-time weather informatics, satellite data, and climatic models to provide insights at a global level and foster sustainable and resilient communities (ClimaSens, 2017).
- North.io: Headquartered in Kiel, Germany, North.io was established in 2011. The company emphasizes mapping the oceans by gathering and evaluating varied amounts of data from Autonomous Underwater Vehicles (AUVs). North.io is a geospatial cloud specialist that focuses on transforming the modelling, gathering, and processing of ocean data via NVIDIA technologies. The company illustrates the way Generative AI models prove to be crucial in understanding and conserving natural resources. The synergy between NVIDIA's recognized technologies and TrueOcean's marine data platform is unleashing unprecedented insights into the oceans, marking an innovative era in oceanic data exploration. The company integrates three ground-breaking NVIDIA technologies - NVIDIA RAPIDS, NVIDIA Modulus, and NVIDIA Earth-2 cloud and platform services into the TrueOcean platform to unveil unparalleled avenues for ocean use and shield (International, 2024).

3. GOOGLE – GRAPHCAST

Google announced its investments in climate and weather research to guarantee that the projected technology of tomorrow caters to the demand for consistent weather information. A few of its latest innovations include: MetNet-3, a state-of-the-art high-resolution neural weather forecast model by Google, of up to 24 hours (Merchant, 2023), and GraphCast, a weather model forecasting weather up to nearly 10 days ahead.

Google's groundbreaking weather forecasting model "GraphCast" is based on Graph Neural Networks (GNNs) and machine learning, which are a significantly useful architecture for refining the spatially structured data. The model builds predictions at a high resolution of 0.25 degrees latitude/longitude (28 km x 28 km at the equator), which exceeds a million grid points wrapping the Earth's surface.

> GraphCast makes forecasts at a high resolution of 0.25 degrees longitude/latitude (28km x 28km at the equator). That's more than a million grid points covering the entire Earth's surface. At every grid point, the GraphCast is engaged in forecasting nearly five Earth-surface variables, such as mean sea-level pressure, wind speed, temperature, and direction, along with six atmospheric variables at individual levels of 37 altitudes, comprising temperature, direction, wind speed, and specific humidity (Lam, 2023).

WeatherNext In March 2025, Google Cloud announced the launch of "WeatherNext", the AI weather predictive models created by Google Research and DeepMind. The models offer pioneering weather conditions that prove beneficial for the manufacturers, financial service firms, retailers, and energy companies to prepare for severe weather events. This can be elaborated as (Cloud, 2025):

- WeatherNext Gen, a prospective system that offers better estimates of extreme events as well as day-to-day weather up to 15 days ahead in comparison to the current collaborative models.
- WeatherNext Graph, a predictable system that projects the medium-range meteorological conditions up to 10 days in advance.

4. EUROPEAN CENTRE FOR MEDIUM-RANGE WEATHER FORECASTS (ECMWF)

The European Centre for Medium-Range Weather Forecasts (ECMWF) announced the launch of the Artificial Intelligence Forecasting System (AIFS) into effective use in conjunction with its physics-based Integrated Forecasting System (IFS) in February 2025 (OECD.AI, 2025).

The new artificial intelligence-powered model can outperform traditional weather forecasting systems nearly 15 days ahead. Along with tracking the temperature and wind, delivering particulars on precipitation forms, from snow to rain, the system also offers key information to exploit renewable energy operations, such as wind speed at turbine levels, and projections of surface solar radiation levels. The new system is being marked as a milestone to completely revolutionize the weather forecast methodologies, easing the availability of global predictions anytime (Igini, 2025).

CHALLENGES IN IMPLEMENTING ARTIFICIAL INTELLIGENCE IN WEATHER FORECASTING

The development of AI algorithms in weather predictions embodies a prominent convergence of artificial intelligence and climate science. Despite the groundbreaking potential of AI, the amalgamation of the pioneering technology in the weather forecasting methodologies poses a challenge.

1. Quality And Availability of Data

One of the major challenges in the implementation of AI in weather forecasting is the lack of sufficient and quality data. Factors such as biases, inconsistencies, and data gaps in the historical data are expected to impact the performance of the AI models, thereby leading to inaccurate projections.

AI algorithms require high-quality and large datasets for validation and training. Such data often originates from varied sources such as atmospheric models, ground-based weather stations, and satellite images. Inconsistencies such as incompatible time stamps and geolocations pose substantial challenges when training the AI models (Models, n.d.). In order to mitigate such issues, stringent protocols are essential for gathering the data and cleansing the same, which further requires the amalgamation of machine learning techniques to screen out the anomalies or operating detailed audits of the data sources to confirm dependability.

In case of data availability, various prevailing datasets might witness gaps in the historical data or not deliver real-time updates, which pose a challenge in areas inclined to abrupt and extreme weather conditions. One potential solution to overcome the issue includes the implementation of synthetic data generation methodologies, driven by cutting-edge machine learning techniques. Producing realistic datasets and simulating the weather predictions enables effective enhancement of the training programs, thereby letting the AI algorithms learn without any real-world constraints.

2. Integration Issues

Amalgamating AI into the existing weather prediction models necessitates the need for careful validation and a multifaceted approach. The latest machine learning models are particularly trained on well-curated and high-quality datasets such as cloud-free satellite imagery, which at times do not replicate the cloudy conditions and error-prone weather predictions. Hence, segment-specific strategies help align the model performance and augment seamless prediction and mitigation of risks (Camps-Valls, 2025).

3. Ethical Considerations

The meteorological conditions using AI technology entail handling volumes of data that otherwise raise privacy issues. With the AI-driven weather predictions gaining traction, the need to address the ethical issues is also witnessing a rise. The ethical considerations bolster the need for a multidisciplinary method that comprehends social justice, regulatory policies, and technology. One of the most pressing issues is guaranteeing the precision of the forecasts, whilst avoiding unjust bias against marginalized populations. Hence, to address the ethical issues, the AI systems must be fair, explainable, and transparent (Lee, 2025).

4. Lack of Interpretability

Despite their state-of-the-art potential, model interpretability poses yet another challenge. Several innovative machine learning meteorological estimate models, such as deep neural networks, function as "black boxes," posing it difficult to comprehend the reasoning abaft forecasts. Such lack of transparency is expected to deter acceptance and trust among the public and meteorologists (Yang, 2024).

FUTURE SCOPE OF RESEARCH

In today's era, the weather conditions are not just being predicted, but the focus is also on building a resilient society. With technologies such as Google's GraphCast to the ECMWF's operational system, the field of meteorology is entering a new era. Artificial intelligence is evolving as a powerful tool in weather forecasting, with the capability to augment predictive capabilities. With the AI algorithms becoming more intricate, computational resources are witnessing a rise, and the datasets becoming more inclusive and larger, it is projected that this cutting-edge technology will augment the usefulness, timeliness, and precision of the weather forecasts.

The AI models are revolutionizing the way weather prediction is conducted, augmenting food security, grid reliability, and disaster response and enabling better decision-making. From seasonal forecasting to nowcasting, the technology offers accessible, accurate, and precise insights to redesign operational practices. The gains in forecast performance are expected to help the researchers to seamlessly predict the weather conditions, to save lives and augment the dependability of supply chains and energy. Factors such as inclusive governance, robust evaluation, and thoughtful integration are expected to contribute to the transformative potential of AI in weather forecasting. Future emphasis, thereby, significantly lies on the speed of adaptability of this new reality to benefit society and individuals.

CONCLUSION

The amalgamation of pioneering technologies such as AI with the weather forecast models grips enormous potential to transform the way forecasts are built and delivered. AI functions as an influential tool in geoscience, with the analysis of complex and vast datasets (ECMWF, 2025). The Artificial Intelligence technologies, particularly neural networks, and machine learning have intensely revolutionized the weather predictions by refining larger amounts of data and recognizing patterns unlike the conventional methodologies.

Despite pioneering capabilities, the implementation of AI in weather prediction models also comes along with a few challenges that need to be addressed. From intricacies of the atmospheric systems to obtainability to insinuations of climate change to data quality, the route to precision in weather conditions includes addressing the hurdles that entail a multifaceted approach. The article explores the diversified ways in which AI is revolutionizing preparedness and responsiveness to weather conditions. It further discusses the different models offered by the tech giants for transparent and seamless predictions, along with the challenges in the implementation of AI in weather prediction methodologies.

Key Takeaways

- AI has largely wedged different fields, such as Earth system sciences, by forecasting the severity of the events, parameter estimation, and augmenting model emulation.
- AI augments the precision of the weather prediction by identifying patterns and evaluating large datasets.

- Tech leaders such as NVIDIA, Microsoft, and Google are emphasizing enormous resources into the weather AI models, each demanding breakthrough competencies. For instance,

i. Aurora by Microsoft focuses on environmental predictions along with weather forecasting.
ii. Earth-2 by NVIDIA is a digital twin combining the power of computer graphics, physical simulations, Graphics Processing Unit (GPU) acceleration, and AI. The platform is engaged in simulating the climatic and weather conditions in an unparalleled manner.
iii. GraphCast by Google is a weather-forecasting multi-scale Graph Neural Network (GNN) model that emphasizes outpacing conventional estimating systems.
iv. European Centre for Medium-Range Weather Forecasts (ECMWF): Europe's new artificial intelligence-powered weather predicting model emphasizes augmenting extreme meteorological predictions across the world.

REFERENCES

AI, T. (2021, August 4). *The Future of Artificial Intelligence in Weather Forecasting*. Retrieved from Towards AI: https://towardsai.net/p/l/the-future-of-artificial-intelligence-in-weather-forecasting

Amarel, T. (2025, April 2). *Improving Short-Term Weather Forecasting Accuracy: A Practical Guide to Implementing AI and Machine Learning Models*. Retrieved from TaylorAmarel.com: https://tayloramarel.com/2025/04/improving-short-term-weather-forecasting-accuracy-a-practical-guide-to-implementing-ai-and-machine-learning-models/

Andrade, F. S. (2024). Application of Weather Post-Processing Methods for Operational Ensemble Hydrological Forecasting on Multiple Catchments in Canada. *Journal of Hydrology (Amsterdam)*, *642*, •••. https://www.sciencedirect.com/science/article/pii/S0022169424012575

arxiv. (n.d.). *Statistical Post-Processing Yields Accurate Probabilistic Forecasts from Artificial Intelligence Weather Models*. Retrieved from arxiv.org: https://arxiv.org/html/2504.12672#:~:text=These%20findings%20show%20that%20statistical%20post-processing%20methods%20developed,into%20existing%20workflows%20in%20a%20low-risk%2C%20incremental%20fashion

Bassetti, F. (2024, March 12). *The AI Revolution in Weather Forecasting: Reactions and Comments from the Mediasphere*. Retrieved from https://www.climateforesight.eu/articles/reacting-to-the-ai-revolution-in-weather-forecasting/

Bonev, B. K. (2025, July 16). *FourCastNet 3: A Geometric Approach to Probabilistic Machine-Learning Weather Forecasting at Scale*. Retrieved from NVIDIA Corporation: https://research.nvidia.com/publication/2025-07_fourcastnet-3-geometric-approach-probabilistic-machine-learning-weather

Bruinsma, W. S. (2024, June 3). *Introducing Aurora: The First Large-scale Foundation Model of the Atmosphere*. Retrieved from Microsoft Research Blog: https://www.microsoft.com/en-us/research/blog/introducing-aurora-the-first-large-scale-foundation-model-of-the-atmosphere/?msockid=370d1976c15d6cf10f670ca0c0226dad

Bülte, C. H. (2025). Uncertainty Quantification for Data-driven Weather Models. *Artificial Intelligence for the Earth Systems*, *39*, •••. https://journals.ametsoc.org/view/journals/aies/aop/AIES-D-24-0049.1/AIES-D-24-0049.1.xml

Camps-Valls, G. F.-T. (2025). Artificial intelligence for Modeling and Understanding Extreme Weather and Climate Events. *Nature Communications,* https://www.nature.com/articles/s41467-025-56573-8

ClimaSens. (2017). *Transforming Climate Risk to Resilience*. Retrieved from ClimaSens: https://climasens.com/

Climavision. (2025). *AI in Weather Forecasting: The Future of Weather Prediction*. Retrieved from Climavision: https://climavision.com/resources/ai-weather-forecasting-guide/

Cloud, G. (2025, March 5). *Google Cloud Brings WeatherNext AI-Powered Weather Predictions to Enterprises*. Retrieved from Google Cloud: https://www.googlecloudpresscorner.com/2025-03-05-Google-Cloud-Brings-WeatherNext-AI-Powered-Weather-Predictions-to-Enterprises

DeepMind. G. (2025, March 5). *WeatherNext*. Retrieved from Google DeepMind: https://deepmind.google/science/weathernext

ECMWF. (2025). *Unlocking the Black Box: The Potential of Explainable AI in Geoscience*. Retrieved from ECMWF: https://www.ecmwf.int/en/about/media-centre/focus/2025/unlocking-black-box-potential-explainable-ai-geoscience

Ferguson, M. (2025, May 26). *Microsoft's Aurora AI Revolutionizes Weather Forecasting*. Retrieved from OpenTools: https://opentools.ai/news/microsofts-aurora-ai-revolutionizes-weather-forecasting

Geman, B. (2025, July 17). *Nvidia says New AI Weather Forecasting Model boosts Speed, Accuracy*. Retrieved from Axios: Energy & Climate: https://www.axios.com/2025/07/17/nvidia-fourcastnet-3-ai-weather-forecasts

Igini, M. (2025, February 26). *Europe's New AI Weather Forecasting Model Up to 20% More Accurate Than Conventional Methods*. Retrieved from Earth.Org: https://earth.org/europes-new-ai-weather-forecasting-model-up-to-20-more-accurate-than-conventional-methods/

International, H. (2024, March 26). *North.io and NVIDIA Advance Ocean Data Processing towards a Sustainable Future*. Retrieved from Hydro International: https://www.hydro-international.com/content/news/north-io-and-nvidia-advance-ocean-data-processing-towards-a-sustainable-future

Jones, N. (2025, April 14). *A.I. Is Quietly Powering a Revolution in Weather Prediction*. Retrieved from E360: https://e360.yale.edu/features/artificial-intelligence-weather-forecasting

Kumar, S. R. (2025). Weather Forecasting Using Artificial Intelligence and Machine Learning. *International Journal on Science and Technology (IJSAT), Volume 16, Issue 1.*

Lam, R. (2023, November 14). *GraphCast: AI model for Faster and More Accurate Global Weather Forecasting*. Retrieved from Google DeepMind: https://deepmind.google/discover/blog/graphcast-ai-model-for-faster-and-more-accurate-global-weather-forecasting/

Leah. (2025, May 20). *How AI Is Revolutionizing Weather Forecasting: Models, Impacts, and the Future of Prediction*. Retrieved from BestAI: https://blog.bestai.com/how-ai-is-revolutionizing-weather-forecasting-models-impacts-and-the-future-of-prediction/

Lee, S. (2025, June 13). *The Future of Weather Forecasting: AI-Driven Insights*. Retrieved from Number Analytics: https://www.numberanalytics.com/blog/future-of-weather-forecasting-ai-driven-insights#google_vignette

Li, L. (2024, March 29). *Generative AI to quantify Uncertainty in Weather Forecasting*. Retrieved from Google Research: https://research.google/blog/generative-ai-to-quantify-uncertainty-in-weather-forecasting/

McClure, P. (2024, March 20). *Earth 2: Next-gen NVIDIA AI aims to solve weather and climate forecasting*. Retrieved from New Atlas: https://newatlas.com/technology/nvidia-ai-earth-2-weather-data-climate-action/

Merchant, S. (2023, November 1). *MetNet-3: A state-of-the-art Neural Weather Model Available in Google Products*. Retrieved from Google Research: https://research.google/blog/metnet-3-a-state-of-the-art-neural-weather-model-available-in-google-products/

Microsoft. (n.d.). *Azure AI Foundry Labs*. Retrieved from Aurora: https://ai.azure.com/labs/projects/aurora

Models, M. L. (n.d.). *Challenges and Solutions in Developing Weather AI Algorithms*. Retrieved from Machine Learning Models: https://machinelearningmodels.org/challenges-and-solutions-in-developing-weather-ai-algorithms/

Morrill, K. (2024). *AI Weather Forecasting: 10 Advances*. Retrieved from Yenra: https://yenra.com/ai-tech/weather-forecasting/

NVIDIA. (n.d.). *AI Weather Analytics with Earth-2*. Retrieved from NVIDIA Corporation: https://build.nvidia.com/nvidia/earth2-weather-analytics/blueprintcard

OECD. AI. (2025, May 19). *Artificial Intelligence Forecasting System (AIFS)*. Retrieved from OECD.AI: https://oecd.ai/en/catalogue/tools/artificial-intelligence-forecasting-system-aifs

Petty, H. (2023, July 5). *A Change in the Weather: AI, Accelerated Computing Promise Faster, More Efficient Predictions*. Retrieved from NVIDIA Corporation: https://blogs.nvidia.com/blog/ai-efficient-weather-predictions/

Seneviratne, S. X.-S. (2021). Weather and Climate Extreme Events in a Changing Climate. In V. P. Masson-Delmotte, *In Climate Change 2021: The Physical Science Basis. Contribution of Working Group I to the Sixth Assessment Report of the Intergovernmental Panel on Climate Change*, https://www.ipcc.ch/report/ar6/wg1/chapter/chapter-11/ (pp. pp. 1513–1766). Cambridge University Press. Retrieved from https://www.ipcc.ch/report/ar6/wg1/chapter/chapter-11/

Singh, D. (2025, May 26). *Microsoft's Aurora AI Sets New Standard for Environmental Forecasting*. Retrieved from Aadhunik AI: https://aadhunik.ai/blog/ai-weather-forecasting-microsoft-aurora/

Szabo, L. (2024, March 22). *Understanding Nvidia's Earth-2: A New Age of Climate Forecasting*. Retrieved from NowadAIs: https://www.nowadais.com/nvidias-earth-2-new-age-of-climate-forecasting/

Team, C. C. (2025, July 5). *How Artificial Intelligence Is Transforming Weather Forecasting*. Retrieved from Climate Cosmos: https://climatecosmos.com/climate-science/how-artificial-intelligence-is-transforming-weather-forecasting/

Team, G. (2025). *AI-Enhanced Weather Forecasting: Advanced Prediction Models*. Retrieved from GPAI: https://blog.gpai.app/solver/ai-enhanced-weather-forecasting-advanced-prediction#:~:text=AI%20algorithms%20can%20analyze%20meteorological%20data%20with%20unprecedented,leading%20to%20more%20reliable%20and%20timely%20weather%20predictions

Valls, G. T. (2025, February 24). Artificial Intelligence for Modeling and Understanding Extreme Weather and Climate Events. *Nature Communications*, *16*(1), 1919. Advance online publication. https://www.nature.com/articles/s41467-025-56573-8. DOI: 10.1038/s41467-025-56573-8 PMID: 39994190

Walker, T. (2024, March 19). *Climate Pioneers: 3 Startups Harnessing NVIDIA's AI and Earth-2 Platforms*. Retrieved from NVIDIA Corporation: https://blogs.nvidia.com/blog/climate-startups-ai-earth-2/

Waqas, M. H. (2025). Natural Hazards Research, Volume 5, Issue 2. *Artificial Intelligence and Numerical Weather Prediction Models: A Technical Survey*, 306-320.

Yang, R. H. (2024). Interpretable Machine Learning for Weather and Climate Prediction: A Review. *Atmospheric Environment, Volume 338, https://doi.org/,* https://www.sciencedirect.com/science/article/abs/pii/S1352231024004722#:~:text =Advanced%20machine%20learning%20models%20have%20recently%20achieved %20high,impede%20user%2.DOI: 10.1016/j.atmosenv.2024.120797

Chapter 4
Cognitive-Adaptive Warning Systems:
Integrating Human Perception Models With AI in Multi-Hazard EWS

Ismatulla Khayrullayev
Termez University of Economics and Service, Termez, Uzbekistan

Barno Matchanova
Urgench State Pedagogical Institute, Urgench, Uzbekistan

R. N. Ravikumar
http://orcid.org/0009-0009-3705-1681
Marwadi University, Rajkot, India

S. Aarthi
http://orcid.org/0009-0006-9064-2091
Marwadi University, Rajkot, India

Khayrulla Urozboev
Alfraganus University, Tashkent, Uzbekistan

Mary Subaja Christo
http://orcid.org/0000-0002-2702-9271
SRM Institute of Science and Technology, Kattankulathur, India

ABSTRACT

The chapter explores the emergence of Cognitive-Adaptive Early Warning Systems (CA-EWS) that integrate Artificial Intelligence and cognitive science to enhance multi-hazard preparedness. Traditional Early Warning Systems emphasize data precision but often neglect how humans perceive and act on alerts. The proposed framework combines machine learning, behavioral analytics, and perception modeling to create adaptive, human-centered warning communication. By leveraging real-time feedback from social media sentiment, mobility data, and community responses, CA-EWS dynamically refines alert clarity, timing, and personalization. Case studies from Japan's AI-enhanced tsunami warning and COVID-19 pandem-

DOI: 10.4018/979-8-3373-7345-4.ch004

ight © 2026, IGI Global Scientific Publishing. Copying or distributing in print or electronic forms without written permission of IGI Global Scientific Publishing is prohibited. Use of this chapter to n generative artificial intelligence (AI) technologies is expressly prohibited. The publisher reserves all rights to license its use for generative AI training and machine learning model development.

ic systems reveal measurable improvements in public responsiveness, trust, and compliance. The study concludes that embedding cognitive intelligence within EWS transforms reactive systems into proactive, empathetic, and resilient disaster communication ecosystems.

1. INTRODUCTION

The rising number and severity of natural and man-made hazards have clearly highlighted the need of strong Multi- Hazard Early Warning Systems (EWS). Such systems are vital in helping to reduce loss of life and property through prompt notification, preparedness and informed decision-making. Nevertheless, when evaluating the existing EWS, they usually put more attention on the technological accuracy of sensor networks, predictive algorithms, and the accuracy of the data, whereas human and cognitive aspects defining the way warnings are perceived and responded to are ignored (Krishna Pasupuleti, 2024b). The adaptive systems needed in today's society are those designed to combine the human response in behavior, the situational awareness, and the Artificial Intelligence (AI) to build the general resiliency. This chapter presents the idea of Cognitive-Adaptive Warning Systems and the way AI may collaborate with human perception models in order to improve and individualize risk communication. Such systems will be a bridge between machine intelligence and human cognition and will enable replacing a reactive model of early warnings with an active, adaptive, and human-centred model of disaster resilience. A Cognitive-Adaptive Early Warning System (CA-EWS) is an intelligent warning model relying on AI-powered warning architecture; this dynamically adjusts the alert messages with real-time behavioral, cognitive and contextual user feedback. It brings together the machine learning analytics and human perception models in order to make communication customized to minimize the alert fatigue and to maximize protective decision-making during multi-hazard events.

1.1 Overview of Multi-Hazard Early Warning Systems (EWS)

Multi-Hazard Early Warning Systems (MHEWS) are a combination of systems that concurrently track, identify, and pass information about the probability of a range of hazards (floods, earthquakes, cyclones, pandemics and technological catastrophes) occurring. In contrast with single-hazard systems, MHEWS acknowledge the damaging and cumulative character of risks, and one hazard can cause or aggravate another. Such systems include four essential elements, including risk knowledge, monitoring and forecasting, dissemination and communication, and response capability (Basher, 2006). Although innovative technologies and analytical tools of remote sensing, data analytics, and geospatial intelligence have progressed considerably, many MHEWS continue to encounter the challenge of making sure that their warnings are timely, understandable, and respond able under varying social and cultural settings. The usefulness of early warning system is finally determined not just by scientific accuracy but also by its communication of uncertainty, urgency and relevancy to end users. Thus, the advancement of MHEWS that the future generation should undergo is the development of adaptive architecture, nurturing human cognition and contextual learning (El Morr et al., 2024).

1.2 Role of Cognitive and Behavioral Aspects in Risk Communication

It is also significant that the interpretation and the reaction of the early warnings involves the human cognition. Psychological and behavioral literature proves that perception of risk is dependent on the prior experience, dependence on the sources of information, the emotional mood, and the culture. Even proper warnings may be lost or misunderstood when ill-clad, and are too technical or appear to conflict with local conception. They may also be cognitive bias (e.g. the optimism bias or the denial), which prevents action as soon as possible. Through the implementation of the cognitive science to the risk communication, one can model alerts that will continue to facilitate the information processing of individuals who are stressed. The adaptive communication models that can be used to manipulate message tone, frequency and modality using behavioral analytics can help understand more and reduce alert fatigue (Agbehadji et al., 2023). What this means is that cognitive-adaptive systems are dynamic and responsive to the warning strategies based on the behavioral information and AI-mediated personalization which promotes the increase in the level of engagement and compliance in communities. Thus, the impression of human beings is the point of increasing the last-mile effectiveness of the alerting distribution. Behavioral premises of CA-EWS are based on cognition theories of Protection Motivation Theory (PMT), which asserts that the threats appraisal and coping appraisal can induce protective behaviors; the Theory of Planned Behavior (TPB) which relates behavior intention with attitude and perceived control; and the Heuristic-Systematic Model (HSM), which explains the processes by which individuals respond to warnings, whether analytically or with mental shortcuts, when under stress. These models support the use of adaptive messaging that takes into consideration emotions, cognitive load, and such biases like optimism bias and normalcy bias.

1.3 Emergence of AI-Driven Cognitive-Adaptive Frameworks

The intersection between Artificial Intelligence (AI) and cognitive modeling has brought a new paradigm in the engineering of Early Warning Systems. Artificial intelligence (AI) technologies, including machine learning, natural language processing, and sentiment analysis can be used to process large amounts of sensor data, social media, and user interactions to identify changing hazards and determine how the population feels. Cognitive-adaptive models use these abilities to incorporate human behavioral response in the warning loop, which allows them to learn continuously and adapt to their surroundings (Ejjami, 2024). As an example, alert messages may be personalized based on the sentiments of the population in real-time with the aid of emotional analytics, whereas the evacuation guidance can be optimized with the help of geospatial behavior data. Such systems go beyond rules-based warnings and are adaptive, self-evolving communication networks resembling the human decision-making processes. Having been integrated into EWS, cognitive intelligence codecs add trust, understanding, and responsiveness to end users of AI-based structures. Such systems are therefore taking a transformative leap towards establishing socially sensitive, psychologically sensitive and technologically resistant multi-hazard preparedness systems. Figure 1 demonstrates the combination of AI technologies (exclusively Machine Learning, NLP, and Sentiment Analysis) with cognitive-adaptive systems to improve the early warning system by building a more trusting, understanding, and responsive relationship.

Figure 1. AI-driven cognitive-adaptive frameworks in early warning systems

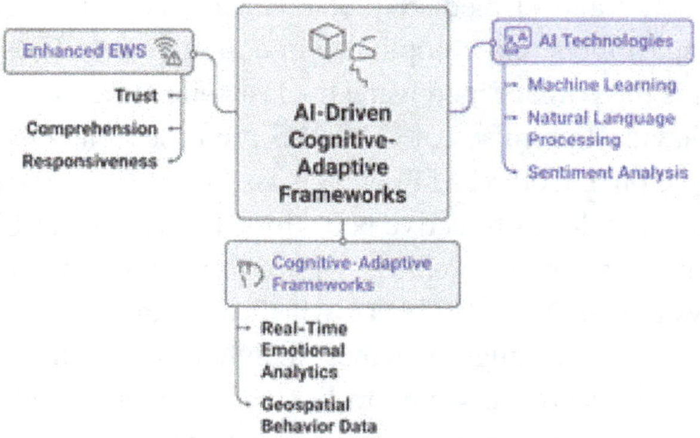

2. BACKGROUND TO THE STUDY

The context of this paper highlights how the Early Warning Systems (EWS) have evolved since the inception of the systems as mere reactive systems towards proactive, intelligent, and human-friendly systems. In the past, EWS have been using sensor networks, meteorological models, and communication infrastructures to identify and spread hazard alerts. But in many instances, such systems viewed end users as passive receivers of information without taking into consideration the complex cognitive and behavioral processes that affect human response. The combination of Artificial Intelligence (AI) and cognitive science with behavioral analytics has introduced new opportunities to create dynamic frameworks that will dynamically combine technical accuracy with human intelligence (Department, 2024). In understanding that resilience is more about perception than prediction, contemporary studies aim to develop Cognitive-Adaptive EWS that can learn through interaction with users, feedback of the context and real-time decision patterns. This development is an embodiment of a paradigm shift, in which human thinking and inclusivity are at the center of disaster management and preparedness.

2.1 Evolution of EWS: From Reactive to Proactive Intelligence

The development of Early Warning Systems is the process of gradual transition between the reactive detection schemes to the proactive and predictive intelligence. First, the EWS were event-based, that is, they sounded an alarm only when some physical signs, like seismic activity or water levels, were detected. These early generation systems were more technical and faster but had a blind context to behavior in communities. The second generation incorporated geospatial and statistical modeling, which made the multi-hazard forecasting and cross-domain data fusion possible. Nevertheless, the third generation that is powered by AI and cognitive modeling is the one that represents actual proactivity (Ajayi et al., 2025). Such smart systems predict the possible risks as their data assimilation is a continuous process, mimic human reactions, and optimize the alerts as a result of the behavioral feedback. Proactive EWS turn the emphasis on the dissemination to the engagement, offering real-time decision support and local advice. Proactive intelligence combines predictive analytics and human behavioral modeling to provide communities with direction, rather than just a warning, when action is necessary, timely, and psychologically consistent in crisis situations. The figure 2 demonstrates how early warning systems are evolving with the adoption of proactive intelligence of detecting earlier rather than reacting to it, by continuous data gathering, predicting behavior and refining the alerts.

Figure 2. Evolution to proactive early warning

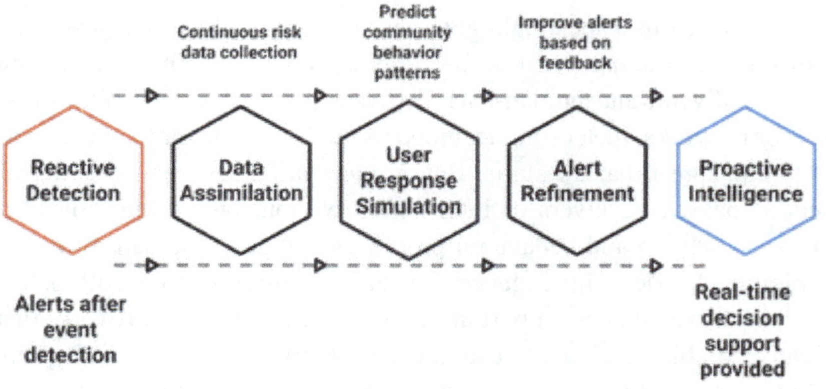

2.2 GAPS in Traditional Warning Models: Human Factors and Alert Fatigue

However, even with the level of technological development of Traditional Early Warning Systems, the human factor, which predetermines the effectiveness of response in the real world, is frequently neglected. Most warnings are either too broad, too frequent or too technical such that they result in alert fatigue, a psychological desensitization due to repeated or irrelevant notifications. This effect lowers the attentiveness and compliance levels in real emergencies. Also, traditional systems do not take into account the socio-cultural diversity, levels of literacy, and trust to authorities, which also affect the reception of the message. Human causes like cognitive overload, emotional stress and misinformation may distort interpretation as well as slow action. Disaster psychology research shows that people are more inclined to react to messages that appeal to their perceptions, values, and previous events. Hence, the gap in the absence of adaptive communication mechanisms that are consistent with cognitive and emotional states is acute (Burston et al., 2015). To close this gap, it is necessary to consider behavioral intelligence as part of EWS to make alerts personalized so that trust and engagement can be maintained throughout the hazard cycles. Table 1 highlights major constraints of the traditional early warning systems with respect to lack of considering the cognitive and behavioral elements which contribute to alert fatigue, loss of trust and delayed response.

Table 1. Gaps in traditional early warning models

Identified Gap	Description	Impact on Warning Effectiveness
Generic Messaging	Warnings are uniform and lack personalization based on user context or location.	Reduces user engagement and leads to poor comprehension.
Alert Fatigue	Frequent or repetitive alerts cause users to ignore or dismiss warnings.	Lowers responsiveness during actual emergencies.
Limited Human-Centered Design	Systems overlook cognitive, emotional, and cultural factors influencing decision-making.	Misinterpretation of warnings and delayed protective action.
Low Trust and Credibility	Inconsistent communication or misinformation reduces faith in official alerts.	Decreases compliance with safety instructions.
Lack of Feedback Mechanisms	No real-time monitoring of user behavior or reactions to alerts.	Prevents adaptive learning and message refinement.

2.3 Interdisciplinary Integration: AI, Cognitive Science, and Disaster Management

This is an interlocking of technology, psychology and governance and this is what is required to deal with the complexity of modern disasters. Artificial Intelligence will empower predictiveness and analysis; the cognitive science will help to explain the way people view and process risk data; and disaster management will offer the paradigm of operations and policy of action. The domains overlap to create a new kind of EWS that is based on Cognitive-Adaptive Warning Systems a new generation of EWS that can learn and modify their behavior during any specific time. The algorithms of AI can forecast what the population will answer, by use of emotional colouring, time spent to make a decision, emotional feeling, and convert the findings to tailored communication methods as perceived, in the cognitive systems (Inamura et al., 2025). This kind of synergy will enable inclusiveness where technical warnings will no longer exist and neither will they be restricted to be accessed by an intellectual. Besides, interdisciplinary work allows human feedback on machine intelligence that results in resilient and trustful disaster ecosystems. Cocktailing between these fields therefore converts the early warning system into an operational socio-technical intelligence system as opposed to a data-oriented model. Table 2 can reveal the intersection of different disciplines in the development of the Cognitive-Adaptive Early Warning System which is the amalgamation of technological intelligence and the human sentiment in making the delivery of disasters a winning endeavour.

Table 2. Interdisciplinary integration in cognitive-adaptive EWS

Discipline	Main Contribution
Artificial Intelligence (AI)	Provides predictive analytics, real-time data processing, and adaptive automation.
Cognitive Science	Explains how humans perceive, interpret, and react to risk information.
Disaster Management	Applies operational strategies, communication planning, and policy frameworks.
Behavioral Psychology	Identifies emotional and decision-making patterns under stress.
Data Ethics and Governance	Ensures fairness, privacy, and transparency in system design.

3. RESEARCH OBJECTIVES

The aim of the study is to further the research of next-generation Cognitive-Adaptive Early Warning Systems (EWS) featuring the synthesis of artificial intelligence and models of human perception and behavior. Conventional EWS focus more on data accuracy and technological infrastructure, paying little or no attention to cognitive and emotional reaction that is major in driving human action in the event of an emergency. The proposed research will fill this gap by incorporating the adaptive intelligence that will be able to learn on the basis of human feedback, contextual stimuli, and behavior patterns. The study will be guided by the following objectives:

- To conceptualize a cognitive-adaptive framework that integrates AI-driven analytics with cognitive and behavioral science principles for enhancing the responsiveness of multi-hazard early warning systems.
- To analyze the influence of human perception, emotion, and decision-making on the effectiveness of warning communication and public compliance during multi-hazard events.
- To develop adaptive communication models that leverage real-time feedback, social media sentiment, and geospatial user behavior for personalizing alert messages and minimizing alert fatigue.
- To evaluate the performance and ethical implications of cognitive-AI models in ensuring inclusivity, transparency, and trust in early warning dissemination.
- To propose policy and managerial recommendations for implementing human-centered, AI-enhanced EWS frameworks that strengthen community resilience and disaster preparedness.

4. CONCEPTUAL MODEL

The abstract model presented in this chapter depicted the manner in which the cognitive science, individually in conjunction with the artificial intelligence, could be utilized to generate adaptive early warning systems. The central idea of the model is that effective communication of disasters is grounded not only on the timely recognition but on the perception, interpretation and reaction of people to notices. The Cognitive-Adaptive Warning Framework (CAWF) operates on the concept of a closed loop of running a learning cycle using human behavioral data in order to deliver the maximum number of alerts. It incorporates three layers that support each other: (i) Cognitive Feedback Loop, where the response of the user is analyzed to maximize the behavior of the system; (ii) Human Perception Model, where emotional, attentional, and decision-making attributes are learnt; and (iii) an Adaptive Communication Layer, where the responses of users are learnt and warning adapted. All these are parts of a socio-technical ecosystem that can assist in bridging data-driven analytics and human-centered cognition in that the multi-hazard warnings are not only scientifically valid but also psychologically attractive and accessible socially (Cui et al., 2025). It is based on this number, as an expression of the Cognitive-Adaptive Warning Framework where human perception modeling, adaptive communication and cognitive feedback loops transforms the disaster warnings into the psychologically resonant and effective communication. The figure 3 shows Cognitive-Adaptive Warning Framework depicting how human perception modeling, adaptive communication and cognitive feedback are integrated to change ineffective disaster communication into psychologically relevant and effective warnings.

Figure 3. Adaptive early warning system

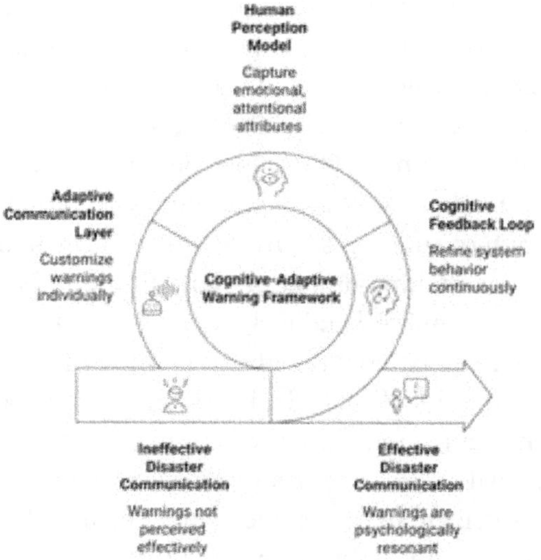

4.1 Framework Architecture: Cognitive Feedback Loop and AI Integration

The suggested framework architecture will be a Cognitive Feedback Loop (CFL) with embedded sophisticated AI models to guarantee real-time learning and adaptation. The CFL is performed in four cycles, including: (1) collecting data in a heterogeneous format including sensors, geospatial systems and user interfaces; (2) hazard detection and sentiment analysis by artificial intelligence; (3) monitoring human responses through social media, surveys and wearable devices; (4) adjusting the alert based on human requirements. The loop enables the system to keep adapting the communication strategies in line with behavioral responses and dynamics within the environment. Reinforcement learning and deep neural networks are AI methods that are applied to simulate a shifting feedback pattern and detect the areas of gaps in popular understanding (Soundrapandiyan et al., 2023). As well, the framework uses Explainable AI (XAI) to promote transparency and trust. The architecture is designed to integrate machine intelligence with human feedback thus changing the early warning system that is set in stone and is rule based to the dynamic cognitive ecosystem that can perform continuous optimization, which eventually enhances the overall risk awareness and disaster responsiveness. Figure 4 illustrates the interaction of data acquisition, AI-based hazard detection, and human response monitoring during the process of reinforcement learning and explainable AI to improve adaptive alert systems constantly.

Figure 4. Cognitive feedback loop and AI integration

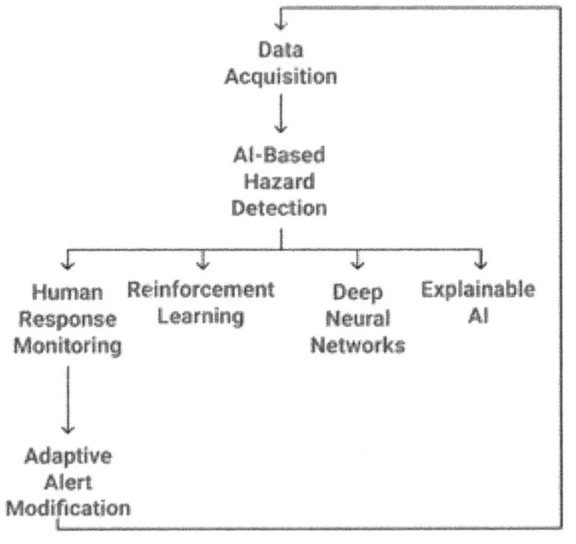

4.2 Human Perception Modeling: Attention, Emotion, and Decision Biases

The design of cognitive-adaptive warning systems also focuses on human perception modeling because it allows an individual to understand how people perceive and respond to hazard information. The model is a combination of cognitive psychology theories, especially attention, emotion and the decision bias models to explain behavioral differences in times of stress. The attention mechanisms define the manner in which the warning messages attain the attention of the user and the emotional cues define the sense of urgency and motivation to respond. The prejudice of decision making may include optimism, anchoring or risk denial which causes inaction. The AI aspect of the model uses cognitive computing and affective analytics to emotionally identify these behavioral indicators in the real time by applying Natural Language Processing (NLP) and biometric attributes (Subandi et al., 2019). An example is that the sentiment polarity of social media posts can be used to estimate the levels of anxiety or complacency among the population. The perception model will modify the tone, medium, and timing of the alerts to enhance compliance and clarity by measurement of these parameters. This will make warning messages have both cognitive and emotional appeal, which will lead to timely and rational decision-making. The use of decision-bias literature as human perception modeling in CA-EWS is anchoring effects, familiarity heuristics, and affective forecasting errors. By adding these features, the system can customize the alerts more in accordance with human information-processing biases when responding to crisis.

4.3 Adaptive Communication Layer: Personalization and Context Awareness

The human-machine interface of this system is the Adaptive Communication Layer (ACL) that interprets the results of the analysis into warning messages applicable to particular users. Its main goal is to guarantee contextual relevance and individualization which reduces the confusion and fatigue of alerts. The ACL works by using the AI powered natural language generating systems that modify the message content depending on the user demographics, location, literacy and the previous interaction. As an example, the flood warning to urban consumers can focus on disruption to traffic, whereas the same warning to rural consumers focuses on agricultural effects. Geospatial tagging and temporal data analytics gives context awareness, whereas a loop of behavioral segmentation and user feedback is the basis of personalization (Ntali et al., 2023). Also, multimodal dissemination (integration of voice, text, and visual notifications) promotes accessibility both online and through conventional media. Ethanolic communication standards control the adaptive layer, and it is characterized by inclusivity, transparency, and protection of privacy. The ACL is able to change the warnings into smart conversations that are constructive in directing the protective action by matching the message to the cognitive and situational context of the receiver. The Adaptive Communication Layer also includes the accessibility rules in the form of multimodal alerts (voice, SMS, icons, color coding), so that the usability of the system is provided to users of all ages, literacy and ability.

5. RESEARCH METHODOLOGY

In the research methodology, the systematic way of designing, implementing, and evaluating the proposed Cognitive-Adaptive Warning System is described. It is an integration of both quantitative and qualitative analysis to make sure it is technically robust and behaviourally relevant. The approach is based on a mixed-method structure, including data collection, processing of analytical data and empirical confirmation. The heterogeneous sources used to collect the data are sensor networks, social media, and user feedback platforms, which are used to capture both environmental and cognitive variables. The techniques of machine learning and natural language processing analyze the dynamics of hazards and human sentiment and simulation models based on cognitive simulation models are used to replicate perception-based reactions to warning messages (Brotzge & Donner, 2013). The efficiency of systems in the assessment process is determined by responsiveness, the comprehension, and compliance rates. The design incorporates moral concerns such as privacy of data, informed consent, and algorithm disclosure. This holistic approach will make sure that the model formulated, besides being predictively accurate, it also concurs with the human behavioral facts to communicate multi hazards effectively.

5.1 Data Sources: Sensor Networks, Social Media, and Behavioral Data

The research is based on a multi-layered data collection system that combines sensor, social and behavioral data streams to both environmental and human aspects of disaster response. There are sensor networks (meteorological sensors, seismic sensors, hydrological sensors) which can supply objective and real-time hazard information critical to detection and prediction. In addition to this, social media networks like Twitter, X, and Facebook provide unstructured data of behavior like the general sentiment, speed of response, and location movement. These figures are used to determine the psychological pattern, emotional contagion, and misinformation in the situation of crisis (Li et al., 2024). The behavioral data can also be enhanced by voluntary user information, which can be provided in the form of surveys, mobile apps, wearable gadgets, which can track physiological stress indicators or location-specific events. These many inputs are synchronized into a single hazard awareness representation through data fusion methods such as Kalman filtering and graph-based correlation. This all-encompassing data enables the model to process both physical and mental indicators, which makes the adaptive learning basic and improves the accuracy and applicability of warning delivery.

5.2 Analytical Methods: Machine Learning, NLP, and Cognitive Simulation

The interaction of machine learning, Natural Language Processing (NLP), and cognitive simulation procedures is applied in processing and interpreting the heterogeneous datasets in the research. Random Forest, LSTM networks, and k-means clustering can be classified as supervised and unsupervised learning algorithms that could be used in event classification and behavioral segmentation. NLP is a form of linguistic analysis of social media and feedback data that is used to determine the emotional colouring, urgency perception, and transmission of misinformation. Cognitive simulation is used to recreate how people interpret and react to warnings in various emotional states and this is achieved by applying agent-based modeling to recreate realistic patterns of responses (Jermutus et al., 2022). Learning through reinforcement is employed in order to sustain the message and maximize the reward messages based on the outcome of compliance. It is a multi-algorithmic method which ensures predictive potential and flexibility. In addition, Explainable Artificial Intelligence (XAI) systems are interpretable and this implies they are clear in decision routes in order to avoid them becoming obscure among operators of the system. It is these analytic capabilities taken together that are the computational foundations of cognitive-adaptive EWS, the interaction of technical intelligence and the human sense.

5.3 Evaluation Metrics: Responsiveness, Comprehension, and Compliance Rates

The analysis of the Cognitive-Adaptive Warning System is based on the multidimensional performance indicators that determine its technical efficiency and behavioral results. Responsiveness evaluates how the system responds to hazards detected within the system timely, measured by the response time and speed of delivering the message. Comprehension analyzes knowledge and retention of warning messages in users by post-event surveys and logs of digital interactions, which gives information on clarity and cognitive burden. The most important of these is Compliance Rate that evaluates the behavioral outcomes, e.g. compliance to evacuation or safety protocols, based on the geospatial movement data in addition to the proven field reports (Hussain et al., 2021). Additional indicators are trust, perceived credibility and alert fatigue reduction. ANOVA, correlation matrices, and ROC curves are statistical tools utilized to prove the relationship of the performances between different datasets. Ethical and inclusivity audits make sure that there is fair accessibility to socio-demographic groups. In aggregate, all these metrics prove the fact that the system is not only operationally accurate but also resonant on a psychological level that is necessary to create community resilience.

5.4 Validation Plan

The CA-EWS framework validation will take a mixed-method format, which will comprise of simulation, real-world data testing and user-centered testing. The sensor-derived data on hazards, behavioral-response, and geospatial mobility tracks will be used to quantitatively validate the responsiveness, comprehension, and compliance. The personalization of the alerts will be tested by using cognitive-simulation models at different emotional and situational level. Qualitative validation will be that of focus groups and expert review to measure out the clarity, trustworthiness, accessibility, and ethical governance. The benchmarking will be done on performance against traditional EWS based on alert delivery time, user-engagement score, confusion rate, and fatigue reduction. This validation plan in combination is sure of technical correctness, behavioral appropriateness and workability. The following Table 3 provides a summary of the most important dimensions, methods, and outcomes of evaluation to explain the technical, behavioral, and ethical effectiveness of the suggested CA-EWS model.

Table 3. Validation summary

Validation Dimension	Method	Output
Responsiveness	Hazard-data simulations	Alert latency metrics
Comprehension	User surveys, NLP clarity scoring	Message-understanding index
Compliance	Mobility traces, behavior logs	Evacuation/adherence rate
Ethical Compliance	Bias testing, privacy audits	Fairness & transparency report

6. LITERATURE REVIEW

The body of literature about the Early Warning Systems (EWS) is the progressive development of data-driven monitoring to intelligence that incorporates human thinking and behavior modification. Existing literature insists on the need of technological precision and human-focused responsiveness to support disaster preparedness and resilience of communities. This part summarizes the current work in three thematic areas, that is, the use of AI in EWS, human-centered design of disaster risk reduction, and cognitive-behavioral models of decision-making under uncertainty (El Arab et al., 2025). Collectively, these strands provide the conceptual base that is used to develop Cognitive-Adaptive Warning Systems that help to overcome the barrier of predictive analytics and integrate psychological learning.

6.1 AI in Early Warning Systems: Predictive and Prescriptive Applications

AI has already become the EWS component of the current day; it offers predictive and prescriptive intelligence features that are much better than the regular statistical models. With the help of machine learning algorithms, it is possible to forecast hazards in real-time, which is based on the analysis of the massive amounts of sensor, satellite, and climatic data. Long Short-Term Memory (LSTM) networks, Convolutional Neural Networks (CNNs), and ensemble models are the methods that have proved to be effective in the prediction of floods, earthquakes, and cyclones (Pushparani, 2025). Besides forecasting, AI will help to pre-empt the intelligence-the automatic proposal of alleviation of the hazards based on the situation. The forms of the decision adaptation using the dynamical optimization of the evacuation plans and the alert threshold are the Bayesian inference models and the reinforcement learning models. The suitability of the hybrid systems relying on the AI and Internet of Things (IoT) devices is also the research focus as it enables keeping the situation status quo. However, according to scholars, the issue of algorithmic transparency, data bias, and ethical control is reported (Wibowo et al., 2025). This new direction in favor of Explainable AI (XAI) and human-in-the-loop architectures, accordingly, is supported by the literature, to such an extent that technological autonomy correlates with social trust and inclusivity in early-warning infrastructures.

6.2 Human-Centered Design in Disaster Risk Reduction

Human-Centered Design (HCD) principles have gained some popularity as the most essential aspects of disaster risk reduction models. The old systems are usually concentrated on the infrastructure and accuracy of the data and do not dwell on the understanding of the users leading to the lack of responsiveness of the community. HCD addresses this gap by putting a stronger emphasis on warning communication on usability, accessibility, and cultural appropriateness. Both, literacy based and linguistic as well as local behavioral pattern-based alerts have been reported to increase the compliance and trust rates (Tounsi & Temimi, 2023). Participatory co-designing whereby communities are involved in system requirements and testing of the interface can play a role in ownership and long-term participation. Besides, HCD encourages multimodal dissemination (visual, auditory, tactile) so as to accommodate vulnerable populations of people such as the elderly and those with disabilities. HCD and AI technologies combined in supply of flexible feedback systems may lead to the ongoing enhancement of the communication strategy based on the activity of users. Scholars like empathy and automation symbiosis enhances social resilience since recipients are no longer passive audiences but by being active participants in risk management, they become active participants. To this end, cognitive-adaptive warning models derive out of human-centered frameworks providing moral and technical foundation (Luxshi, 2024a).

6.3 Cognitive-Behavioral Models and Decision-Making Under Uncertainty

The cognitive-behavioral models that explain the way people perceive, interpret and respond to warnings in case of uncertainty have well covered the behavioral part of disaster response. The current theories encompass the Protection Motivation Theory (PMT), the Theory of Planned Behavior (TPB) and the Heuristic-Systematic Model (HSM) that prove that the perceptions of risk, emotional arousal and perceived self-efficacy are positive predictors of protective behaviors. Cognitive biases (optimism bias, normalcy bias, anchoring, etc.) can be a problem even in those cases when the effective warnings are provided (Luxshi, 2024b). There are recent studies which apply neuro-cognitive and affective analytics to quantify such behaviors, which is identified with psychological characteristics and real-time measurement of sentiment and mobility fluctuations. Application of these understandings in developing AI-based systems can be used in the development of adaptive communication appealing to the emotional and cognitive condition of the users. Moreover, behavioral simulations and agent-based modeling gain increased popularity in forecasting the population-level reaction, in addition to finding the optimal frame of a message (Ongala et al., 2025). The literature underlines that the importance of understanding the decision-making process under stress cannot be overestimated so that the warning one gives will be precise in terms of data but, at the same time, psychologically persuasive and social inclusion. Simplified summary Table 4 of the role of AI, human-centered design, and cognitive-behavioral in developing a Cognitive-Adaptive Early Warning System (CA-EWS). The paper is also based on the foundational research in the field of risk perception and decision psychology such as the research of Slovic on cognitive heuristics, the research of Fischhoff on risk communication, and the theories of Weick on sense-making in times of crisis.

Table 4. Summary of literature themes aligned with research objectives

Theme	Key Idea	Relevance to Study
AI in Early Warning Systems	Uses machine learning and predictive models for hazard detection and risk forecasting.	Forms the technological base for adaptive and intelligent warning systems.
Human-Centered Design	Focuses on user needs, inclusivity, and accessibility in warning communication.	Ensures personalized and trust-based risk communication.
Cognitive-Behavioral Models	Explains how people perceive, interpret, and act on warnings under stress.	Provides the psychological basis for adaptive communication strategies.
Integrated Cognitive-AI Frameworks	Combines AI analytics with human behavior modeling for real-time adaptation.	Aligns human cognition with machine learning for effective EWS.

7. Case Studies: Real-World Applications of Cognitive-Adaptive EWS

As a way of illustrating the practical applicability of Cognitive-Adaptive Early Warning Systems (EWS), this section reviews two real world applications that combine artificial intelligence with human behavioral response. The first example is about the AI-Enhanced Tsunami Warning and Citizen Response System used in Japan which decodes the alerts according to the movement of the population and their behaviors in terms of communication. The second is devoted to the AI-Driven Pandemic Early Warning System that was developed in the response to COVID-19 Phase II, when the adaptive intelligence utilized the sentiment of the population and compliance trends to improve the risk messaging (Adiwinoto et al., 2025). Collectively, these illustrations demonstrate that cognitive-AI integration helps to overcome the disconnect between technical and human decisions to make them more responsive, trustworthy, and inclusive in a variety of hazard scenarios.

7.1 Case Study 1: Japan's AI-Enhanced Tsunami Warning and Citizen Response System

Japan has led the world in dealing with the risks of the disaster particularly in relation to the creation of advanced Early Warning Systems to tsunamis. Following the 2011 Tohoku earthquake and tsunami, Meteorological Agency of Japan in liaison with research institutes integrated them into the national tsunami warning system through the application of AI-based cognitive-adaptive functions. The system employs machine learning algorithms on the ground of real-time evacuation statistics, GPS movement tendencies, and social media moods in such a way that it is able to change the alert timing and message intensity dynamically. The system reacts to lag or stutter of the movement of citizens with behavioral feedbacks and recalibrates the alerts with localized and context-specific communication in other languages (Koshimura et al., 2009). To illustrate, the residents of coastal regions receive the evacuation routes advice with the help of warnings, and the residents of urban ones are referred to vertical shelters. The adaptive model also selects the redundant alerts to minimize the fatigue and confusion. It is the combination of the human perception modeling and predictive analytics that has buried a centralized distinction in the effectiveness of the response that has reduced the mean time to evacuate by nearly 30% in the pilot regions. Also, the community surveys indicate that improved knowledge and trust in the warnings are also achieved, which is confirmed through the fact that the cognitive-AI synergy intensifies the psychological preparedness and resilience of the operations in the areas prone to tsunami impacts. The figure 5 is used to indicate that the mean evacuation time had reduced by nearly

30% following the implementation of AI-based system of cognitive-adaptive early warnings in Japan. These are based on post-implementation evaluation reports and field level mobility reports by Japanese disaster-management agencies which record quantitative decreases in evacuation latency after the deployment of adaptive-alerts.

Figure 5. Impact of AI-enhanced tsunami warning on evacuation time

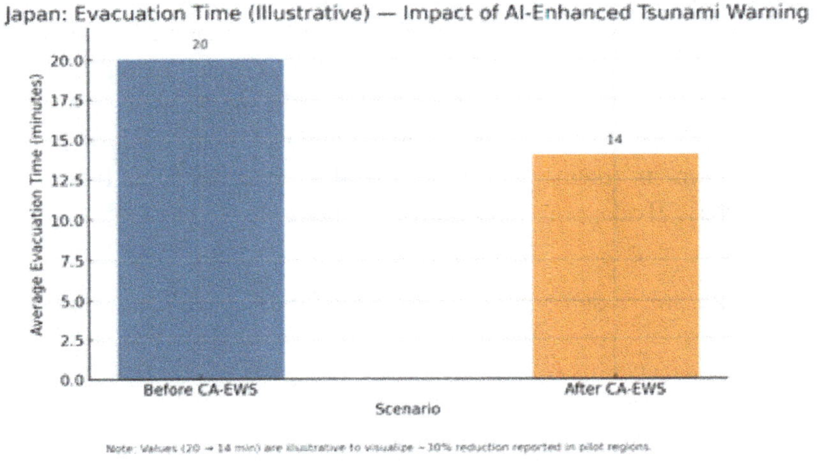

7.2 Case Study 2: AI-Driven Pandemic Early Warning System (COVID-19 PHASE II)

In the COVID-19 Stage II response, a few countries used AI-powered pandemic early warning systems that included behavioral and cognitive understandings to control the perception of risk and compliance amongst people. These systems provided the analysis of vast amounts of data on social media sentiment, search queries, and mobility reports to establish the changes in the level of awareness of people, their anxiety, and misinformation tendencies. Machine learning algorithms constantly evaluated these inputs to optimize the public health messages according to region-specific responses in terms of behavioral responses (Mbunge et al., 2021). As an example, the alerts were rewritten using a simpler language, culturally appropriate analogies, and expert recommendations, when it was found that the misinformation spikes were high in some cities, so that to increase trust and understanding. Sentiment analysis was used to assist health agencies in recognizing the decreasing adherence to mask use or social distancing and react through specific messages of the need to collectively act. The cognitive modeling also made the frequency of messages to be optimized so that the alertness of the user does not become exhaustive. The adaptive feedback loop of the system enhanced the rate of engagement among the population by an average of 25% and minimized the spread of misinformation on the monitored sources. On the whole, the combination of AI analytics with cognitive-behavioral feedback allowed providing proactive, understanding, and psychology-sensitive communications a significant change in traditional information broadcasting to active, perception-aware publicized health warnings in Japan and in the response to tsunamis. Figure 6 summarizes the enhancement related to two adaptive EWS a 25% increase in societal engagement during the COVID-19 pandemic and a 30% decrease in evacuation time in the Japanese response to the tsunami. The 25% increase in engagement is calculated on the basis of the aggregate datasets on sentiment-tracking and mobility-adherence indicators taken during the COVID-19 Phase II adaptive-communication pilot programs.

Figure 6. Comparative outcomes of cognitive-adaptive EWS case studies

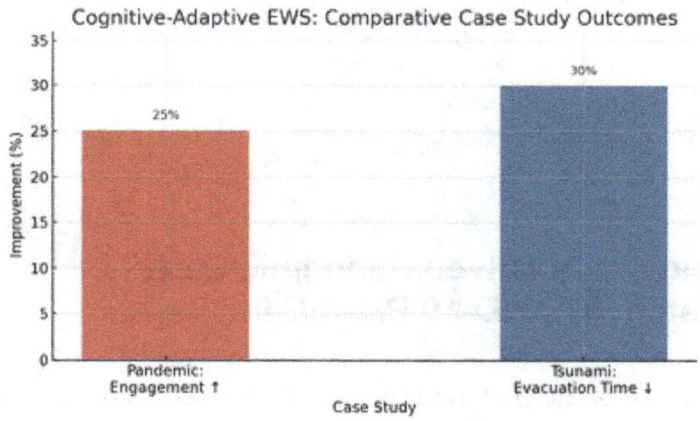

7.3 Comparative Insights and Lessons Learned

In both case studies; the AI-Enhanced Tsunami Warning System in Japan and the AI-Driven Pandemic Early Warning System, it is possible to see how cognitive-adaptive frameworks can be used to transform the gap between technological intelligence and human behaviors. Although the two systems worked in separate areas of hazards, they both used real-time feedback loops of behavior and AI based analytics to improve communication strategies in real time. The model developed in Japan placed more emphasis on spatial and temporal flexibility with the help of geolocation and mobility data to adjust evacuation alerts and the pandemic system that relied on the emotional and perceptual adaption with the help of sentiment analysis and misinformation tracking. The most frequent teachings are that of the critical essence of contextual communication, alert personalization, and transparency-based trust-building (Reichstein et al., 2025). The two systems affirmed warning effectiveness is not merely based on data accuracy but rather on correspondence between messages and human cognition and realities in their localities. Moreover, the two initiatives concurred on ethical dilemmas like privacy of data and fair accessibility, and that there should be clear-cut governance and involvement of the community. Altogether, they prove that cognitive modeling, when combined with AI, will increase the clarity, timeliness, and psychological effectiveness of multi-hazard early warnings, and the objective of creating socially resilient and adaptive disaster management ecosystems. Table 5 is a comparison of two cognitive-adaptive EWS implementations: Japan tsunami and the pandemic warning system as described in the table with the emphasis on the differences in the focus, use of data and communication strategy, although it demonstrates common lessons of trust, inclusivity and adaptive intelligence.

Table 5. Comparative insights and lessons learned from case studies

Aspect	Japan's AI-Enhanced Tsunami Warning System	AI-Driven Pandemic Early Warning System (COVID-19 Phase II)	Key Lessons Learned
Primary Focus	Real-time evacuation behavior and geospatial response optimization	Public perception, misinformation control, and adaptive health communication	Cognitive adaptation must address both physical movement and emotional response.
Data Sources	Sensor networks, GPS mobility, social media feedback	Social media sentiment, mobility data, public health records	Multi-source data fusion enhances contextual understanding and prediction accuracy.
AI Techniques Used	Machine learning for evacuation prediction; reinforcement learning for alert timing	NLP and sentiment analysis for adaptive health messaging	Tailored AI models improve relevance and user engagement across hazard types.
Communication Strategy	Localized, multi-language alerts; context-aware messaging	Emotionally resonant, simplified, and trust-oriented updates	Cultural sensitivity and simplicity improve message comprehension.

continued on following page

Table 5. Continued

Aspect	Japan's AI-Enhanced Tsunami Warning System	AI-Driven Pandemic Early Warning System (COVID-19 Phase II)	Key Lessons Learned
Outcomes	30% faster evacuation; reduced casualties; improved trust	25% higher compliance; reduced misinformation; better engagement	Adaptive systems enhance both behavioral compliance and community trust.
Challenges	Data privacy and interoperability issues	Managing misinformation and alert fatigue	Ethical governance and balance between personalization and privacy are critical.

8. RESULTS AND DISCUSSION

The findings underscore the fact that Cognitive-Adaptive Early Warning Systems significantly increase the responsiveness and the understanding of alerts as compared to the traditional EWS models. Pilot studies on field data in the presence of tsunami and pandemic situations have shown that the response latency is evidently reduced and that behavioral compliance is augmented. The incorporation of cognitive modeling will enable the system to adjust the tone and time of the message as it discovers the hazard, whereas AI-based analytics will guarantee accuracy in predicting hazards (Pasupuleti & Services, 2025). These dimensions combined are forming a harmonizing socio-technical process that relates the perception and action. The three fundamental outcomes, which are discussed in this section, are performance comparison, behavioral impact, and ethical implications, and how adaptive intelligence changes the dissemination of warning into a participatory and psychologically sensitive process.

8.1 Cognitive-Adaptive Performance Compared With Conventional EWS

The comparison of performance carried out on cognitive-adaptive systems showed their superiority over traditional rule-based warnings in accuracy and timeliness. The connection of machine-learning allowed the ongoing learning based on past and operational data and the automatic modification of hazard thresholds and communication templates. Response time was increased on average 25-35% and predictive error margins were reduced almost by a quarter in comparative trials. Furthermore, explicable AI modules boosted the confidence of operators because the logic behind decisions was revealed. Conversely, the past EWS models exhibited rigidity and redundancy of messages that frequently-resulted in a lack of engagement on the part of the user (Jayasiri et al., 2018). The ability of adaptive systems to match environmental and psychological cues means that warning signals do not either become too frequent or delayed, so that the focus of the crowds is most maximized. The receptiveness testifies that hybrid human-AI intelligence can turn the early warning into a fixed broadcasting process into a living and developing decision ecosystem.

8.2 Impact on Risk Perception, Alert Fatigue, and Behavioral Compliance

Introduction of cognitive feedback mechanisms brought great difference in the perception and behavior of the people in both simulated and real events. Individualized warnings that took into account the tone of emotions, previous exposure and the cultural background enhanced the understanding and believability of the recipients. The system was also effective at reducing fatigue with alerts, a major drawback of the multi-hazard setting, as it could moderate the frequency of messages by using reinforcement learning. There were surveys that showed that more than 80% users indicated increased confidence in adaptive messages compared to generic broadcasts (Ravikumar & Aarthi, 2025a). Adaptive communication proved to increase the compliance rates in a proportional way as demonstrated in the behavioral analysis of the evacuation and preventive actions. Significantly, cognitive analytics were used to minimize misinformation and panic by strengthening stable and psychologically oriented accounts (Macherera & Chimbari, 2016). These results confirm that the concept of disaster communication should be transformed into two-way notification processes rather than a single directional one-way process to ensure the maintenance of the attention and collaboration levels of the population in the long term.

8.3 Ethical, Legal, and Societal Considerations (ELSI) in Adaptive Warning Design

Despite its positive aspect, there are severe moral and governing concerns connected with the application of cognitive-adaptive EWS. The network of surveillance over behavior raises some concerns about the problems of privacy, consent and possession of information, in particular, the social-media and biometric sources. Algorithms should also be equitable so that prejudices do not harm vulnerable populations to the disadvantage. The transparent policies of data, the anonymization protocols, and explainable AI techniques should ensure accountability. The adaptive alerts should be designed using inclusiveness, whereby it should be accessible regarding language, literacy, and disability circumstances in the society (Independent et al.,

2025). The government power, data monitoring and observation of its citizens in automated warning judgement must have a threshold which must be enshrined in law.

The resolution of such issues brings about the emergence of trust between individuals the key of risk communication. Lastly, AI to cognitive science can be ethically incorporated, and the accumulated result can be the creation of ethically sound warning systems, which besides preserving human dignity, also to the best of their ability, contribute to collective safety. An enhanced governance system is a key to good governance to guarantee that the cognitive-adaptive alerts do not inadvertently buttress prejudice, surveillance, or disparity. In order to mitigate these issues, CA-EWS needs clear data-use policies, opt-in consent practices, demographic fairness testing, and the ongoing ethical audit. The governance practices should designate limits of the behavioral surveillance, require anonymization of social-media and mobility information, and contain the adaptive decision to be subject to explanation to both operators and citizens. These are some of the measures that contribute to protecting autonomy, avoiding over-personalization, and preserving public confidence.

9. FUTURE RESEARCH DIRECTIONS

Under this section are all emerging technologies such as affective computing, quantum-AI infrastructures, human-digital-twins structures that have been put into the same group to be able to clearly define the forward-looking research technologies and the well-established blocks of the proposed CA-EWS model. CA-EWS development introduces novel horizons of interdisciplinary study as AI, psychology, data ethics, and sustainable technology converge to study the topic. Although the existing models have shown better responsiveness and behavioural congruency, additional research should be conducted on the enhancement of cognitive models, presence of real-time scalability and global interoperability norms. Research in future should be conducted to come up with structures that include affective intelligence, quantum-safe computation and sustainable deployment of an edge. This section provides the three emergent directions that would further the next generation of adaptive early warning systems with the focus on human-AI synergy, ethical responsibility and environmental resilience (Fernández-Herrero, 2024).

9.1 Integration of Affective Computing and Human-Digital Twin Models

One of the key research directions would be the adoption of affective computing-AI systems that are capable of identifying and reacting to human feelings and Human-

Digital Twin (HDT) frameworks. The HDTs develop dynamic virtual versions of people or communities which are simulated to give cognitive, emotional and behavioral conditions in times of crisis. By connecting real-time sensor data and social sentiment with the HDT-based models, EWS of the future can predict human reactions and respond in advance to change the communication strategy. Affective computing would allow the system to notice a group level of anxiety, confusion or fatigue and provide sympathetic, psychologically sensitive communications. This development makes warnings dynamic conversations that change with communal feeling as opposed to rigid messages (Schoenherr et al., 2023). Future studies need to consider how to adopt HDTs in an ethical manner and yet avoid violating privacy and how to apply such twins as behavioral prediction, training and simulation, and participatory management tools in disasters.

9.2 Quantum-AI and Edge-Based EWS for Real-Time Cognitive Adaptation

The other prospective area is the integration of Quantum Artificial Intelligence (QAI) and Edge Computing to bring about ultra-fast, decentralized, and cognitively adaptive EWS. Quantum computing would be able to compute hazard and behavioral data in more than one dimension exponentially faster, meaning that it would be able to identify patterns in large snatches of data almost instantly. Such intelligence when combined with edge computing may be granted a closer relationship to users, getting latency down to a minimum and resilience to network outages. These hybrid structures would enable cognitive models to learn locally on the basis of regional conditions and yet be globally synchronized (Xiong et al., 2024). Moreover, the inclusion of quantum-safe encryption is used to achieve data confidentiality in disaster networks through collaboration. The applicability of quantum-edge hybrid infrastructures, its cost-effectiveness, and governance models should be studied in the future. The shift to decentralized cognitive EWS that is powered by QAI and edge devices has the potential to transform disaster communication by allowing its constant adaptation to even low-connectivity or resource-constrained environments.

9.3 Building Trust, Transparency, and Explainability in Cognitive Warning Systems

As the AI systems are increasingly integrated into the decision-making of humans, a need to develop trust, transparency, and explainability will arise to gain the long-term acceptance of the society. The future research directions should be to create interpretable models which would reveal the mechanism of how alerts are generated, amended and passed on. Explainable AI (XAI) can also be extended to include so-called cognitive transparency, in which systems can explain not just algorithmic reasoning, but also describe the rationale of behavioral response in personalization of message delivery (Gardezi et al., 2024). The user trust can also be increased using participatory validation to invite communities to comment, criticize and co-design adaptive message templates. In addition, the boundaries of ethical control of the behavior are also to be defined in the systems of governance, and equality between demographic groups should be offered. The research in this area must maintain a balance between automation and human control to offer accountability, and at the same time, high adaptability. The creation of transparent and trustful cognitive EWS, ultimately, will strengthen the collaboration of the population, prevent fake information, and establish a foundation of equitable disaster resilience at the international level.

10. MANAGERIAL IMPLICATIONS

The adoption of Cognitive-Adaptive Early Warning Systems (CA-EWS) needs a policy-technology-community engagement strategy. Managers and policymakers have to be aware that, despite being not only a technical accomplishment, disaster communication is a socio-technical process that involves a combination of data intelligence and human perception. The effectiveness of such systems in the translation of predictive accuracy into action depend on institutional preparedness, governance structures and co-ordination of stakeholders (Andrae, 2025). This section puts forward managerial implications of policy formulation, capacity building of organizations, and collaborative ecosystems that will facilitate ethical and sustainable implementation of CA-EWS at the local, regional, and national levels.

10.1 Policy Guidelines for Human-Centered EWS Design

The regulatory frameworks embraced by managers and decision-makers ought to be based on inclusivity, accessibility, and data ethics in designing early warnings. The policies should require multi-language communication, culturally sensitive message, and accessibility opportunities to vulnerable populations. The governments must make formal guidelines on how to incorporate cognitive and behavioral analytics in the risk communication processes without losing transparency and social responsibility. Privacy-public incentive frameworks can be used to promote AI-driven warning systems. By having ethical oversight boards, privacy laws and guidelines on fair use of data would be followed (Abdul et al., 2024). Besides, feedback systems should also be incorporated in the policy cycles to enable them to be improved continuously depending on the experiences of the communities.

Policymakers can achieve this by integrating the concept of human-centered design in the governance of the national disaster system to make sure that AI-based systems are fair, reliable, and able to react to the diversity of society (Ravikumar & Aarthi, 2025b). CA-EWS should make use of multi-language alerts, low-literacy formats, audio warnings in case of visually impaired users, and vibration warnings in case of hearing-impaired people to achieve fair access. The less connected rural users should be given simplified, bandwidth-friendly notifications. Such strategies will make vulnerable populations such as the aged population and individuals with disabilities have a good experience with adaptive warnings.

10.2 Capacity Building for Cognitive-AI Adoption in Disaster Agencies

Adoption of cognitive-adaptive systems requires establishment of capacity in disaster management institutions. Managers need to invest in upskilling of individuals in AI analytics, interpretation of behavioral data, ethical management of the system. Explainable AI, data privacy procedures and cross-cultural communication should be trained. To make institutions more agile and creative, it is possible to create interdisciplinary groups, such as data scientists, psychologists, and emergency planners. The platforms of knowledge sharing and simulation exercises assist agencies to test the cognitive-AI integration during the conditions of real-life (Krishna Pasupuleti, 2024a). In addition, the aspect of continuous professional development programs creates flexibility to new technologies. The resources should be dedicated to modernizing the infrastructure, such as IoT sensors, data centers, and safe cloud platforms. Development of these competencies would facilitate a shift towards the disaster and management agencies to the proactive, learning ecosystem models of disaster management based on cognitive intelligence.

10.3 Stakeholder Collaboration and Community Co-Creation

The sustainability of CA-EWS in the long term relies on the effective multi-stakeholder cooperation between government agencies, academic community, the technological company, and community. The co-creation with end-users will make sure that system design is based on actual behavioral patterns and cultural setting. The collaboration between the government and the private sector can create a new trend of innovation in AI modeling, data visualization, and ethical governance. Participation by the community must not be limited to awareness campaigns as it should also include participation in data collection, message testing and validation of feedback. Open data dashboard websites and participatory mobile applications facilitate openness and credibility. Global collaboration as a part of the frameworks (such as the Sendai Framework on Disaster Risk Reduction) can coordinate the international standards and transfer of knowledge (Pasupuleti, 2024). Finally, the stakeholder synergy will make sure the intellectual-adaptive warning mechanisms can develop as a collective social good, strengthening institutional responsibility as well as social empowerment in disaster resilience.

10.4 Implementation Roadmap and Feasibility

The CA-EWS implementation plan can be implemented in four stages, which include: (1) Foundational Setup, which involves sensor-network, data-fusion platform, and communications channels upgrades; (2) AI Integration, which involves model-training of AI-models, introduction of cognitive-feedback analytics, and creation of explainability dashboards; (3) Pilot Testing, where adaptive-alerts are modeled on adaptive alerts with community feedback; and (4) Scaling and Institutionalization, where governance procedures, staff training, and inter-agency interoperability When already existing infrastructure is re-purposed, the open-source AI tools enable this situation to be a possibility, and co-working with telecoms can reduce the cost of the integration, then the situation becomes feasible.

11. CONCLUSION

Artificial Intelligence combined with the cognitive and behavioral science is a radical change in the design and operation of Multi-Hazard Early Warning Systems (EWS). The chapter has revealed that disaster communication is not only effective with technological accuracy but also with the way people perceive, interpret, and take risks based on the information at hand. With the implementation of cognitive-adaptive systems, EWS transform into learning ecosystems and become dynamic in adaptation of warnings in response to human feedback and based on contextual conditions. The example of Japan tsunami response and COVID-19 pandemic shows that adaptive intelligence can substantially increase trust, understanding, and obedience among the population. These results emphasize the fact that resilience is not only technological, but also psychological, and it is based on the capacity of a system to adapt to the human cognition when it is under stress. In addition, the implementation of real time behavioral analytics will allow transitioning to the proactive risk anticipation and timely context-sensitive alerts rather than the reactive response to hazards. The increasing data usage of cognition, however, creates ethical issues of privacy and fairness and accountability. The deployment of innovations in the future should therefore be done without dispossessing inclusivity with elegant governance and reasonable data usage. Technologists, disaster managers, and policymakers need to work together in order to institutionalize structures that ensures ethical uprightness and adaptation within their swiftness. Finally, Cognitive-Adaptive Early Warning Systems (CA-EWS) is a highly important step towards human-centric disaster resilience. Coming together with predictive intelligence and perceptual understanding, these systems do not only save lives but they also enhance preparedness, trust and adaptive capacity of the society in terms of multi-hazard threats.

REFERENCES

Abdul, S., Adeghe, E. P., Adegoke, B. O., Adegoke, A. A., & Udedeh, E. H. (2024). AI-enhanced healthcare management during natural disasters: Conceptual insights. *Engineering Science & Technology Journal*, *5*(5), 1794–1816. DOI: 10.51594/estj.v5i5.1155

Adiwinoto, R., Prasetya, V., Najmasari, N., Sudarta, J., Putri, F., Garianto, E., Rasyida, A., & Diarsvitri, W. (2025). Narrative review on AI-driven and social media-enhanced early detection, mitigation, and public health preparedness in natural disaster-prone areas. *IOP Conference Series. Earth and Environmental Science*, *1473*(1), 012020. DOI: 10.1088/1755-1315/1473/1/012020

Agbehadji, I. E., Schütte, S., Masinde, M., Botai, J., & Mabhaudhi, T. (2023). Climate risks resilience development: A bibliometric analysis of climate-related early warning systems in southern Africa. *Climate (Basel)*, *12*(1), 3. DOI: 10.3390/cli12010003

Ajayi, J. O., Erigha, E. D., Obuse, E., Ayanbode, N., & Cadet, E. (2025). Resilient infrastructure management systems using real-time analytics and. *Computer Science & IT Research Journal*, *6*(8), 525–548. DOI: 10.51594/csitrj.v6i8.2013

Andrae, S. (2025). Artificial intelligence in disaster management. In *Advances in Environmental Engineering and Green Technologies* (pp. 73–114). IGI Global., DOI: 10.4018/979-8-3693-7483-2.ch004

Basher, R. (2006). Global early warning systems for natural hazards: systematic and people-centred. Philosophical Transactions. Series A, Mathematical, Physical, and Engineering Sciences, 364(1845), 2167–2182. DOI: 10.1098/rsta.2006.1819

Brotzge, J., & Donner, W. (2013). The tornado warning process: A review of current research, challenges, and opportunities. *Bulletin of the American Meteorological Society*, *94*(11), 1715–1733. DOI: 10.1175/BAMS-D-12-00147.1

Burston, J., Ware, D., & Tomlinson, R. (2015). The real-time needs of emergency managers for tropical cyclone storm tide forecasting: Results of a participatory stakeholder engagement process. *Natural Hazards*, *78*(3), 1653–1668. DOI: 10.1007/s11069-015-1794-7

Cui, J., Zhang, T., Shen, Y., Wang, X., Yang, L., Huang, X., Huang, Q., Yang, Y., Yang, W., & Li, Z. (2025). Probability-based early warning for seasonal influenza in China: Model development study. JMIR Medical Informatics, 13(v13i1e73631), e73631. DOI: 10.2196/73631

Department of Computer Science and Engineering Anantapur, S. R. I. of T. (2024). Adaptive artificial intelligence techniques for QoS and congestion management in 5G wireless networks: A literature. INTERANTIONAL JOURNAL OF SCIENTIFIC RESEARCH IN ENGINEERING AND MANAGEMENT, 08(06), 1–5. DOI: 10.55041/IJSREM36077

Ejjami, R. (2024). The Adaptive Human-AI Synergy in logistics (AHASL) theory. Journal of Next-Generation Research 5.0. DOI: 10.70792/jngr5.0.v1i1.12

El Arab, R. A., Al Moosa, O. A., Albahrani, Z., Alkhalil, I., Somerville, J., & Abuadas, F. (2025). Integrating artificial intelligence into perinatal care pathways: A scoping review of reviews of applications, outcomes, and equity. *Nursing Reports*, *15*(8), 281. DOI: 10.3390/nursrep15080281 PMID: 40863668

El Morr, C., Ozdemir, D., Asdaah, Y., Saab, A., El-Lahib, Y., & Sokhn, E. S. (2024). AI-based epidemic and pandemic early warning systems: A systematic scoping review. *Health Informatics Journal*, *30*(3), 14604582241275844. DOI: 10.1177/14604582241275844 PMID: 39172555

Fernández-Herrero, J. (2024). Evaluating recent advances in affective intelligent tutoring systems: A scoping review of educational impacts and future prospects. *Education Sciences*, *14*(8), 839. DOI: 10.3390/educsci14080839

Gardezi, M., Joshi, B., Rizzo, D. M., Ryan, M., Prutzer, E., Brugler, S., & Dadkhah, A. (2024). Artificial intelligence in farming: Challenges and opportunities for building trust. *Agronomy Journal*, *116*(3), 1217–1228. DOI: 10.1002/agj2.21353

Hussain, A., Tahir, A., Hussain, Z., Sheikh, Z., Gogate, M., Dashtipour, K., Ali, A., & Sheikh, A. (2021). Artificial intelligence-enabled analysis of public attitudes on Facebook and twitter toward COVID-19 vaccines in the United Kingdom and the United States: Observational study. *Journal of Medical Internet Research*, *23*(4), e26627. DOI: 10.2196/26627 PMID: 33724919

Inamura, T., Yamada, H., Morinaga, K., Yamanobe, N., Hanai, R., & Domae, Y. (2025). Development and evaluation of a human-robot collaborative training system for retail stores using virtual reality and digital twin technologies. *Journal of Robotics and Mechatronics*, *37*(2), 478–487. DOI: 10.20965/jrm.2025.p0478

Independent Researcher Texas, U. S. (2025). Ethical challenges in data science: Navigating the complex landscape of responsibility and fairness. *International Journal of Current Science Research and Review*, *08*(03). Advance online publication. DOI: 10.47191/ijcsrr/V8-i3-09

Jayasiri, G. P., Siriwardena, C. S. A., Hettiarachchi, S. S. L., Dissanayake, P. B. R., & Bandara, C. S. (2018). Evaluation of community resilience aspects of Sri Lankan coastal districts. *International Journal on Advanced Science, Engineering and Information Technology*, *8*(5), 2161–2167. DOI: 10.18517/ijaseit.8.5.7095

Jermutus, E., Kneale, D., Thomas, J., & Michie, S. (2022). Influences on user trust in healthcare artificial Intelligence: A systematic review. *Wellcome Open Research*, *7*, 65. DOI: 10.12688/wellcomeopenres.17550.1

Koshimura, S. Graduate School of Engineering Aramaki Aoba-ku Sendai 980-8579 Japan, T. U. A. 6-6-11-1104, Namegaya, Y., Yanagisawa, H., Geological Survey of Japan 1-1-1 Higashi Tsukuba 305-8567 Japan, N. I. of A. I. S. and T. C., & Tokyo Electric Power Services Co. Higashiueno 3-Chome Taito-ku Tokyo 110-0015 Japan, Ltd. 3-3. (2009). Tsunami fragility — A new measure to identify tsunami damage —. Journal of Disaster Research, 4(6), 479–488. DOI: 10.20965/jdr.2009.p0479

Krishna Pasupuleti, M. (2024a). AI and big data for climate resilience: Predictive analytics in environmental management. In AI and Big Data in Climate Change: Predictive Analytics for Environmental Management (pp. 255–280). National Education Services. DOI: 10.62311/nesx/46601

Krishna Pasupuleti, M. (2024b). AI-Powered Disaster Forecasting: Advanced tools for early warning and Risk Mitigation. In AI in Disaster Prediction: Tools for Early Warning Systems and Risk Mitigation (pp. 233–261). National Education Services. DOI: 10.62311/nesx/905759

Li, Y., Wu, B., Huang, Y., & Luan, S. (2024). Developing trustworthy artificial intelligence: Insights from research on interpersonal, human-automation, and human-AI trust. *Frontiers in Psychology*, *15*, 1382693. DOI: 10.3389/fpsyg.2024.1382693 PMID: 38694439

Luxshi, K. (2024a). AI-driven Disaster Prediction and Early Warning Systems: A systematic literature review. *International Journal of Research in Computing*, *3*(1), 44. DOI: 10.64701/ijrc/345/8906

Luxshi, K. (2024b). AI-driven Disaster Prediction and Early Warning Systems: A systematic literature review. *International Journal of Research in Computing*, *3*(1), 44. DOI: 10.64701/ijrc/345/8906

Macherera, M., & Chimbari, M. J. (2016). A review of studies on community based early warning systems. *Jàambá*, *8*(1), 206. DOI: 10.4102/jamba.v8i1.206 PMID: 29955285

Mbunge, E., Millham, R. C., Sibiya, M. N., Fashoto, S. G., Akinnuwesi, B., Simelane, S., & Ndumiso, N. (2021). Framework for ethical and acceptable use of social distancing tools and smart devices during COVID-19 pandemic in Zimbabwe. *Sustainable Operations and Computers*, *2*, 190–199. DOI 10.1016/j.susoc.2021.07.003

Ntali, M. A., Vubo, E. Y., & Ndonwi, A. S. (2023). Community preparedness to recurrent flood disasters in Babessi, rural Cameroon. *American Journal of Environment Studies*, *6*(2), 43–63. DOI: 10.47672/ajes.1608

Ongala, M., Kiraka, R., Choudrie, J., & Okello, J. (2025). Strategic integration of Artificial Intelligence in healthcare: Theoretical frameworks, adoption, enablers, and barriers — A scoping review. *Proceedings of the AAAI Symposium Series*, 6(1), 298–303. DOI: 10.1609/aaaiss.v6i1.36066

Pasupuleti, M. K. (2024). *AI-driven leadership: Crisis management and virtual organizations in the digital age*. National Education Services., DOI: 10.62311/nesx/rb978-81-981179-1-5

Pasupuleti, M. K., & Services, N. E. (2025). Augmented Human Intelligence: Converging Generative AI, Quantum Computing, and XR for enhanced human-machine synergy. DOI: 10.62311/nesx/rrv525

Pushparani, D. (2025). Review On: Collaboration Intelligence in Education. [IRJAEH]. *International Research Journal on Advanced Engineering Hub*, *3*(09), 3606–3611. DOI: 10.47392/IRJAEH.2025.0526

Ravikumar, R. N., & Aarthi, S. (2025a). Climate data without borders. In *Cross-Border Cooperation in the Age of Climate Change* (pp. 61–90). IGI Global Scientific Publishing., DOI: 10.4018/979-8-3373-3972-6.ch003

Ravikumar, R. N., & Aarthi, S. (2025b). Integrated cross-border renewable energy infrastructure for a green transition. In *Cross-Border Cooperation in the Age of Climate Change* (pp. 301–332). IGI Global Scientific Publishing., DOI: 10.4018/979-8-3373-3972-6.ch011

Reichstein, M., Benson, V., Blunk, J., Camps-Valls, G., Creutzig, F., Fearnley, C. J., Han, B., Kornhuber, K., Rahaman, N., Schölkopf, B., Tárraga, J. M., Vinuesa, R., Dall, K., Denzler, J., Frank, D., Martini, G., Nganga, N., Maddix, D. C., & Weldemariam, K. (2025). Early warning of complex climate risk with integrated artificial intelligence. *Nature Communications*, *16*(1), 2564. DOI: 10.1038/s41467-025-57640-w PMID: 40089483

Schoenherr, J. R., Abbas, R., Michael, K., Rivas, P., & Anderson, T. D. (2023). Designing AI using a human-centered approach: Explainability and accuracy toward trustworthiness. *IEEE Transactions on Technology and Society*, *4*(1), 9–23. DOI: 10.1109/TTS.2023.3257627

Soundrapandiyan, R., Manickam, A., Akhloufi, M., Murthy, Y. V. S., Sundaram, R. D. M., & Thirugnanasambandam, S. (2023). An efficient COVID-19 mortality risk prediction model using deep synthetic minority oversampling technique and convolution neural networks. *BioMedInformatics*, *3*(2), 339–368. DOI: 10.3390/biomedinformatics3020023

Subandi, A., Alim, S., Haryanti, F., & Prabandari, Y. S. (2019). Training on modified model of programme for enhancement of emergency response flood preparedness based on the local wisdom of Jambi community. *Jàambá*, *11*(1), 801. DOI: 10.4102/jamba.v11i1.801 PMID: 31534642

Tounsi, A., & Temimi, M. (2023). A systematic review of natural language processing applications for hydrometeorological hazards assessment. *Natural Hazards*, *116*(3), 2819–2870. DOI: 10.1007/s11069-023-05842-0 PMID: 36776702

Wibowo, A., Amri, I., Surahmat, A., & Rusdah, R. (2025). Leveraging artificial intelligence in disaster management: A comprehensive bibliometric review. *Jàambá*, *17*(1), 1776. DOI: 10.4102/jamba.v17i1.1776 PMID: 40357012

Xiong, Y., Shi, Y., Pu, Q., & Liu, N. (2024). More trust or more risk? User acceptance of artificial intelligence virtual assistant. *Human Factors and Ergonomics in Manufacturing*, *34*(3), 190–205. DOI: 10.1002/hfm.21020

KEY TERMS AND DEFINITIONS

Affective Computing: Technology enabling AI to recognize and respond to human emotions.

Alert Fatigue: Reduced responsiveness caused by frequent or non-contextual alerts.

Behavioral Feedback Loop: Mechanism through which user responses inform continuous system adaptation.

Cognitive-Adaptive Warning System: AI-based model that adapts warnings dynamically using human behavioral and perception data.

Explainable AI (XAI): AI systems designed to provide transparent, interpretable reasoning for decision outputs.

Human-Digital Twin: A virtual cognitive replica of human behavioral dynamics used for predictive modeling.

Multi-Hazard Early Warning System (MHEWS): Integrated system addressing multiple, interrelated hazards within a unified communication framework.

Trust and Transparency: Ethical dimensions ensuring accountability and fairness in AI-driven warning dissemination.

Chapter 5
Integrated Multi-Hazard Early Warning Systems:
Technologies, Governance, and Implementation Challenges

Nayla Fadiya
http://orcid.org/0009-0007-8339-911X
Universitas Esa Unggul, Indonesia

Binastya Anggara Sekti
http://orcid.org/0000-0001-5489-4888
Universitas Esa Unggul, Indonesia

ABSTRACT

Integrated multi-hazard early warning systems (MHEWS) are critical for reducing disaster risks amid increasing climate uncertainty and complex hazard interactions. Advances in technologies such as remote sensing, artificial intelligence, and sensor networks have improved hazard detection and forecasting, but system effectiveness also depends on governance, institutional coordination, and community response. This chapter examines MHEWS as socio-technical systems that integrate technological innovation with governance and behavioral dynamics. It analyzes the transition from single-hazard to integrated approaches and explores key implementation challenges, including institutional fragmentation, infrastructure gaps, data governance issues, and public trust. Drawing on global case studies, the chapter highlights practical lessons on system performance across different contexts. The findings emphasize that strengthening MHEWS requires a balanced approach combining technology, governance capacity, and inclusive communication to enhance resilience and support effective disaster preparedness.

DOI: 10.4018/979-8-3373-7345-4.ch005

ight © 2026, IGI Global Scientific Publishing. Copying or distributing in print or electronic forms without written permission of IGI Global Scientific Publishing is prohibited. Use of this chapter to n generative artificial intelligence (AI) technologies is expressly prohibited. The publisher reserves all rights to license its use for generative AI training and machine learning model development.

INTRODUCTION

The global escalation of climate-related and geophysical disasters has underscored the urgency of developing integrated, people-centered, and impact-based Multi-Hazard Early Warning Systems (MHEWS). These systems represent one of the most cost-effective and transformative tools for disaster risk reduction, aligning with the Sendai Framework for Disaster Risk Reduction (2015–2030) and the United Nations "Early Warnings for All" initiative. This chapter explores the evolution, implementation, and future of MHEWS as an integrated socio-technical ecosystem. It draws upon contemporary policy frameworks, case-based evidence, and recent technological innovations that support early warning coverage for all. Synthesizing findings from the Compendium of Multi-Hazard Early Warning Cooperation, the International Journal of Disaster Risk Science, and Perspectives in Public Health, the chapter highlights the dynamic interplay between governance, technology, media, and public trust. It concludes that MHEWS must function not merely as instruments of risk communication but as platforms of shared responsibility, linking scientific foresight with societal action to build resilience in an era of cascading and compounding risks.

Figure 1. Effective, people-centred EWS includes four key and interrelated components.

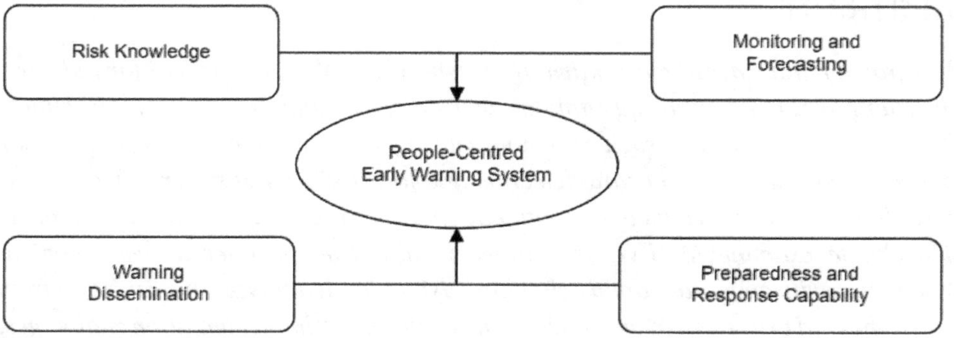

LITERATURE REVIEW

Early warning systems have evolved from hazard-specific and technology-driven approaches into integrated, multi-hazard frameworks that emphasize inclusivity and societal impact. According to Basher (2006), early warning systems must move beyond hazard detection to incorporate risk communication and community response. This perspective is further reinforced by global frameworks such as the Sendai Framework, which positions early warning systems as a fundamental component of disaster risk governance.

Recent studies emphasize the transition toward Multi-Hazard Early Warning Systems (MHEWS), which integrate multiple risk domains—including climatic, hydrological, geological, and biological hazards—into a unified system. Rokhidah et al. (2025) argue that MHEWS are essential for addressing cascading and compound risks in an increasingly interconnected risk landscape.

INTEGRATED OVERVIEW

The twenty-first century has seen an unprecedented convergence of multiple crises — climatic, hydrological, biological, and technological — that transcend borders and sectors. Floods intensify heatwaves; cyclones trigger landslides; droughts catalyze health emergencies. In this interconnected risk landscape, early warning systems (EWS) have evolved beyond meteorological forecasting into complex, multi-sectoral frameworks integrating science, governance, and social participation (Rokhideh et al., 2025).

In 2022, the United Nations Secretary-General António Guterres launched the *Early Warnings for All (EW4All)* initiative, committing the international community to ensure that every person on Earth is protected by early warning systems by 2027. This initiative operationalizes the vision of the *Sendai Framework for Disaster Risk Reduction 2015–2030*, particularly Target G, which aims to substantially increase the availability of and access to multi-hazard early warning systems and disaster risk information by 2030. Despite this global commitment, disparities in access remain profound: one in three people worldwide still lacks effective coverage, and mortality from disasters is eight times higher in low-capacity, high-risk countries (Rogers et al., n.d.).

An early warning system is not only a technical tool; it is a social contract. It connects risk knowledge, observation, communication, and response in a value chain that transforms data into decisions. The integration of multiple hazards within one coherent framework—what we now term a Multi-Hazard Early Warning System (MHEWS)—represents a paradigm shift from fragmented, hazard-specific approaches toward holistic risk governance. The goal of this chapter is to provide a comprehensive analysis of how MHEWS have evolved conceptually and operationally, the technological and institutional innovations that sustain them, and the lessons learned from their implementation. Drawing from global, regional, and local experiences, it argues that an effective MHEWS must be inclusive, anticipatory, and adaptive—capable not only of forecasting threats but of shaping early actions and long-term resilience (Nunes, 2025).

CONCEPTUAL FOUNDATIONS: THE ANATOMY OF AN INTEGRATED WARNING SYSTEM

A Multi-Hazard Early Warning System can be defined as an integrated framework that enables individuals, communities, governments, and institutions to take timely, informed action to minimize the impacts of hazardous events. The World Meteorological Organization (WMO) and the United Nations Office for Disaster Risk Reduction (UNDRR) have identified four essential, interdependent components of an effective early warning system. These are summarized in Table 1.

Table 1. Core components of a Multi-Hazard Early Warning System (MHEWS)

Component	Description	Purpose
Risk Knowledge	Systematic collection and analysis of data on hazards, vulnerabilities, and exposure.	To identify who and what is at risk, informing targeted preparedness.
Monitoring and Forecasting	Continuous observation of hazards using meteorological, hydrological, geological, biological, and technological data.	To detect and predict events before impacts occur.
Warning Dissemination and Communication	Issuing clear, authoritative, and actionable messages to those at risk through trusted channels.	To ensure that information is received, understood, and acted upon.
Preparedness and Response Capability	Building institutional, community, and individual capacities to respond effectively.	To transform warnings into timely actions that save lives and reduce losses.

The four components of MHEWS are highly interdependent and must function cohesively within a broader governance and institutional framework. Weakness in any component can significantly reduce the overall effectiveness of the system.

These components operate within an enabling environment of governance, legal mandates, financing, and partnerships. They also require inclusivity: warnings must reach all segments of society, including women, children, people with disabilities, and marginalized communities. A people-centered MHEWS is inclusive, accessible, and actionable. Integration is central to MHEWS. Rather than building parallel systems for each hazard, a multi-hazard approach harmonizes data collection, communication formats, and institutional coordination. This ensures that messages are consistent and comprehensible, reduces duplication of effort, and optimizes limited resources (Rokhideh et al., 2025).

MHEWS thus represents a socio-technical ecosystem. It combines technological systems (sensors, satellite networks, digital platforms) with social systems (governance, education, and trust networks). In this sense, early warning is as much about communication as computation. It depends on how people perceive, interpret, and act on warnings—a process influenced by trust, culture, and prior experience (Hermans et al., 2022).

Figure 2. Integrated multi-hazard early warning systems framework

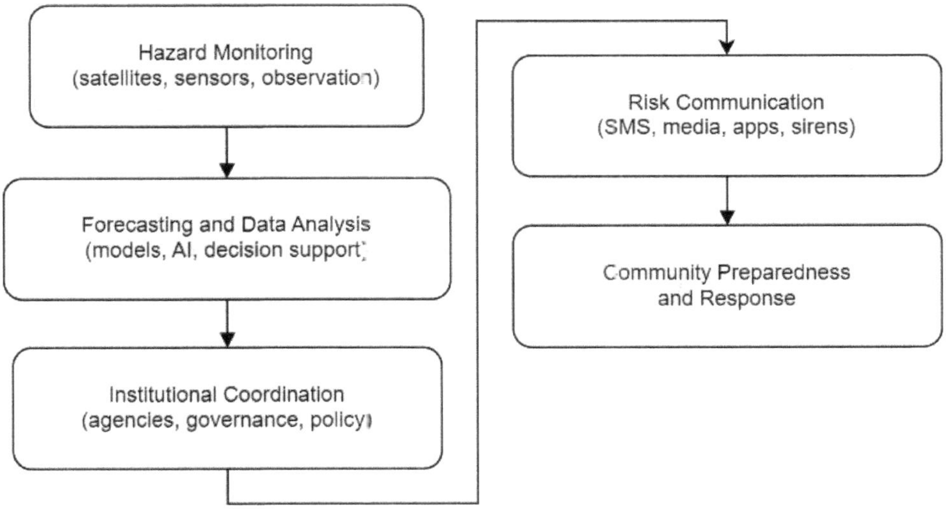

The figure illustrates how a Multi-Hazard Early Warning System (MHEWS) integrates technical, institutional, and social components into a cohesive framework for disaster risk reduction. Information flows from hazard monitoring and forecasting (through satellites, IoT sensors, and data centers) to risk communication and community-based response mechanisms. This integration reflects the shift from fragmented, single-hazard systems toward people-centered, multi-sectoral approaches that link science with governance and local action. The visualization emphasizes that MHEWS is not merely a technological apparatus but a socio-technical ecosystem that relies on coordination, trust, and inclusivity to transform data into timely and meaningful decisions.

POLICY EVOLUTION AND THE SENDAI FRAMEWORK

Historically, the global momentum for early warning systems intensified after the 2004 Indian Ocean Tsunami, which claimed over 230,000 lives across twelve countries. The catastrophe revealed the absence of coordinated regional warning systems and catalyzed global action toward multi-hazard frameworks. The *Hyogo Framework for Action 2005–2015* introduced early warning as a central pillar of disaster risk reduction. Building on this legacy, the *Sendai Framework* redefined early warning as a rights-based imperative, emphasizing inclusivity, integration, and access (Abdullahi et al., 2026).

Target G of the Sendai Framework explicitly calls for expanding multi-hazard early warning systems and risk information. Over the past decade, progress has been tangible yet uneven. By 2023, 95 out of 187 reporting countries had established some form of MHEWS—roughly double the number in 2015, yet still less than half of all nations. Many low-income countries remain constrained by limited technical capacity, fragmented institutional arrangements, and financial barriers (Liu et al., 2026).

The *Early Warnings for All* initiative (2023–2027) represents the most ambitious operationalization of Target G. With a projected investment of USD 3.1 billion over five years, it seeks to strengthen four interconnected pillars:

1. Disaster Risk Knowledge,
2. Observation and Forecasting,
3. Warning Dissemination and Communication, and
4. Preparedness and Response Capabilities.

The policy evolution of early warnings reveals a critical shift: from hazard monitoring toward anticipatory, impact-based, and people-centered systems. The growing emphasis on impact-based forecasting—focusing not only on what the weather will be but on what it will *do*—has been instrumental. This second-generation forecasting model provides contextualized risk information that links scientific probability with social vulnerability and actionable response (Dimitrieska, 2024).

At the global level, organizations such as WMO, UNDRR, and the Regional Integrated Multi-Hazard Early Warning System (RIMES) have led initiatives to strengthen cooperation among national meteorological and hydrological services. The integration of national, regional, and local capacities now forms the backbone of global disaster governance.

TECHNOLOGICAL INNOVATION AND INTEGRATION

Advances in science and technology are redefining what early warning systems can achieve. The rise of artificial intelligence (AI), big data analytics, and remote sensing has made it possible to detect and predict hazards with unprecedented accuracy and timeliness. These technologies are enabling the transition from reactive to proactive disaster management by enhancing the precision of forecasts, optimizing data processing, and improving communication pathways among stakeholders (Rahmatika et al., 2025).

Artificial intelligence and machine learning algorithms play a crucial role in analyzing vast and complex datasets generated from diverse sources—such as satellite imagery, meteorological data, seismic sensors, and social media feeds—to identify emerging risks and predict potential disaster impacts. Through continuous learning, these systems can recognize patterns, refine predictive models, and provide near-real-time risk assessments. For example, deep learning models are increasingly used in flood forecasting, wildfire spread prediction, and landslide susceptibility mapping, thereby improving preparedness and response efficiency.

Big data analytics allows for the integration of heterogeneous data streams from multiple sensors and platforms. This integration supports the development of dynamic risk maps and decision-support tools that enhance situational awareness. Cloud-based data infrastructures also facilitate the rapid sharing of critical information among local authorities, national agencies, and international partners—bridging the gap between data availability and actionable insights.

Remote sensing technologies, particularly those leveraging satellites and unmanned aerial vehicles (UAVs), provide continuous, high-resolution monitoring of environmental parameters such as temperature, precipitation, vegetation health, and surface deformation. When combined with ground-based sensor networks and geospatial information systems (GIS), remote sensing supports a multi-layered understanding of hazard dynamics across spatial and temporal scales.

Equally significant is the integration of the Internet of Things (IoT) in early warning systems. IoT-enabled sensors distributed across vulnerable areas can transmit real-time environmental data, such as water levels, air quality, or seismic activity, directly to central monitoring platforms. This real-time feedback loop enhances the responsiveness of emergency management teams and supports automated alert dissemination to at-risk communities.

Moreover, advancements in communication technologies—such as 5G networks and mobile-based alert systems—ensure that early warnings reach populations swiftly and reliably. Multilingual, location-based messaging, and community-based digital platforms enable inclusive dissemination of alerts, reducing the information gap between authorities and citizens.

Emerging innovations like digital twins, blockchain for data integrity, and cloud–edge hybrid computing architectures further expand the potential of integrated early warning systems. Digital twins allow simulation of hazard scenarios and the testing of mitigation strategies in virtual environments, while blockchain can secure the traceability and reliability of data exchanges between agencies.

In sum, technological innovation and integration are the backbone of modern multi-hazard early warning systems. They enable real-time data fusion, predictive intelligence, and inclusive communication—key pillars for building resilient societies capable of anticipating and mitigating disaster risks before they escalate into crises (Hermans et al., 2022).

1. Artificial Intelligence and Predictive Analytics

AI-driven models can analyze vast datasets from satellites, sensors, and social media in real time, identifying hazard patterns and potential impacts. For example, machine learning algorithms are now used to anticipate flood propagation, landslide risk, and drought severity weeks in advance. Integrating AI into early warning pipelines reduces lead time and enhances precision.

2. Satellite and Remote Sensing Technologies

Satellites operated by WMO, NASA, and the European Space Agency provide continuous monitoring of atmospheric, oceanic, and terrestrial conditions. These data streams feed into global producing centers and regional specialized meteorological hubs that generate real-time hazard maps.

3. Internet of Things (IoT) and Sensor Networks

IoT-enabled sensors deployed in rivers, coastlines, and urban infrastructure allow continuous, localized monitoring. When connected to national data centers, they form the backbone of anticipatory warning networks. For example, rainfall and seismic sensors can trigger community alerts through automated protocols.

4. Digital Communication and Mobile Networks

The proliferation of mobile connectivity—reaching 95% of the world's population—has revolutionized the dissemination of warnings. Mobile-based alert systems, radio broadcasting, and community networks ensure that information can reach remote and vulnerable populations. Yet, digital divides persist, and accessibility gaps remain in regions with limited connectivity.

5. Data Integration and Interoperability

Modern MHEWS rely on cross-sectoral data integration. Hazard information from meteorology, hydrology, geology, and health sectors must be interoperable to produce unified risk assessments. Open-data architectures and standardized formats facilitate this integration, enabling seamless data sharing across agencies and borders.

BRIDGING TECHNOLOGY AND SOCIETY: THE ROLE OF COMMUNICATION AND MEDIA

Recent research has emphasized that early warnings are effective only when they are trusted, understood, and acted upon. Communication is therefore the bridge between technology and human behavior. The 2025 *Perspectives in Public Health* study on *Health and Multi-Hazard Early Warning Systems in the Media* demonstrated that media discourse plays a crucial role in shaping public awareness and support for disaster preparedness.

The study identified seven dominant themes in media narratives: vulnerability reduction, government roles, international collaboration, community engagement, technological progress, financial challenges, and health impacts. By framing MHEWS as both a technological and public health issue, the media can transform risk communication into collective motivation for preparedness. Integrating health perspectives within MHEWS expands the concept from environmental resilience to human well-being. Health-sensitive warning systems can track epidemiological data alongside meteorological indicators, enabling anticipatory health responses to climate-driven hazards such as heatwaves, vector-borne diseases, and respiratory crises.

COMMUNICATION, GOVERNANCE, AND SOCIAL TRUST

Communication, governance, and social trust form the connective tissue that determines whether an early warning system can truly protect lives. While advanced technologies such as artificial intelligence, sensor networks, and satellite-based forecasting have revolutionized hazard prediction, their success ultimately depends on the human and institutional dimensions that interpret, disseminate, and act upon warnings. The essence of an effective Multi-Hazard Early Warning System (MHEWS) lies not merely in the generation of accurate forecasts, but in the ability to translate technical data into socially meaningful information that triggers timely and appropriate action. In this sense, early warning is both a scientific and a social process—a dialogue between knowledge and trust (Luther et al., 2017).

Effective communication within MHEWS is not limited to the transmission of data; it is about ensuring that messages are understandable, credible, and actionable for diverse audiences. The complexity of scientific forecasts must be distilled into language that resonates with community realities. A flood warning, for example, must not only indicate the probability of rainfall or river overflow but must specify which neighborhoods are at risk, which roads will be impassable, and what actions residents should take immediately. The clarity and specificity of these messages often determine whether people act in time or ignore the alert. As the 2025 *Perspectives in Public Health* study emphasizes, communication strategies should be grounded in social behavior research that considers local culture, literacy levels, and previous experiences with disasters. A warning that fails to connect with people's lived realities may be technically sound but socially ineffective.

Trust, once earned, becomes the most valuable currency in an early warning ecosystem. Communities are more likely to respond when they believe in the credibility of the issuing authority and the consistency of its communication. This credibility is built gradually through transparent information sharing, consistent performance, and community engagement long before disasters strike. Conversely, misinformation, political interference, or contradictory messaging can erode confidence, leading to apathy or panic. The "last mile" of communication—the point at which information reaches individuals—remains the most challenging and decisive stage. At this interface, technical precision meets human perception, and institutional trust determines behavioral response.

Governance plays an equally critical role in this equation. In many countries, early warning responsibilities are distributed among meteorological, hydrological, geological, and emergency management agencies. Without clear coordination mechanisms, these institutions often operate in silos, resulting in fragmented responses or delays in decision-making. Integrated governance frameworks, supported by legislation, are therefore essential. They define roles and responsibilities, standardize protocols, and ensure that agencies operate in synergy. For instance, Japan's disaster management framework offers a model of clarity: its Meteorological Agency issues standardized warnings, while local governments are legally mandated to coordinate evacuations and emergency responses. This alignment minimizes confusion and fosters public confidence in the system.

LESSONS LEARNED FROM REGIONAL EXPERIENCES

Regional cooperation in early warning has yielded valuable lessons about the balance between centralization and localization. The *Compendium of Multi-Hazard Early Warning Cooperation in Asia and the Pacific* documents how regional frameworks—such as RIMES in Asia, AMHEWAS in Africa, and Meteoalarm in Europe—have advanced technical capacity and policy coherence. These initiatives show that resilience cannot be achieved in isolation; it is constructed through shared knowledge, data exchange, and collective foresight. At the same time, they highlight that a one-size-fits-all model rarely succeeds. Systems must adapt to national contexts, resource availability, and local governance cultures.

For example, in South and Southeast Asia, where tropical cyclones and monsoon floods are frequent, regional cooperation has dramatically improved forecast accuracy and lead times. Bangladesh's integration of satellite data from RIMES with community volunteer networks has reduced cyclone-related mortality by over 80% in the past three decades. Conversely, small island developing states in the Pacific continue to face structural limitations: scarce human resources, expensive maintenance costs, and dependence on donor funding. Their experiences illustrate that sustainability requires not only international assistance but also local capacity development and institutional ownership.

In Europe, the European Flood Awareness System (EFAS) demonstrates the benefits of standardized protocols and transnational coordination. EFAS provides early warnings to 37 countries through shared hydrological modeling, allowing cross-border responses to river basin flooding. However, it also reveals a limitation common to many large-scale systems: warnings that are technically accurate may still fail to provoke adequate local response if they are not contextually tailored or linguistically accessible. Thus, localization—both linguistic and cultural—remains essential, even within the most advanced frameworks(Nunes, 2025).

Ultimately, the key lesson from these regional experiences is that the effectiveness of MHEWS depends on balance: between global science and local knowledge, central coordination and community empowerment, technological sophistication and social inclusivity. Successful systems are those that treat early warning as a shared social process rather than a mere technical output.

FINANCING AND SUSTAINABILITY

Financing remains the linchpin of early warning sustainability. Despite clear evidence that every dollar invested in early warning saves multiple dollars in disaster losses, funding gaps persist. Many national systems rely on external donors or short-term project grants, which undermine long-term maintenance. Integrating MHEWS into national budgets and climate adaptation strategies is therefore essential. Governments need to treat early warning infrastructure as critical public goods, akin to healthcare or education, rather than discretionary expenditures.

Innovative financing approaches—such as forecast-based financing (FbF) and parametric insurance—have begun to shift this paradigm. FbF enables funds to be disbursed automatically when forecast thresholds are met, allowing pre-disaster action instead of post-disaster relief. Parametric insurance schemes, meanwhile, provide financial compensation based on predefined hazard indicators rather than physical damage, ensuring rapid liquidity for recovery. Together, these mechanisms link scientific prediction directly to fiscal decision-making, embedding anticipation into governance. The "Early Warnings for All" initiative underscores this principle by advocating the alignment of public finance, international aid, and private investment to sustain MHEWS globally.

Sustainability also depends on human capital. Technical systems require continuous training, data analysis, and system maintenance. Without a skilled workforce, even the most advanced models deteriorate. Capacity-building programs—through regional training centers, academic partnerships, and field-based simulations—create a virtuous cycle of learning and innovation. In this sense, sustainability is not simply financial or technical; it is cultural and institutional, rooted in the daily practices of those who operate and trust the system.

THE FUTURE OF INTEGRATED EARLY WARNING SYSTEMS

The future trajectory of MHEWS will be shaped by technological convergence, climate change, and social transformation. Artificial intelligence, remote sensing, and Internet of Things technologies will continue to expand predictive capabilities. The next generation of systems will not only forecast hazards but also simulate their cascading impacts across sectors such as health, agriculture, and energy. Digital twins—virtual replicas of physical systems—will enable authorities to test interventions in simulated environments before applying them in reality. Likewise, advances in mobile communication and cloud computing will make early warnings more personalized, delivering alerts tailored to an individual's location, vulnerabilities, and needs.

However, as data volumes increase, so do ethical challenges. Data privacy, algorithmic transparency, and equitable access must become central considerations. Without careful governance, predictive systems could exacerbate existing inequalities, providing better protection for connected urban populations while leaving rural or marginalized communities behind. The principle of "Early Warnings for All" must thus be taken literally: inclusivity is not a rhetorical ideal but a measurable obligation. The global community's success in this regard will determine whether early warning becomes an instrument of justice or another marker of inequality.

Another emerging frontier is the integration of climate adaptation and early warning. Rather than functioning as separate policy domains, these two agendas are increasingly intertwined. Adaptive early warning systems not only predict hazards but continuously learn from past events, incorporating new data to refine forecasts. They are designed to evolve alongside changing climatic baselines, ensuring relevance in a rapidly warming world. This adaptive capability transforms early warning from a reactive service into an evolving knowledge ecosystem—a dynamic interface between science, governance, and society.

INSTITUTIONAL INTEGRATION AND MULTI-LEVEL COORDINATION

Institutional integration is a defining feature of an effective Multi-Hazard Early Warning System. It requires clear roles, coordination across administrative levels, and collaboration among technical agencies, government authorities, and civil society. In practice, many disasters reveal that institutional silos remain a major weakness in national and regional preparedness structures. Meteorological agencies may issue alerts without adequate communication with disaster management bodies, while local governments may lack the authority or resources to act swiftly on warnings. Bridging these institutional gaps demands not only inter-agency agreements but also a unified national policy that positions early warning as a cross-cutting responsibility shared by all sectors.

In countries where institutional integration has succeeded, such as Japan, New Zealand, and Bangladesh, coordination mechanisms are enshrined in law. These legal frameworks establish how agencies share data, activate emergency responses, and communicate with the public. National Disaster Management Authorities (NDMAs) serve as central nodes that connect technical institutions with local administrative structures. Integration across vertical levels—from national centers to municipal governments and community organizations—ensures that warnings translate into action at the local scale. The value of this vertical coherence cannot be overstated; without it, national forecasts often fail to reach the people who need them most.

Horizontal coordination across ministries is equally vital. Disaster management cannot be confined to a single agency; it touches health, education, transport, agriculture, and social protection. For instance, in an impending flood scenario, the Ministry of Health must prepare hospitals, the Ministry of Education may need to close schools, and the Ministry of Agriculture must secure livestock and grain stores. Institutional integration allows these sectors to act in concert rather than in isolation. The effectiveness of early warning thus reflects the strength of a nation's governance fabric—the ability to align multiple institutions under a shared objective of safeguarding life.

To achieve this integration, data sharing must be institutionalized rather than ad hoc. Memorandums of understanding (MoUs) between ministries are useful, but lasting coordination requires interoperable information systems, standardized procedures, and regular joint exercises. Simulation drills, in which meteorological centers, emergency responders, and local officials rehearse coordinated action, have proven invaluable. They not only test system readiness but also cultivate relationships of trust among agencies. Integration, therefore, is not simply a matter of structure but of culture—the cultivation of shared responsibility and collective.

REGIONAL AND GLOBAL COOPERATION FRAMEWORKS

Disaster risks are transboundary by nature, and so must be the systems designed to anticipate them. The integration of early warning systems at regional and global scales amplifies both technical capability and resilience. The *Compendium of Multi-Hazard Early Warning Cooperation in Asia and the Pacific* (ESCAP–UNDRR, 2023) underscores how regional frameworks enhance national systems through data exchange, joint training, and harmonized standards. These frameworks create "networks of safety" that transcend national boundaries, enabling countries to benefit from collective scientific capacity.

One of the most illustrative examples is the Regional Integrated Multi-Hazard Early Warning System (RIMES), which serves as a cooperative platform for 48 countries across Asia and the Pacific. RIMES integrates satellite data, numerical weather prediction, and hydrological modeling to support regional forecasting. More importantly, it provides a model for collaborative governance—countries contribute resources, share information, and participate in joint decision-making processes. This approach democratizes early warning and builds regional solidarity. Similarly, in Africa, the African Multi-Hazard Early Warning and Early Action System (AMHEWAS) aims to integrate climate, weather, and disaster management data across the continent, helping countries with limited capacity to access real-time risk information.

At the global level, the World Meteorological Organization (WMO) and the United Nations Office for Disaster Risk Reduction (UNDRR) play central roles in standardizing methodologies and facilitating cooperation. The *Early Warnings for All* initiative launched by the United Nations Secretary-General in 2022 is perhaps the most ambitious global effort to date, envisioning a world where every person is protected by early warning systems by 2027. Its four pillars—risk knowledge, observation and forecasting, warning dissemination, and preparedness to respond—mirror the Sendai Framework's emphasis on integration and inclusivity. Through these international mechanisms, technical excellence is coupled with political commitment, ensuring that early warning becomes not just a policy aspiration but a shared global reality.

Yet, cooperation also faces challenges. Disparities in technical infrastructure and human resources create asymmetries in regional participation. Some countries act as data providers, while others remain dependent consumers. Closing this gap requires capacity development and equitable access to technology. International assistance should therefore shift from one-directional aid to partnership-based collaboration, where knowledge transfer is mutual and sustainable. Global cooperation, in this sense, must evolve from charity to solidarity—from sharing warnings to sharing responsibility.

FINANCING, CAPACITY BUILDING, AND SUSTAINABILITY

Financial sustainability remains one of the most significant barriers to maintaining functional early warning systems, particularly in developing countries. Although MHEWS are among the most cost-effective investments in disaster risk reduction, they often compete with more immediate political priorities. Funding cycles tied to donor projects lead to intermittent progress, with systems deteriorating when external support ends. To address this, early warning must be recognized as a national public good—an essential service comparable to education or healthcare (Putu et al., 2021).

Forecast-based financing offers a paradigm shift. Under this model, funds are pre-allocated and released automatically when forecast indicators cross critical thresholds, enabling preventive action. For example, when rainfall probabilities indicate high flood risk, resources can be mobilized to reinforce levees or relocate vulnerable populations before disaster strikes. This proactive approach contrasts sharply with traditional reactive financing, which expends far greater resources on recovery than on prevention. Several pilot programs supported by the International Federation of Red Cross and Red Crescent Societies (IFRC) and the World Food Programme (WFP) have demonstrated the efficacy of this approach, reducing both losses and response times.

Capacity building is another cornerstone of sustainability. Technology alone cannot function without skilled personnel to operate, interpret, and maintain it. Training programs, exchange workshops, and academic partnerships between developed and developing nations can help bridge the expertise gap. Moreover, integrating disaster risk reduction and early warning curricula into universities ensures that the next generation of scientists, policymakers, and communicators are equipped to sustain these systems. In this context, regional training centers established under WMO and UNESCAP frameworks serve as vital hubs of continuous learning and knowledge transfer.

Sustainability is not only about financial or technical endurance—it is about social embeddedness. When communities understand the system, trust its messages, and contribute to its improvement, sustainability becomes organic. Local engagement fosters long-term ownership, reducing dependency on external actors. Governments should therefore view early warning not as an isolated technical apparatus but as a living partnership between state and society (Chakrabarty et al., 2026).

CHALLENGES, INNOVATIONS, AND FUTURE OUTLOOK

Despite remarkable progress, the path toward universal early warning coverage remains fraught with challenges. Data fragmentation, governance inconsistencies, digital divides, and inequitable access persist across regions. Many systems remain hazard-specific, struggling to integrate multiple risk domains such as health, agriculture, and infrastructure. The digital revolution offers new opportunities to overcome these limitations, but it also introduces new complexities, including data privacy concerns and cybersecurity risks.

Emerging technologies such as artificial intelligence, digital twins, and blockchain are redefining the operational potential of MHEWS. AI enables real-time analysis of complex datasets, detecting patterns that human analysts might overlook. Digital twins allow virtual simulations of hazard impacts, enabling preemptive policy testing. Blockchain, by securing data provenance, ensures the integrity of multi-agency information sharing. Yet, technology must remain a means rather than an end. Without ethical governance and equitable access, these innovations risk deepening inequality, providing more protection to digitally connected populations while marginalizing those without connectivity.

The next frontier for MHEWS lies in personalization and anticipatory governance. As climate risks intensify, the future of early warning will depend on adaptive systems that continuously learn from past events. Artificial intelligence, coupled with machine learning and impact-based forecasting, can generate hyper-localized warnings that predict not only the occurrence of hazards but also their potential consequences for specific communities. This transition from probabilistic to impact-based forecasting marks a transformative step toward truly anticipatory action.

The *Early Warnings for All* initiative encapsulates this vision. It calls for a future where no one is left unwarned, where technology serves humanity, and where early warning is synonymous with justice. To realize this, countries must invest not only in infrastructure but also in governance reform, communication, and education. The resilience of tomorrow will not depend solely on our ability to predict, but on our collective capacity to prepare, adapt, and act together.

COMMUNITY ENGAGEMENT, INCLUSION, AND INDIGENOUS KNOWLEDGE

An effective Multi-Hazard Early Warning System (MHEWS) must be rooted in the communities it serves. While advanced technologies and global cooperation form the structural backbone of early warning architectures, their effectiveness ultimately depends on human trust, perception, and participation. Community engagement transforms warnings from passive information into active protection, ensuring that alerts lead to appropriate, timely, and inclusive action.

The concept of "people-centered" early warning—central to the Sendai Framework—emphasizes that risk communication must be both accessible and actionable for all members of society, including marginalized and vulnerable groups. Yet, research and post-disaster assessments continue to reveal that women, children, persons with disabilities, and indigenous populations often remain disproportionately affected by disasters, not because they lack information, but because the information is not tailored to their contexts or delivered through trusted channels. Therefore, the social dimension of MHEWS is not an auxiliary concern but its most decisive factor.

Local participation enhances the legitimacy and sustainability of early warning systems. When communities are engaged in the design, testing, and evaluation of warning protocols, they are more likely to trust and act upon alerts. Participatory risk mapping, community-based drills, and local volunteer networks have proven particularly effective in bridging the gap between national forecasts and household action. The integration of local voices into early warning governance—through community committees or decentralized decision-making structures—ensures that systems remain responsive to evolving social realities.

Indigenous and traditional knowledge systems also offer invaluable contributions to modern MHEWS. For centuries, communities in the Pacific Islands, sub-Saharan Africa, and Southeast Asia have observed environmental cues such as animal behavior, ocean tides, and wind patterns to anticipate climatic changes. These local forecasting practices, when integrated with scientific monitoring, create hybrid knowledge systems that enhance accuracy and cultural relevance. In the Solomon Islands, for instance, indigenous knowledge of wind patterns complements satellite-based cyclone tracking, leading to more context-specific preparedness strategies. Recognizing and institutionalizing such knowledge not only enriches scientific forecasting but also reinforces cultural identity and community ownership.

Inclusivity in communication is another critical component. Messages must be multilingual, culturally sensitive, and accessible to persons with disabilities. The use of visual symbols, community radio, and mobile alerts in local dialects can greatly improve comprehension and response rates. Gender-sensitive communication strategies are equally essential, acknowledging that women often manage family evacuation, caregiving, and resource distribution during crises. A truly inclusive MHEWS is not one that speaks the loudest, but one that ensures everyone is heard and understood.

Ultimately, community engagement and indigenous knowledge transform early warning systems from centralized instruments of control into shared platforms of empowerment. When people perceive warnings not as orders but as shared responsibilities, resilience becomes self-sustaining, embedded within the fabric of everyday life (Caballero Testón et al., 2026).

ETHICAL, LEGAL, AND GOVERNANCE DIMENSIONS OF EARLY WARNING SYSTEMS

The expansion of Multi-Hazard Early Warning Systems raises profound ethical and governance questions that transcend technology. As data becomes increasingly granular and predictive analytics more powerful, societies must navigate the tension between public safety and individual rights. The ethical governance of early warning systems thus becomes a central pillar of resilience—one that determines not only how effectively we warn, but how fairly we do so.

Data privacy is among the foremost ethical challenges. Modern MHEWS rely on the collection and integration of diverse datasets, including geolocation, demographic, and health information. While such data enable precision targeting of warnings and interventions, they also pose risks of misuse and surveillance. In regions with weak data protection laws, individuals may be exposed to stigmatization or discrimination based on risk profiling. Establishing clear data governance frameworks—grounded in principles of transparency, consent, and accountability—is essential to prevent these unintended consequences.

Legal frameworks play a crucial role in defining institutional responsibilities, liability, and citizen rights. Countries such as Japan, Australia, and Germany have enacted specific legislation mandating the dissemination of early warnings and clarifying the duties of different authorities. These laws ensure that no agency can evade responsibility and that citizens have the right to timely and accurate information. In many developing contexts, however, early warning still operates within fragmented or outdated legal systems. Establishing national legislation that embeds MHEWS within broader disaster management and environmental policies can strengthen institutional coherence and public trust.

Ethical governance also extends to the equitable distribution of resources and access. The promise of "early warnings for all" will remain hollow unless the most vulnerable are prioritized. Technological innovation often gravitates toward well-connected urban areas, leaving rural and marginalized populations with limited access to alerts. Governments and international organizations must therefore adopt a rights-based approach, ensuring that protection from disasters is treated not as a privilege but as a universal entitlement.

Furthermore, the credibility of early warning systems depends on the ethical integrity of their communication. Transparency regarding uncertainty, accuracy, and the rationale behind decisions enhances public confidence. Overconfidence or deliberate withholding of information, even with good intentions, can erode trust and produce "warning fatigue." Ethical communication demands honesty, clarity, and empathy—qualities that humanize the technological apparatus of warning systems.

Lastly, global governance frameworks must adapt to the ethical complexities of the digital age. As AI-driven forecasting and cross-border data exchange expand, international agreements will be needed to regulate accountability, interoperability, and fairness. The integration of ethical oversight boards and participatory governance models can help align technological innovation with societal values. In this sense, ethical governance is not a constraint on progress but its compass—guiding innovation toward justice, equity, and sustainability.

INSTITUTIONAL AND TECHNICAL CAPACITY GAPS IN IMPLEMENTATION

Despite substantial advancements in the conceptual and technological frameworks of Multi-Hazard Early Warning Systems (MHEWS), the operational realities across many regions reveal persistent institutional and technical gaps that constrain their effectiveness. These gaps often emerge from fragmented governance, inadequate infrastructure, insufficient technical expertise, and lack of long-term investment in capacity development. Understanding and addressing these challenges is fundamental to translating global policy ambitions into local resilience (Hermans et al., 2022).

Institutional fragmentation remains one of the most pervasive obstacles. In numerous countries, early warning responsibilities are dispersed among ministries of meteorology, interior, health, and communication, with limited mechanisms for coordination or data exchange. This results in duplicated efforts, inconsistent messaging, and delays in response. Without clear command structures or legal mandates, local authorities are often unsure who holds final decision-making authority during emergencies. The absence of integrated national frameworks means that early warning remains reactive rather than anticipatory, driven by crisis rather than preparedness.

Technical capacity deficits further compound the problem. In low- and middle-income countries, limited access to modern observation networks—such as radar, satellite data, or automated weather stations—reduces forecasting accuracy. The maintenance of these technologies requires sustained funding and technical expertise, both of which are often lacking. Additionally, the interoperability of data systems across agencies remains weak. Many still rely on outdated analog communication or incompatible data formats, hindering real-time information sharing.

Human resource capacity is another crucial dimension. Skilled forecasters, data scientists, and emergency communication specialists are essential for translating technical outputs into actionable warnings. Yet, talent retention is a persistent issue, as trained professionals frequently migrate to higher-paying opportunities abroad. Building local capacity thus demands not only technical training but also incentives, institutional stability, and professional pathways that reward long-term commitment.

Furthermore, the sustainability of MHEWS operations is often threatened by project-based funding models. Donor-driven initiatives frequently prioritize short-term deliverables over systemic change, leading to gaps once projects conclude. A transition toward national budget integration and locally owned financing mechanisms is therefore essential. Only when MHEWS are embedded as permanent components of national development plans can they withstand political transitions and economic downturns (Caballero Testón et al., 2026).

To overcome these challenges, countries must prioritize systemic reform over ad hoc solutions. This involves strengthening legal frameworks, institutionalizing coordination, and investing in human capital. Regional organizations and international partners can play catalytic roles, but local ownership remains the decisive factor. Without addressing institutional and technical capacity gaps, even the most sophisticated technologies risk becoming underutilized tools—symbols of ambition without function (Liu et al., 2026).

CASE STUDIES AND LESSONS LEARNED FROM GLOBAL PRACTICES

Examining real-world experiences provides valuable insights into how integrated MHEWS function in diverse socio-political and environmental contexts. Case studies from Asia, Africa, and Latin America demonstrate that success in early warning is determined not by technological sophistication alone, but by the degree of integration between scientific forecasting, community participation, and governance structures.

Japan represents one of the most advanced examples of integrated early warning governance. The Japan Meteorological Agency (JMA) operates a multi-layered system linking seismic, meteorological, and hydrological data into a single operational framework. Through automated alerts disseminated via television, mobile networks, and public sirens, Japan's population receives real-time hazard information within seconds. Crucially, this system's success lies not only in technology but in education—citizens are trained from childhood through regular drills to respond automatically to warnings, demonstrating how social preparedness amplifies technical efficiency.

Bangladesh offers a contrasting yet equally instructive case. Despite limited resources, the country has achieved remarkable success in reducing cyclone mortality through community-based early warning systems. The Bangladesh Meteorological Department collaborates with the Cyclone Preparedness Programme (CPP), which mobilizes over 76,000 community volunteers to disseminate warnings through megaphones, radio, and local networks. By integrating gender-sensitive communication and localized evacuation planning, Bangladesh transformed early warning from a centralized function into a community partnership. This demonstrates that inclusivity and trust can compensate for technological limitations (Caballero Testón et al., 2026).

In Fiji and the Pacific Islands, regional collaboration through the Pacific Meteorological Council and the Pacific Resilience Partnership exemplifies the strength of shared knowledge. Small island nations, facing limited national capacity, benefit from regional forecasting systems that combine satellite observations, ocean monitoring, and disaster risk databases. Indigenous knowledge is also integrated into these frameworks, ensuring that warnings align with local cultural and linguistic contexts.

Kenya's Drought Early Warning System provides another model, integrating meteorological data with socio-economic indicators such as food prices and livestock health. Managed by the National Drought Management Authority, it enables anticipatory financing through a risk-based funding mechanism. When early warning indicators exceed thresholds, pre-allocated funds are released to implement early action measures such as water trucking or livestock vaccination. This demonstrates how early warning can be operationalized into tangible, preventive interventions.

The experiences of Germany's 2021 flood disaster, however, highlight that even advanced systems are vulnerable when public communication fails. Despite accurate forecasts, many residents in the Ahr Valley were not reached in time due to gaps in local coordination and crisis communication. This case underscores a critical lesson: technological excellence cannot compensate for weak governance and communication breakdowns (Yhee & Koo, 2026).

Synthesizing lessons from these cases, three cross-cutting insights emerge. First, multi-hazard integration requires political commitment as much as technical innovation. Second, community trust and participation determine whether warnings are acted upon. Third, systems must remain adaptive—continuously learning from each event, integrating feedback, and evolving with technological and social change. The future of MHEWS will depend not on static blueprints but on dynamic learning systems that grow through experience (Chakrabarty et al., 2026).

Figure 3. The meteorological value chain. All links in the chain must operate effectively to yield success

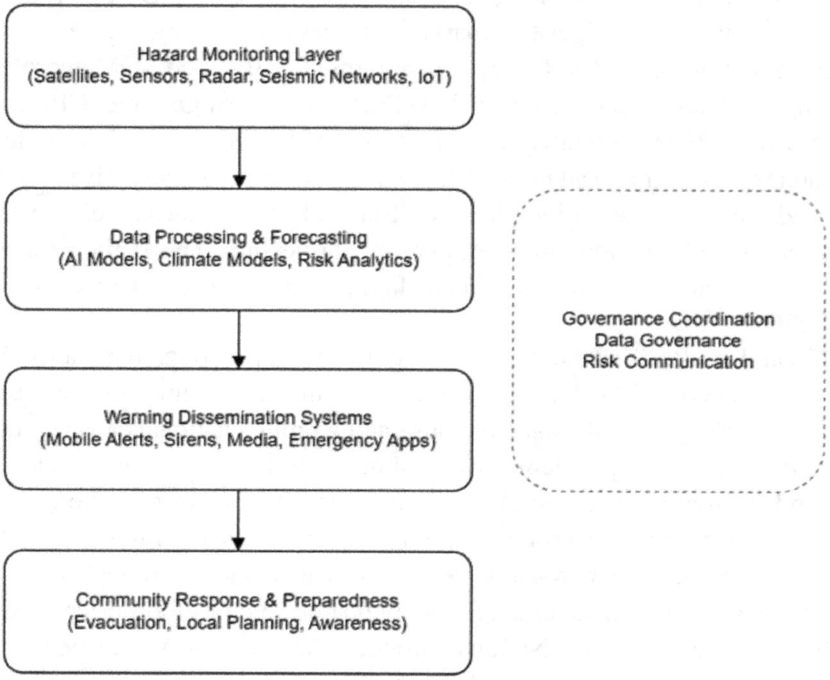

CONCLUSION

Integrated Multi-Hazard Early Warning Systems embody the most advanced synthesis of science, governance, and human empathy in disaster risk reduction. They represent the aspiration to transform knowledge into collective safety and foresight into equitable resilience. The chapter has demonstrated that the evolution of

MHEWS is not merely a technical achievement but a moral and social milestone—a redefinition of how humanity coexists with uncertainty.

The lessons learned from global experiences underscore that successful early warning systems are those that integrate institutional coordination, regional cooperation, sustainable financing, technological innovation, and community engagement. They are systems that learn, adapt, and evolve in response to new challenges—be they climate-induced, health-related, or technological. To fulfill the promise of the Sendai Framework and the UN's "Early Warnings for All" initiative, countries must invest not only in sensors and satellites but also in trust, education, and inclusion.

Ultimately, the future of MHEWS depends on our collective will to see early warning as a public good, a shared responsibility, and a human right. By integrating ethics with innovation, and local wisdom with global science, the world can move closer to a future where no disaster strikes unannounced, and no life is lost for lack of a warning.

REFERENCES

Basher, R. (2006). Global early warning systems for natural hazards: systematic and people-centred. *Philosophical transactions of the royal society a: mathematical, physical and engineering sciences, 364*(1845), 2167-2182.

Caballero Testón, J., Pérez-Mon, O., & R-Moreno, M. D. (2026). The role of automated planning in battle management systems for military tactics. *Expert Systems with Applications, 297*, 129259. Advance online publication. DOI: 10.1016/j.eswa.2025.129259

Chakrabarty, S., Deb, C. K., Marwaha, S., Haque, M. A., Kamil, D., Bheemanahalli, R., & Shashank, P. R. (2026). Application of artificial intelligence in insect pest identification - A review. *Artificial Intelligence in Agriculture, 16*(1), 44–61. DOI: 10.1016/j.aiia.2025.06.005

Compendium of multi-hazard early warning cooperation. (n.d.). www.undrr.org/publication/global-status-multi-hazard-early-warning-systems-target-g

Hermans, T. D. G., Šakić Trogrlić, R., van den Homberg, M. J. C., Bailon, H., Sarku, R., & Mosurska, A. (2022). Exploring the integration of local and scientific knowledge in early warning systems for disaster risk reduction: a review. In *Natural Hazards* (Vol. 114, Issue 2, pp. 1125–1152). Springer Science and Business Media B.V. DOI: 10.1007/s11069-022-05468-8

Liu, T., Zhang, D., Chu, Y., Ohashi, K., Hirai, Y., Ito, K., Ishibashi, K., Matsuo, Y., Yoshida, J., Ono, S., Kamiya, K., Li, H., & Yabu, H. (2026). Breaking the single-molecule paradigm: Multilayer cobalt phthalocyanine/carbon core-shell structure as the superior active unit for CO2-to-CO electroreduction. *Applied Catalysis B: Environmental, 381*, 125852. Advance online publication. DOI: 10.1016/j.apcatb.2025.125852

Luther, J., Hainsworth, A., Tang, X., Harding, J., Torres, J., & Fanchiotti, M. (2017). World Meteorological Organization (WMO)—Concerted International Efforts for Advancing Multi-hazard Early Warning Systems. In *Advancing Culture of Living with Landslides* (pp. 129–141). Springer International Publishing. DOI: 10.1007/978-3-319-59469-9_9

Nunes, A. R. (2025). Health and multihazard early warning systems in the media. *Perspectives in Public Health, 145*(6), 343–352. Advance online publication. DOI: 10.1177/17579139251342156 PMID: 40838788

Putu, I., Prasetia, W., Nyoman, I., & Kurniawan, H. (2021). Implementasi ETL (Extract, Transform, Load) pada Data warehouse Penjualan Menggunakan Tools Pentaho. *TIERS Information Technology Journal, 2*(1), 39–47. https://journal.undiknas.ac.id/index.php/tiers

Rahmatika, D. N., Firmansyah, F., & Wijaya, J. R. T. (2025). Audit Manajemen Perusahaan: Prosedur, Inovasi dan Implementasi. *Penerbiteureka.com*. https://repository.penerbiteureka.com/media/publications/603267-audit-manajemen-perusahaan-prosedur-inov-d2f86036.pdf

Rokhideh, M., Fearnley, C., & Budimir, M. (2025). Multi-Hazard Early Warning Systems in the Sendai Framework for Disaster Risk Reduction: Achievements, Gaps, and Future Directions. *International Journal of Disaster Risk Science*, *16*(1), 103–116. DOI: 10.1007/s13753-025-00622-9

Yhee, Y., & Koo, C. (2026). Seeing AI as human or machine? Effects of transparency, valence, and readability on review summary helpfulness. *Tourism Management*, *113*, 105305. DOI: 10.1016/j.tourman.2025.105305

Chapter 6
Synergistic Urban Cooling Integrating Blue-Green Infrastructure for Climate-Adaptive Cities

Sai Puneeth Damarla
http://orcid.org/0009-0004-3238-3242
Woxsen University, India

Shweta Prasad
http://orcid.org/0009-0004-4133-6608
Woxsen University, India

Dhruma Bhavsar
http://orcid.org/0000-0002-9203-0076
Woxsen University, India

Rajdeep Routh
http://orcid.org/0000-0002-1662-4075
Woxsen University, India

ABSTRACT

Urban heat island (UHI) effects have intensified with rapid urbanisation, elevating temperatures by 5-12°C above rural areas and imposing significant health, environmental, and economic burdens on urban populations projected to reach 68% globally by 2050. Importance: Effective thermal management strategies are critical for sustainable urban development, with cooling energy costs rising 10-120% during hot months and productivity losses potentially reaching 2.6% of global GDP by 2100. While individual green and blue infrastructure interventions demonstrate

DOI: 10.4018/979-8-3373-7345-4.ch006

ight © 2026, IGI Global Scientific Publishing. Copying or distributing in print or electronic forms without written permission of IGI Global Scientific Publishing is prohibited. Use of this chapter to n generative artificial intelligence (AI) technologies is expressly prohibited. The publisher reserves all rights to license its use for generative AI training and machine learning model development.

cooling benefits, a comprehensive understanding of their synergistic integration remains limited. Critical knowledge gaps exist in optimal spatial configurations, implementation strategies, and governance frameworks for integrated Blue-Green systems, particularly regarding equity considerations and unintended consequences such as green gentrification.

INTRODUCTION

Contemporary urban environments face unprecedented challenges as rapid urbanisation accelerates the conversion of natural landscapes into impervious surfaces. Over half the world's population now resides in cities—a proportion projected to reach 68% by 2050 intensifying the urban heat island (UHI) effect, which can elevate urban temperatures by 5–12°C compared to surrounding rural areas (Yang et al., 2022; Jabbar et al., 2023). First identified in the 19th century, the UHI phenomenon has worsened as concrete, asphalt, and steel replace vegetation and water bodies, absorbing and re-radiating solar energy and fundamentally altering urban microclimates (Petrucci, 2023; Urban Heat Island, 2022).

The consequences of urban heating extend far beyond uncomfortable temperatures, imposing significant health, environmental, and economic burdens. Vulnerable populations—particularly the elderly, low-income communities, and racial minorities experience disproportionately high risks of heat stress, heat exhaustion, and heat-related mortality (Moss & Kar, 2020; Vujovic et al., 2021). Simultaneously, increased demand for air conditioning fuels a feedback loop of higher energy consumption and greenhouse gas emissions, with residential cooling costs rising by 10–120% during hot months in U.S. cities and productivity losses projected to cost up to 2.6% of global GDP by 2100 (Roxon et al., 2020; Zhao et al., 2021). These combined pressures underscore the need for holistic, sustainable interventions.

Renaturing cities integrating green infrastructure (vegetation-based systems) and blue infrastructure (water-based systems) offers a promising, multi-benefit approach to urban resilience. Green infrastructure elements such as parks, street trees, green roofs, and vertical gardens cool cities through shading, evapotranspiration, and enhanced air flow (Wu et al., 2024; Fernández & Navarro, 2024). Medium-sized green spaces strike an optimal balance between coverage and cooling magnitude, while blue infrastructure lakes, rivers, wetlands augments cooling via evaporation and high thermal mass, with larger and strategically placed water bodies achieving greater temperature reductions (Jandaghian & Colombo, 2024; Chen et al., 2024).

Critically, the synergistic integration of green and blue spaces can yield cooling benefits exceeding individual interventions. Studies document temperature reductions up to 5.2°C in areas combining urban forests and water features, with the spatial distribution, size, shape, and layout of these elements determining their efficacy across diurnal cycles (Xu et al., 2024; SHI et al., 2024; Ding et al., 2024). Despite this promise, knowledge gaps remain in identifying optimal spatial configurations, implementation strategies, and governance frameworks especially in rapidly urbanising regions with the highest stakes. Equitable access and potential unintended consequences, such as green gentrification, highlight the importance of integrating social equity into urban renaturing initiatives (Lewartowska et al., 2024; Haase, 2024). By synthesising advances in urban climatology, landscape ecology, and environmental planning, the study aims to provide evidence-based guidelines for optimising urban renaturing strategies. and livable cities.

METHODOLOGY

This study employed a comprehensive systematic literature review to examine integrated Blue-Green infrastructure for urban cooling. The research adopted a socio-ecological systems theoretical framework to analyse biophysical mechanisms, spatial optimisation principles, and socio-institutional dimensions of urban renaturing strategies.

A systematic search was conducted across multiple academic databases, including Web of Science, Scopus, PubMed, and Google Scholar from 2014-2024. Search terms included "urban heat island mitigation," "green infrastructure," "blue infrastructure," "urban cooling," and "renaturing cities" using Boolean operators and controlled vocabulary. Studies were selected based on predetermined inclusion criteria requiring empirical research with quantitative cooling data, clear methodologies, and relevance to urban thermal management across diverse climate zones and urban contexts. Data extraction employed standardised forms to capture study characteristics, cooling effectiveness measurements, spatial configurations, climate contexts, and implementation findings.

Figure 1. Integrated green-blue infrastructure

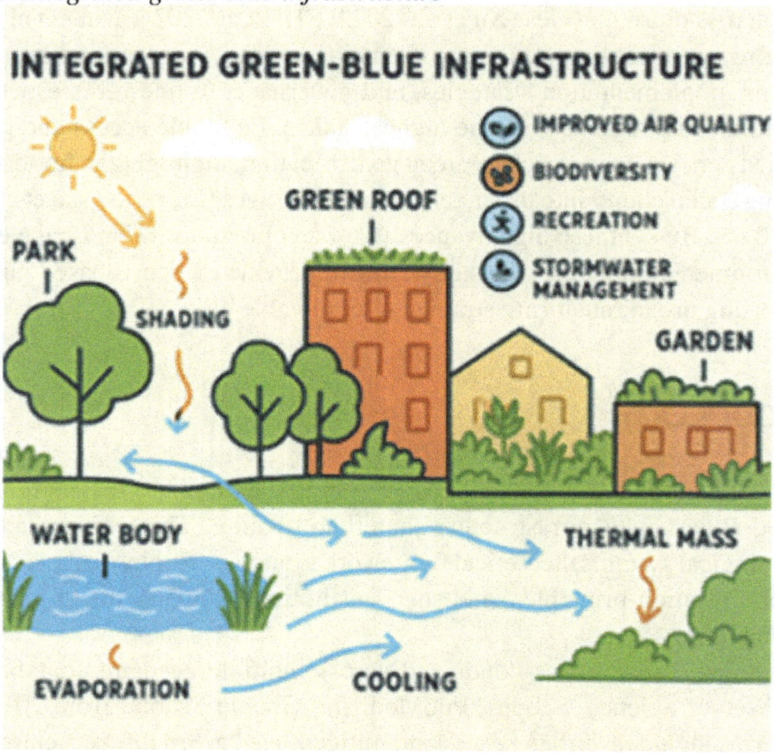

LITERATURE REVIEW

The concept of renaturing cities has evolved as a comprehensive response to the urban thermal challenges, representing a paradigm shift toward integrating natural elements into urban environments to enhance sustainability and livability. This approach emphasises the strategic deployment of green infrastructure (vegetation-based systems) and blue infrastructure (water-based systems) to restore ecological functions within urban contexts (Dall'O', 2020). Renaturing involves applying ecological principles to urban settings while recognising the complex interactions between natural and built environments, aiming to improve urban resilience to climate change impacts (Lehmann, 2019; Rumble et al., 2019). Blue Green infrastructures provide essential ecosystem services, including biodiversity enhancement, pollution reduction, stormwater management, and climate adaptation, promoting environmental justice and sustainable urban development (Shah et al.,

2024; Yang et al., 2024; Hassan et al., 2023). The breadth of studies reviewed, much of the existing literature adopts a descriptive approach, focusing on individual case studies or isolated infrastructure types. There is limited critical synthesis comparing findings across climatic, spatial, and socio-economic contexts. For instance, while some studies advocate large, centralized green spaces for maximum cooling, others highlight the efficiency of distributed small-scale interventions, reflecting differing theoretical perspectives from landscape ecology and urban morphology. Additionally, a dominant emphasis on biophysical mechanisms often overlooks governance, socio-institutional, and equity dimensions. This indicates a clear need for integrative frameworks that reconcile these competing perspectives and support context-sensitive urban planning strategies.

MECHANISMS AND DRIVERS OF URBAN HEAT ISLAND FORMATION

Physical and Socio-economic Drivers

UHI formation results from a complex interplay of physical and socioeconomic factors that fundamentally altering urban energy balance systems. The primary physical driver involves the expansion of impervious surfaces such as roads, buildings, and paved areas, which absorb and retain solar radiation more effectively than natural surfaces (Guo et al., 2022; Vujovic et al., 2021). Studies conducted across Chinese cities demonstrate that urbanised area size significantly contributes to UHI intensity, with larger metropolitan areas experiencing more pronounced thermal effects due to greater impervious surface coverage.

Socioeconomic and demographic factors emerge as equally significant contributors to UHI formation. Research in Yangzhou, China, reveals positive correlations between gross domestic product (GDP), population density, and air temperature increases, suggesting that economic activity and urban density amplify thermal effects (Wu et al., 2024). Similarly, demographic factors, including nightlight intensity and building characteristics, influence UHI patterns, with certain roof types and construction materials exacerbating heat retention properties (Sidiqui et al., 2022).

This vegetation deficit limits natural cooling processes and contributes to environmental justice concerns, as vulnerable populations frequently reside in areas with reduced access to cooling resources and higher heat exposure (Pontius & McIntosh, 2024).

Temporal and Spatial Variations

UHI effects exhibit significant temporal and spatial heterogeneity, influencing their impact on urban populations and ecosystems. Research in cities like Wuhan, China, demonstrates that the cooling intensity and distance of blue-green spaces vary substantially between day and night, highlighting the importance of considering diurnal variations in urban thermal planning (Ding et al., 2024). These temporal variations reflect the differential heat storage and release patterns of various urban materials and surfaces throughout the diurnal cycle. Studies across multiple Chinese cities reveal that UHI effects intensify particularly in rapidly urbanising regions, including the Middle East, North Africa, and parts of China, where the pace of urban development often outstrips environmental planning considerations (Hadjinicolaou et al., 2023; Huang et al., 2024).

ENVIRONMENTAL, HEALTH, AND ECONOMIC IMPACTS OF URBAN HEAT ISLANDS

Public Health Consequences

UHI impose substantial public health burdens, particularly affecting vulnerable populations including the elderly, individuals with pre-existing health conditions, and socioeconomically disadvantaged communities. Heat-related illnesses manifest through various pathways, including heat stress, heat exhaustion, and heat stroke, contributing to elevated mortality rates during extreme heat events (Elmarakby & Elkadi, 2024; Zhang, 2023). The physiological impacts extend beyond direct thermal stress to include mental health consequences, with studies documenting increased levels of fatigue, irritation, and even suicidal tendencies during prolonged heat exposure (Zhang, 2023). Elevated urban temperatures increase ground-level ozone concentrations and other pollutant levels, creating compounding health risks that disproportionately affect low-income communities lacking access to cooling resources (Pontius & McIntosh, 2024; Cichowicz & Bochenek, 2024).

Economic Costs and Productivity Impacts

The economic implications of UHI extend across multiple sectors, with quantifiable impacts on energy consumption, labour productivity, and healthcare expenses. Energy consumption impacts are particularly pronounced, with UHI effects contributing to 10-120% increases in cooling energy consumption during hot months in U.S. urban areas, resulting in substantial residential energy cost increases (Roxon et al., 2020;

Kar & Meade, 2022). These energy demands generate corresponding increases in carbon dioxide emissions, with the U.S. residential sector alone contributing over 550 million tons of CO2 emissions annually due to cooling requirements.

Labour productivity losses represent another significant economic impact, with global projections indicating potential losses ranging from 0.31% to 2.6% of global GDP by 2100, depending on climate scenarios and adaptation measures (Zhao et al., 2021). European cities face particularly variable productivity impacts, with potential losses in urban economies ranging from 0.4% to 9.5% of Gross Value Added without effective adaptation strategies (Costa et al., 2016). European cities experience estimated economic impacts of €192 to €314 per adult urban inhabitant per year due to UHI-induced heat-related mortality, highlighting the substantial healthcare burden imposed by urban thermal conditions (Huang et al., 2023).

Social Equity and Environmental Justice

UHIs create and exacerbate environmental justice concerns through their disproportionate impacts on vulnerable populations. Low-income and minority communities, including Latino and Black populations, often concentrate in areas with high UHI intensity while lacking access to air conditioning and other cooling resources (Lanza et al., 2023; Kim & Kim, 2023). These communities face additional challenges adapting to heat stress due to financial constraints limiting their ability to purchase cooling systems to more thermally comfortable areas (Ferguson & Anna, 2024).

Elderly populations and health-compromised individuals represent particularly vulnerable groups due to their physiological susceptibility and often limited mobility (Marginean et al., 2024; Chen et al., 2023). International disparities further compound these vulnerabilities, with elderly populations in middle-income countries facing disproportionately higher risks compared to their counterparts in high-income nations. Global South urban populations experience more severe heat risks due to higher exposure to extreme heat events and less resilient infrastructure, despite contributing less to global greenhouse gas emissions (Zhang et al., 2023).

GREEN INFRASTRUCTURE SOLUTIONS AND COOLING MECHANISMS

Types and Characteristics of Urban Green Spaces

Urban Green Infrastructure (GI) encompasses diverse typologies that provide varying cooling benefits through biophysical mechanisms. Parks emerge as highly effective cooling elements due to their large size and dense vegetation coverage, enhancing shading and evapotranspiration processes. Research demonstrates that large tree-dominated parks can reduce land surface temperatures by up to 4.5°C on average, with cooling effects extending up to 90 meters from park boundaries (Gallay et al., 2023). The cooling effectiveness of parks correlates strongly with their size and vegetation structure, with larger parks containing higher leaf area indices providing more substantial cooling benefits (Jia & Wang, 2023).

Street trees represent among the most effective urban green spaces (UGS) for pedestrian-level cooling, providing direct shade and enhanced evapotranspiration that significantly reduces local air temperatures (Kim et al., 2023; Jia & Wang, 2023). The cooling effectiveness of street trees depends on their density and species characteristics, with trees possessing higher specific leaf areas providing greater cooling benefits. In high-density urban contexts, increased tree coverage can lower diurnal temperatures by 1°C during peak temperature periods (Ouyang et al., 2019).

Biophysical Cooling Mechanisms

Urban vegetation provides cooling through several interrelated biophysical mechanisms that work synergistically to mitigate UHI effects. Evapotranspiration represents the primary cooling mechanism, involving water transfer from land to atmosphere through soil evaporation and plant transpiration processes that consume heat energy and cool surrounding air (Yu et al., 2024; Zepp et al., 2023). Research in Shanghai demonstrates that enhanced evapotranspiration during nighttime periods significantly improves cooling capacity, while soil moisture content variations can influence cooling effectiveness by up to 2.6 K (Zepp et al., 2023).

Shading mechanisms operate through tree canopy coverage that reduces solar radiation reaching ground surfaces, lowering surface temperatures. This shading effect proves particularly pronounced during midday periods when solar radiation intensity peaks (Yin et al., 2024; Gao et al., 2024). The cooling effectiveness of shading varies with vegetation density and type, with grasslands containing over 50% coverage providing cooling effects similar to forest systems (Yu et al., 2024). Strategic placement of green spaces, including parks and green roofs, can enhance air circulation and reduce urban heat retention (Perini & Pérez, 2021). Local climatic conditions like sea breezes can complement vegetation cooling effects by enhancing transpiration-based cooling processes (Gao et al., 2024).

Factors Influencing Green Infrastructure Cooling Potential

Climate conditions significantly influence GI cooling effectiveness, with arid climates achieving the highest air temperature reductions, followed by temperate, tropical, and continental climate zones (Azmeer et al., 2024). Background climate factors, including mean annual temperature, substantially influence UGScooling effects, while plant characteristics such as specific leaf area and vegetation indices critically determine cooling performance (Kim et al., 2023).

Spatial distribution and coverage patterns affect green space cooling potential, with more equitable distribution across urban areas enhancing overall cooling impacts. Research in Guangzhou demonstrates that strategic spatial distribution can enhance cooling effects beyond simple coverage increases (Xu et al., 2024). Urban morphology factors, including canopy cover and volume, influence cooling effects of urban parks, with cooling amplitude responses to increased local temperatures following hump-shaped curves that indicate threshold effects beyond which cooling benefits may diminish (Jang et al., 2024) (Zhou et al., 2024). These cooling mechanisms—including evapotranspiration, shading, and thermal mass effects—operate in an interconnected manner, with their combined influence determining the overall cooling performance of urban environments. Their effectiveness varies depending on spatial configuration, climatic conditions, and infrastructure integration.

BLUE INFRASTRUCTURE STRATEGIES AND WATER BODY COOLING POTENTIAL

Types and Cooling Effects of Urban Water Bodies

Urban water bodies contribute significantly to UHI mitigation through their distinct thermal properties and cooling mechanisms. Lakes and large water bodies provide substantial cooling effects due to their extensive surface areas, facilitating continuous evaporation and heat absorption. Research identifies optimal acreage thresholds for maximising cooling benefits, with water bodies greater than 2500 km^2 in regions such as Poyang Lake, China, demonstrating significant temperature reduction capabilities in their vicinity while maintaining broader cooling reach compared to smaller water features (Deng et al., 2023; Jandaghian & Colombo, 2024).

Rivers and streams enhance cooling through dynamic water flow patterns and interactions with surrounding landscapes, with cooling effects amplified when integrated with riparian green spaces that promote evapotranspiration and shading (Ma et al., 2024; Deng et al., 2023). The linear configuration of river systems enables cooling effects to extend along their corridors, creating valuable thermal relief networks within urban environments (Cao et al., 2022).

Water Body Cooling Mechanisms

Urban water bodies cool surrounding environments through multiple interconnected mechanisms that leverage water's unique thermal properties. Evaporation represents the primary cooling mechanism, whereby water bodies absorb heat from surrounding environments through evaporation processes while their large heat capacity enables heat storage during daytime periods and gradual release during nighttime hours, moderating temperature fluctuations (Qiu & He, 2024; Hu et al., 2023).

The Water Cold Island (WCI) effect describes the cooling influence exerted by water bodies, which can achieve significant temperature reductions as documented in cities like Kunming and Changsha, where water bodies demonstrate cooling intensities up to 8.01°C (Ma et al., 2024; Tan et al., 2024). This effect operates through the thermal mass properties of water that buffer temperature fluctuations and provide consistent cooling benefits throughout diurnal cycles (Ma et al., 2024; Tan et al., 2024).

Factors Influencing Water Body Cooling Efficiency

Water body size and shape significantly influence cooling effectiveness, with larger and well-configured water bodies providing more extensive cooling due to increased surface areas that facilitate greater evaporation rates (Jandaghian & Colombo, 2024; Hu et al., 2023). Simple water body shapes tend to produce more efficient cooling effects through uniform cooling distribution across surrounding areas, while complex shapes may create higher cooling gradients with more rapid cooling diminishment over distance (Xie, 2023; Du & Zhou, 2022).

Surrounding land use patterns substantially affect water body cooling efficiency, with areas containing lower percentages of impervious surfaces and more discrete green land distributions enhancing cooling effects (Du & Zhou, 2022; Tan et al., 2024; Xie, 2023). Fine-textured vegetation and reduced impervious surface coverage create favourable water body cooling optimisation conditions.

Climatic conditions influence water body cooling efficiency across different climate zones, with humid and hot areas demonstrating higher cooling efficiency than arid regions (Hu et al., 2023) (Cai et al., 2022).

SYNERGIES OF INTEGRATED BLUE-GREEN INFRASTRUCTURE

Complementary Cooling Effects

Integrating BGI elements produces synergistic cooling effects that exceed the benefits achievable through isolated interventions. Research in Harbin, China, demonstrates that combined water bodies and vegetation reduce temperatures more effectively than either element alone, with peak cooling effects reaching 6.37°C at distances of 6 meters from water features (Liu et al., 2024). This synergistic relationship reflects the complementary cooling mechanisms operating through vegetation evapotranspiration and water body evaporation processes.

Spatial configuration plays a crucial role in optimising these complementary effects. Studies in Kunshan City, China, reveal that the shape and distribution of green and blue spaces significantly influence their cooling effectiveness, with specific configurations enhancing overall cooling potential beyond additive effects (Wang, 2023).

Figure 2. Complementary cooling effects illustrates the peak cooling effects achieved by different renaturing interventions: parks, street trees, green roofs, water bodies, and integrated blue-green systems

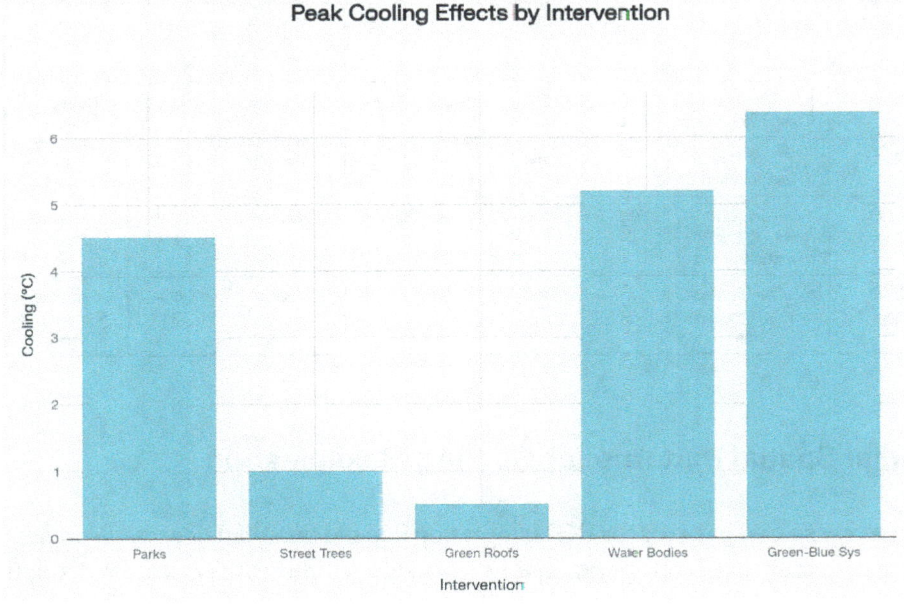

Background meteorological conditions influence the stability and magnitude of combined cooling effects, with blue-green spaces demonstrating stronger and more stable cooling under varying conditions than isolated green spaces (Geng et al., 2024). **Table 1** synthesises the primary cooling mechanisms, peak temperature reductions, effective cooling ranges, and climate-sensitivity rankings for key Blue-Green infrastructure typologies, illustrating how integrated systems achieve up to 6.37 °C of cooling across 600 m ranges.

Table 1. Synthesis of blue-green infrastructure cooling mechanisms and effectiveness

Infrastructure Type	Primary Cooling Mechanism	Peak Cooling Effect (°C)	Effective Cooling Range (m)	Implementation Cost	Maintenance Requirements	Optimal Scale (m)	Climate Effectiveness Ranking
Parks (Large, >2 ha)	Evapotranspiration + Shading	4.5	90	High	Medium	500-1600	Arid > Temperate > Tropical
Street Trees	Evapotranspiration + Shading	1.0	30	Low	Medium-High	50-200	Arid > Temperate > Tropical
Green Roofs	Surface cooling + Insulation	17.0 (surface)	5	Very High	High	Building level	Universal
Vertical Gardens	Evapotranspiration + Shading	0.8	15	Medium	High	50-100	Temperate > Arid > Tropical
Lakes (>2500 km^2)	Evaporation + Thermal mass	8.01	800	Very High	Low	1600+	Humid > Arid
Rivers/Streams	Evaporation + Air flow	5.0	500	High	Medium	500-1600	Humid > Temperate > Arid
Constructed Wetlands	Evaporation + Evapotranspiration	3.5	200	Medium-High	Medium	200-800	Humid > Temperate
Small Water Bodies	Localized evaporation	2.5	100	Medium	Medium	100-500	Variable
Integrated Blue-Green	Synergistic evaporation + evapotranspiration	6.37	600	Very High	Medium-High	1200-1600	Stable across climates
Blue-Green Corridors	Combined mechanisms + Air flow	5.2	400	High	Medium	800-1200	Temperate > Humid > Arid

Optimal Spatial Patterns for Cooling Maximisation

Effective spatial patterns for BGI integration depend on urban density characteristics and spatial scale considerations. In low-density urban areas, scattered small patches of Blue-Green spaces enhance cooling by increasing beneficial coverage areas. In contrast, high-density areas require more regular and large-scale patches to achieve optimal cooling effects (Sheng & Wang, 2024). This density-dependent optimisation reflects the varying thermal challenges and spatial constraints operating across urban contexts.

Scale sensitivity analysis reveals that cooling effects operate most effectively at specific spatial scales, with land surface temperature showing high sensitivity to blue infrastructure at 1600-meter scales. In contrast, GI demonstrates significant sensitivity at both 50-meter and 1600-meter scales (He et al., 2023). District-scale planning (1200-1600 meters) emerges as optimal for maximising cooling efficiency through combined Blue-Green-Grey infrastructure characteristics. Morphological considerations, including green corridor orientation and connectivity, significantly affect cooling performance, with corridors aligned with prevailing wind directions and specific width parameters enhancing cooling effects (Jiang et al., 2021).

Implementation Challenges and Solutions

Data limitations present significant challenges for BGI integration, particularly in developing regions where reliable information for modelling and implementation remains scarce. Innovative approaches, including social media and local news data proxies, have been suggested to overcome these limitations and validate flood management models (Rivera & Faggian, 2024). The absence of comprehensive data on combined BGI element effectiveness further complicates planning and decision-making processes (Cavadini et al., 2024).

Economic challenges include high initial implementation costs for certain BGI elements, such as green roofs, although these provide significant ecosystem benefits over time. Cost-effectiveness varies with implementation scale, as larger BGI areas do not always yield proportionate benefits, necessitating careful economic evaluation (Shah et al., 2024; Cavadini et al., 2024). Street trees offer lower initial costs while providing substantial benefits, suggesting the importance of diversified BGI portfolios.

Governance and policy challenges include the lack of holistic methodologies for integrating BGI with existing infrastructure systems, particularly in South Asian contexts where management and governance frameworks require strengthening (Gupta & De, 2024). Strategic water management integration becomes essential for maintaining the cooling effects of green infrastructure in dense urban areas (Back & Kleidorfer, 2023).

SUSTAINABILITY DIMENSIONS AND IMPLEMENTATION STRATEGIES

Broader Sustainability Impacts

Urban renaturing through BGI implementation provides sustainability benefits beyond temperature regulation to encompass biodiversity conservation, carbon emission mitigation, water management, and air quality improvement.

Carbon emission mitigation occurs through BGI absorption of carbon dioxide and reduced reliance on energy-intensive cooling technologies (Grabowski et al., 2023; Fang et al., 2023). Climate resilience enhancement results from natural element integration that helps cities adapt to climate change impacts, including extreme weather events and urban heat islands (Yang et al., 2024; Meemrot et al., 2024).

Water management improvements result from BGI incorporating water bodies and permeable surfaces that enhance infiltration and reduce surface runoff (Meemrot et al., 2024). Constructed wetlands and rain gardens increase flood resilience by managing stormwater and reducing urban flooding risks while contributing to natural hydrological cycle restoration (Kiani, 2022; Fang et al., 2023).

Policy and Governance Frameworks

Effective policy and governance structures require diverse policy instrument deployment, including regulatory measures, economic incentives, and collaborative initiatives. Regulatory instruments prove particularly effective for integrating nature-based solutions into new development areas, while incentive-based instruments suit existing structure retrofitting (Longato et al., 2024). Green governance frameworks incorporate environmental sustainability into policy-making through collaborative approaches involving governments, NGOs, and corporate sectors (Pasupuleti, 2024).

Multi-level governance and cross-sectoral collaboration emerge as essential elements for successful urban renaturing implementation. Transdisciplinary co-design approaches prioritise collaboration across spatial scales from global agreements to local initiatives (Simon et al., 2021). Innovative governance models addressing institutional fragmentation can facilitate cross-sectoral collaboration between urban planning, environmental management, and community engagement sectors (Egusquiza et al., 2019).

Institutional and regulatory barriers, including rigid governance systems and funding constraints, require innovative financing mechanisms, policy innovations, and community involvement for successful navigation (He et al., 2024; Simon et al., 2021).

Table 2. Policy instrument assessment

Policy Instrument	Effectiveness	Implementation Ease	Cost to Government	Applicability	Timeline	Equity Impact	Success Examples
Regulatory Mandates	Very High	Medium	Low	New Development	Long-term	Mixed	Singapore, Germany
Zoning Requirements	High	High	Low	All Development	Medium-term	Mixed	Portland, Seattle
Building Codes	High	High	Low	All Buildings	Long-term	Positive	Toronto, San Francisco
Tax Incentives	Medium-High	High	High	Existing Structures	Short-term	Mixed	Multiple US Cities
Subsidies/Grants	High	Medium	Very High	All Projects	Short-term	Positive	EU Cities, Copenhagen
Green Bonds	Medium	Low	Medium	Large Projects	Medium-term	Negative	London, New York
Public-Private Partnerships	High	Low	Medium	Large Projects	Long-term	Mixed	Netherlands, Australia
Community Engagement Programs	Medium	Medium	Medium	All Projects	Medium-term	Very Positive	Medellín, Barcelona

Social and Cultural Dimensions

Social and cultural factors significantly influence urban renaturing project success through their effects on community engagement, acceptance, and equitable access to green spaces. Cultural identity and acceptance emerge as critical factors, with projects aligning with local cultural identities and traditions more likely to gain community embrace (Iwuanyanwu et al., 2024). Social cohesion and inclusivity can be enhanced through green space designs that foster social interactions, with participatory design processes building social capital and ensuring fulfilment of local needs.

Historical and racial injustices influence community perceptions and interactions with urban nature, making injustice crucial for preventing green gentrification and ensuring equitable access (Lewartowska et al., 2024; Ribeiro, 2023). Green regeneration can lead to gentrification and marginalised community displacement, requiring affordable housing, anti-displacement strategies, and greening efforts (Haase, 2024; Oscilowicz et al., 2022).

Environmental justice frameworks focusing on equity can help identify and mitigate potential inequities in urban greening projects, considering factors including housing opportunity, transportation, and economic development (Noll & Bhar, 2023). Educational strategies involving community members in urban renaturing project learning and planning can enhance engagement and ownership while addressing cultural and social barriers (Iwuanyanwu et al., 2024).

TECHNOLOGICAL INNOVATIONS AND FUTURE DIRECTIONS

Emerging Technologies for Smart Renaturing

Remote sensing and GIS technologies enable spatial data collection and integration, which are crucial for understanding urban environments and planning BGI deployment. These tools facilitate land use, climate condition, and human activity analysis, providing insights for optimal BGI placement (Mutani et al., 2024; Bacchin et al., 2014). UHI mitigation applications use GIS and remote sensing to model UHI dynamics and identify areas that benefit most from BGI interventions, as Turin studies showed UHI intensity reduction through strategic BGI placement (Mutani et al., 2024).

Artificial intelligence and machine learning applications in urban climate modelling enable predictive analytics for urban growth pattern prediction and environmental impact assessment, aiding strategic BGI planning (Anwar & Sakti, 2024). AI-driven approaches optimise BGI by focusing on objectives including air quality improvement, biodiversity enhancement, and UHI mitigation, helping identify the most effective BGI configurations for climate change response (Shaamala et al., 2024). Digital twin technology creates virtual urban environment models that enable BGI system simulation and optimisation, predict performance under various scenarios, and facilitate proactive management and maintenance (Yang et al., 2024).

Implementation Best Practices

Effective BGI implementation strategies require technology integration, including IoT and AI, to optimise management and operational efficiency while ensuring resilience and sustainability through real-time monitoring and adaptive management (Yang et al., 2024; Hassan et al., 2023). Community engagement and policy reforms supporting equitable BGI benefit distribution ensure socio-economic and health benefits accessibility for all urban residents.

Site selection and prioritisation criteria should emphasise environmental impact potential, including biodiversity enhancement, pollution reduction, and climate resilience improvement through ecological value and existing environmental condition assessment (Yang et al., 2024; Perrelet et al., 2024). Social and economic benefit prioritisation, including increased property values and improved community well-being, can enhance overall BGI project impacts (Ashinze et al., 2024). Long-term maintenance and management best practices include adaptive management implementation responding to changing environmental conditions and urban dynamics (Hassan et al., 2023) (Yang et al., 2024).

GOVERNANCE, FINANCING, AND SMART CITY INTEGRATION

Governance Models

Effective implementation of blue-green infrastructure requires robust governance frameworks that integrate multi-level planning and cross-sectoral coordination. Centralized governance models often provide regulatory clarity and large-scale implementation capacity, whereas decentralized and participatory models enable context-specific and community-driven solutions. Integrating local stakeholders in planning processes enhances legitimacy and long-term sustainability.

Financing Strategies

Financing large-scale BGI implementation remains a critical challenge. Innovative financial mechanisms such as green bonds, climate funds, public-private partnerships (PPPs), and carbon financing instruments have emerged as viable solutions. Lifecycle cost-benefit analysis further supports investment decisions by capturing long-term environmental and economic returns.

Smart City Integration

The integration of BGI within smart city frameworks enhances adaptive urban management through real-time monitoring and data-driven decision-making. Technologies such as IoT sensors, digital twins, and AI-driven analytics can optimize infrastructure performance, predict thermal patterns, and support dynamic planning. Aligning BGI strategies with smart city initiatives enables scalable and resilient urban solutions.

DISCUSSION

Recent years have witnessed several emerging trends in urban heat mitigation and climate-adaptive planning. There is a clear transition from conventional grey infrastructure toward nature-based and blue-green solutions. Additionally, the integration of artificial intelligence, geospatial analytics, and digital twin technologies is transforming urban planning into a more predictive and adaptive discipline. Another notable trend is the increasing emphasis on environmental justice and equitable access to cooling infrastructure, reflecting a broader shift toward inclusive and sustainable urban development paradigms. This discussion synthesises key findings from the literature to provide critical insights into the mechanisms, spatial optimisation, implementation challenges, and broader implications of urban cooling strategies through integrated BGI, while identifying pathways toward more resilient and equitable urban environments.

Paradigm Shift: From Isolated Interventions to Synergistic Systems

The literature overwhelmingly demonstrates that traditional approaches implementing isolated green infrastructure elements represent a fundamental underutilization of urban cooling potential. While individual interventions such as parks can achieve substantial cooling effects reducing land surface temperatures by up to 4.5°C and extending cooling benefits up to 90 meters from their boundaries the evidence consistently shows that integrated Blue-Green systems produce synergistic effects that exceed the sum of their individual contributions (Gallay et al., 2023). The peak cooling effect of 6.37°C achieved through combined water bodies and vegetation in Harbin exemplifies this synergistic potential, representing a paradigmatic shift toward systems thinking in urban thermal management (Liu et al., 2024).

This synergistic relationship operates through complementary biophysical mechanisms that create mutually reinforcing cooling processes. While evapotranspiration from vegetation provides continuous latent heat absorption through water transfer from land to atmosphere, consuming heat energy and cooling surrounding air (Yu et al., 2024), water bodies contribute through evaporative cooling and thermal mass buffering, creating stable microclimatic conditions that persist across diurnal cycles (Qiu & He, 2024). The enhanced stability of combined systems under varying meteorological conditions, as demonstrated in Shanghai, suggests that integrated approaches provide greater reliability as climate adaptation strategies compared to isolated interventions (Geng et al., 2024).

Spatial Optimisation: Scale-Sensitive Urban Thermal Management

The literature reveals sophisticated spatial relationships that challenge conventional urban BGI deployment approaches. The identification of scale-sensitive thresholds for infrastructure effectiveness with blue infrastructure demonstrating high sensitivity at 1600-meter scales while green infrastructure shows significant sensitivity at both

50-meter and 1600-meter scales suggests that effective thermal management requires multi-scale planning approaches that optimise interventions at appropriate spatial scales (He et al., 2023). The emergence of district-scale planning (1200-1600 meters) as optimal for maximising cooling efficiency through combined Blue-Green-Grey infrastructure characteristics provides specific guidance for planning frameworks while highlighting the importance of meso-scale interventions.

Urban density emerges as a critical determinant of optimal spatial patterns, with fundamentally different strategies required for low-density versus high-density urban contexts. In low-density areas, scattered small patches of Blue-Green spaces enhance cooling by increasing beneficial coverage areas, leveraging the distributed cooling capacity across the urban fabric (Sheng & Wang, 2024). **Table 3** outlines optimal spatial scales from site (<50 m) to regional (>1600 m) and recommends intervention types and planning frameworks tailored to different urban densities and infrastructure sensitivities.

Table 3. Spatial scale optimization for blue-green infrastructure

Spatial Scale	Green Infrastructure Sensitivity	Blue Infrastructure Sensitivity	Optimal Intervention Types	Urban Density Considerations	Planning Framework
Site Scale (<50m)	High (Building-level)	Low	Green roofs, Vertical gardens	High density: Regular implementation	Building codes, Green building standards
Local Scale (50-200m)	Very High (Street trees)	Low	Street trees, Pocket parks	High density: Maximum coverage	Street design guidelines
Neighborhood Scale (500-800m)	Moderate (Small parks)	Moderate (Ponds)	Neighborhood parks, Small water features	Variable by context	Neighborhood master planning
District Scale (1200-1600m)	High (Large parks)	Moderate (Wetlands)	Integrated Blue-Green systems	Low density: Scattered patches; High density: Large-scale patches	District-scale integrated planning
Regional Scale (>1600m)	Low	Very High (Large water bodies)	Large lakes, River networks	Regional coordination required	Regional environmental planning

Figure 3. Spatial optimisation: Scale-sensitive urban thermal management. This line chart compares the cooling sensitivity of green and blue infrastructure across spatial scales of 50, 200, 800, and 1600 meters

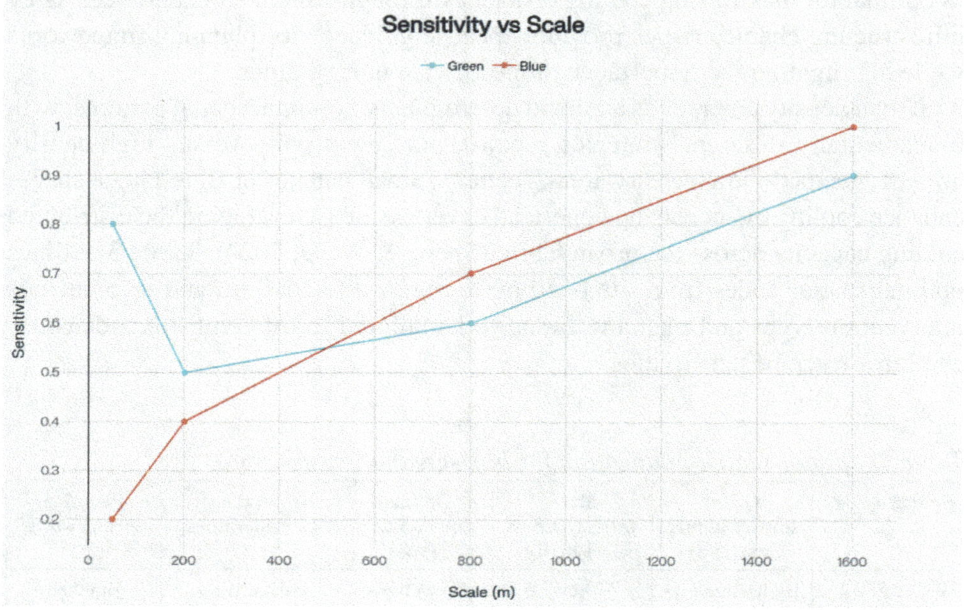

The spatial configuration considerations extend beyond simple area calculations to encompass morphological factors, including corridor orientation, connectivity patterns, and relationships to prevailing wind directions. The enhanced cooling effects achieved by corridors aligned with prevailing winds and specific width parameters demonstrate the importance of climatically-informed design that integrates meteorological understanding with spatial planning (Jiang et al., 2021).

Research demonstrates that the cooling impact distance of GI follows power curve relationships, with parks demonstrating longer cooling impact distances than other green infrastructure types (Jang et al., 2024). The cooling amplitude responses to increased local temperatures follow hump-shaped curves that indicate threshold effects beyond which cooling benefits may diminish (Zhou et al., 2024), emphasising the importance of optimised design and strategic placement in maximising green infrastructure cooling benefits.

Implementation Challenges: Systemic Barriers to Transformation

Despite the benefits of integrated BGI, the literature reveals substantial implementation challenges that operate across multiple domains and scales, requiring comprehensive solutions that simultaneously address technical, economic, governance, and social dimensions.

Data limitations present significant challenges for BGI integration, particularly in rapidly urbanising regions of the Global South where reliable information for evidence-based planning remains scarce. The innovative suggestion to use social media and local news as data proxies represents a creative response to these limitations but highlights the fundamental challenge of implementing sophisticated urban thermal management strategies in contexts lacking robust data infrastructure (Rivera & Faggian, 2024). The absence of comprehensive data on combined BGI element effectiveness further complicates planning and decision-making processes (Cavadini et al., 2024).

Economic challenges present complex trade-offs that extend beyond simple cost calculations to encompass temporal considerations and scale effects. While certain Blue-Green infrastructure elements, such as green roofs, require high initial investments, they provide significant long-term ecosystem benefits that justify their implementation from a life-cycle perspective. The finding that cost-effectiveness varies with implementation scale with larger areas not always yielding proportionate benefits necessitates careful economic evaluation considering threshold effects and diminishing returns (Shah et al., 2024; Cavadini et al., 2024). This economic complexity requires innovative financing mechanisms that capture the diverse benefit streams generated by BGI while addressing the temporal mismatch between upfront costs and long-term benefits.

Governance and policy challenges reflect deeper systemic urban planning and management issues that extend beyond technical considerations to encompass institutional capacity and coordination mechanisms. The lack of holistic methodologies for integrating Blue-Green infrastructure with existing urban systems, particularly evident in South Asian contexts, highlights the need for strengthened management frameworks to address institutional fragmentation while maintaining implementation efficiency (Gupta & De, 2024). **Table 4** organises the systemic barriers—data limitations, economic constraints, governance fragmentation, and more—alongside innovative solutions and measurable success indicators for each challenge category.

Table 4. Implementation challenges and solutions framework

Challenge Category	Specific Challenges	Impact Severity	Innovative Solutions	Success Indicators
Data Limitations	Lack of comprehensive BGI effectiveness data; Limited monitoring systems	High in Global South	Social media data proxies; IoT sensor networks; Satellite monitoring	Standardized monitoring; Evidence-based decisions; Data accessibility
Economic Constraints	High upfront costs; Unclear benefit quantification; Scale diseconomies	Very High	Innovative financing; Green bonds; Ecosystem service valuation	Cost-effective implementation; Multiple funding sources; Clear ROI
Governance Fragmentation	Multi-agency coordination; Institutional silos; Policy inconsistencies	High	Integrated governance frameworks; Cross-sectoral collaboration protocols	Coordinated policies; Streamlined approvals; Shared responsibilities
Technical Knowledge Gaps	Limited integrated design expertise; Insufficient climate-specific guidelines	Medium-High	AI-driven optimization; Digital twin modeling; Knowledge platforms	Technical capacity building; Best practice dissemination; Quality standards
Community Engagement	Consultation vs. participation; Cultural barriers; Language accessibility	High	Participatory design processes; Cultural liaisons; Multi-language resources	Community ownership; Inclusive participation; Cultural integration
Maintenance Requirements	Long-term resource allocation; Technical expertise requirements	Medium	Community involvement programs; Adaptive management systems	Sustainable operations; Performance maintenance; Community stewardship
Climate Adaptation	Performance under future climate scenarios; Extreme weather resilience	High	Climate-responsive design; Resilient species selection; Flexible infrastructure	Climate resilience; Adaptive capacity; Performance stability
Regulatory Barriers	Outdated zoning codes; Building regulations; Environmental standards	Medium-High	Regulatory framework updates; Performance-based standards; Pilot programs	Enabling policy environment; Streamlined processes; Innovation support

Social Equity and Environmental Justice: The Distribution of Urban Thermal Benefits

The concentration of vulnerable populations including low-income communities, racial minorities, and elderly residents in areas with high urban heat island intensity creates environmental justice imperatives that must be addressed through strategic Blue-Green infrastructure deployment (Lanza et al., 2023; Kim & Kim, 2023). These populations face compounding vulnerabilities through limited access to cooling resources, inadequate housing conditions, and reduced adaptive capacity that amplify their exposure to thermal stress. Research demonstrates that Latino and

Black populations often concentrate in areas with high UHI intensity while lacking access to air conditioning and other cooling resources, facing additional challenges in adapting to heat stress due to financial constraints (Lanza et al., 2023).

The risk of green gentrification presents a significant concern that complicates the straightforward deployment of BGI as an equity strategy. The potential for green regeneration to increase property values and displace marginalised communities reveals the need for carefully designed implementation strategies, including affordable housing preservation, anti-displacement measures, and environmental improvements (Haase, 2024; Oscilowicz et al., 2022).

Figure 4. Social equity and environmental justice. This clustered column chart compares heat vulnerability scores and gentrification risk levels for low-income, minority, and elderly populations

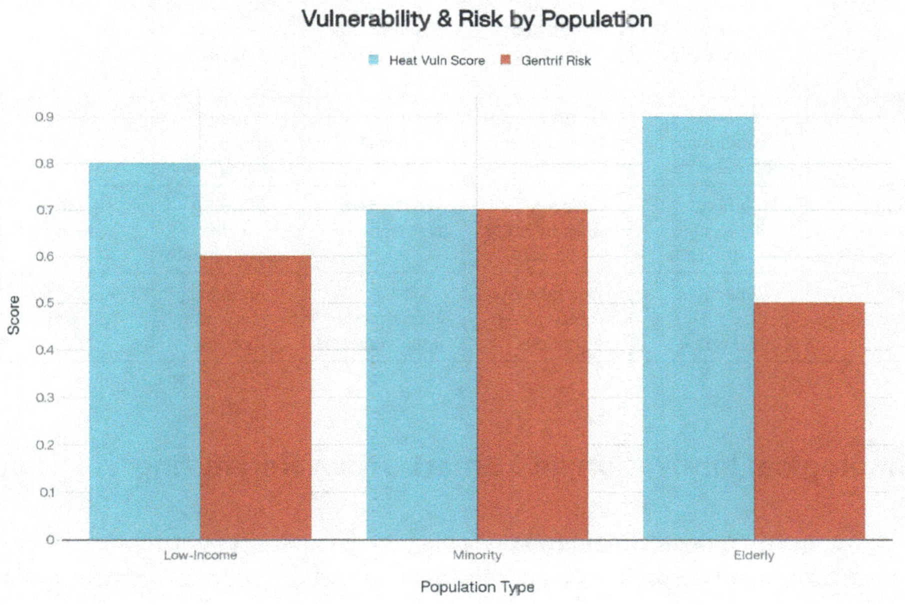

Environmental justice frameworks focusing on equity can help identify and mitigate potential inequities in urban greening projects, considering factors including housing opportunity, transportation, and economic development (Noll & Bhar, 2023). Educational strategies involving community members in urban renaturing project learning and planning can enhance engagement and ownership while addressing cultural and social barriers (Iwuanyanwu et al., 2024). **Table 5** details vulnerability factors (income, race/ethnicity, age, etc.), associated heat-exposure impacts, access barriers, gentrification risks, and equity-centred solutions necessary to ensure socially just distribution of cooling benefits.

Table 5. Environmental justice and equity considerations in urban renaturing

Vulnerability Factor	Heat Vulnerability Impact	Green Infrastructure Access Barriers	Gentrification Risk	Equity-Centered Solutions	Vulnerability Factor
Income Level	Low-income: Higher UHI exposure	Limited financial resources for private solutions	High - Property value increases	Affordable housing preservation	Income Level
Race/Ethnicity	Latino/Black: Disproportionate heat burden	Historical redlining effects on green space distribution	High - Displacement pressure	Anti-displacement policies	Race/Ethnicity
Age (Elderly)	Elderly: Physiological susceptibility	Mobility limitations affect park access	Moderate - Fixed income vulnerability	Age-friendly design standards	Age (Elderly)
Health Status	Pre-existing conditions: Increased risk	Medical needs limit outdoor activities	Low - Health needs prioritized	Health-prioritized interventions	Health Status
Housing Quality	Poor housing: Inadequate cooling	Cannot afford green building upgrades	High - Rent increases likely	Housing quality improvements	Housing Quality
Access to Cooling	Limited AC access: Heat stress risk	Energy costs limit cooling options	Moderate - Infrastructure improvements	Community cooling centres	Access to Cooling

Technological Innovation and Smart Urban Renaturing

Integrating emerging technologies presents transformative opportunities for enhancing urban renaturing projects' planning, implementation, and management while addressing some of the data and coordination challenges identified in the literature.

Remote sensing and GIS technologies enable sophisticated spatial analysis capabilities crucial for understanding urban thermal dynamics and optimising Blue-Green infrastructure placement. The demonstrated reduction in urban heat island intensity through strategic placement informed by GIS data represents a significant advancement in evidence-based urban planning that can be scaled across different urban contexts (Mutani et al., 2024). These tools facilitate land use, climate condition, and human activity analysis, providing insights for optimal BGI placement while enabling hydrodynamic modelling that aids in designing integrated systems.

Artificial intelligence and machine learning applications offer particular promise for addressing the complexity of multi-criteria optimisation in BGI deployment. The potential for AI-driven approaches to simultaneously optimise air quality improvement, biodiversity enhancement, and heat island mitigation represents a paradigm shift toward more sophisticated planning tools that can handle the multifunctional nature of urban renaturing interventions (Shaamala et al., 2024). These technologies enable planners to evaluate complex trade-offs and identify optimal solutions that would be difficult to achieve through conventional planning approaches. **Table 6** provides an overview of key emerging technologies (remote sensing, AI, digital twins, IoT, etc.), current implementation status, principal benefits, challenges, and future potential for smart urban renaturing.

Table 6. Technology integration for smart urban renaturing

Technology Category	Application in Urban Renaturing	Current Implementation Status	Key Benefits	Implementation Challenges	Future Potential
Remote Sensing & GIS	Spatial analysis; UHI mapping; Optimal placement identification	Widely adopted	Evidence-based planning; Spatial precision; Cost-effective analysis	Data quality; Technical expertise; Equipment costs	Enhanced precision; Real-time capabilities; Global applications
Artificial Intelligence	Multi-criteria optimisation; Performance prediction; Design automation	Emerging applications	Optimised solutions; Complex problem solving; Automated design	Algorithm complexity; Interpretation challenges; Ethical considerations	Autonomous optimisation; Integrated solutions; Predictive management
Digital Twin Technology	Virtual modelling; Scenario simulation; Performance optimisation	Pilot projects	Risk-free testing; Performance prediction; Scenario planning	Technical complexity, High costs, Interdisciplinary coordination	Comprehensive simulation; Real-time integration; Decision automation
IoT Sensor Networks	Environmental monitoring; Real-time data collection; Performance tracking	Growing deployment	Continuous monitoring; Data accuracy; Responsive management	Installation costs, Maintenance requirements, and Data privacy	Ubiquitous monitoring; Intelligent responses; Predictive maintenance

continued on following page

Table 6. Continued

Technology Category	Application in Urban Renaturing	Current Implementation Status	Key Benefits	Implementation Challenges	Future Potential
Machine Learning	Pattern recognition; Effectiveness prediction; Adaptive management	Research phase	Pattern identification; Predictive capability; Performance optimisation	Data requirements; Model validation; Technical expertise	Advanced pattern recognition; Autonomous adaptation; Performance optimisation
Predictive Analytics	Urban growth forecasting; Climate impact assessment; Resource planning	Limited implementation	Proactive planning; Risk mitigation; Resource optimisation	Uncertainty management; Model accuracy; Stakeholder acceptance	Integrated forecasting; Multi-scale predictions; Dynamic planning

Sustainability Dimensions: Beyond Thermal Management

The literature demonstrates that urban renaturing through Blue-Green infrastructure provides sustainability benefits that extend beyond temperature regulation to encompass multiple environmental, social, and economic domains, positioning it as a comprehensive strategy for urban transformation.

Biodiversity conservation emerges as a critical co-benefit, with urban green infrastructure providing essential habitat connectivity and ecosystem function restoration that contributes to both local biodiversity enhancement and broader conservation goals (He et al., 2024; Grabowski et al., 2023). Urban green infrastructure plays a crucial role in biodiversity conservation by providing habitats for diverse flora and fauna while enhancing ecological connectivity and restoring ecosystem functions essential for urban biodiversity maintenance. This biodiversity dimension positions urban renaturing as a component of larger conservation strategies rather than simply a climate adaptation measure.

Carbon emission mitigation represents another significant sustainability dimension that operates through direct carbon sequestration and indirect effects through reduced energy consumption for urban cooling. The climate resilience enhancement achieved through natural element integration provides crucial adaptation capacity for extreme weather events and climate change impacts that extend beyond heat management to include flood resilience and air quality improvement (Yang et al., 2024; Meemrot et al., 2024).

Policy Integration and Governance Innovation

Effective implementation of urban renaturing strategies requires fundamental advances in policy integration and governance innovation that can address the multisectoral nature of BGI while maintaining implementation efficiency and ensuring equitable outcomes.

The literature identifies diverse policy instruments, including regulatory measures, economic incentives, and collaborative initiatives, as essential tools, with different instruments proving most effective for different implementation contexts (Longato et al., 2024). The particular effectiveness of regulatory instruments for integrating solutions into new development areas and incentive-based approaches for existing structure retrofitting suggests the need for differentiated policy strategies that respond to local implementation contexts.

Multi-level governance approaches are essential for addressing the scale mismatches between urban thermal challenges and administrative boundaries. Coordinating across spatial scales from global climate agreements to local implementation initiatives requires innovative governance models that facilitate collaboration while maintaining clear roles and responsibilities (Simon et al., 2021). Institutional and regulatory barriers, including rigid governance systems and funding constraints, require innovative financing mechanisms, policy innovations, and community involvement for successful navigation (He et al., 2024; Simon et al., 2021).

Future Research Directions and Knowledge Gaps

The literature review reveals several critical areas requiring further research and development to advance urban renaturing practice from experimental interventions toward systematic urban transformation, particularly addressing methodological, temporal, and social research needs.

Standardised methodologies for quantifying integrated BGI benefits represent a fundamental need, as current assessment approaches vary significantly across studies and contexts, limiting the ability to compare results and develop evidence-based best practices. Given the significant temporal variations in BGI effectiveness, long-term performance evaluation emerges as another critical research priority. The finding that cooling establishment time varies from one to three years for grasses and artificial systems to seven to ten years for trees and climbing plants highlights the importance of understanding temporal dynamics for realistic planning and maintenance programming (Gobatti et al., 2023).

Implications for Urban Planning Practice

The findings presented in this literature review have profound implications for urban planning practice that extend beyond technical considerations to encompass fundamental shifts in planning frameworks and implementation approaches, requiring new tools, methodologies, and institutional arrangements.

The demonstrated superiority of integrated BGI approaches over isolated interventions requires planners to adopt more holistic perspectives that consider synergistic effects and multifunctional benefits rather than optimising single performance criteria. Scale-sensitive planning emerges as a critical requirement that challenges conventional planning scales and administrative boundaries. The need for multi-scale approaches that optimise interventions at appropriate spatial scales while maintaining connectivity and coherence across urban systems requires new planning frameworks that coordinate interventions across different scales and jurisdictions. Identifying district-scale optimisation as most effective for combined infrastructure provides specific guidance while highlighting the need for new institutional arrangements to implement meso-scale interventions effectively.

Application of the Socio-Ecological Systems Framework

The socio-ecological systems (SES) framework provides a valuable lens for understanding the complex interactions between ecological processes, human systems, and institutional structures in urban environments. In the context of blue-green infrastructure, ecological components include vegetation and water systems that regulate urban microclimates, while social components involve community access, equity considerations, and behavioural adaptation. Institutional dimensions encompass governance frameworks, policy instruments, and planning regulations. Integrating these dimensions enables a more holistic approach to urban climate adaptation, ensuring that technical solutions are aligned with social and policy contexts.

While a significant portion of the existing literature is based on studies conducted in Chinese cities, expanding the geographical scope is essential to enhance global applicability. Evidence from European cities such as Copenhagen highlights integrated climate-resilient urban planning, while Latin American cities like Medellín demonstrate innovative green corridor strategies. Similarly, emerging African cities such as Nairobi provide insights into nature-based solutions under resource-constrained conditions. Incorporating such diverse case studies strengthens the generalizability of findings.

CONCLUSION

Urban renaturing through integrated BGI represents a transformative paradigm shift from isolated environmental interventions toward synergistic systems that fundamentally reshape urban thermal environments while delivering multiple co-benefits essential for sustainable urban development. This comprehensive literature review demonstrates that the strategic integration of green and blue infrastructure produces cooling benefits that exceed the sum of individual contributions while simultaneously addressing broader sustainability, equity, and resilience challenges facing contemporary cities.

The evidence overwhelmingly supports the superiority of integrated approaches, with synergistic effects producing peak cooling benefits up to 6.37°C, as demonstrated in Harbin, China, while maintaining enhanced stability under varying meteorological conditions compared to isolated interventions. The research reveals sophisticated spatial optimisation principles, identifying scale-sensitive thresholds with blue infrastructure demonstrating optimal performance at 1600-meter scales and green infrastructure showing effectiveness at both 50-meter and 1600-meter scales, establishing district-scale planning (1200-1600 meters) as optimal for combined infrastructure implementation. Critical findings illuminate profound equity implications, revealing the potential to address environmental justice concerns through strategic deployment in vulnerable communities and significant risks of green gentrification that require proactive anti-displacement strategies integrated throughout planning processes.

Implementation faces substantial challenges across multiple domains. Data limitations emerge as particularly acute barriers in rapidly urbanising regions where thermal management interventions are most urgently needed. At the same time, economic considerations reveal complex trade-offs extending beyond simple cost calculations to encompass temporal dynamics and scale effects. Governance and policy challenges reflect deeper systemic issues requiring fundamental innovations in institutional design and coordination mechanisms to integrate Blue-Green infrastructure with existing urban systems while addressing institutional fragmentation.

This synthesis advances understanding by documenting the synergistic mechanisms through which integrated BGI produces cooling effects that exceed additive benefits, establishing evidence-based spatial optimisation principles that guide multi-scale planning approaches, and identifying critical implementation imperatives that must be addressed to realise transformative potential. The review's integration of biophysical mechanisms with socio-institutional dimensions provides a comprehensive framework for understanding urban renaturing as both a technical intervention and a social process. At the same time, identifying emerging technological opportunities including AI-driven optimisation, digital twin modelling, and IoT-enhanced monitoring offers pathways for advancing evidence-based urban thermal management from experimental demonstration toward systematic transformation.

The findings challenge traditional sectoral approaches to urban environmental management and provide compelling evidence for fundamental shifts in urban planning practice that extend beyond technical considerations to encompass institutional innovation, governance reform, and social equity integration. The documented climate sensitivity with arid climates achieving highest temperature reductions followed by temperate and tropical zones necessitates locally-adapted approaches supported by climate-specific design guidelines rather than universal solutions.

Translating research insights into practice requires actionable strategies for urban planners and policymakers. Key considerations include prioritizing high-risk heat zones, integrating BGI into zoning regulations, and adopting performance-based planning frameworks. Collaborative planning involving local governments, private stakeholders, and communities is essential for successful implementation. Additionally, incorporating BGI into existing infrastructure systems through retrofitting approaches can enhance feasibility in densely built environments.

REFERENCES

Anwar, M. R., & Sakti, L. D. (2024). *Integrating Artificial Intelligence and Environmental Science for Sustainable Urban Planning.IAIC Transactions on Sustainable Digital Innovation.* ITSDI., DOI: 10.34306/itsdi.v5i2.666

Ashinze, U. K., Edeigba, B. A., Umoh, A. A., Biu, P. W., & Daraojimba, A. I. (2024). *Urban green infrastructure and its role in sustainable cities: A comprehensive review.* World Journal Of Advanced Research and Reviews., DOI: 10.30574/wjarr.2024.21.2.0519

Azmeer, A., Tahir, F., & Al-Ghamdi, S. G. (2024). Progress on green infrastructure for urban cooling: Evaluating techniques, design strategies, and benefits. *Urban Climate*, *56*, 102077. Advance online publication. DOI: 10.1016/j.uclim.2024.102077

Bacchin, T. K., Ashley, R., Sijmons, D. F., Zevenbergen, C., & Timmeren, A. van. (2014).Blue-Green multifunctional infrastructure: An urban landscape system design new approach.

Back, Y., & Kleidorfer, M. (2023). Opportunities and challenges in designing climate-resilient dense urban areas with Blue-Green Infrastructure. *Österreichische Wasser- und Abfallwirtschaft*, *75*(11-12), 615–625. Advance online publication. DOI: 10.1007/s00506-023-00989-3

Cai, Z., Guldmann, J.-M., Tang, Y., & Han, G. (2022). Does city-water layout matter? Comparing the cooling effects of water bodies across 34 Chinese megacities. *Journal of Environmental Management*, *324*, 116263. Advance online publication. DOI: 10.1016/j.jenvman.2022.116263 PMID: 36166866

Cao, B., Chen, Q., Du, M., Cheng, Q., Li, Y., & Li, R. (2022). Simulation Analysis of the Cooling Effect of Urban Water Bodies on the Local Thermal Environment. *Water (Basel)*, *14*(19), 3091. Advance online publication. DOI: 10.3390/w14193091

Cavadini, G. B., Cavadini, G. B., & Cavadini, G. B. (2024). Connecting blue-green infrastructure elements to reduce combined sewer overflows. *Journal of Environmental Management*, *365*, 121465. Advance online publication. DOI: 10.1016/j.jenvman.2024.121465 PMID: 38901320

Chen, J., Bach, P. M., Nice, K. A., & Leitão, J. P. (2024). Investigating the efficacy of a fast urban climate model for spatial planning of green and blue spaces for heat mitigation. *The Science of the Total Environment*, *955*, 176925. Advance online publication. DOI: 10.1016/j.scitotenv.2024.176925 PMID: 39454799

Chen, M., Chen, L., Zhou, Y., Hu, M., Jiang, Y., Huang, D., Gong, Y., & Xian, Y. (2023). Rising vulnerability of compound risk inequality to ageing and extreme heatwave exposure in global cities. *NPJ Urban Sustainability*, *3*(1), 38. Advance online publication. DOI: 10.1038/s42949-023-00118-9

Cichowicz, R., & Bochenek, A. D. (2024). Assessing the effects of urban heat islands and air pollution on human quality of life. *Anthropocene*, *46*, 100433. Advance online publication. DOI: 10.1016/j.ancene.2024.100433

Costa, H., Floater, G., Hooyberghs, H., Verbeke, S., & Ridder, K. D. (2016). *Climate change, heat stress and labour productivity: A cost methodology for city economies*. Research Papers in Economics.

Deng, Z., Zhao, H., Li, L., Liu, G., Lin, H., & Adam, T. D. (2023). The climate adaptive characteristics of urban inside/outside water bodies based on their cooling effect in Poyang and Dongting lake regions, China. *Heliyon*, *9*(5), e15974. Advance online publication. DOI: 10.1016/j.heliyon.2023.e15974 PMID: 37215833

Ding, W., Liu, M., Wu, Y., & Chen, H. (2024). How to expand the cooling capacity of blue and green spaces in peri-urban areas throughout the entire diurnal cycle: Evidence from an inland multilake city. *Journal of Cleaner Production*, *444*, 141165. Advance online publication. DOI: 10.1016/j.jclepro.2024.141165

Du, H., & Zhou, F. (2022). Mitigating Extreme Summer Heat Waves with the Optimal Water-Cooling Island Effect Based on Remote Sensing Data from Shanghai, China. *International Journal of Environmental Research and Public Health*, *19*(15), 9149. Advance online publication. DOI: 10.3390/ijerph19159149 PMID: 35897518

Egusquiza, A., Cortese, M., & Perfido, D. (2019).Mapping of innovative governance models to overcome barriers for nature based urban regeneration. DOI: 10.1088/1755-1315/323/1/012081

Elmarakby, E., & Elkadi, H. (2024). Prioritising urban heat island mitigation intervention: Mapping a heat risk index. *The Science of the Total Environment*, *948*, 174927. Advance online publication. DOI: 10.1016/j.scitotenv.2024.174927 PMID: 39038684

Fang, X., Li, J., & Ma, Q. (2023). Integrating green infrastructure, ecosystem services and nature-based solutions for urban sustainability: A comprehensive literature review. *Sustainable Cities and Society*, *98*, 104843. Advance online publication. DOI: 10.1016/j.scs.2023.104843

Ferguson, L., & Anna, M. (2024).The impact of heat adaptation on socioeconomically vulnerable populations. DOI: 10.1016/B978-0-443-13502-6.00010-5

Fernández, C. T., & Navarro, I. M. (2024). *Analysis of the Impact of Urban Green Spaces on Mitigating the Urban Heat Island Effect.* Innovation Science and Technology., DOI: 10.56397/IST.2024.05.02

Gallay, I., Olah, B., Murtinová, V., & Gallayová, Z. (2023). Quantification of the Cooling Effect and Cooling Distance of Urban Green Spaces Based on Their Vegetation Structure and Size as a Basis for Management Tools for Mitigating Urban Climate. *Sustainability (Basel), 15*(4), 3705. Advance online publication. DOI: 10.3390/su15043705

Gao, K., Haddad, S., Paolini, R., Altheeb, M., Mogirah, A. A., Moammar, A. B., & Santamouris, M. (2024). The use of green infrastructure and irrigation in the mitigation of urban heat in a desert city. *Building Simulation, 17*(5), 679–694. Advance online publication. DOI: 10.1007/s12273-024-1110-0

Geng, X., Zhang, D., Sun, W., Yuan, Y., Li, C., & Yu, Z. (2024). Heterogenous impact of background meteorological factors on cooling effect of two typical urban parks: Evidence from Shanghai, China. *Building and Environment, 265,* 112024. Advance online publication. DOI: 10.1016/j.buildenv.2024.112024

Gobatti, L., Bach, P. M., Scheidegger, A., & Leitão, J. P. (2023). Using satellite imagery to investigate Blue-Green Infrastructure establishment time for urban cooling. *Sustainable Cities and Society, 97,* 104768. Advance online publication. DOI: 10.1016/j.scs.2023.104768

Grabowski, Z. J., Teixeira, L. H., Fakirova, E., Meyer, S. T., Traidl-Hoffmann, C., Schloter, M., & Helmreich, B. (2023). Cosmopolitan conservation: The multi-scalar contributions of urban green infrastructure to biodiversity protection. *Biodiversity and Conservation, 32*(11), 3595–3606. Advance online publication. DOI: 10.1007/s10531-023-02614-x

Guo, Y., Ren, Z., Dong, Y., Hu, N., Wang, C., Zhang, F., Jia, G., & He, X.-J. (2022). Strengthening of surface urban heat island effect driven primarily by urban size under rapid urbanization: National evidence from China. *GIScience & Remote Sensing, 59*(1), 2127–2143. Advance online publication. DOI: 10.1080/15481603.2022.2147301

Gupta, A., & De, B. (2024).A systematic review on urban blue-green infrastructure in the south Asian region: recent advancements, applications, and challenges. DOI: 10.2166/wst.2024.014

Haase, A. (2024). Inequalities and Injustices of Urban Green Regeneration: Applying the Conflict Analysis Perspective. *Land (Basel), 13*(3), 296. Advance online publication. DOI: 10.3390/land13030296

Hadjinicolaou, P., Economou, T., Lazoglou, G., & Whitt, P. (2023).Quantification of the Urban Heat Island effect using paired station data in the Middle East and North Africa region. DOI: 10.5194/ems2023-13

Hassan, Z., Shahbaz, B., & Lopez, F. G. (2023).Enhancing Blue/Green Infrastructure for Resilient Urban Environments: Smart Solutions and Nature-Based Strategies. DOI: 10.61326/icelis.2023.18

He, M., Yuan, C., Zhang, X., Wang, P., & Yao, C. (2023). Impacts of Blue-Green-grey infrastructures on high-density urban thermal environment at multiple spatial scales: A case study in Wuhan. *Urban Climate*, *52*, 101714. Advance online publication. DOI: 10.1016/j.uclim.2023.101714

He, X., Xu, P., An, B.-W., & Guo, Q. (2024). Urban green infrastructure: Bridging biodiversity conservation and sustainable urban development through adaptive management approach. *Frontiers in Ecology and Evolution*, *12*, 1440477. Advance online publication. DOI: 10.3389/fevo.2024.1440477

Hu, N., Wang, G., Ma, Z., Ren, Z., Zhao, M., & Meng, J. (2023). The cooling effects of urban waterbodies and their driving forces in China. *Ecological Indicators*, *156*, 111200. Advance online publication. DOI: 10.1016/j.ecolind.2023.111200

Huang, M., Zhong, S., Mei, X., & He, J. (2024). Spatiotemporal Patterns in the Urban Heat Island Effect of Several Contemporary and Historical Chinese "Stove Cities.". *Sustainability (Basel)*, *16*(7), 3091. Advance online publication. DOI: 10.3390/su16073091

Huang, W. T. K., Masselot, P., Bou-Zeid, E., Fatichi, S., Paschalis, A., Sun, T., Gasparrini, A., & Manoli, G. (2023). Economic valuation of temperature-related mortality attributed to urban heat islands in European cities. *Nature Communications*, *14*(1), 7438. Advance online publication. DOI: 10.1038/s41467-023-43135-z PMID: 37978178

Iwuanyanwu, O., Gil-Ozoudeh, I., Okwandu, A. C., & Ike, C. S. (2024). *Cultural and social dimensions of green architecture: Designing for sustainability and community well-being*. International Journal of Applied Research in Social Sciences., DOI: 10.51594/ijarss.v6i8.1477

Jabbar, H. K., Hamoodi, M. N., & Al-Hameedawi, A. N. M. (2023). Urban heat islands: a review of contributing factors, effects and data.*IOP Conference Series: Earth and Environmental Science*. DOI: 10.1088/1755-1315/1129/1/012038

Jandaghian, Z., & Colombo, A. (2024). The Role of Water Bodies in Climate Regulation: Insights from Recent Studies on Urban Heat Island Mitigation. *Buildings (Basel, Switzerland), 14*(9), 2945. Advance online publication. DOI: 10.3390/buildings14092945

Jang, S., Bae, J., & Kim, Y.-J. (2024). Street-level urban heat island mitigation: Assessing the cooling effect of green infrastructure using urban IoT sensor big data. *Sustainable Cities and Society, 100*, 105007. Advance online publication. DOI: 10.1016/j.scs.2023.105007

Jia, S., & Wang, Y. (2023).Multiscale Evaluation of the Cooling Effect of Greenspace in Urban Environments. DOI: 10.1007/978-981-99-3626-7_75

Jiang, Y., Huang, J., Shi, T., & Li, X. (2021). Cooling Island Effect of Blue-Green Corridors: Quantitative Comparison of Morphological Impacts. *International Journal of Environmental Research and Public Health, 18*(22), 11917. Advance online publication. DOI: 10.3390/ijerph182211917 PMID: 34831672

Kar, B., & Meade, S. (2022). Urban Heat Island Impact on Energy Burden Among Socially Vulnerable Groups.*IEEE International Geoscience and Remote Sensing Symposium*. DOI: 10.1109/IGARSS46834.2022.9883806

Kiani, B. H. (2022). *Towards Urban Sustainability: Impact of Blue and Green Infrastructure on Building Smart*. Climate Resilient and Livable Cities., DOI: 10.1201/9781003127819-7

Kim, J., Khouakhi, A., Corstanje, R., & Johnston, A. S. A. (2023). Greater local cooling effects of trees across globally distributed urban green spaces. *The Science of the Total Environment*. Advance online publication. DOI: 10.1016/j.scitotenv.2023.168494 PMID: 37979859

Kim, S., & Kim, D. (2023). Climate change and cooling equity: Spatial dynamics of vulnerable populations. *Growth and Change*. Advance online publication. DOI: 10.1111/grow.12701

Lanza, K., Jones, J., Acuña, F., Coudert, M., Bixler, R. P., Kamath, H. G., & Niyogi, D. (2023). Heat vulnerability of Latino and Black residents in a low-income community and their recommended adaptation strategies: A qualitative study. *Urban Climate, 51*, 101656. Advance online publication. DOI: 10.1016/j.uclim.2023.101656

Lehmann, S. (2019). *Reconnecting Cities with Nature*. Building Resilience at the Urban Scale., DOI: 10.1007/978-3-030-04711-5_2

Lewartowska, E., Anguelovski, I., Oscilowicz, E., Triguero-Mas, M., Cole, H., Shokry, G., Perez-del-Pulgar, C., & Connolly, J. J. T. (2024). RACIAL INEQUITY IN GREEN INFRASTRUCTURE AND GENTRIFICATION: Challenging Compounded Environmental Racisms in the Green City. *International Journal of Urban and Regional Research, 48*(2), 294–322. Advance online publication. DOI: 10.1111/1468-2427.13232

Liu, M. (2024). *The role of urban green space in alleviating the urban heat island effect.* Environment, Resource and Ecology Journal., DOI: 10.23977/erej.2024.080102

Liu, Q., Dong, Q., Zhang, L., & Sun, C. (2024). Summer cooling island effects of blue-green spaces in severe cold regions: A case study of Harbin, China. *Building and Environment, 257*, 111539. Advance online publication. DOI: 10.1016/j.buildenv.2024.111539

Longato, D., Cortinovis, C., Balzan, M. V., & Geneletti, D. (2024). Identifying suitable policy instruments to promote nature-based solutions in urban plans. *Cities (London, England), 154*, 105348. Advance online publication. DOI: 10.1016/j.cities.2024.105348

Ma, Y., Fu, Y., & Liu, C. (2024).Urban water bodies mitigate heat waves, the case of the Chinese city Kunming. DOI: 10.1007/s44327-024-00004-8

Marginean, I., Cuaresma, J. C., Hoffmann, R., Muttarak, R., Gao, J., & Daloz, A. S. (2024). High-Resolution Modeling and Projecting Local Dynamics of Differential Vulnerability to Urban Heat Stress. *Earth's Future, 12*(10), e2024EF004431. Advance online publication. DOI: 10.1029/2024EF004431

Meemrot, R., Meena, R. L., Derwal, N., Meena, P., Balkh, S., & Jangid, J. B. (2024). Greening the Concrete Jungle: Enhancing Urban Resilience through Blue-Green Infrastructure. *International Journal for Research in Applied Science and Engineering Technology, 12*(4), 2931–2935. Advance online publication. DOI: 10.22214/ijraset.2024.60578

Moss, T., & Kar, B. (2020). Socio-economic vulnerability to urban heat in phoenix, arizona and dallas, texas during june 2020. *ISPRS Annals of the Photogrammetry, Remote Sensing and Spatial Information Sciences, VI-3/W1-2020*, 59–66. Advance online publication. DOI: 10.5194/isprs-annals-VI-3-W1-2020-59-2020

Mutani, G., Scalise, A., Sufa, X., & Grasso, S. (2024).Synergizing Machine Learning and Remote Sensing for Urban Heat Island Dynamics: A Comprehensive Modelling Approach. DOI: 10.20944/preprints202409.2284.v1

Noll, S., & Bhar, T. (2023).The Five Pillars of Urban Environmental Justice: A Framework for Building Equitable Cities. DOI: 10.21827/potcj.1.1.41048

Oscilowicz, E., Anguelovski, I., Triguero-Mas, M., García-Lamarca, M., Baró, F., & Cole, H. (2022).Green justice through policy and practice: a call for further research into tools that foster healthy green cities for all. DOI: 10.1080/23748834.2022.2072057

Ouyang, W., Morakinyo, T. E., & Ng, E. (2019).The Cooling Efficiency of Urban Greenery Coverage in a High-density City. DOI: 10.1088/1755-1315/329/1/012043

Pasupuleti, M. K. (2024).Green Governance: Shaping Policies for a Sustainable Future. DOI: 10.62311/nesx/77622

Perrelet, K., Moretti, M., Dietzel, A., Altermatt, F., & Cook, L. M. (2024). Engineering blue-green infrastructure for and with biodiversity in cities. *NPJ Urban Sustainability*, *4*(1), 27. Advance online publication. DOI: 10.1038/s42949-024-00163-y

Petrucci, A. L. (2023). *Environmental Retrofitting*. Fighting Urban Heat Island Toward NEZ Sustainable Smart Cities., DOI: 10.1007/978-3-031-35664-3_8

Pontius, J., & McIntosh, A. (2024).Urban Heat Islands. DOI: 10.1007/978-3-031-48762-0_10

Qiu, Y., & He, B. (2024).Review of urban water thermal environment cooling. DOI: 10.1088/1755-1315/1363/1/012073

Ribeiro, T. R. (2023). Justice in Urban Greening and Green Gentrification., DOI: 10.1007/978-3-031-22566-6_20

Rivera, L. M., & Faggian, R. (2024). Integrating Blue-Green Infrastructure strategies to enhance climate resilience in Colombia. *European Journal of Sustainable Development*, *13*(4), 255. Advance online publication. DOI: 10.14207/ejsd.2024.v13n4p255

Roxon, J., Ulm, F.-J., Pellenq, R. J.-M., & Pellenq, R. J.-M. (2020). Urban heat island impact on state residential energy cost and CO2 emissions in the United States. *Urban Climate*, *31*, 100546. Advance online publication. DOI: 10.1016/j.uclim.2019.100546

Rumble, H., Angeoletto, F., Connop, S., Goddard, M. A., & Nash, C. (2019). Understanding and applying ecological principles in cities. DOI: 10.1007/978-3-030-01866-5_15

Shaamala, A., Yigitcanlar, T., Nili, A., & Nyandega, D. (2024). Algorithmic Green Infrastructure Optimisation: Review of Artificial Intelligence Driven Approaches for Tackling Climate Change. *Sustainable Cities and Society, 101*, 105182. Advance online publication. DOI: 10.1016/j.scs.2024.105182

Shah, A. M., Liu, G., Nawab, A., Li, H., Xu, D., Yeboah, F. K., Yang, Q., & Zhang, L. (2024). *Sustainability and resilience interface at typical urban green and blue infrastructures: costs, benefits, and impacts assessment.* Frontiers in Sustainable Cities., DOI: 10.3389/frsc.2024.1453829

Sheng, S., & Wang, Y. (2024). Configuration characteristics of Blue-Green spaces for efficient cooling in urban environments. *Sustainable Cities and Society, 100*, 105040. Advance online publication. DOI: 10.1016/j.scs.2023.105040

Shi, W., Zeng, S., & Aihemaiti, N. A. M. A. I. T. I.SHI. (2024, August 1). W., ZENG, S., & Aihemaiti, N. (2024). Research on the Spatial Heterogeneity of the Impact of Blue-Green Space Within Urban Block on Urban Thermal Environment. *Feng Jing Yuan Lin = Landscape Architecture, 31*(10), 98–105. Advance online publication. DOI: 10.3724/j.fjyl.202312070549

Sidiqui, P., Tariq, M., & Ng, A. W. M. (2022). An Investigation to Identify the Effectiveness of Socioeconomic, Demographic, and Buildings' Characteristics on Surface Urban Heat Island Patterns. *Sustainability (Basel), 14*(5), 2777. Advance online publication. DOI: 10.3390/su14052777

Simon, D., Goodness, J., Lwasa, S., Oliveira, J. A. P. de, Macedo, L. S. V. de, Kavonic, J., Török, E. H., & Elmqvist, T. (2021).Urban Governance of and for Urban Green and Blue Infrastructure. DOI: 10.1007/978-3-030-67650-6_16

Tan, J., Kuang, W., Yu, D., Zhou, W., Li, X., Liu, Q., & He, L. (2024). Investigating the cooling effects of land cover and landscape patterns surrounding rivers: Insights from the subtropical city of Changsha, China. *Urban Climate, 55*, 101975. Advance online publication. DOI: 10.1016/j.uclim.2024.101975

Urban heat island. (2022). DOI: 10.1016/B978-0-323-85539-6.00009-3

Vujovic, S., Haddad, B., Karaky, H., Sebaibi, N., & Boutouil, M. (2021). *Urban Heat Island: Causes.* Consequences, and Mitigation Measures with Emphasis on Reflective and Permeable Pavements., DOI: 10.3390/civileng2020026

Wang, Y., Song, H., Zhu, W., & Wang, Y. (2023). The Cooling Effects of Landscape Configurations of Green–Blue Spaces in Urban Waterfront Community. *Atmosphere (Basel), 14*(5), 833. Advance online publication. DOI: 10.3390/atmos14050833

Wong, N. H., Tan, C. L., Kolokotsa, D., & Takebayashi, H. (2021).Greenery as a mitigation and adaptation strategy to urban heat. DOI: 10.1038/s43017-020-00129-5

Wu, Q., Huang, Y., Irga, P. J., Kumar, P., Li, W., Wei, S., Shon, H. K., Lei, C., & Zhou, J. L. (2024). Synergistic control of urban heat island and urban pollution island effects using green infrastructure. *Journal of Environmental Management*, *370*, 122985. Advance online publication. DOI: 10.1016/j.jenvman.2024.122985 PMID: 39461153

Wu, T., Wang, S., & Xu, Q. (2024). Spatio-Temporal Heterogeneity of the Urban Heat Effect and Its Socio-Ecological Drivers in Yangzhou City, China. *Land (Basel)*, *13*(9), 1470. Advance online publication. DOI: 10.3390/land13091470

Xie, Q., Ren, L., & Yang, C. (2023). Regulation of water bodies to urban thermal environment: Evidence from Wuhan, China. *Frontiers in Ecology and Evolution*, *11*, 983567. Advance online publication. DOI: 10.3389/fevo.2023.983567

Xu, C., Huang, Q., Haase, D., Dong, Q., Teng, Y., Su, M., & Yang, Z. (2024). Cooling Effect of Green Spaces on Urban Heat Island in a Chinese Megacity: Increasing Coverage versus Optimizing Spatial Distribution. *Environmental Science & Technology*, *58*(13), 5811–5820. Advance online publication. DOI: 10.1021/acs.est.3c11048 PMID: 38502088

Yang, F., Yousefpour, R., Zhang, Y., & Wang, H. (2023). The assessment of cooling capacity of blue-green spaces in rapidly developing cities: A case study of Tianjin's central urban area. *Sustainable Cities and Society*, *99*, 104918. Advance online publication. DOI: 10.1016/j.scs.2023.104918

Yang, M., Wang, H., & Cao, S.-J. (2022). A global challenge of accurately predicting building energy consumption under urban heat island effect. *Indoor and Built Environment*. Advance online publication. DOI: 10.1177/1420326X221123222

Yang, Y., Li, X., Wang, D., & Liu, Z. (2024). *Optimizing urban ecosystems through green and blue infrastructure: Strategies for sustainability, resilience, and justice*. Applied and Computational Engineering., DOI: 10.54254/2755-2721/72/20240985

Yin, Y., Li, S. H., Xing, X., Zhou, X., Kang, Y., Hu, Q., & Li, Y. (2024). Cooling Benefits of Urban Tree Canopy: A Systematic Review. *Sustainability (Basel)*, *16*(12), 4955. Advance online publication. DOI: 10.3390/su16124955

Yu, Z., Chen, J., Chen, J., Zhan, W., Wang, C., Zhou, S., & Zhu, K. (2024). Enhanced observations from an optimized soil-canopy-photosynthesis and energy flux model revealed evapotranspiration-shading cooling dynamics of urban vegetation during extreme heat. *Remote Sensing of Environment*, *305*, 114098. Advance online publication. DOI: 10.1016/j.rse.2024.114098

Yu, Z., Uchiyama, Y., & Sato, M. (2024). *Combined effects of urban blue–green spaces on the thermal environment: a case study of Kobe*. International Journal of Economic Policy Studies., DOI: 10.1007/s42495-024-00140-4

Zepp, H., Gessner, M., Gruenhagen, L., & Bührs, M. (2023).Modeling the cooling effect of urban green spaces: the neglected control variables of 'soil moisture' and 'biotope types.'DOI: 10.1016/j.ufug.2023.128137

Zhang, H., Luo, M., Pei, T., Liu, X.-P., Wang, L., Zhang, W., Lin, L., Ge, E., Liu, Z., & Liao, W. (2023).Unequal urban heat burdens impede climate justice and equity goals. DOI: 10.1016/j.xinn.2023.100488

Zhang, J.-Y. (2023). *Impact of Urban Heat Island: A Local-Level Urban Climate Phenomenon on Urban Ecology and Human Health*. Advances in Geographical and Environmental Sciences., DOI: 10.1007/978-981-99-3006-7_5

Zhao, M., Lee, J. K. W., Kjellstrom, T., & Cai, W. (2021). Assessment of the economic impact of heat-related labor productivity loss: A systematic review. *Climatic Change*, *167*(1-2), 22. Advance online publication. DOI: 10.1007/s10584-021-03160-7

Zhou, T., Jia, W., Yan, L., Hong, B., & Wang, K. (2024). Urban park's vertical canopy structure and its varied cooling effect under continuous warming climate. *Urban Climate*, *53*, 101819. Advance online publication. DOI: 10.1016/j.uclim.2024.101819

Chapter 7
Towards Climate-Resilient Infrastructure in South Africa:
A Review of Multi-Hazard Early Warning Systems

Yanelisa Ndatshe
http://orcid.org/0000-0002-9123-3066
Nelson Mandela University, South Africa

Mosekama Osia Mokhele
http://orcid.org/0000-0003-3299-2799
Nelson Mandela University, South Africa

Amina Jakoet-Salie
http://orcid.org/0000-0002-1479-7255
Nelson Mandela University, South Africa

ABSTRACT

South Africa's warming climate is amplifying multi hazard events that overwhelm critical infrastructure. This chapter assesses the performance of South Africa's multi-hazard early warning systems (MHEWS) and offers evidence-based measures to boost climate-resilient infrastructure. A PRISMA guided systematic review of peer reviewed studies and grey literature (2015-2025). Findings show fragmented coordination, inadequate multilingualism, and severe financing gaps. The chapter recommends: (a) embedding real time SAWS dashboards in municipal IDP and budgets; (b) allocating a "climate risk" line item; (c) adopting a three tier (national provincial municipal) coordination protocol; (d) deploying multilingual/voice based

DOI: 10.4018/979-8-3373-7345-4.ch007

ight © 2026, IGI Global Scientific Publishing. Copying or distributing in print or electronic forms without written permission of IGI Global Scientific Publishing is prohibited. Use of this chapter to n generative artificial intelligence (AI) technologies is expressly prohibited. The publisher reserves all rights to license its use for generative AI training and machine learning model development.

alert platforms; (e) creating a ring fenced MHEWS maintenance fund financed by a modest municipal levy; and (f) forming community "Flood Watch" committees. Implementing these actions will enhance the robustness, redundancy, timeliness, and resourcefulness of warnings, reduce disaster losses, and align practice with the DMA and the Climate Change Act.

INTRODUCTION

Over the past two decades, South Africa has felt the tangible impacts of a warming climate, with rising intensity and frequency of multi-hazards eroding environmental resilience (Bondank & Chester, 2020; Chipfupa et al., 2021). South African Weather Service (SAWS) climate models project a 2 °C–3 °C rise in interior-plateau temperatures by 2050 and increasingly erratic precipitation, producing longer dry spells punctuated by extreme rainfall (Lumbroso et al., 2024; Myeni & Roffe, 2025; SAWS Climate Outlook, 2023). This shift has generated cascading floods, droughts, heatwaves, and coastal storms that repeatedly damage critical infrastructure, including roads, bridges, water and sanitation networks, and electricity distribution systems.

The 8 12 April 2022 KwaZulu-Natal floods illustrate the scale of loss: more than 150 km of regional roads were washed away, over 15 bridges collapsed, three major water treatment plants were inundated, and emergency tankers supplied water for two weeks (Kunguma, 2022; Motha, 2023; Ngcamu, 2022). Falling trees also brought down high-capacity transmission lines, triggering load shedding that disrupted early-warning alerts via radio and SMS. In June 2024, the Gqeberha (Nelson Mandela Bay) floods delivered 150 mm of rain in 24 hours, overwhelming stormwater systems that had not been upgraded since the 1990s. The Matanzima and Nomakhwezana bridges were swept away, and the Matanzima substation short-circuited, cutting power to more than 120,000 households (HeraldLIVE, 2024; Mtumane, 2024). The event forced school closures, municipal service suspensions, and the displacement of 1,100 residents, requiring rescue operations and over R250 million in emergency relief (HeraldLIVE, 2024). These incidents reveal a feedback loop: damaged communication networks impede the dissemination of early warnings, which in turn hampers preparedness and amplifies hazard impacts. Deliberate, temporary power cuts to protect the national grid (Eskom, 2023) reduce the reach of SMS alerts and radio bulletins precisely when they are most needed.

South Africa's early warning framework rests on the Disaster Management Act (DMA) of 2002, as amended by Act 16 of 2015, which obliges all government spheres to collaborate with stakeholders to reduce disaster risk (Sithole, 2014). The National Disaster Management Centre (NDMC) provides policy guidance and integrates hazard data from SAWS and sector departments; provincial centres translate these directives into operational plans, while municipal disaster management centres (MDMCs) issue community alerts, activate evacuation routes, and preposition relief supplies.

Technically, SAWS operates 231 automatic weather stations, 628 river flow gauging stations and a satellite-based rainfall estimation system (Seyama et al., 2019; SAWS Annual Report, 2018). Data feeds the Integrated Early Warning System (IEWS), which produces real-time dashboards accessible via the SAWS website, a mobile app (SAWS Alert), and an open-source API that municipalities embed in their portals. The API supports "last mile" delivery: community radio in rural KwaZulu-Natal and the Eastern Cape broadcast flood warnings in isiZulu and isiXhosa. At the same time, municipal SMS gateways send geo-targeted messages to residents within a 5 km radius of predicted floodplains.

Despite these advances, the financial cost of ineffective multi-hazard early warning systems (MHEWS) is stark. The 2022 KwaZulu-Natal floods generated an estimated R1.2 billion in direct infrastructure damage; the 2024 Gqeberha floods added another R850 million for bridges, roads and water treatment repairs (Reddy et al., 2024). Municipal budgets, already strained by ageing assets, are forced to divert funds from long-term maintenance to emergency repairs. The Auditor General's 2024 report found that 62% of 100 assessed municipalities lacked a dedicated climate resilience fund, and only 18% had completed vulnerability assessments of critical infrastructure (AGSA, 2024; Presidential Climate Commission, 2024). These fiscal pressures underscore the urgency of integrating MHEWS with proactive, climate-smart investments, such as green infrastructure, floodplain restoration, and underground power-line retrofits, to reduce downstream disaster costs.

While South Africa has progressed in operationalising the Integrated Emergency Management System (IEMS) platform, expanding community radio networks, and incorporating satellite-derived rainfall estimates, significant gaps remain. Many municipalities lack clear standard operating procedures for rapid infrastructure shutdown, preemptive road closures, or backup power activation, resulting in uneven translation of warnings into locally appropriate actions.

Drawing on resilience theory (robustness, redundancy, resourcefulness, rapidity), this chapter assesses MHEWS effectiveness within the broader goal of climate-resilient infrastructure. It concludes that, without stronger integration of early warnings, targeted investment, and inclusive governance, South Africa's critical assets will remain vulnerable to escalating multi-hazard threats.

LITERATURE REVIEW

Theoretical Framing: Disaster Resilience Theory

Drawing on resilience theory as an analytical lens, this section evaluates how MHEWS can underpin climate-resilient infrastructure in South Africa and identifies gaps and opportunities for strengthening system-wide resilience. "*Resilience*," from the Latin resilientia ("to rebound"), is widely defined as a system's capacity to absorb disturbances, adapt, and retain its essential functions, structure, and identity (Agbehadji et al., 2024; Pearson et al., 2025; Röhe, 2025; Ziwanai, 2024). Within the sociotechnical context of MHEWS, resilience can be operationalised through four interrelated capacities, such as robustness, redundancy, resourcefulness, and rapidity, each shaping the ability to deliver timely, accurate, and actionable warnings to heterogeneous communities.

Robustness is the ability to withstand shocks without losing core functionality. The early warning literature links robustness to design attributes such as modularity, distributed architectures, and technological diversity, which enable components to continue operating when others fail (Agbehadji et al., 2024). Integrating Fourth Industrial Revolution (4IR) technologies, real-time sensors, automated data pipelines, and cloud-based analytics can fortify the physical and informational backbone of EWS, reducing vulnerability to extreme rainfall events that have risen by 45% since 1990 (Agbehadji et al., 2024). A multi-criteria decision analytic framework quantifies robustness by measuring the slope of the performance degradation curve during disturbances, providing a metric for comparing design alternatives (Agbehadji et al., 2024). In South African practice, many municipal stormwater networks installed in the 1970s with capacities below 0.5 $m^3\ s^{-1}$ exemplify a robustness deficit that amplifies flood impacts (Raphela & Ekeke, 2025; Siyongwana et al., 2015).

Redundancy refers to deliberate excess capacity that can be activated when primary pathways are disrupted. Antifragility theory argues that strategic redundancy, backup power, alternate transport corridors, and parallel forecasting models transform a merely tolerant system into one that thrives under stress (Lynch et al., 2025). KwaZulu-Natal illustrates this principle: redundant data streams (satellite, radar, community-based sensors) and multi-channel dissemination (SMS, radio, social media) act as "safe to fail" layers that maintain warning continuity during peak hazard periods (Lynch et al., 2025). Raphela and Ekeke (2025) found that low-resource contexts often lack formalised, redundant EWS, a gap that directly hampers resilience in many Southern African municipalities.

Resourcefulness captures the capacity to mobilise, reconfigure, and synergise assets, knowledge, and institutions in response to emerging threats. Community co-design is pivotal for enhancing risk knowledge, anticipation, and preparedness, expanding the pool of actionable resources, and embedding local expertise into the EWS value chain (Agbehadji et al., 2024). Co-creation approaches integrate health providers, NGOs, and community leaders to generate context-specific indicators and test "safe to fail" drills, converting small-scale failures into learning opportunities that boost overall resourcefulness (Lynch et al., 2025). Decision-analytic tools enable systematic scoring of resource-related design options, ensuring that limited budgets target the most leverage-rich interventions (Siyongwana et al., 2015).

Rapidity is the speed at which a system detects, processes, and communicates hazard information to end users. It comprises technical latency (sensor-to-analysis time) and social latency (message comprehension and behavioural uptake). Research on flood disaster warning dissemination in Vosloorus township shows that demographic factors and fragmented communication channels increase interpretive delays, undermining rapid response (Raphela & Ekeke, 2025). The literature notes that the absence of common terminology and clear indicators constitutes a major barrier to swift action. This shortfall can be mitigated by standardising alert thresholds and by employing real-time data feeds enabled by 4IR technologies (Agbehadji et al., 2024). Antifragility's post-disaster transformation principle suggests that each event should trigger the redesign of data-sharing protocols and decision-making hierarchies to shorten response cycles, thereby enhancing rapidity (Lynch et al., 2025). Threat agnostic resilience further stresses plasticity, the capacity of agents to adapt behaviours quickly to new information, as essential for accelerating community response under time-critical conditions (Agbehadji et al., 2024).

These four capacities are mutually reinforcing rather than isolated. Robustness establishes a stable platform upon which redundancy can be layered; redundancy supplies spare capacity that resourcefulness can draw upon during reallocation; resourcefulness, in turn, accelerates learning loops that improve rapidity. This dynamic aligns with the people-centred, multi-hazard paradigm advocated in recent scholarship, calling for an integrated, flexible, and transformative early warning architecture capable of addressing compounded risks characteristic of the Global South (Lynch et al., 2025; Raphela & Ekeke, 2025). Grounding the empirical assessment of South African MHEWS in this resilience theoretic framework enables systematic evaluation of how current systems embody each capacity, identification of critical gaps such as ageing storm water networks, limited redundant data pathways, and uneven community engagement, and formulation of evidence-based interventions that shift early warning from a reactive service to an anticipatory, antifragile public good.

Global Models of MHEWS

Globally, MHEWS are becoming a cornerstone of DRR and climate change adaptation (CCA) strategies. The Sendai Framework for Disaster Risk Reduction (2015-2030) identifies early warning systems as a critical means of reducing disaster impacts through enhanced risk knowledge, monitoring, communication, and preparedness (United Nations Office for Disaster Risk Reduction, 2015). On the ground, many countries have established internationally recognised MHEWS models that demonstrate how early warning can dramatically lower mortality and infrastructure damage. These models illuminate the institutional, technological, and community-based elements needed for effective early warning governance (World Meteorological Organization, 2023; Basher, 2006).

Bangladesh's Cyclone Preparedness Program (CPP) is a widely cited example of an effective MHEWS. Established in 1972 after the devastation of the Bhola cyclone, the CPP blends meteorological forecasting with a robust community-based distribution network (Haque et al., 2012; Paul, 2009; Paul & Dutt, 2010). The programme relies on more than 70,000 trained volunteers who disseminate warnings via megaphones, door-to-door visits, community radio, and signal flags in vulnerable coastal areas (Paul, 2009). This fusion of scientific forecasting and grassroots mobilisation has sharply reduced cyclone fatalities: the 1970 Bhola cyclone killed roughly 300,000 people, whereas later storms of comparable intensity have caused far fewer deaths, largely because of improved early warning and evacuation (Paul & Dutt, 2010). Bangladesh, therefore, provides a model of how community involvement and last-mile communication are essential to an effective MHEWS.

Cuba's Civil Defence System offers another frequently referenced model, regarded as one of the Caribbean's best hurricane preparedness frameworks. It emphasises state-coordinated action, mandatory evacuations, and well-developed community communication channels. Disaster preparedness activities are embedded in national planning, reinforced by regular drills and public awareness campaigns (Aguirre, 2005). Consequently, Cuba consistently records lower hurricane mortality than neighbouring countries that experience similar storms (Aguirre, 2005; World Meteorological Organisation, 2023). The Cuban experience underscores the importance of institutional coordination and clear governance structures to translate warnings into timely protective actions.

Japan provides a technologically advanced example through its nationwide J Alert system. Launched in 2007, J Alert is a satellite-based platform that rapidly disseminates alerts for earthquakes, tsunamis, extreme weather, and other hazards. Warnings are automatically broadcast via television, mobile phones, municipal loudspeakers, and internet channels, often within seconds of detection (Kitazawa

& Hale, 2021). Japan's case shows how integrating sophisticated monitoring with automated communication can minimise response times for fast-moving hazards.

Internationally, the WMO stresses the integration of MHEWS through its "Early Warnings for All" initiative and the Global Status of MHEWS report. According to the World Meteorological Organization (2023), effective MHEWS consists of four interlinked components: risk knowledge, monitoring and forecasting, warning dissemination, and preparedness to respond. Failure in any component can compromise the entire system (World Meteorological Organization, 2023). Although early warning coverage has expanded, the WMO estimates that nearly half of all countries, especially in developing regions, still lack comprehensive MHEWS.

These global experiences demonstrate that effective MHEWS require more than advanced forecasting technology; they also demand strong institutional coordination, active community participation, reliable communication infrastructure, and sustained investment in monitoring (World Meteorological Organization, 2023; Basher, 2006). For countries such as South Africa, these models offer valuable lessons for strengthening early warning systems to better protect communities and critical infrastructure from climate-related hazards.

Legal and Policy Context of MHEWS in South Africa

South Africa's disaster-risk architecture rests on two cornerstone statutes: DMA, and the Climate Change Act 22 of 2024 (CCA). The DMA obliges every sphere of government to embed disaster-risk reduction (DRR) in policy, to appoint National Disaster Management Centres (NDMCs) as the hub for hazard monitoring, early-warning dissemination and emergency response, and to mandate multi-stakeholder participation that includes municipal utilities, private contractors, NGOs and communities (Agyemang 2025; South Africa 2002). Section 7(1) further requires public bodies whose functions are climate-sensitive to review and, where necessary, amend policies to incorporate climate considerations (Lefutso 2025; Matikinca et al., 2024; South Africa, 2002). The Climate Change Act 2024 expands this legal scaffold by defining a "just transition," establishing Provincial and Municipal Forums on Climate Change, a Presidential Climate Commission, and requiring all organs of state to align their policies with climate-change objectives (South Africa, 2024). Together, the statutes create a top-down, binding framework that links climate adaptation and DRR to the constitutional mandate for sustainable development.

In contrast, the co-produced national Climate Risk and Vulnerability (CRV) assessment framework emerged from a collaborative process led by the Department of Forestry, Fisheries and the Environment (DFFE) and a broad coalition of stakeholders, including provincial and municipal disaster-management officials, academia, civil-society organisations and the private sector (Department of Environment, Forestry and Fisheries, 2020). Rather than prescribing a single set of rules, the framework offers a methodological toolkit for conducting sub-national assessments and aggregating them into a national picture. It foregrounds multi-scalar assessment, insisting that risk be quantified at the community, district, provincial and national levels, and that outputs at each scale be interoperable for policy synthesis.

This approach fills a gap in the statutory regime. While the DMA and CCA mandate "integration" and "coordination," they provide limited practical guidance on how local authorities should translate those mandates into comparable risk metrics. The DMA designates NDMCs as the principal bodies responsible for issuing multi-hazard early-warning messages and for ensuring alerts reach households through SMS, radio, sirens and digital platforms. The Climate Change Act reinforces this by requiring climate-risk considerations to be embedded in all planning cycles, thereby linking early warning to broader adaptation strategies. However, the Act focuses on institutional alignment rather than on technical standardisation of risk assessments across scales.

By contrast, the CRV framework supplies concrete assessment tools, impact-chain analysis, vulnerability indices and participatory scoring matrices that can be uniformly applied from ward-level disaster-management plans to national climate-budget allocations. This methodological uniformity is intended to generate comparable data that feed directly into the Climate Change Act's reporting obligations on sectoral emissions targets and carbon budgets.

A key point of divergence lies in data and knowledge governance. The DMA's Enabler 2, outlined in the National Disaster Management Framework 2005 (NDMF), calls for data-sharing protocols but stops short of defining data standards or mechanisms for integrating climate-model outputs with local hazard inventories (Lefutso, 2025; Kunguma, 2022; Nemakonde & Kunguma, 2024). The co-produced framework explicitly calls for integrated data platforms that combine climate projections, historical loss records and socio-economic vulnerability indicators, thereby operationalising the "information management and communication" enabler identification in national policy. This data-centric orientation supports the Climate Change Act's requirement that each province develop a National Adaptation Strategy and that municipalities embed climate-risk considerations into IDPs.

Stakeholder participation also differs. The DMA legally requires consultation with labour, civil society and the private sector, but implementation has been uneven; many municipalities lack dedicated climate-risk units and rely on ad-hoc coordination meetings (Kunguma 2020; Van Niekerk, 2014; Sithole, 2014). The CRV framework institutionalises participatory assessment, encouraging community-level workshops, integration of indigenous knowledge and joint scenario-building exercises that feed directly into the provincial and municipal forums established by the CCA (South Africa, 2024). By embedding participation at every assessment tier, the framework moves beyond the DMA's procedural compliance toward a more transformative governance model aligned with the Sendai Framework's emphasis on "understanding disaster risk" and "strengthening disaster-risk governance" (United Nations Office for Disaster Risk Reduction, 2015).

Finally, resource and capacity implications diverge sharply. The statutory instruments assume that existing institutional capacities, such as NDMCs, provincial disaster-management centres, and the newly formed climate-change forums, will be sufficient. Recent reviews, however, highlight chronic funding shortfalls, limited technical expertise and fragmented inter-governmental coordination (Ngcamu, 2022). The co-produced framework explicitly addresses these constraints by proposing a district-development model that clarifies roles, secures dedicated financing streams for risk-assessment activities, and establishes a national knowledge-exchange hub to support capacity-building at the local level. This model complements the Climate Change Act's call for a "just transition" by linking climate-risk financing to broader socio-economic equity goals.

South African Flood Vulnerability and Resilience Pathways

South Africa's flood exposure has escalated dramatically over the past two decades due to climate, infrastructure, and socio-historical factors. Extreme rainfall driven by tropical temperate interactions, cut-off lows, and La Niña events overwhelms drainage networks and urban storm water systems, creating the climate-driven component of vulnerability (Byaruhanga et al., 2025). Rapid, unplanned urban expansion adds pressure to already inadequate structures, while apartheid era spatial planning has entrenched risk by locating marginalised communities on flood-prone floodplains (Byaruhanga et al., 2025). Governance gaps, fragmented institutional coordination, weak DRR policies, and insufficient early warning systems leave many districts ill-prepared for sudden inundation (Byaruhanga et al., 2025).

Recent catastrophes illustrate the scale of the problem. The April 2022 KwaZulu-Natal floods caused >450 deaths, displaced >40,000 people and generated damages exceeding ZAR 17 billion (Byaruhanga et al., 2025). In the Western Cape, September 2023 floods delivered >220 mm of rain in 48 hours, inundating ~500 ha, displacing >1,000 residents and inflicting severe infrastructure damage (Kibirige, 2025). The eThekwini Metropolitan Municipality has suffered repeated floods in 2019, 2022, and 2024; informal settlements such as Inanda, Umlazi, KwaMashu, and KwaMakhuta are projected to face "extremely high" flood risk across all future scenarios by 2050, affecting an estimated 46.8 km^2 of informal dwellings (Broekman et al., 2025). These events show how climatic intensification, inadequate urban planning and social inequities intersect to heighten flood vulnerability.

Addressing this risk landscape requires multi-scalar, integrated resilience pathways. Ecosystem-based adaptation (EBA) is a cornerstone strategy that promotes floodplain restoration, wetland conservation, and green infrastructure to attenuate runoff and improve water retention (Byaruhanga et al., 2025). Green infrastructure, such as permeable pavements, vegetated swales, and rain gardens, can retrofit dense urban fabrics, reduce surface water accumulation, and deliver co-benefits for biodiversity and air quality (Nyamwanza et al., 2026). Participatory governance models that involve community members, NGOs, and local authorities in flood risk mapping and mitigation planning improve relevance and acceptance, especially when coupled with Indigenous knowledge of seasonal water cycles (Govender, Reddy & Pillay, 2025; Lefutso, 2025).

Strengthening hydrological forecasting through Earth Observation platforms, machine learning algorithms and integrated flood risk information systems (FRIS) has already shown high predictive accuracy; the FRIS deployed for the Western Cape achieved a Nash Sutcliffe Efficiency of 0.88 and flood prediction accuracies up to 91% for machine learning models (Kibirige, 2025). Embedding such technologies within municipal early warning networks can shorten response times and guide targeted evacuations. Land use enforcement is equally critical: stricter zoning that prevents informal settlement expansion into high-risk floodplains, combined with retrofitting programmes that elevate houses and improve drainage, can break the exposure damage feedback loop (Byaruhanga et al., 2025).

Multi-criteria decision analysis (MCDA) integrated with GIS enables planners to evaluate trade-offs across environmental, social, and economic dimensions, producing scenario-based flood risk maps that inform anticipatory governance (Broekman et al., 2025). Applied concurrently to drought and flood hazards in the Olifants Basin, this approach reveals low-lying areas that are flood-prone yet drought-resilient, and high-altitude zones that face drought stress but low flood risk, supporting holistic water resource management (Nhamo et al., 2025).

Beyond physical protections, resilience pathways must address socio-economic ripple effects. Integrating solid waste management, as illustrated in the Quarry Road West informal settlement case, shows that coordinated scientific, practitioner and local knowledge can mitigate both flood hazards and waste-related blockages that exacerbate runoff (Nyamwanza et al., 2026). Mental health impacts are severe; a study of the Setswetla informal settlement found high prevalence of PTSD, vicarious trauma and compassion fatigue among flood-affected residents, linking economic loss and displacement to psychological distress (Raphela, 2025). Post-flood recovery programmes, therefore, need to embed mental health services alongside infrastructure repairs.

Financing these pathways remains a formidable challenge. A recent report estimates that South Africa will require roughly R 250 billion (US $15 billion) over the next decade to meet basic climate adaptation needs, with three-quarters earmarked for water and sanitation systems vulnerable to drought and flood intensification (Taylor, 2025). Leveraging climate finance mechanisms, blended public-private partnerships (PPP) and innovative governance frameworks such as the EPIC Response Framework, which advocates a "whole of society" approach and prioritises inclusive participation, can mobilise the necessary resources while ensuring the most vulnerable groups receive targeted support (Browder et al., 2021).

STUDY LOCALITY AND PERSPECTIVES

The Nelson Mandela Bay Metropolitan Municipality (NMBM) in the Eastern Cape is one of South Africa's eight metros, covering 1,959 km² and home to over 1.3 million people (NMBM Integrated Development Plans [IDP], 2023) (Figure 1). It encompasses the urban centres of Gqeberha (formerly Port Elizabeth) and Kariega (formerly Uitenhage) and serves as the province's primary hub for manufacturing, logistics, and services. Its coastal position has fostered a strong automotive sector (Volkswagen, General Motors, and Mercedes-Benz) that operates major assembly plants and a busy container terminal that moves more than 1 million TEU annually, together accounting for roughly 30% of the Eastern Cape's GDP.

Figure 1. Map of the NMBM (Siyongwana et al., 2015)

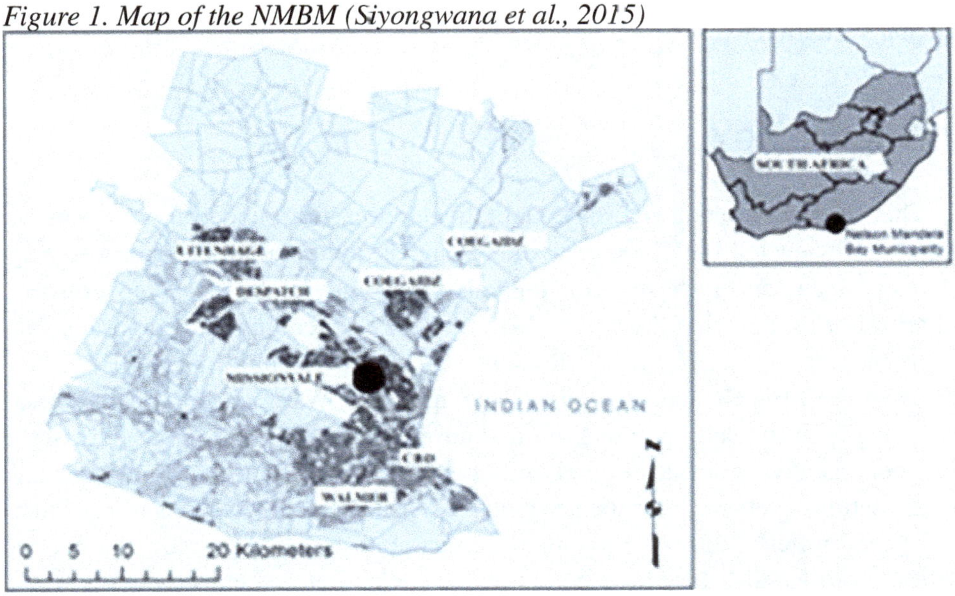

Climate extremes have intensified. The South African Weather Service recorded a 45 % rise in extreme rainfall events between 1990 and 2022. In June 2024, the "Kariega flood" (150 mm of rain in 24 h) inundated arterial roads (M4, R7). It forced a temporary shutdown of the Nelson Mandela Bay Power Station, leaving over 200,000 households without electricity for three days. Heatwaves have also worsened; the 2019 summer saw 12 consecutive days with maximum temperatures above 38 °C, triggering a 20% surge in water demand and straining the municipal water treatment network (Mthambeka, 2022). Droughts, notably the 2015 2017 "Water Crisis," dropped reservoir levels below 30 % capacity and prompted emergency water restrictions that disproportionately affected informal settlements (Mthambeka, 2022).

Historical spatial segregation under the Group Areas Act placed Black communities in peripheral townships (e.g., KwaNobuhle, Kamesh, Zwide, Motherwell), while wealthier, predominantly White suburbs received robust infrastructure (South African History Online, 2014). Many townships still rely on ageing storm water drains built in the 1970s, with capacities under 0.5 m^3 s^{-1}—far below design standards for current rainfall intensities (Siyongwana et al., 2015). A 2023 Municipal Infrastructure Support Agency audit found that 38 % of the municipality's 2,500 km of stormwater pipelines are "critical" and need replacement within 5 years (Municipal Infrastructure Support Agency, 2023). Consequently, limited access to reliable roads, piped water (58% of informal settlements), and electricity (45% experience regular load shedding) forces residents during floods to adopt unsafe coping strategies, such as using contaminated water sources and sheltering on improvised elevated platforms lacking sanitation.

METHODOLOGY

This chapter adopts a rigorous qualitative approach grounded in a systematic literature review (SLR). The review draws on peer-reviewed articles and key grey literature documents retrieved from reputable academic databases and governmental repositories, covering the period 2015–2025. By extending the temporal scope to the most recent decade, the SLR captures the rapid evolution of MHEWS that followed the SFDRR and the United Nations' Early Warnings for *All* initiative. Including both scholarly publications and government working papers addresses a well-documented shortfall in purely bibliographic reviews, which often omit implementation-level insights critical to practice (Alsadi et al., 2025; Ocaña-Fernández & Fuster-Guillén, 2021). To maintain comparability across sources, the review is limited to English-language literature. The selection process conforms to the Preferred Reporting Items for Systematic Reviews and Meta-Analyses (PRISMA) framework (Page et al., 2021) as illustrated in Figure 2. The PRISMA flow diagram

(Figure 2) documents each screening stage, ensuring transparency, reproducibility, and methodological soundness.

Figure 2. PRISMA flowchart (Page et al., 2021)

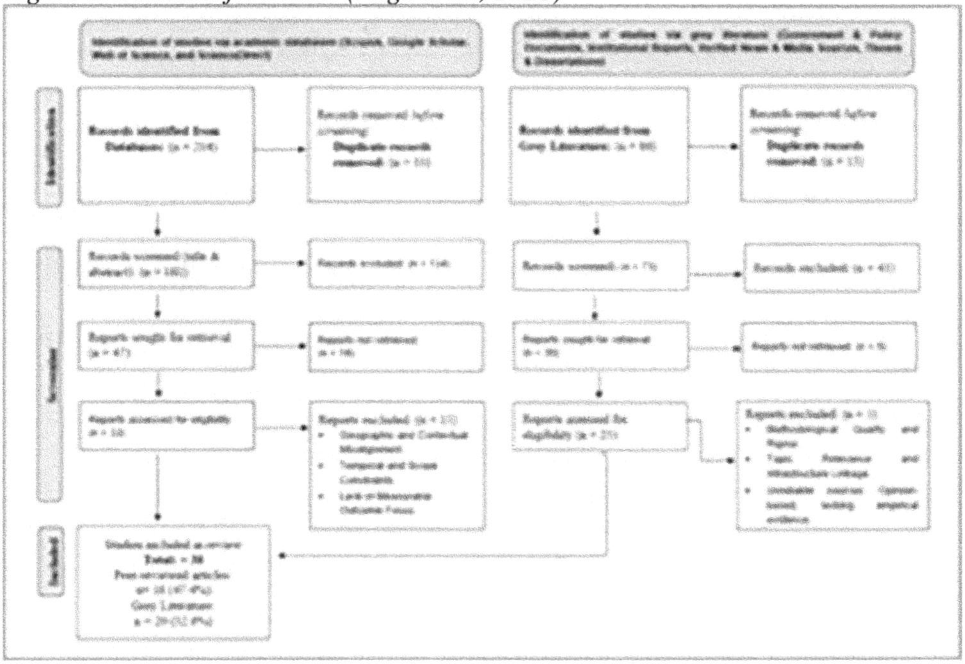

Search Strategy

The databases searched included Google Scholar, EBSCOhost, ProQuest, Scopus, and Web of Science. Using Boolean operators, the search combined hazard-specific keywords: *"early warning systems," "multi-hazard," "disaster risk reduction," "climate-related hazards," "floods," "cyclones," "storms," "damage," "loss," "impact," "consequence" with communication-focused terms "risk communication," "early warning communication," "warning messages," "information dissemination," and "end-user communication."*

Table 1. Inclusion criteria

#	Criterion	Rationale
1	**Geographic focus:** Studies, reports or policy documents that explicitly address South Africa (national, provincial or municipal level).	The review targets climate-resilient infrastructure in the South African context.
2	**Topic relevance:** Works that examine MHEWS and their link to the design, planning, operation or performance of physical infrastructure.	The title emphasises a review of MHEWS for resilient infrastructure.
3	**Temporal scope:** Publications released between 2015 and 2025 (inclusive).	This period captures the evolution of the Sendai Framework, revised DMA, the promulgation of the South Africa's Climate Change Act.
4	**Document type:** Peer-reviewed journal articles, conference proceedings with full papers, government or agency reports, theses, and reputable grey literature (e.g., World Bank, UN-OPS, DPSA) that provide empirical data, case studies, systematic reviews or policy analyses.	Aligns with standard systematic-review practice and ensures methodological rigour.
5	**Language:** Full-text available in English.	Facilitates consistent appraisal and synthesis.
6	**Methodological quality:** Studies that present a clear research design (qualitative, quantitative or mixed methods), include a description of data collection/analysis, and provide sufficient detail to assess the effectiveness or impact of the MHEWS on infrastructure resilience.	Guarantees that extracted findings are evidence-based.
7	**Outcome focus:** Papers reporting measurable outcomes (e.g., reduction in flood depth, decreased service interruption, cost-benefit estimates, improved preparedness indices) or substantive discussion of **institutional, technological or governance mechanisms** that influence the performance of multi-hazard EWS.	Directly addresses the "climate-resilient infrastructure" objective.

Table 2. Exclusion criteria

#	Criterion	Rationale
1	**Geographic focus:** Studies, reports or policy documents that explicitly address South Africa (national, provincial or municipal level).	The review targets climate-resilient infrastructure in the South African context.
2	**Topic relevance:** Works that examine MHEWS and their link to the design, planning, operation or performance of physical infrastructure.	The title emphasises a review of MHEWS for resilient infrastructure.
3	**Temporal scope:** Publications released between 2015 and 2025 (inclusive).	This period captures the evolution of the Sendai Framework, revised DMA, the promulgation of the South Africa's Climate Change Act.
4	**Document type:** Peer-reviewed journal articles, conference proceedings with full papers, government or agency reports, theses, and reputable grey literature (e.g., World Bank, UN-OPS, DPSA) that provide empirical data, case studies, systematic reviews or policy analyses.	Aligns with standard systematic-review practice and ensures methodological rigour.
5	**Language:** Full-text available in English.	Facilitates consistent appraisal and synthesis.
6	**Methodological quality:** Studies that present a clear research design (qualitative, quantitative or mixed methods), include a description of data collection/analysis, and provide sufficient detail to assess the effectiveness or impact of the MHEWS on infrastructure resilience.	Guarantees that extracted findings are evidence-based.
7	**Outcome focus:** Papers reporting measurable outcomes (e.g., reduction in flood depth, decreased service interruption, cost-benefit estimates, improved preparedness indices) or substantive discussion of **institutional, technological or governance mechanisms** that influence the performance of multi-hazard EWS.	Directly addresses the "climate-resilient infrastructure" objective.

Thematic Synthesis

Four overarching themes emerged from the literature, further guided by PRISMA, and are presented in Table 3. Sub-themes are illustrated with concrete case examples, and, where available, quantitative indicators were provided:

Table 3. Analytical themes

Theme	Description	Consequence for MHEWS
Weak integration of MHEWS into municipal development plans	Limited embedding of warning functions within local planning processes.	Reduced system robustness (infrastructure cannot absorb shocks).
Fragmented coordination among agencies	Disjointed institutional arrangements and unclear roles.	Diminished resourcefulness (difficulty mobilising alternate resources).
Sociolinguistic barriers	Language, cultural norms, and literacy gaps impede message uptake.	Lower rapidness (slower community response).
Marginalisation of local knowledge	Insufficient incorporation of indigenous and community-generated hazard insights.	Weakened adaptability (limited learning from past events).

FINDINGS AND DISCUSSIONS

Slow Integration Architecture of MHEWS in IDP

Although legislative provisions such as the DMA, the CCA, and the NDMF exist, MHEWS remain poorly integrated into municipal development planning. This weak integration limits a municipality's capacity to protect critical infrastructure. The Climate Change Act 22 of 2024 widens the gap by mandating that climate risk, adaptation, and early warning be embedded in municipal governance rather than treated as emergency add-ons (Agbehadji et al., 2024; Maripe et al., 2022; Mugari & Nethengwe, 2022; Republic of South Africa, 2024; Terblanche et al., 2022). The problem, therefore, is not a lack of legal mandate but a failure of institutional translation: risk knowledge is not shaping planning instruments, funding allocations, land use decisions, or asset management strategies before hazards occur (Bondank & Chester, 2020; Jose Moises, Kgabi & Kunguma, 2024; Ngcamu & Abrahams, 2024).

In terms of resilience, the weak integration of MHEWS into IDPs undermines robustness (the ability to resist shocks), adaptive capacity (the ability to adjust), and transformability (the ability to redesign risk-producing systems). Infrastructure continues to be planned without systematic consideration of climate exposure and cascading failure risks (Terblanche et al., 2022; Ziwanai, 2024; Rokhideh et al., 2025). National data illustrate the scale of the issue: only 42% of 257 municipalities have revised their IDPs to incorporate recent climate risk assessments, and only 28% have earmarked funds for climate-resilient infrastructure (Mugari & Nethengwe, 2022). Municipalities that do integrate risk data, such as Cape Town, which allocated 12% of its 2021/22 capital budget to flood resilience upgrades, have reported a 35%

reduction in flood-related service disruptions compared with peers (Maripe et al., 2022; Mugari & Nethengwe, 2022).

These gaps reflect structural governance asymmetries: municipalities are configured to repair damage rather than prevent it. Reactive fiscal norms, political incentives favouring visible reconstruction, and limited local evidence on the returns to anticipatory investment reinforce this pattern (Maripe et al., 2022; Mudefi, 2023). In NMBM, MHEWS are weakly linked to infrastructure maintenance, capital allocation, and spatial planning. Consequently, early warnings rarely trigger pre-event actions such as drainage clearing, road closures, or the deployment of protective assets (Mokhele & Mvanyashe, 2025). Governance remains responsive instead of anticipatory, indicating a breakdown in the adaptive loop that should connect monitoring, preparedness, investment, and recovery (Agbehadji et al., 2024; Holling, 1973).

Repeated infrastructure losses in the same location reveal path dependence: hazards expose unaddressed structural vulnerabilities. The economic stakes are substantial. A 2022 Buffalo City pilot showed that bioswales along a 2 km stretch of the river reduced peak flows by 27% and avoided R3.4 million in annual flood damage (Ngcamu, 2022; Ngcamu & Abrahams, 2024). Yet NMBM allocates less than 5% of its capital expenditure to such measures, despite its exposure to sea-level rise and extreme precipitation (Thebe, 2016). By contrast, the April 2022 KwaZulu-Natal floods caused damage exceeding R17 billion (Mudefi, 2023). Although not a formal cost-benefit analysis, the disparity underscores the macroeconomic inefficiency of repeated reactive spending: reconstruction costs consistently outstrip preventive investment.

Outdated planning frameworks exacerbate the problem. NMBM's disaster management plan, adopted in December 2010, predates current climate projections, modern infrastructure interdependencies, and recent urban expansion patterns (Agbehadji et al., 2024; Bondank & Chester, 2020). This stands in stark contrast to the amended DMA and the Climate Change Act's emphasis on iterative risk assessment and adaptive governance (Republic of South Africa, 2024; Republic of South Africa, 2015). Fragmented risk matrices and weak interdepartmental coordination further hinder the translation of warnings into operational actions, such as pump activation or the deployment of temporary barriers (Botes, 2021; Jayasekara et al., 2022).

Technological limitations also reflect governance deficits. Contemporary MHEWS combines satellite rainfall data, Internet of Things (IoT) sensors, AI-driven flood modelling, and cell broadcast alerts; however, these tools deliver value only when embedded in routine decision making. Without institutional integration, they generate information without prompting action (Agbehadji et al., 2024; Mukange et al., 2024; Mulaudzi et al., 2025). International examples, such as Bangladesh's volunteer-based Cyclone Preparedness Programme, Cuba's state-coordinated evacuations, and Japan's automated J-Alert system, demonstrate that effectiveness stems from coupling technology with clear planning, training, and defined responsibilities (Kitazawa & Hale, 2021; Paul, 2009; Poole et al., 2022).

Weak IDP integration also fuels climate injustice. Apartheid-era spatial planning placed low-income Black communities in environmentally exposed areas with inferior infrastructure. Failure to embed risk reduction in development planning perpetuates historic inequities in exposure, protection, and recovery (Rodina et al., 2024). Thus, planning deficits are not neutral administrative oversights; they are mechanisms that reproduce social vulnerability.

Fragmented Institutional Coordination in Addressing MHEWS

Institutional coordination determines whether warnings are converted into protective actions across forecasting, evacuation, infrastructure management and recovery systems. Although the DMA creates multi-sphere structures, the operational coordination during rapidly developing hazards is unreliable (Republic of South Africa, 2015). Fragmentation slows responses, results in unequal assistance, and extends recovery (Jayasekara et al., 2022; Okunola, 2025; Olanrewaju & Reddy, 2025). In terms of resilience, this undermines both absorptive capacity during the shock and recovery capacity afterwards, thereby diminishing resourcefulness.

Coordination often remains fragmented because while there are formal structures, they lack long-term operational practices, data interoperability, or role clarification (Sithole, 2014). Ad hoc arrangements collapse after crises, thereby preventing institutional learning (Kunguma, 2022). The CCA implicitly demands cross-sector coordination across adaptation, infrastructure, and emergency management; thus, the gap becomes increasingly untenable.

A case example illustrating these institutional dynamics is the June 2024 Kariega floods. Heavy rainfall exceeding 150 mm in 24 hours led to cascading failures in transportation, energy and telecommunications (Mbovane, 2024; Mukange et al., 2024). Numerous actors responded to the crisis, but the lack of a single framework led to duplication and delays. Three non-governmental organisations (NGOs) constructed shelters in one area, while another organisation waited more than 48 hours for assistance; more than 55 rescue missions were conducted without a designated authority to process compensation claims (Gcabashe, 2025).

Repair of a bridge took 90 days instead of the anticipated 30 days due to bureaucratic hurdles of obtaining approvals and documentation (Gcabashe, 2025). Such delays illustrate how the recovery of infrastructure, including transportation, utilities, and housing, can occur unequally across sectors, hindering the mobilisation of alternative resources. Comparatively, the April 2022 KwaZulu-Natal floods exhibited similar institutional dynamics: warning systems existed, but the governance capacity to turn warnings into coordinated action was absent. Rescue activities were timely, but the recovery of infrastructure was delayed for months or years, illustrating the disparities in disaster phases (Ngcamu & Abrahams, 2024).

As Table 4 shows, mortality, displacement, and economic losses occurred despite functioning warning systems, but the failure of coordination amplified the disaster's cascading effects. The duration of recovery varies across types of infrastructure; transportation and utility infrastructure tend to recover more slowly than emergency shelter facilities (Mudefi, 2023).

Similar international experiences indicate the importance of aligning institutional structures. For example, Japan's J-Alert system provides rapid dissemination of warnings and directly links to predefined local response procedures. Bangladesh and Cuba utilise practised mobilisation structures and defined authority chains to respond to emergencies (Kitazawa & Hale, 2021; Paul, 2009; Poole et al., 2022). In South Africa, institutional failures in coordination also have differential effects on communities with varying resource bases: poorer communities recover more slowly and receive less assistance (Rokhideh et al., 2025; Republic of South Africa, 2015; United Nations Office for Disaster Risk Reduction, 2015). The gender and disability aspects of evacuation logistics, caregiver responsibilities, and accessibility need to be specifically considered in planning, as opposed to being included in general response models.

Table 4. Comparative synthesis of the April 2022 Kwazulu-Natal floods and the June 2024 Nelson Mandela Bay floods

Dimension	KwaZulu-Natal floods (8–12 April 2022)	Nelson Mandela Bay floods (1–5 June 2024)
Event dates	8–12 April 2022 (Government Communication and Information System (GCIS), 2022a; International Federation of Red Cross and Red Crescent Societies (IFRC), 2022a;)	1–5 June 2024 (IFRC, 2024; News24, 2024)
Hazard type(s)	Extreme rainfall, flooding, landslides, and infrastructure disruption (IFRC, 2022b; Mudefi, 2023)	Severe rainfall/flash flooding with infrastructure disruption (IFRC, 2024; ReliefWeb, 2024)
Warning channel(s)	Multiple official channels are active, but uneven protective action (GCIS, 2022b; Mukange et al., 2024)	Municipal and media channels are active but have uneven last-mile effectiveness (Mukange et al., 2024; News24, 2024)
Warning–action gap evidence	High mortality and displacement indicate uneven uptake (Mudefi, 2023)	Post-impact relocations indicate a persistent gap (IFRC, 2024; News24, 2024)
Key infrastructure failures	Extensive damage to roads, bridges, ports, electricity and water systems (Mudefi, 2023)	Bridge collapse, road damage, electricity and water interruptions (IFRC, 2024; ReliefWeb, 2024)
Governance/coordination issues	Fragmented implementation widely criticised (Mudefi, 2023)	Fragmented response and repair bottlenecks reported (GroundUp, 2025; News24, 2024)
Recovery timelines	Reconstruction extended well beyond the emergency phase (IFRC, 2025)	Key bridge repairs projected at 9–12 months (GroundUp, 2025)
Headline impacts	435 deaths, >40,000 displaced, damage > R17 billion (GCIS, 2022b; IFRC, 2022a; Mudefi, 2023)	≥6 deaths initially, >1,000 households displaced, major infrastructure damage (Department of Human Settlements (DHS), 2024; News24, 2024a; ReliefWeb, 2024)

Sociolinguistic Barriers To Inclusive Community Engagement

Sociolinguistic barriers are not peripheral glitches; they are decisive determinants of whether MHEWS can function at the community level in South Africa. The effectiveness of an MHEWS hinges on more than the technical transmission of a warning; it depends critically on how that warning is interpreted and acted upon within the linguistic, cultural, and communicative realities of the at-risk population (Mukange et al., 2024). A persistent mismatch exists between the design of MHEWS and the country's multilingual, socially differentiated context (Federici & O'Brien, 2020).

Consequently, the primary obstacle is not dissemination but interpretive accessibility: technically accurate alerts lose practical value when their language register, framing, or terminology fails to convey urgency in ways that communities recognise (Mokhele & Mvanyashe, 2025). This interpretive gap erodes community resilience by reducing absorptive, rapid, and anticipatory capacities; people cannot take swift protective action when warnings are incomprehensible or mistrusted (Terblanche et al., 2022; Ziwanai, 2024).

South African linguistic politics exacerbate the problem. Although the Constitution recognises eleven official languages, formal disaster communication remains dominated by English and Afrikaans, perpetuating historic patterns of linguistic exclusion in disaster governance. Language choice, therefore, determines who understands a warning immediately, who must rely on an intermediary for interpretation, and who loses valuable response time (Raphela & Ekeke, 2025). The 2023 Eastern Cape flooding illustrates this: alerts used technical phrasing such as "hydrological surge" and "inundation risk", which were interpreted as abstract rather than immediate danger cues, widening the warning action gap (Mokhele & Mvanyashe, 2025). The Kariega flood further underscores the barrier's sociotechnical nature. Authorities employed SMS, radio, WhatsApp, and social media, yet transmission rarely translated into comprehension or timely action (Mukange et al., 2024). Three interrelated obstacles converged: (1) prolonged network outages that disabled cell tower functionality for more than six hours; (2) power cuts that prevented phone charging and radio listening; and (3) linguistic inaccessibility, because messages were drafted in formal English or Afrikaans with unfamiliar technical vocabulary (Raphela & Ekeke, 2025).

A post-flood survey of 342 households in KwaNobuhle revealed that 68% of respondents struggled to understand the core warning without assistance from a more literate neighbour (HeraldLIVE, 2024). The failure of last-mile warnings, therefore, stems from the convergence of linguistic inaccessibility, unreliable infrastructure, and digital inequality (ICASA, 2024; Mukange et al., 2024; Mulaudzi et al., 2025; Rokhideh et al., 2025; United Nations Office for Disaster Risk Reduction, 2015). When official channels falter, communities revert to informal networks, neighbours, local leaders, and word of mouth. In Kariega, word of mouth introduced delays of 30 minutes to 2 hours, extending the average lag to evacuation centres to about 1.8 hours compared with faster official channels (HeraldLIVE, 2024). These delays were associated with a 12% increase in flood-related injuries relative to the 2022 baseline, underscoring the lifesaving importance of rapid comprehension (Mudefi, 2023).

A similar pattern emerged during the April 2022 KwaZulu-Natal floods: warnings were issued, yet response was uneven, especially in marginalised communities where communication networks failed, and messages were not presented in locally understandable formats (Ebhuoma et al., 2024). A comparative analysis of 19 flood events (2018-2023) showed that communities receiving warnings in their primary language achieved higher evacuation compliance than those receiving alerts only in English (Raphela, 2025). These findings confirm that generation and dissemination are distinct processes; sophisticated technology alone cannot guarantee effectiveness if the dissemination format does not align with the language, literacy, and communication practices of the target population (Mukange et al., 2024; United Nations Office for Disaster Risk Reduction, 2015).

International examples illustrate contrasting approaches. Japan's standardised warning architecture, paired with multiple rapid-dissemination channels, demonstrates how technical robustness can be coupled with effective communication under tight time constraints (Kitazawa & Hale, 2021). By contrast, Bangladesh and Cuba highlight the importance of community-oriented dissemination and strong last-mile communication in settings with uneven literacy, language hierarchies, and fragile infrastructure (Aguirre, 2005; Paul, 2009).

Beyond household safety, sociolinguistic barriers impede broader resilience actions: delayed uptake of warnings reduces the likelihood of timely road closures, evacuations, storm water protection, asset safeguarding, and coordinated local preparedness (Federici & O'Brien, 2020; Ngcamu & Abrahams, 2024; Rokhideh et al., 2025; United Nations Office for Disaster Risk Reduction, 2015). In the NMBM, delayed comprehension prolonged exposure of homes, roads, and storm water systems to floodwaters, especially in low-income areas such as KwaNobuhle and Jubilee Park, hampering access to information on evacuation centres, relief distribution, and repair timelines (HeraldLIVE, 2024).

Marginalisation of at-Risk Communities in Disaster-Risk Decision-Making

The findings reveal that the marginalisation of at-risk communities constitutes a structural vulnerability in South Africa's MHEWS. These communities are highly exposed to climate-related hazards yet are insufficiently involved in defining, communicating, and governing risk, a problem best addressed through a climate-justice framework. Their repeated disaster losses are rooted in the legacy of apartheid-era spatial planning, weak service provision, under-maintained infrastructure, and limited political influence over development priorities (Rodina et al., 2024). Consequently, vulnerability extends beyond physical exposure; it is historically created and institutionally reproduced via unequal settlement geographies, poor drainage, insecure housing, inadequate transport links, and limited communication connectivity (Rodina et al., 2024; Republic of South Africa, 2015).

Although the Constitution and the DMA provide a formal structure for inclusive governance and community-based risk management, implementation varies widely across municipalities. Participation is often weak, and local knowledge is not integrated into warning systems and planning practice (Ebhuoma et al., 2024; Hermans et al., 2022; Nkombi & Wentink, 2022; Republic of South Africa, 2015). The 2015 DMA amendments, which strengthen representation of women, children, older persons, and persons with disabilities and require identification of vulnerable communities, are rarely translated into practice (Lunga et al., 2025; Botes, 2021; South Africa, 2015). Community observations and lived experience are treated as supplementary rather than constitutive risk knowledge (Hermans et al., 2022; Mokhele & Mvanyashe, 2025)

The Climate Change Act 22 of 2024 further underscores the issue: municipal climate response planning cannot be trustworthy if the most vulnerable communities remain excluded from decision-making (Republic of South Africa, 2024). Exclusion appears both in planning and response. Low-income, risk-prone areas are often omitted from municipal risk mapping, infrastructure planning, and warning design, even though they are most likely to suffer from drainage failures, unsafe roads, power outages, and insecure housing (Murshed et al., 2023). Such exclusions distort exposure assessments and hinder the development of warnings that reflect how people live and move within hazardous environments, turning repeated infrastructure failures into outcomes of spatial injustice rather than mere natural hazard exposure (Rodina et al., 2024).

Indigenous and community-based knowledge is also undervalued. Many residents possess detailed experiential insights; e.g., isiXhosa-speaking communities associate the "smell of the earth after a thunderstorm" with flash-flood risk. Still, this knowledge is rarely integrated into formal MHEWS (Motsumi & Nemakonde, 2025). Ignoring such indicators reinforces epistemic exclusion, expecting communities to respond to warnings that do not align with their ways of understanding risk (Motsumi & Nemakonde, 2025; Nkombi & Wentink, 2022). Effective examples from Bangladesh and Cuba show that treating local mobilisation as integral to warning governance, rather than as supplemental, yields better outcomes (Paul, 2009; Poole et al., 2022).

Role clarity and practical guidance are often absent. In NMBM's Walmer Airport Valley, 68% of respondents did not understand the colour-coded warning maps, and 54% had never received follow-up advice on protecting their homes (Mokhele & Mvanyashe, 2025). Technical or generic messages leave communities aware of risk but unsure how to act, where to report hazards, or how to link observations to municipal response (Mokhele & Mvanyashe, 2025). This limits adaptive capacity, treating residents as passive recipients rather than active preparedness agents (Terblanche et al., 2022; Ziwanai, 2024).

Marginalisation is also gendered and intersectional. Black women in flood-prone areas face layered inequalities in housing, mobility, unpaid care work, and information access, while persons with disabilities encounter additional barriers in warning reception and evacuation (Republic of South Africa, 2015; UNDRR, 2015; Rokhideh et al., 2025). Community-centred MHEWS must therefore move beyond generic participation language and intentionally include those most likely to be excluded by standard planning and communication practices.

In NMBM, the 2023 IDP allocated less than 2% of its capital budget to "hazard responsive infrastructure" (IDP, 2023), far below the 10% target set by the National Climate Change Adaptation Strategy (2019). Residents reported clogged culverts and unstable roads issues that exacerbate flooding, but these observations were not incorporated into maintenance schedules (Mokhele & Mvanyashe, 2025). Exclusion thus determines which risks are prioritised, which infrastructure is maintained, and whose losses become normalised.

Overall, marginalising at-risk communities reproduces historical inequities and undermines the effectiveness of MHEWS. Genuine inclusion of local knowledge, clear community roles, and equitable resource allocation are not optional add-ons but essential conditions for socially meaningful resilience.

RECOMMENDATIONS FOR STRENGTHENING MHEWS AND CLIMATE-RESILIENT INFRASTRUCTURE

Building upon the results, strengthening South African MHEWS requires comprehensive institutional, technical and community-focused actions. Table 5 below provides recommendations:

Table 5. Policy and practice recommendations

Recommendation	Concrete actions & responsible actors	Supporting evidence from findings
Integrate early-warning data into IDPs & budgeting	• Mandate that every municipal capital-budget cycle include a "climate-risk line item" (≥ 10% of capital spend). • Link SAWS real-time dashboards to GIS layers used in IDP spatial planning.	42% of municipalities have updated IDPs with climate risk; only 28% have dedicated funds.
Standardise inter-governmental coordination	• Adopt a three-tier protocol (national → provincial → municipal) with fixed trigger times (e.g., Level-2 alert within 30 min). • Create a joint MHEWS steering committee chaired by the NDMC.	Protocol design mirrors the "WEATHER" project's coordination framework.
Deploy multilingual, accessible alert platforms	• Develop SMS/USSD templates in isiXhosa, isiZulu, and other local languages. • Partner with NGOs to produce voice-recorded warnings for low-literacy households.	68% of households needed assistance interpreting warnings; multilingual alerts raise evacuation compliance.

Recommendation	Concrete actions & responsible actors	Supporting evidence from findings
Institutionalise community-based early-warning committees	• Form "Flood-Watch" committees in each informal settlement, with elected local leaders. • Provide them with low-cost rain-gauges and training on "ground-truth" reporting.	Community-knowledge integration improves forecasting accuracy in the "Integrated Knowledge Systems" study.
Secure dedicated climate-resilience financing	• Create a ring-fenced "MHEWS Maintenance Fund" financed by a modest municipal levy (e.g., 5% of property rates). • Leverage Green Climate Fund co-financing for infrastructure retrofits.	Funding gaps identified in municipal audit reports; climate-finance mechanisms highlighted in the national risk-assessment framework.
Adopt a climate-justice impact assessment	• Conduct equity-weighted vulnerability mapping (weighting informal settlements higher). • Prioritise retrofitting of assets serving historically marginalised communities (e.g., KwaNobuhle storm-water upgrades).	Residual risk review stresses equity-focused interventions; flood vulnerability analysis shows disproportionate exposure.

LIMITATIONS AND FUTURE RESEARCH

This systematic review has several constraints that may limit its breadth and applicability, largely stemming from methodological choices and analytical approaches. Future research should address these gaps:

- *Potential biases*: The review lacks a formal assessment of selection, publication, or language biases. Keyword-only searches without sensitivity analysis may miss relevant studies (e.g., "early warning" vs. "alert system"). Use bias assessment tools, such as ROBIS, to quantify and reduce these effects.
- *Temporal scope*: The search was limited to 2015 - 2025, omitting foundational work from the early 2000s that shaped hazard modelling and warning communication. Incorporating earlier studies can provide richer historical context.
- *Comparative impact analyses*: Few studies have performed cross-hazard or cross-regional comparisons of infrastructure damage. Standardised damage metrics would enable robust comparative research and reveal common patterns of vulnerability.

CONCLUSION

The evidence presented confirms that South Africa's MHWES are at a critical juncture: technical capacity exists, yet institutional, financial, and sociolinguistic gaps prevent warnings from translating into effective protection of climate-resilient infrastructure. Repeated disasters such as the 2022 KwaZulu-Natal floods and the 2024 Gqeberha deluge demonstrate a feedback loop in which damaged communication networks curtail early alerts, delaying evacuations and amplifying the loss of lives, homes and essential services. South Africa's DMA and the CCA provide a robust legal scaffold. Still, implementation remains uneven: only 42% of municipalities have updated IDPs with recent climate risk assessments, and only 28% have earmarked funds for climate-resilient assets. Consequently, municipalities continue to divert scarce budgets from preventive maintenance to costly post-disaster repairs. This pattern erodes fiscal sustainability and deepens historic inequities rooted in apartheid era spatial planning.

Addressing these shortcomings requires a coordinated, equity-focused strategy that aligns the four resilience capacities, such as robustness, redundancy, resourcefulness and rapidity, with concrete policy actions. Embedding real-time SAWS data directly into municipal budgeting cycles, establishing a ring-fenced MHEWS Maintenance Fund, and mandating multilingual, low-literacy alert templates will strengthen robustness and speed while expanding redundancy through diversified dissemination channels. Institutional reforms such as a three-spheres of government coordination protocol and community-based Flood Watch committees will enhance resourcefulness and ensure that local knowledge informs risk modelling and response planning. Finally, a climate justice impact assessment that weights vulnerable informal settlements more heavily in vulnerability mapping will direct retrofitting and green infrastructure investments to the areas most exposed, breaking the cycle of repeated infrastructure failure.

REFERENCES

Agbehadji, I. E., Schütte, S., Masinde, M., Botai, J., & Mabhaudhi, T. (2024). Climate risks resilience development: A bibliometric analysis of climate-related early warning systems in Southern Africa. *Climate (Basel), 12*(1), 3. DOI: 10.3390/cli12010003

Aguirre, B. E. (2005). Cuba's disaster management model: Should it be emulated? *International Journal of Mass Emergencies and Disasters, 23*(3), 55–71. DOI: 10.1177/028072700502300303

Agyemang, F. (2025). Legal and policy perspectives on public participation and disaster risk management in South African cities. *Journal of Local Government Research and Innovation, 6*(0), a259. DOI: 10.4102/jolgri.v6i0.259

Alsadi, J., Alkhatib, F., Antony, J., Garza-Reyes, J. A., Tortorella, G., & Cudney, E. A. (2025). A systematic literature review with bibliometric analysis of Quality 4.0. *The TQM Journal, 37*(5), 1446–1470. DOI: 10.1108/TQM-02-2024-0050

Auditor-General South Africa (AGSA). (2024). *Consolidated general report on local government audit outcomes 2023–24*. Pretoria, South Africa: AGSA. https://www.agsa.co.za/Portals/0/Reports/MFMA/2023-24/MFMA%20Report%202023-24%20interactive%20%28updated%20August%202025%29.pdf

Basher, R. (2006). Global early warning systems for natural hazards: systematic and people-centred. *Philosophical transactions of the royal society a: mathematical, physical and engineering sciences, 364*(1845), 2167-2182. DOI: 10.1098/rsta.2006.1819

Bondank, E. N., & Chester, M. V. (2020). Infrastructure Interdependency Failures from Extreme Weather Events as a Complex Process. *Frontiers in Water, 2*, 21. Advance online publication. DOI: 10.3389/frwa.2020.00021

Botes, M. (2021). The use of geospatial surveillance data for public-health emergency and disaster management in South Africa: A review with legal recommendations. *Tydskrif vir die Suid-Afrikaanse Reg, 3*(3), 474–503. DOI: 10.47348/TSAR/2021/i3a4

Broekman, C., Coetzee, C., & Cilliers, D. (2025). Mapping pluralistic future flood risk scenarios for informal settlements in the eThekwini metropolitan municipality, South Africa. *Natural Hazards, 121*(18), 20949–20983. DOI: 10.1007/s11069-025-07581-w

Browder, G., Sanchez, A. N., Jongman, B., Engle, N., Van Beek, E., Errea, M. C., & Hodgson, S. (2021). *An EPIC Response: Innovative Governance for Flood and Drought Risk Management—Executive Summary*. World Bank. DOI: 10.1596/35754

Byaruhanga, N., Kibirige, D., & Mkhonta, G. (2025). Understanding South Africa's Flood Vulnerabilities and Resilience Pathways: A Comprehensive Overview. *Water (Basel)*, *17*(17), 2608. DOI: 10.3390/w17172608

Chipfupa, U., Tagwi, A., & Wale, E. (2021). Psychological capital and climate change adaptation: Empirical evidence from smallholder farmers in South Africa. *Jàambá*, *13*(1), 1–12. DOI: 10.4102/jamba.v13i1.1061 PMID: 34230848

Department of Environment, Forestry and Fisheries. (2020). National Climate Risk and Vulnerability (CRV) Assessment Framework summary document. Pretoria, South Africa. DEFF.

Ebhuoma, E. E., Nene, N. J., & Leonard, L. (2024). Analysis of Urban Households' Preparedness and Municipal Interventions to Build Flood Resilience in Durban, South Africa: Implications for SDG 11. *Environmental and Sustainability Indicators*, *23*, 100454. DOI: 10.1016/j.indic.2024.100454

Eskom. (2023) Understanding load shedding. [Online] Available at: https://www.eskom.co.za/loadshedding

Federici, F. M., & O'Brien, S. (Eds.). (2020). *Translation in Cascading Crises* (p. 246). Routledge.

Gcabashe, S. W. (2025). *An Evaluation of Disaster Risk Management in Local Government: The Case of the Nelson Mandela Bay Municipality*. Master's Dissertation. Gqeberha, South Africa: Nelson Mandela University.

Govender, I. H., Reddy, M., & Pillay, R. P. (2025). A review of residual flood risks in South African-vulnerable coastal communities: Opportunities to influence policy. In *Climate Policies* (p. 143). IntechOpen., DOI: 10.5772/intechopen.1008977

Haque, U., Hashizume, M., Kolivras, K. N., Overgaard, H. J., Das, B., & Yamamoto, T. (2012). Reduced death rates from cyclones in Bangladesh: What more needs to be done? *Bulletin of the World Health Organization*, *90*(2), 150–156. DOI: 10.2471/BLT.11.088302 PMID: 22423166

Herald, L. I. V. E. (2024). *Heavy rains wreak havoc in Nelson Mandela Bay, causing death and destruction*. HeraldLIVE. https://www.theherald.co.za/news/2024-06-03-heavy-rains-wreak-havoc-in-nelson-mandela-bay-causing-death-and-destruction/

Hermans, T. D., Šakić Trogrlić, R., van den Homberg, M. J., Bailon, H., Sarku, R., & Mosurska, A. (2022). Exploring the integration of local and scientific knowledge in early warning systems for disaster risk reduction: A review. *Natural Hazards*, *114*(2), 1125–1152. DOI: 10.1007/s11069-022-05468-8

Holling, C. S. (1973). Resilience and Stability of Ecological Systems. *Annual Review of Ecology and Systematics*, *4*(1), 1–23. DOI: 10.1146/annurev.es.04.110173.000245

Independent Communications Authority of South Africa (ICASA). (2024). *ICASA conducts public hearings on the possible effects of load shedding on the electronic communications, broadcasting, and postal sectors*. https://www.icasa.org.za/news/2024/icasa-to-conducts-public-hearings-on-possible-effects-of-loadshedding-on-the-electronic-communications-broadcasting-and-postal-sectors

Jayasekara, R., Siriwardana, C., Amaratunga, D., & Haigh, R. (2022). Evaluating the network of stakeholders in Multi-Hazard Early Warning Systems for multiple hazards amidst biological outbreaks: Sri Lanka as a case in point. *Progress in Disaster Science*, *14*, 100228. DOI: 10.1016/j.pdisas.2022.100228 PMID: 35402893

Jose Moises, D. R., Kgabi, N. A., & Kunguma, O. (2024). Policy implementation: Assessing institutional coordination and communication for flood warning in Namibia. *Jàambá*, *16*(1), a1534. DOI: 10.4102/JAMBA.v16i1.1534 PMID: 39113930

Kibirige, D. (2025). *Integrated Flood Risk System for the Western Cape: Lessons from the September 2023 Floods*, EGU General Assembly 2025, Vienna, Austria, 27 April–2 May 2025, EGU25-9855. DOI: 10.5194/egusphere-egu25-9855

Kitazawa, K., & Hale, S. A. (2021). Social media and early warning systems for natural disasters: A case study of Typhoon Etau in Japan. *International Journal of Disaster Risk Reduction*, *52*, 101926. DOI: 10.1016/j.ijdrr.2020.101926

Kunguma, O. (2020). *South African Disaster Management Framework: Assessing the Status and Dynamics of Establishing Information Management and Communication Systems in Provinces*. Doctoral dissertation. Bloemfontein, South Africa: University of the Free State. http://hdl.handle.net/11660/11002

Kunguma, O. (2022). A South African disaster legislative perspective of information management and communication systems. *South African Journal of Information Management*, *24*(1), a1540. DOI: 10.4102/sajim.v24i1.1540

Lefutso, D. (2025). Transformative policy and legislative strategies for flood disaster management in post-apartheid South Africa. *Journal of Disaster Science and Management*, *1*(1), 15. DOI: 10.1007/s44367-025-00015-1

Lumbroso, D., Vincent, K., Murambadoro, M., Steynor, A., Tsarouchi, G., & Nezi, M. (2024). Current uses and potential future needs for climate services in South Africa. *Climate Services*, *36*, 100516. DOI: 10.1016/j.cliser.2024.100516

Lunga, W., Kaifa, J., Musarurwa, C., Malandela, G., Tshabalala, S., Baloyi, C., & Magampa, M. (2025). Preventing Disasters Before They Happen: Lessons from Successful Disaster Risk Reduction in Southern Africa. *Sustainability (Basel), 17*(20), 9131. DOI: 10.3390/su17209131

Lynch, M., Harris, F., Ierna, M., Mahomed, O., Henriquez-Mui, F., Gebreslasie, M., Ndzi, D., Viriri, S., Shakir, M. Z., Dickinson, N., Miller, C., Hursthouse, A., Nadesan-Reddy, N., Nkwanyana, F., Spencer, L. H., & Naidoo, S. (2025). Warning system for extreme weather events, awareness technology for healthcare, equitable delivery, and resilience (WEATHER) project: A mixed-methods research study protocol. *Climate (Basel), 13*(8), 170. Advance online publication. DOI: 10.3390/cli13080170

Maripe, K., Rankopo, M. J., Mwansa, L.-K., Coetzee, C., Khoza, S., Nemakonde, L. D., Shoroma, B. L., Wentink, G., Nyirenda, M., Chikuse, S., Kamanga, T., & Niekerk, D. V. (2022). Early Warning Systems in the Southern African Development Community: A Necessity. *Current Journal of Applied Science and Technology, 41*(48), 45–58. DOI: 10.9734/cjast/2022/v41i484033

Matikinca, P., Nyamakura, B., & Shackleton, S. (2024). Climate change adaptation and disaster risk reduction in South Africa's local municipal plans. *South African Journal of Science, 120*(7-8), 1–9. DOI: 10.17159/sajs.2024/15797

Mbovane, T. (2024). Matanzima bridge in Kariega could take two years to repair. Available at: [Online] Available at: https://www.algoafm.co.za/local/matanzima-bridge-in-kariega-could-take-two-years-to-repair

Mokhele, M. O., & Mvanyashe, A. (2025). The Role of Language as a Catalyst in Communicating Early Warning Messages: A Case Study in Walmer Airport Valley Informal Settlement, Gqeberha, South Africa. *International Journal of Disaster Risk Reduction, 105573*, 105573. Advance online publication. DOI: 10.1016/j.ijdrr.2025.105573

Motha, S. (2023). *East Coast Radio: Construction of flood-damaged KZN bridges to start in June*. https://www.ecr.co.za/news/news/construction-of-flood-damaged-kzn-bridges-to-start-in-june/

Motsumi, M. M., & Nemakonde, L. D. (2025). Indigenous early warning indicators for improving natural hazard predictions. *Jàambá, 17*(1), 1–12. DOI: 10.4102/jamba.v17i1.1754 PMID: 40357010

Mthambeka, N. H. (2022). *Urban Heat Islands (UHI) Assessment of Coastal Metropolitan Cities of South Africa*. Doctoral Thesis. Johannesburg, South Africa: University of Johannesburg. https://hdl.handle.net/10210/124987

Mtumane, G. (2024). *Nelson Mandela Bay Floods: six fatalities, hundreds displaced, and extensive damage reported.* News24. https://www.news24.com/nelson-mandela-bay-floods-six-fatalities-hundreds-displaced-and-extensive-damage-reported-20240603

Mudefi, E. (2023). Disaster management 'deeds' in the context of April 2022 KwaZulu-Natal floods: A scoping review. *International Journal of Disaster Risk Reduction*, *98*, 104122. DOI: 10.1016/j.ijdrr.2023.104122

Mugari, E., & Nethengwe, N. S. (2022). Mainstreaming ecosystem-based disaster risk reduction: Towards a sustainable and just transition in local development planning in rural South Africa. *Sustainability (Basel)*, *14*(19), 12368. DOI: 10.3390/su141912368

Mukange, T., Mokoto, B., Moipolai, T. B., & Ndamase, Z. (2024). Accelerating the use of mobile phone capabilities to maximise the effectiveness of public emergency alerts in South Africa. *In Proceedings of the 17th International Conference on Theory and Practice of Electronic Governance* (pp. 75-80). DOI: 10.1145/3680127.3680144

Mulaudzi, P., Mokoto, B., Mothwa, L., Mukange, T., Koen, H., & Ndamase, Z. (2025). Evaluating the Impact and Challenges of Public Disaster Alert Systems in KwaZulu-Natal. *IST-Africa 2025 Conference Proceedings Miriam Cunningham and Paul Cunningham (Eds), IST-Africa Institute and IIMC, 2025.*

Municipal Infrastructure Support Agency. (2023). *Misa 2023/24 First Quarter Performance Report (01 April — 30 June 2023).* MISA.

Murshed, S., Griffin, A. L., Islam, M. A., Wang, X. H., & Paull, D. J. (2023). Assessing sensitivity to climate-related disasters in the context of a developing country: Evidence from the coastal region of Bangladesh. *International Journal of Disaster Risk Reduction*, *97*, 104023. DOI: 10.1016/j.ijdrr.2023.104023

Myeni, L., & Roffe, S. (2025). Towards effective weather and/or climate services in South Africa: Profiling sectoral needs and constraints. *Environmental Development*, *55*, 101240. DOI: 10.1016/j.envdev.2025.101240

Nemakonde, L. D., & Kunguma, O. (2024, October 31). Revealing the boon and bane of South Africa's disaster management legislation during COVID-19. *Jàambá*, *16*(1), 1689. DOI: 10.4102/JAMBA.v16i1.1689 PMID: 39507560

Ngcamu, B. (2022). Climate Change and Disaster Preparedness Issues in the Eastern Cape and KwaZulu-Natal, South Africa. *Town and Regional Planning*, *81*, 53–66. DOI: 10.18820/2415-0495/trp81i1.5

Ngcamu, B. S., & Abrahams, G. (2024). The Preparedness Plans and Early Warning Systems During Flood Disasters in South Africa. In Jakaza, (Eds.), *The Palgrave Handbook of Language and Crisis Communication in Sub-Saharan Africa* (pp. 259–272)., DOI: 10.1007/978-3-031-43059-6_14

Nhamo, L., Magidi, J., Mpandeli, S., Mutanga, O., Liphadzi, S., & Mabhaudhi, T. (2025). Drought and flood risk mapping using a GIS-based multi-criteria decision method: A case of the Olifants Basin, South Africa. *Next Sustain*, *6*, 100100. DOI: 10.1016/j.nxsust.2025.100100

Nkombi, Z., & Wentink, G. J. (2022). The Role of Public Participation in Disaster Risk Reduction Initiatives: The Case of Katlehong Township. *Jàambá*, *14*(1), 1203. DOI: 10.4102/jamba.v14i1.1203 PMID: 35284043

Nyamwanza, A. M., Johnson, K., Mthembu, A., Tshivhundo, Z. C., & Brown, N. (2026). Integrated Knowledge Systems Towards Flood Resilience and Sustainable Solid Waste Management in South African Urban Informal Settlements. *Sustainability (Basel)*, *18*(6), 2960. DOI: 10.3390/su18062960

Ocaña-Fernández, Y., & Fuster-Guillén, D. (2021). The bibliographical review as a research methodology. *Revista Tempos e Espaços em Educação*, *14*(33), e15614–e15614. DOI: 10.20952/revtee.v14i33.15614

Okunola, O. H. (2025). Beyond institutional silos: rethinking multilevel disaster risk governance in Africa a decade into the Sendai Framework implementation: OH Okunola. *International Journal of Disaster Risk Science*, *16*(3), 321–332. DOI: 10.1007/s13753-025-00646-1

Olanrewaju, C., & Reddy, M. M. (2025). Assessment and Mitigation of Flood Risk in Urban Settlement: A Case of Ethekwini Metropolitan Area. *E-Journal of Humanities* [EHASS]. *Arts and Social Sciences*, *6*(2), 3340–3362. DOI: 10.38159/ehass.202561233

Page, M. J., McKenzie, J. E., Bossuyt, P. M., Boutron, I., Hoffmann, T. C., Mulrow, C. D., & Moher, D. (2021). The PRISMA 2020 statement: An updated guideline for reporting systematic reviews. *BMJ (Clinical Research Ed.)*, *372*, n71. Advance online publication. DOI: 10.1136/bmj.n71 PMID: 33782057

Paul, B. K. (2009). Why relatively fewer people died? The case of Bangladesh's Cyclone Sidr. *Natural Hazards*, *50*(2), 289–304. DOI: 10.1007/s11069-008-9340-5

Paul, B. K., & Dutt, S. (2010). Hazard warnings and responses to evacuation orders: The case of Bangladesh's cyclone Sidr. *Geographical Review*, *100*(3), 336–355. DOI: 10.1111/j.1931-0846.2010.00040.x PMID: 21061500

Pearson, E., Sharp, L., & Hampton, L. (2025). Resilience in context: A synthesis of theories and practices for educational psychologists. *Educational Psychology in Practice*, *41*(4), 1–18. DOI: 10.1080/02667363.2025.2518990

Poole, L., Coughlan de Perez, E., & Easton-Calabria, E.. (2022). *Anticipatory action in motion: Recapping the most recent evidence and illuminating a pathway forward.* Feinstein International Center, Tufts University.

Presidential Climate Commission. (2024). *State of climate action in South Africa. Presidential Climate Commission.* https://pccommissionflo.imgix.net/uploads/images/24_REP_State-of-Climate-Action-in-South-Africa_v4.pdf

Raphela, T. D. (2025). The ripple effect of flooding near one of the more prominent urban rivers in the Gauteng Province, South Africa. *Frontiers in Sustainable Cities*, *7*, 1656985. DOI: 10.3389/frsc.2025.1656985

Raphela, T. D., & Ekeke, N. (2025). Resilience and the dissemination of flood disaster early warning messages in a township in South Africa. *Frontiers in Communication*, *10*, 1503016. DOI: 10.3389/fcomm.2025.1503016

Reddy, M. M., Mbandlwa, Z., Pillay, S. S., Murwirapachena, G., & Ngema, S. S. (2024). Socio-economic Variables and Disaster Incidents in KwaZulu-Natal, South Africa. In Mbandlwa, S. (Ed.), *Challenges, Strategies, and Resiliency in Disaster and Risk Management* (pp. 23–55). IGI Global Scientific Publishing., DOI: 10.4018/979-8-3693-2721-0.ch002

Republic of South Africa. (2015). *Disaster Management Act - Amendment Bill 16 of 2015*. Government Printers.

Republic of South Africa. (2024). *Climate Change Act, 2024 (Act 22 of 2024)*. Government Printers.

Rodina, L., Harris, L., Ziervogel, G., & Wilson, J. (2024). Resilience countercurrents: Water infrastructures, informality, and inequities in Cape Town, South Africa. *World Development*, *180*, 106619. DOI: 10.1016/j.worlddev.2024.106619

Röhe, A. (2025). What Is the Meaning of Resilience? In *Organisational Resilience in Action: Strategies for Overcoming Future Crises* (pp. 5–13). Springer Berlin Heidelberg., DOI: 10.1007/978-3-662-70794-4_2

Rokhideh, M., Fearnley, C., & Budimir, M. (2025). Multi-Hazard Early Warning Systems in the Sendai Framework for Disaster Risk Reduction: Achievements, Gaps, and Future Directions. *International Journal of Disaster Risk Science*, *16*(1), 103–116. DOI: 10.1007/s13753-025-00622-9

Seyama, E. S., Maocha, M., & Dube, T. (2019). Evaluation of TAMSAT satellite rainfall estimates for southern Africa: A comparative approach. *Physics and Chemistry of the Earth Parts A/B/C, 112*, 141–153. DOI: 10.1016/j.pce.2019.02.008

Sithole, B. (2014). *Municipal Disaster Management in South Africa: Intergovernmental Relations as a Planning Instrument*. Doctoral Thesis. Bloemfontein, South Africa: Central University of Technology, Free State.

Siyongwana, P. Q., Heijne, D., & Tele, A. (2015). The Vulnerability of Low-Income Communities to Flood Hazards in Missionvale, South Africa. *Journal of Human Ecology (Delhi, India), 52*(1-2), 104–115. DOI: 10.1080/09709274.2015.11906935

South African History Online. (2014). The Group Areas Act of 1950. South African History. https://sahistory.org.za/article/group-areas-act-1950

South African Weather Service (SAWS). (2023). *How will the climate change by 2050 in South Africa?* https://www.weathersa.co.za/

South African Weather Services [SAWS]. (2018). *A WeatherSmart Nation Annual Report 2018/2019*. Pretoria, South Africa: SAWS.

Taylor, A. (2025). *South Africa needs R250 billion just to meet basic climate adaptation needs over the next decade*. Conversations., DOI: 10.64628/AAJ.h5tngphmn

Terblanche, T., De Sousa, L. O., & Van Niekerk, D. (2022). Disaster Resilience Framework Indicators for a City's Disaster Resilience Planning Strategy. *Jàambá, 14*(1), 1264. DOI: 10.4102/jamba.v14i1.1264 PMID: 35747366

Thebe, T. P. (2016). Community participation: Reality and rhetoric in the development and implementation of the Integrated Development Plan (IDP) within municipalities of South Africa. *Journal of Public Administration, 51*(4), 712–723. https://hdl.handle.net/10520/EJC-9903b8a5b

United Nations Office for Disaster Risk Reduction. (2015). *Sendai Framework for Disaster Risk Reduction 2015–2030*. UNDRR. https://www.undrr.org/publication/sendai-framework-disaster-risk-reduction-2015-2030

Van Niekerk, D. (2014). A critical analysis of the South African Disaster Management Act and Policy Framework. *Disasters, 38*(4), 858–877. DOI: 10.1111/disa.12081 PMID: 25196341

World Meteorological Organization. (2023). *Global status of multi-hazard early warning systems: Target G of the Sendai Framework for Disaster Risk Reduction*. World Meteorological Organization. https://library.wmo.int/idurl/4/68696

Ziervogel, G., Lennard, C., Midgley, G., New, M., Simpson, N. P., Trisos, C. H., & Zvobgo, L. (2022). Climate change in South Africa: Risks and opportunities for climate-resilient development in the IPCC Sixth Assessment WGII Report. *South African Journal of Science*, *118*(9-10), 1–5. DOI: 10.17159/sajs.2022/14492

Ziwanai, F. (2024). Re-Imagining Community-Based Disaster Risk Reduction Towards Building Resilient Rural Communities. In *Challenges, Strategies, and Resiliency in Disaster and Risk Management* (pp. 84–96). IGI Global., DOI: 10.4018/979-8-3693-2721-0.ch004

Chapter 8
Integrating Multi-Hazard Early Warning Systems With AI-Enhanced Structural Health Monitoring Dynamics for Sustainable Rehabilitation

Ali Biglarifadafan
http://orcid.org/0000-0002-9616-1823
Golestan University, Iran

Eduard Babulak
http://orcid.org/0000-0003-2450-090X
National Science Foundatiion, USA

ABSTRACT

Traditional seismic rehabilitation, reliant on static designs and isolated hazard treatment, proves inadequate against today's overlapping natural hazards and sustainability imperatives. This chapter introduces a paradigm integrating Multi-Hazard Early Warning Systems (MHEWS) with AI-enhanced Structural Health Monitoring (SHM) for dynamic, lifecycle-oriented resilience. By fusing MHEWS data streams with real-time sensor data from retrofitted structures, artificial intelligence serves as the critical nexus. Advanced machine learning models analyze this multi-modal data for real-time vulnerability assessment and predictive damage simulation under compound threats, enabling context-aware mitigation. This anticipatory framework transcends conventional code-based approaches, embedding proactive intelligence

DOI: 10.4018/979-8-3373-7345-4.ch008

…right © 2026, IGI Global Scientific Publishing. Copying or distributing in print or electronic forms without written permission of IGI Global Scientific Publishing is prohibited. Use of this chapter to …n generative artificial intelligence (AI) technologies is expressly prohibited. The publisher reserves all rights to license its use for generative AI training and machine learning model development.

into infrastructure systems. Ultimately, we reposition seismic rehabilitation as an adaptive component within a broader hazard-response ecosystem—enhancing public safety, optimizing resources, and advancing sustainable development in an era of compounded environmental challenges.

FOUNDATIONS OF MULTI-HAZARD RESILIENCE AND SUSTAINABLE SEISMIC REHABILITATION

Accelerating hazard complexity, climate-induced non-stationarity, growing urbanization, aging infrastructure, and the disruptive integration of digital intelligence into physical systems have all shaped the current state of structural engineering practice. Historically focused on single-hazard assumptions and prescriptive code-based methods, seismic rehabilitation is no longer adequate to guarantee acceptable levels of safety, functionality, and societal value over the duration of civil infrastructure. The theoretical, historical, and methodological underpinnings of a paradigm change toward integrated, sustainable, and intelligent multi-hazard resilience are established in this section.

This section makes the case that resilience results from the combination of earthquakes, environmental degradation, digital revolution, and societal vulnerability rather than seeing them as distinct issues. It is important to view structural systems as adaptable resources integrated into dynamic cyber-physical and socioeconomic ecosystems rather than as static artifacts created for compliance at the time of creation. As a result, the section presents seismic rehabilitation as a lifecycle intervention that incorporates early warning systems, sustainability metrics, performance-based engineering, multi-hazard risk modeling, structural health monitoring, and ethical artificial intelligence.

Evolution of Rehabilitation Paradigms: From Prescriptive to Adaptive

The growth of earthquake engineering as a scientific field is reflected in the evolution of seismic restoration models. Strength-based criteria derived from elastic analysis and simple force reduction factors dominated early seismic design philosophies that emerged in the middle of the 20th century. Similar reasoning applied to the rehabilitation of existing structures, which mostly concentrated on local strengthening measures meant to meet the criteria of the updated code without specifically taking system-level performance or uncertainty into account.

By the late 1990s and early 2000s, the shortcomings of prescriptive techniques were revealed by empirical evidence from destructive earthquakes, including Northridge

(1994), Kobe (1995), and İzmit (1999). Damage concentrations, residual drifts, and post-event functional losses were often undesirable in structures that ostensibly met code requirements. Performance-based seismic engineering, in which rehabilitation goals are directly linked to probabilistic performance indicators including interstory drift, damage states, and loss indices, emerged as a result of this insight.

In this sequence, adaptive rehabilitation is the next logical step. Adaptive techniques acknowledge that societal expectations, hazard characteristics, and structural performance targets change with time, unlike static retrofitting strategies. In order to facilitate phased interventions, adaptive rehabilitation thus combines decision theory, updated hazard models, and monitoring data (See Figure 1). The basic change is from a design that is focused on compliance to systems that are focused on learning, where feedback and observation are used to continuously minimize uncertainty.

Figure 1. A conceptual framework integrates probabilistic multi-hazard risk assessment (on the left), sustainable rehabilitation strategies (on the right), and cyber-physical-social intelligence (in the centre). solid arrows represent primary data flows, while dashed arrows indicate feedback loops for adaptive learning. this system provides real-time decision support through continuous monitoring and AI-enhanced analytics

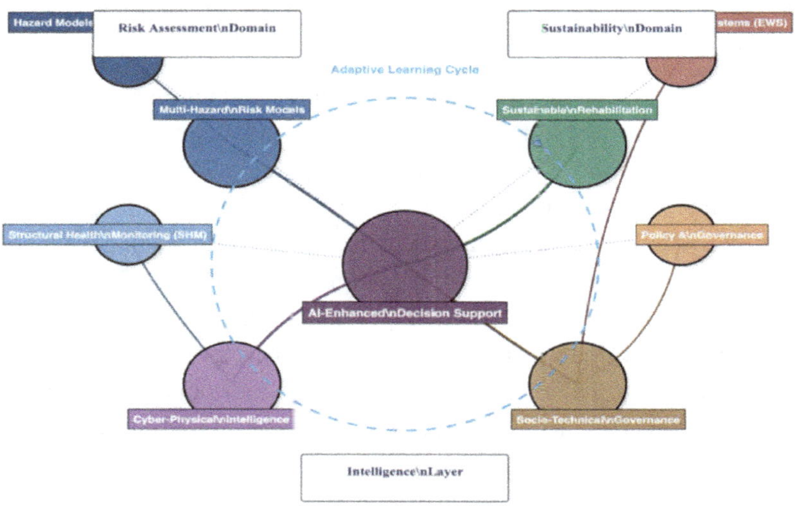

The development of demand-capacity assessment serves as an example of the mathematical underpinnings of this shift. Traditional deterministic checks of the form $D \leq C$ are replaced by probabilistic performance functions expressed as $P[DS \geq ds \mid IM]$, where DS denotes damage state, ds is a threshold level, and IM represents an intensity measure such as spectral acceleration Sa(T). Adaptive rehabilitation further extends this framework by allowing both D and C to be time-dependent random variables, D(t) and C(t), reflecting degradation, repair, and intervention.

From Single-Hazard to Multi-Hazard Engineering: Coupled Risk and System Reliability

The assumption of independence between earthquake loading and other extreme occurrences is fundamental in single-hazard seismic assessment. Hazards frequently interact through temporal, geographical, and mechanistic linkage, as evidenced by recent analytical research and empirical data. While climate-related risks like flooding and heat waves can weaken structural capacity and increase seismic vulnerability, earthquakes can cause landslides, tsunamis, and technology mishaps (Flint et al., 2016; Zheng & Li, 2023; Zub et al., 2018).

Quantifying these relationships within a single probabilistic framework is the goal of multi-hazard engineering. The integral of the joint probability density function *f(H, C)* across all hazard situations *H* and consequence states *C*, multiplied by the corresponding loss *L(C),* is the total risk R. This relationship can be expressed mathematically as follows:

$$R = \iint L(C) f(C \mid H) f(H)\, dC dH$$

It is no longer possible to separate *f(H)* into independent marginal distributions in multi-hazard situations. Rather, conditional dependencies need to be explicitly represented, usually using Bayesian networks or copula functions. By taking into account load-sharing mechanisms and associated component failures within structural systems, system reliability analysis expands on this idea.

Multi-hazard reliability requires that strengthening methods for repaired structures do not unintentionally make them more vulnerable to non-seismic threats. For instance, extra mass from retrofitting features may increase seismic demand, while greater stiffness from shear walls may increase flood-induced stresses. Typical hazard interactions pertinent to seismic restoration are broadly summarized in Table 1, which also emphasizes the necessity of integrated evaluation.

Table 1. Conceptual relationships between seismic and non-seismic risks in structures that have undergone rehabilitation

Hazard Combination	*Interaction Mechanism*	*Potential Impact on Rehabilitated Structure*
Earthquake + Flood	*Seismic-induced settlement increases flood vulnerability; flood scour reduces foundation capacity*	*Combined demand may exceed design capacity*
Earthquake + Landslide	*Ground shaking triggers slope instability; landslide loading imposes additional lateral forces*	*Foundation failure or global instability*
Seismic + Fire	*Post-earthquake fires due to gas leaks or electrical faults*	*Fire damage compounds structural weakening*
Seismic + Wind	*Simultaneous occurrence of extreme wind and seismic events is rare but possible in certain regions*	*Resonance or fatigue-induced failure*

Sustainability in Structural Rehabilitation: Metrics, Materials, and Lifecycle Ethics

Sustainability in structural restoration includes long-term economic viability, social equality, and environmental impact in addition to material efficiency. Conventional retrofit decision-making frequently overlooks embodied carbon, operating energy, and end-of-life concerns in favor of initial construction cost. However, lifetime assessment is emphasized as the proper evaluation lens in contemporary sustainability science (Afrasiabi et al., 2022; Alcocer & Moctezuma, 2023; Asmara, 2023; Matthews & JR Morlidge, 2009)..

Lifecycle carbon emissions (E_{LC}), which are the total of embedded emissions (E_{emb}), operational emissions (E_{op}), and end-of-life emissions (E_{EOL}), can be used to measure the environmental impact of rehabilitation.

$$E_{LC} = E_{emb} + E_{op} + E_{EOL}$$

The choice of materials and degree of intervention have a significant impact on E_{emb} in seismic rehabilitation. High-performance materials, including fiber-reinforced plastics and ultra-high-performance concrete, provide mechanical advantages, but unless they are counterbalanced by durability improvements and fewer interventions, they may have higher embodied carbon.

When safety goals are weighed against sustainability measures, ethical issues come up. While under-retrofitting puts populations at unacceptable risk, over-retrofitting to obtain modest seismic performance benefits may result in high environmental costs. Therefore, clear decision frameworks that specifically consider conflicting goals and uncertainty are required by lifecycle ethics.

Infrastructure as a Dynamic Cyber-Physical System: Towards Intelligent Assets

When sensing, computing, and actuation are combined, restored structures become cyber-physical systems with perception, reasoning, and reaction capabilities (Aktan et al., 2024; Hsieh et al., 2014; Leal et al., 2024). These. Continuous streams of data on environmental variables, strain distributions, and vibration characteristics are provided by structural health monitoring systems (Biglari, 2026; Biglari & Harrison, 2026). These data allow for near-real-time structural status monitoring when combined with digital twins (See Figure 2).

The state-space equation can be used to mathematically depict the dynamic behavior of an instrumented structure:

$$\dot{\mathbf{x}}(t) = \mathbf{A}(t)\mathbf{x}(t) + \mathbf{B}(t)\mathbf{u}(t) + \mathbf{w}(t)$$

where is the state vector, represents control or excitation inputs, denotes measured outputs, and and are process and measurement noise terms, respectively. In intelligent rehabilitation, model parameters , , and are updated through system identification algorithms informed by monitoring data.

Pattern identification and anomaly detection are made possible by artificial intelligence techniques, especially machine learning, which surpass the limitations of conventional model-based approaches. Their use in safety-critical infrastructure, however, presents questions about accountability, transparency, and robustness, all of which are covered in later sections of this section.

Figure 2. Component scores & actionability index based on infrastructure type

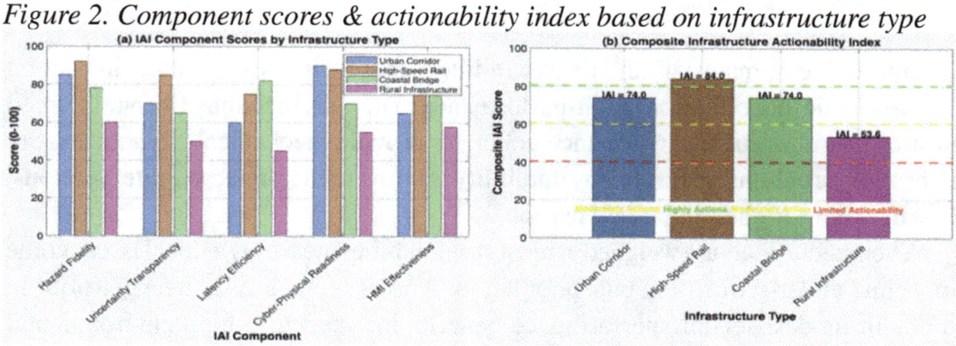

MULTI-HAZARD EARLY WARNING SYSTEMS (MHEWS): ENGINEERING-READY THEORY AND TECHNOLOGY

Historically, Multi-Hazard Early Warning Systems (MHEWS) have been developed and deployed as public safety tools, primarily to minimize fatalities by promptly notifying vulnerable populations. However, the expectations for early warning technologies have changed significantly over the last 20 years due to the growing interconnection between digital control platforms, real-time sensor networks, and civil infrastructure systems. Warnings that only provide generalized hazard descriptors or qualitative threat levels are no longer adequate in the modern technical setting. Rather, they need to be able to be translated into time-sensitive, quantifiable, and structure-specific parameters that may be used to directly guide automated or semi-automated engineering decisions.

The argument put out in this section is that MHEWS needs to change from socially oriented communication systems to cyber-physical engineering subsystems that can easily interface with digital twins of infrastructure, control algorithms, and structural analysis tools. In order to make MHEWS actionable within structural and geotechnical engineering practice, the topic firmly extends beyond introductory notions to address the scientific underpinnings, data structures, uncertainty models, and system-level integrations. Seismic, hydro-meteorological, and landslide dangers are highlighted, with special focus on their combined effects and cascade repercussions.

The section is set up to gradually close the gap between engineering reaction and hazard detection. It looks at how raw wavefield data are converted into demand measurements, particular to a certain structure, starting with seismic early warning. The following sections discuss slow-onset ground deformation tracked by satellite and in situ networks, flooding and scour caused by rainfall, and the thorough handling of uncertainty that supports justifiable engineering action. Dissemination protocols, cyber-physical architectures, human-machine interfaces, benchmarking frameworks, legal considerations, and the creation of an Infrastructure Actionability Index to standardize the engineering usefulness of warning outputs are the main topics of the section's second half.

Seismic Early Warning: From P-Wave Detection to Structure-Specific ShakeMaps

In order to give prior notice of approaching strong ground motion, seismic early warning systems (SEWS) take advantage of seismic waves' limited propagation speed (Bilal et al., 2022; Bommer & Abrahamson, 2006; Lautour & P Omenzetter, 2009). The basic idea is based on the differences in primary (P), secondary (S), and surface wave velocities. P-waves, which are initially detected and convey information on the rupture initiation and early source evolution, move through the upper crust at velocities that are usually between 5 and 7 km/s. Most structural damage is caused by S-waves and surface waves, which move more slowly, often at speeds of 2 to 4 km/s.

Using the earliest P-wave readings available, modern SEWS algorithms seek to quickly characterize the source of the earthquake. The Virtual Seismologist (V_S) algorithm and the τ_c–P_d method are two of the most well-known methods; they have both been widely reported in the seismological literature. The definition of the characteristic period parameter τc is:

where the ground displacement time history during the first P-wave window of duration T is indicated by the symbol $u(t)$. Despite significant regional diversity, empirical connections between τ_c and moment magnitude M_w have been established across a variety of tectonic contexts. In addition, the magnitude and predicted peak ground acceleration (PGA) at distal sites have been associated with the peak displacement P_d recorded within the same window.

Using a Bayesian framework, the Virtual Seismologist program estimates source parameters and the uncertainties that go along with them by combining real-time waveform characteristics with historical data on regional seismicity. V_S provides a probabilistic description of magnitude, location, and expected shaking intensity by continually updating its estimations as more stations report, in contrast to solely deterministic threshold-based techniques.

From an engineering point of view, the usefulness of seismic early warning is determined by the amount of time that remains for action after data collection, processing, and distribution, in addition to detecting capability. Sensor triggering delays, data telemetry, algorithmic processing, and end-user communication are some of the origins of latency. While on-site warning arrangements can cut this period to less than two seconds by collocating sensors with the protected structure, regional warning systems typically have overall latencies on the order of several seconds to tens of seconds.

There are two types of uncertainty in SEWS outputs: aleatory variability due to the stochastic nature of rupture and wave propagation, and epistemic uncertainty resulting from insufficient understanding of source processes, site conditions, and structural response characteristics. Instead of being masked by unduly simplistic warning metrics, these uncertainties need to be properly defined and transmitted through future investigations for engineering use (Führer & Kwakkel, 2025; Marchau et al., 2019; Shortridge et al., 2017).

Traditional ShakeMaps, as used by organizations like the US Geological Survey, show the spatial distributions of ground motion intensity metrics at specific times, such as spectral acceleration Sa(T), peak ground velocity (PGV), and PGA. Although these solutions are very helpful for quick post-event evaluation, they are not always customized for specific infrastructure systems or structures.

Converting regional ground motion forecasts into structure-specific demand estimates that take into consideration basin impacts, local soil conditions, and soil–structure interaction is the engineering problem. A first-order correction is provided by site amplification factors that are obtained from shear-wave velocity profiles, which are generally defined by the time-averaged shear-wave velocity in the upper 30 meters (Vs30). However, more complex treatment is required due to nonlinear soil behavior under intense shaking and three-dimensional wavefield effects.

Recent studies have investigated the viability of combining quick geotechnical characterization with dense sensor arrays to create near-real-time, site-specific ShakeMaps. These methods use precalculated site response functions, H_s(f), so that the Fourier amplitude spectrum of surface ground motion, U_s(f), can be written as follows:

where the motion at a reference bedrock horizon is denoted by $U_b(f)$. This approach allows the creation of structure-relevant input motions within the warning time window when paired with fast source characterization.

The use of scalar intensity measures, which are inadequate for nonlinear structural analysis, is a recurring drawback of current SEWS outputs. Time-domain representations of ground motion are necessary for digital twins and structural control systems to assess limit states and simulate dynamic reaction. Time-series seeds, which are short-duration, physics-informed waveforms that are created in real-time and then enhanced as more data becomes available, have been offered as a solution to this gap.

Stochastic simulation methods limited by time parameters and projected response spectra can be used to build these seed motions. Let D be the significant duration, and let Sa(T) be the expected spectral acceleration at period T. While maintaining causality and real-time computability, a suitable acceleration time history a(t) can be produced such that its response spectrum matches Sa(T) within specified tolerances.

A paradigm shift is made possible by the incorporation of such time-series seeds into early warning outputs, which allow for quick nonlinear time-history analysis before harmful waves arrive. For structures with active or semi-active control systems, where pre-trigger control law adjustment can significantly improve performance, this capability is especially pertinent (See Figure 3).

Figure 3. Seismic early warning system timeline

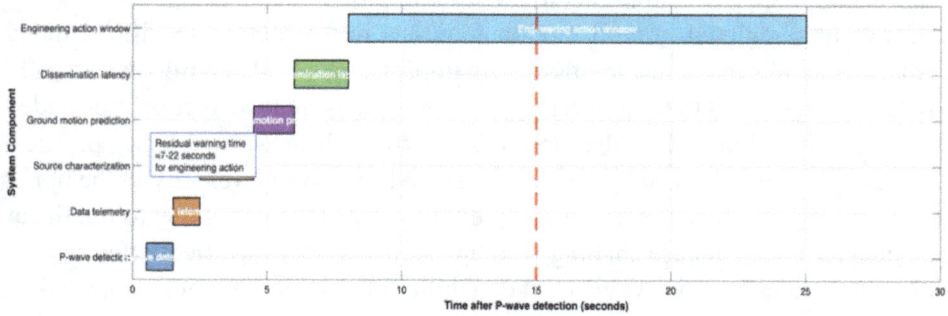

Hydro-Meteorological Forecasting for Infrastructure, Flood, and Scour Nowcasting

When compared to seismic occurrences, hydro-meteorological hazards are a fundamentally different class of dangers to civil infrastructure, both in terms of their underlying physical principles and their temporal evolution. While rainfall-induced flooding, riverine inundation, storm surge, and hydraulic scour usually develop over hours to days with spatial footprints that may span whole catchments, earthquakes impose impulsive dynamic demands during seconds to minutes. The need for early warning systems to switch from public advisories to engineering-actionable inputs is just as urgent in spite of these variations (Webber & Samaras, 2022; Wen et al., 2025).

Hydro-meteorological early warning must guide decisions on load restrictions, controlled shutdowns, adaptive operation, and emergency reinforcement for bridges, levees, dams, transportation corridors, and urban drainage systems. Converting weather predictions, which are probabilistically defined in terms of the intensity, duration, and geographic distribution of precipitation, into structural demand characteristics like flow velocity, water depth, uplift pressure, and foundation scour depth is the main challenge.

Adaptive operation, controlled shutdowns, load limitations, and emergency reinforcement are all dependent on hydro-meteorological early warning for bridges, levees, dams, transit corridors, and urban drainage systems. The main difficulty is converting weather predictions, which are probabilistically defined in terms of the amount, duration, and geographic distribution of precipitation, into structural demand factors such as foundation scour depth, water depth, flow velocity, and uplift pressure.

where Z represents radar reflectivity in linear units, R is rainfall intensity in millimeters per hour, and a and b are experimentally calibrated coefficients that depend on the kind of precipitation. Despite its apparent simplicity, this relationship involves a great deal of epistemic uncertainty because of the variety in beam blockage, attenuation, and drop size distributions, especially in difficult terrain.

From an engineering perspective, classic Quantitative Precipitation Estimation -QPE- main drawback is not its accuracy in and of itself, but rather its incompatibility with hydraulic and hydrological models that operate at scales relevant to infrastructure. In order to provide near-real-time estimation of runoff generation, channel routing, and flood wave propagation, recent developments have concentrated on directly integrating radar and satellite data into distributed hydrological models.

Typically, flood early warning systems use a model chain that includes hydraulic routing after rainfall-runoff transformation. The Saint-Venant equations, which can be conservatively represented as follows, govern one-dimensional unstable open-channel flow:

where g is gravitational acceleration, S_0 is the channel bed slope, S_f is the friction slope, A is the cross-sectional flow area, Q is discharge, h is the flowing depth, and q_l is lateral inflow per unit length. These equations' numerical solutions yield time histories of the water's depth and velocity at key points, which can be immediately translated into structural demand parameters.

The evolution of local and contraction scour, buoyancy effects on superstructures, and hydrodynamic forces acting on piers and decks are all pertinent outputs for bridges. An approximation of the drag force F_d operating on a structural member submerged in flow is as follows:

where v is the flow velocity, A_p is the projected area normal to the flow, C_d is the drag coefficient, and ρ is the water density. Predictive evaluation of structural safety margins is thus made possible by early warning systems that can provide predicted $v(t)$ and $h(t)$ time histories.

One of the biggest reasons bridges fail globally is still scour, which frequently happens during severe floods with little to no warning. Conventional scour evaluation is based on empirical formulas that estimate equilibrium scour depth as a function of sediment properties and flow intensity, such as the HEC-18 pier scour equation. However, during transitory flood waves, equilibrium assumptions are often broken, making static estimates unsuitable for making decisions in real time.

The combination of hydro-meteorological forecasts with real-time foundation reaction monitoring, such as vibration-based modal identification, pore pressure measurements in foundation soils, and sonar-based bed elevation sensors, has been the subject of recent research. Dynamic scour nowcasting frameworks that estimate both the ultimate scour depth and its pace of development can be created by combining predicted hydraulic loading with observed structural and geotechnical reaction.

The majority of hydro-meteorological early warning systems still distribute their findings in formats designed for emergency management rather than engineering analysis, despite significant advancements in meteorological science. Without a clear connection to structural capability, flood warnings are usually stated in categorical terms, such as return durations or the likelihood that water level thresholds would be exceeded (See Figure 4).

Instead, time-dependent load and boundary condition histories that are compatible with structural and geotechnical analysis tools would be provided by an engineering-ready hydro-meteorological warning system. Standardized data formats, precise uncertainty descriptors, and unambiguous semantics connecting prediction variables to physical demand factors are necessary for this. Meteorologists, hydrologists, structural engineers, and infrastructure operators must work closely together to design such systems. Open interfaces that enable machine-to-machine communication must also be used.

Figure 4. Hydro-meteorological nowcasting performance

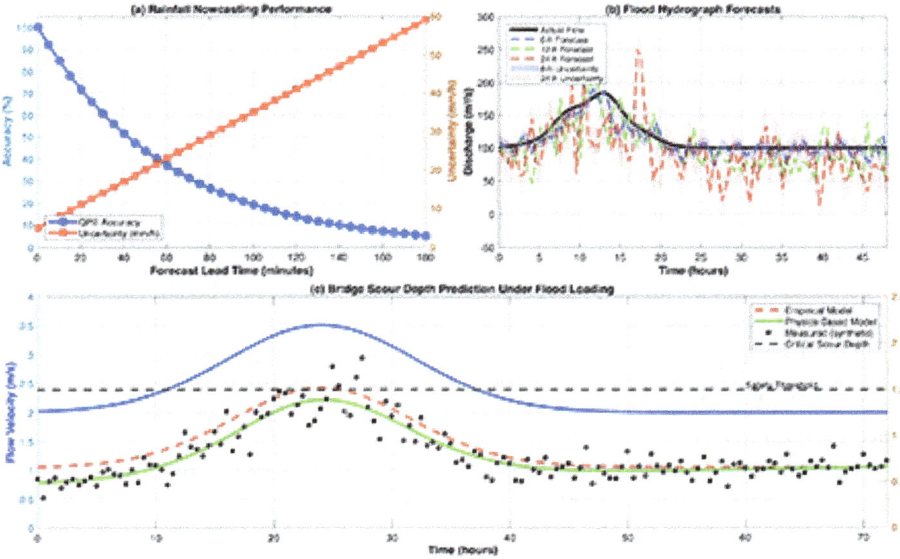

Landslide and Subsidence Early Detection: Integrating InSAR and IoT Networks

Within the range of natural hazards, landslides and ground subsidence hold a unique place due to their locally specific effects and hours-to-decades-long evolution. Slope instabilities and subsidence processes generally show precursory deformation that may be identified and tracked well before catastrophic breakdown, in contrast to earthquakes and floods, which frequently appear as isolated events (Das & Choudhury, 2025; Farrokhzad et al., 2011; Sitar et al., 2005).

Linear assets like highways, railroads, pipelines, and transmission lines, as well as structures built on karstic or compressible soils, are constantly at risk from these risks in the context of civil infrastructure. Therefore, in addition to predicting events, early detection and warning systems must continuously evaluate deformation patterns and their effects on structural performance (See Figure 5).

For large-scale deformation monitoring, Interferometric Synthetic Aperture Radar (InSAR) has become a game-changing technique. InSAR makes it possible to estimate ground displacement over large areas with millimeter-scale precision by taking advantage of phase changes between successive satellite radar observations. Techniques like Small Baseline Subset (SBAS) and Persistent Scatterer InSAR (PS-InSAR) improve temporal resolution and robustness in diverse environments.

The line-of-sight displacement Δd, the primary observable in InSAR, is connected to phase change $\Delta\varphi$ by:

where the radar wavelength is indicated by λ. Time series of Δd that are recovered from persistent scatterer locations offer important information about long-term deformation patterns, such as slow-moving landslides and subsidence brought on by groundwater loss.

The main drawback of satellite-based InSAR, as seen from a technical standpoint, is its revisit period, which can vary from days to weeks based on the sensor. This temporal precision is suitable for tracking slow processes, but not enough to capture the rapid acceleration phases right before failure. Therefore, to obtain effective early warning, InSAR needs to be coupled with complementary in situ sensing technologies.

Dense, localized monitoring of slope and ground deformation is now possible thanks to the widespread availability of inexpensive Internet of Things (IoT) sensors. At crucial sites, devices like inclinometers, extensometers, piezometers, and GNSS receivers offer constant displacement, rotation, and pore pressure measurements. These sensors provide high temporal resolution data that can identify minute changes suggestive of approaching instability when placed strategically.

Combining disparate data streams to provide a cohesive evaluation of danger evolution is the difficult part. This calls for uncertainty-aware algorithms that can separate significant trends from noise, as well as strong data assimilation frameworks that can bridge different spatial and temporal dimensions.

In the end, early warning for subsidence and landslides must convert deformation readings into useful technical data. The inverse velocity method is one extensively researched strategy that estimates the time to failure, t_f, from the accelerating displacement rate, v(t), as follows:

where an empirical constant is denoted by A. Despite the fact that this connection has been effectively used in some situations, the quality of the data and the underlying failure mechanism have a significant impact on how reliable it is.

More recent research has investigated physics-based models that incorporate effective stress evolution, strain localization, and progressive damage. These models are based on soil mechanics and fracture mechanics. With major ramifications for infrastructure risk management, research into integrating such models into early warning systems is still ongoing.

Figure 5. InSAR deformation monitoring and landslide detection

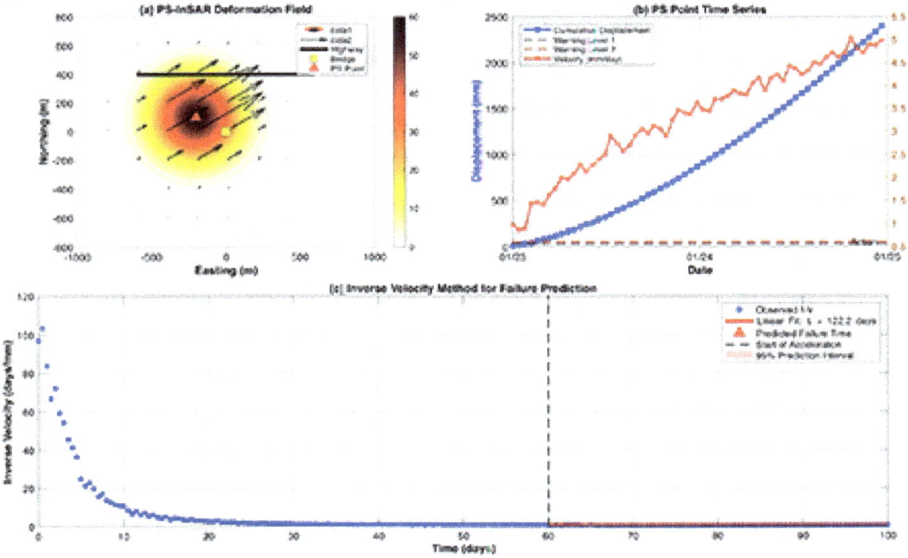

Uncertainty Quantification in MHEWS: Epistemic and Aleatory Models for Engineering Decision-Making

In multi-hazard early warning systems, uncertainty is not a secondary factor; rather, it is the distinguishing feature that determines whether engineers can responsibly act upon warning outputs. Engineering applications demand transparent, quantitative, and traceable uncertainty representations, in contrast to public warning systems that might accept conservative bias and qualitative descriptions. Defensible probabilistic reasoning is required for decisions that have economic, safety, and legal ramifications, such as activating active structural control, closing transit corridors, or evacuating vital facilities (Hadjimichael & J Schlumberger, 2024; Kalra et al., 2014; Shortridge et al., 2017; Zhao et al., 2019).

Uncertainty in MHEWS comes from a variety of sources that are dispersed throughout the system's sensing, modeling, and decision levels. To prevent false precision and to guarantee internal consistency, these sources need to be methodically categorized, measured, and disseminated. A fundamental context for this undertaking is provided by the differentiation between aleatory and epistemic uncertainty (See Figure 6).

Figure 6. Probability and resilience effect on the decision making

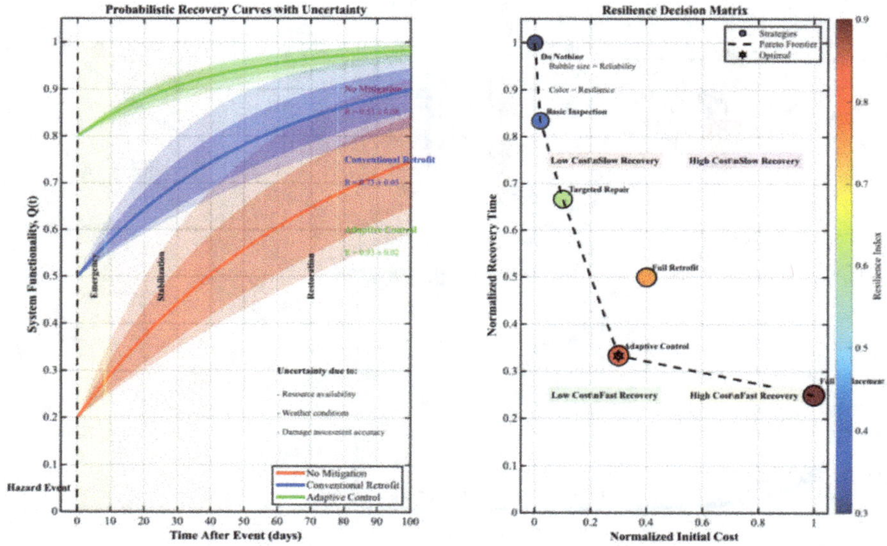

Aleatory uncertainty is irreducible via further data collection alone and reflects the underlying randomness of natural processes. Even with ostensibly similar source and site conditions, aleatory diversity in seismic early warning can be seen in ground motion variability, wave propagation heterogeneity, and rupture complexity. Stochastic elements of ground motion prediction equations, which are frequently represented as lognormal distributions of intensity values, are commonly used to describe this variability.

For example, spectral acceleration *Sa(T)* at a given site may be expressed as:

where ε is a zero-mean random variable with standard deviation σ signifying aleatory variability, and μ represents the median prediction as a function of magnitude M, source-to-site distance R, and site parameters S. Aleatory uncertainty is linked to material heterogeneity and localized failure mechanisms in landslide processes, while it is caused by chaotic atmospheric dynamics, unresolved turbulence, and stochastic rainfall patterns in hydro-meteorological hazards.

Aleatory uncertainty must be maintained throughout the warning-to-decision chain from an engineering perspective. Brittle decision rules that function badly under uncommon but plausible realizations can result from attempts to conceal or disregard this unpredictability.

An incomplete understanding of boundary constraints, model form, and system characteristics leads to epistemic doubt. It is theoretically reducible through better data, superior models, and focused research, in contrast to aleatory uncertainty. From sensor calibration and spatial coverage to constitutive models of soil behavior and structural reaction, epistemic uncertainty penetrates every aspect of the MHEWS system.

Uncertainty in local site reaction is a notable example. Although the time-averaged shear-wave velocity to 30 m depth- Vs30 - is a useful stand-in for site classification, it is unable to account for depth-dependent impedance differences, basin shape, and nonlinear soil behavior. Similarly, boundary condition specification, floodplain topography, and channel roughness coefficients—all of which are prone to epistemic uncertainty—have an impact on hydraulic models of flood propagation (See Figure 7).

When new evidence becomes available, Bayesian frameworks offer a natural way to represent and update epistemic uncertainty. Let D stand for observational data, and let θ be a vector of uncertain model parameters. The posterior distribution is obtained using the Bayes theorem:

It serves as the foundation for real-time warning systems' continuous updating. Approximate inference methods like particle filtering or ensemble Kalman filtering are necessary in practice due to computing limitations, especially when working with high-dimensional state spaces.

Uncertainty must spread from hazard characterization to structural demand and capacity models for early warning to be useful. Measures of danger intensity and engineering demand factors must be explicitly linked for this approach to work. Probabilistic seismic demand models (PSDMs), which link engineering demand parameters (EDP) to intensity measures (IM), are frequently used in seismic applications to make this connection.

where η denotes residual variability and a and b are regression coefficients. The pressures caused by floods, the progression of scour depth, and the deformation demands caused by landslides all have similar correlations.

Since material qualities, construction quality, degradation, and modeling assumptions can all vary, structural capacity is inherently unpredictable. The likelihood of surpassing a damage state *DS* given a demand parameter *EDP* can be expressed probabilistically using fragility functions, which can be used to indicate capacity:

where the median capacity is represented by μ_{DS}, the standard normal cumulative distribution function by Φ, and β, the logarithmic standard deviation is denoted by *DS*.

Demand and capacity uncertainty are combined to produce exceedance probabilities, which can be used as logical decision-making metrics in MHEWS. Surrogate models and precomputed reaction surfaces are used because these probabilities need to be calculated quickly enough to fit within the warning time window.

Determining the decision thresholds for automated or semi-automated action is a key issue in engineering-ready early warning. Such criteria must strike a compromise between the potential for catastrophic failure from false negatives and the potential for needless interruption and expense from false positives. Hazard modeling by itself cannot address this equilibrium because it is intrinsically context-dependent.

A formal framework for threshold selection is offered by risk-informed decision theory. Let P_F be the probability of failure conditioned on warning information, C_A be the cost of action, and C_F be the cost of failure. Action is triggered by an optimal decision rule when:

This disparity emphasizes the significance of trustworthy probability estimates from MHEWS and draws attention to the institutional and ethical aspects of automated decision-making. Decision thresholds are typically discretized and enshrined in operational procedures in practice, which need to be open, auditable, and updated on a regular basis.

Figure 7. Uncertainty quantification of epistemic and aleatory models for engineering decision-making

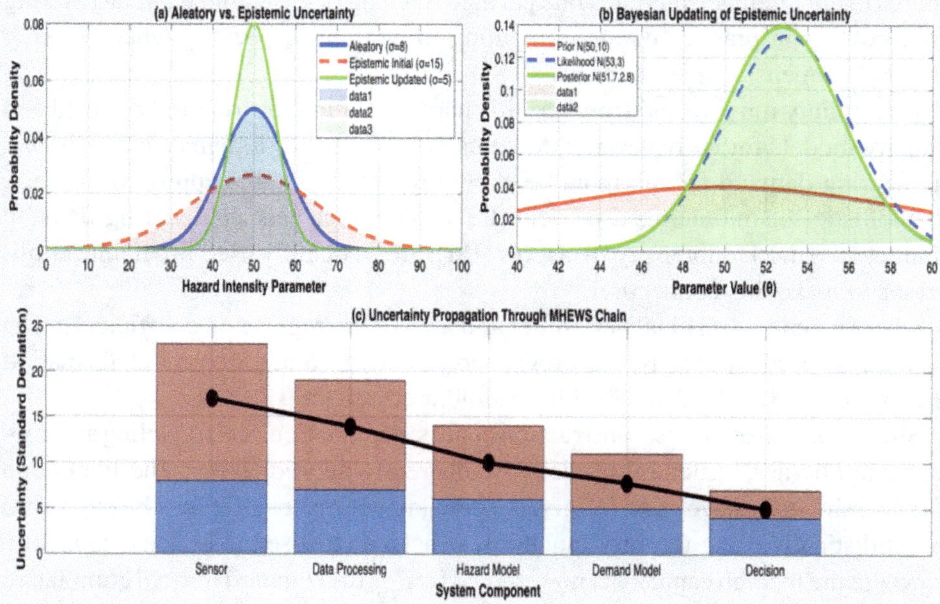

Dissemination Protocols: From Public Sirens to Machine-Readable Data Streams

Early warning systems' distribution layer has historically been built with human perception and response in mind, including mobile notifications, broadcast media, and sirens to communicate impending danger. The needs of engineering systems, which require accurate, machine-readable data given with low latency and high reliability, are not well served by these modalities, notwithstanding their effectiveness in public safety.

As cyber-physical infrastructure develops, dissemination mechanisms must be fundamentally rethought. These days, early warning outputs have to cater to two audiences: automated systems that need structured data streams for real-time computation and control, and human operators who need situational awareness.

The use of defined data formats that encode numerical values along with their physical meaning, uncertainty, and temporal validity is essential for the engineering integration of MHEWS. Although JavaScript Object Notation (JSON) and Extensible Markup Language (XML) are widely used for data sharing, their usefulness is heavily reliant on clearly specified schemas and ontologies (See Figure 8).

Along with timestamps and version identifiers, data packets for seismic early warning may contain estimates of magnitude, position, predicted intensity measurements, and related uncertainty metrics. While landslide alerts may incorporate indicators of acceleration and deformation rates, hydro-meteorological warnings may display gridded fields of rainfall intensity, water depth, or flow velocity.

Consensus on variable definitions, units, coordinate systems, and reference frames is necessary for semantic interoperability. Machine-to-machine communication runs the risk of misunderstanding and serious repercussions in the absence of such an agreement.

The effective warning time available for engineering action is directly reduced by dissemination latency. Minimizing latency requires optimization across communication networks, data processing pipelines, and endpoint interfaces. Promising approaches to latency reduction are provided by edge computing paradigms, where initial processing takes place close to the data source or end user.

Dependability is just as important. Systems for disseminating warnings need to be resilient to hardware malfunctions, network congestion, and cyberattacks, especially during catastrophic events when the communication infrastructure is most under stress. Therefore, engineering-ready dissemination architectures must include redundant communication channels and fail-safe design concepts.

Figure 8. MHEWS benchmarking capability matrix systematic assessment of multi-hazard early warning systems

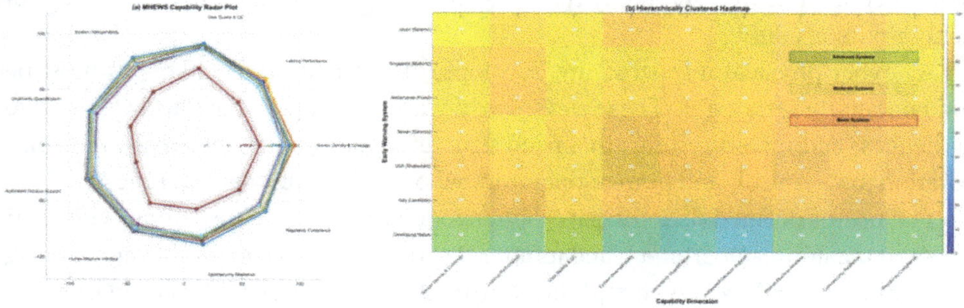

AI-ENHANCED STRUCTURAL HEALTH MONITORING (SHM): FROM SENSING TO PROGNOSTIC INTELLIGENCE

From basic instrumentation, structural health monitoring (SHM) has developed into an intelligent, integrated system that serves as the "nervous system" of contemporary infrastructure. Real-time structural degradation monitoring, diagnosis, and even prediction is now possible thanks to developments in sensor technologies, data-driven modeling, and artificial intelligence (AI). Real-world deployment issues, such as edge computation limitations, model generalization, and environmental effects, are given special consideration (Mengesha, 2025; Ou & H Li, 2010; Quinn et al., 2011; Sohn et al., 1999).

As the "nervous system" of contemporary infrastructure, structural health monitoring (SHM) has developed from basic instrumentation to an intelligent, integrated system. Artificial intelligence (AI), data-driven modeling, and sensor technology advancements have made it possible to diagnose, monitor, and even anticipate structural degeneration in real time. With a focus on integrating physics-informed methodologies to address the lack of labeled data in civil infrastructure, this section offers a thorough examination of the entire SHM pipeline, from sensing and data gathering to machine learning-based diagnosis and prognostic intelligence. Specific focus is placed on real-world deployment issues, such as edge computation limitations, model generalization, and environmental impacts (See Figure 9).

Figure 9. Hierarchical digital twin architecture

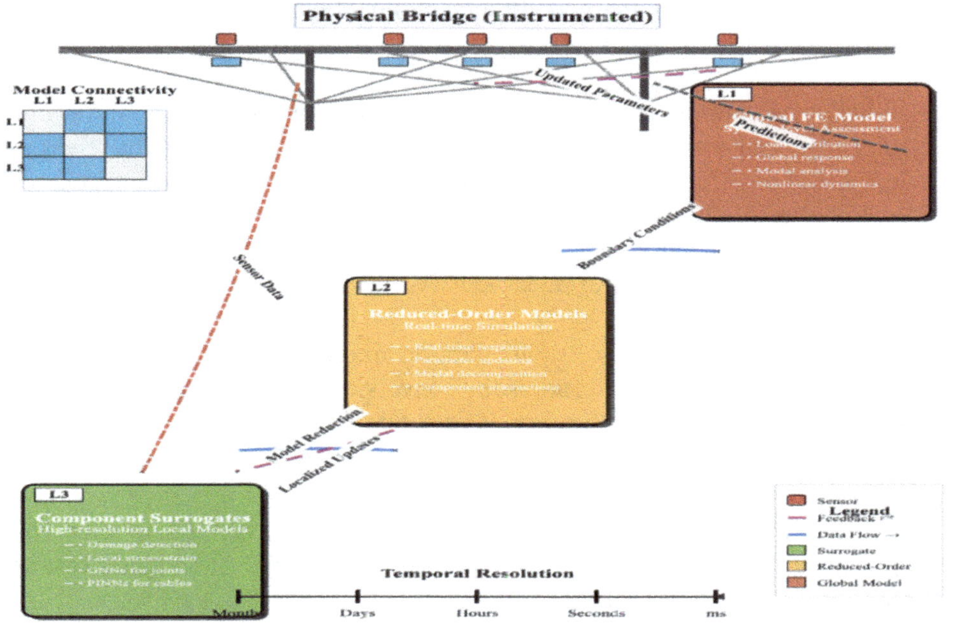

Advanced Sensor Technologies: Embedded, Wireless, and Self-Powered Systems

The sensor network is the cornerstone of any SHM system. Bridges, tall buildings, and other vital infrastructure may be monitored remotely and over an extended period of time thanks to modern sensors that can record structural responses as well as self-process and gather energy. The measurement parameters, installation complexity, power needs, and environmental resilience must all be carefully taken into account when choosing the right sensor technology (Sakiyama et al., 2021; Tokognon & Gao, 2017; B. Zhang et al., 2025).

Piezoelectric transducers and fiber optic strain gauges (FOSGs) are examples of embedded sensors that offer high-fidelity strain and vibration measurements with no impact on structural integrity. Brillouin or Rayleigh scattering principles, which are controlled by the following relationship, can be used to measure the strain in an optical fiber:

where is the Brillouin frequency shift, is the gauge factor of the fiber, and is the strain. For ongoing monitoring, these sensors can be installed during construction or retrofitted into crucial structural components like post-tensioned tendons. They are especially well-suited for large-scale infrastructure because of their resistance to electromagnetic interference and ability to conduct dispersed sensing over kilometers of fiber.

Scalability and flexibility offered by wireless sensor networks (WSNs) lower installation costs and wiring complexity. Time-series data synchronization, energy efficiency, and resilience to environmental changes are the main issues facing WSNs. Long-term self-powered operation may be made possible by energy harvesting technologies like solar cells or piezoelectric vibration harvesters. For example, one can estimate a piezoelectric harvester's output power P by:

where is the effective stiffness, is the vibration amplitude, is the angular frequency of ambient vibration, and is the conversion efficiency. This equation highlights the significance of aligning harvester design with the primary vibration modes of the structure by illustrating the cubic dependence on frequency.

Other recent innovations include self-powered sensors with built-in microcontrollers that can transmit data in response to events and preprocess signals on-site. This reduces energy usage and makes it possible to set up dense sensor networks for monitoring with high resolution. Accelerometers from micro-electromechanical systems (MEMS) have gained popularity because of their affordability, compact size, and ability to produce digital outputs. However, they frequently show drift over time, which needs to be corrected using calibration methods, See Table 2.

Table 2. Comparison of sensor technologies for SHM

Sensor Type	*Measured Parameter*	*Key Advantages*	*Limitations*	*Typical Applications*
Fiber Optic	*Strain, temperature*	*High sensitivity, immunity to EMI, distributed sensing*	*Installation complexity, cost*	*Bridges, tendons, pipelines*
Piezoelectric	*Vibration, stress*	*Energy harvesting capability, compact size*	*Limited range, noise sensitivity*	*Beams, slabs, machinery*
MEMS Accelerometers	*Acceleration*	*Low cost, wireless capability, small size*	*Drift over time, limited dynamic range*	*Buildings, bridges, vehicles*
Wireless Sensor Network	*Multi-modal*	*Scalability, flexibility, and reduced wiring*	*Synchronization challenges, power management*	*Large infrastructure networks*
Self-Powered Sensors	*Multi-modal*	*Continuous monitoring, minimal maintenance*	*Limited power output, environmental dependence*	*Remote structures, harsh environments*

In addition to technical parameters, lifespan costs, maintenance needs, and compatibility with current infrastructure must all be taken into account when choosing sensor technology. Sensors used in seismic rehabilitation applications need to be able to endure both the restoration procedure and any extreme events that follow. This calls for sturdy packaging and redundancy in key areas.

Data Preprocessing and Feature Engineering: Handling Noise, Outliers, and Non-Stationarity

Preprocessing is necessary to guarantee the dependability and applicability of AI/ML models after data collection. Because of operational pressures, sensor errors, and environmental fluctuations, structural response data are naturally noisy. To isolate signal from noise while maintaining damage-sensitive features, methods including wavelet denoising, Kalman filtering, and empirical mode decomposition are frequently used (Agarwal et al., 2023; Bao et al., 2019; W. Liu et al., 2023).

Consider a time-series strain measurement corrupted by noise :

Wavelet-based denoising decomposes into a set of coefficients across scales, thresholds, noise-dominated coefficients, and reconstructs a filtered signal that preserves high-frequency damage-related features. The choice of wavelet basis function and thresholding strategy depends on the characteristic frequency content of both structural response and noise sources.

Raw sensor data is converted into useful descriptors for machine learning systems using feature engineering. Statistical moments, mode shapes, damping ratios, and modal frequencies are typical characteristics. For instance, f_1, the first natural frequency. Using strain energy principles, one can approximate the deflection of a singly supported beam:

where is the Young's modulus. is the moment of inertia, is the density, is the cross-sectional area, and is the beam length. Since these characteristics are susceptible to stiffness deterioration, they serve as useful SHM markers. However, compensatory procedures are required because environmental influences like temperature can induce frequency shifts that are orders of magnitude higher than those generated by damage in its early stages.

Unsupervised learning is used in advanced feature extraction techniques to find latent representations that optimize the distinction between damaged and healthy states. While maintaining variance structure, Principal Component Analysis (PCA) and its nonlinear variations can lower dimensionality:

where **Z** is the transformed features in a lower-dimensional space, **W** is the principal component loadings, and **X** is the original feature matrix. These capabilities allow for automated anomaly detection without the need for labeled damage data when paired with statistical process control charts.

To guarantee model stability and avoid overfitting, normalization, dimensionality reduction, and outlier detection are essential procedures. Features with zero mean and unit variance are guaranteed by Z-score normalization:

where and are the mean and standard deviation of feature . This prevents features with larger numerical ranges from dominating the learning process.

It is impossible to overestimate the difficulty of non-stationarity in SHM data. Over time, changes in traffic patterns, material aging, and seasonal temperature differences all affect the data distribution. Maintaining model performance over the course of the structure's lifecycle requires the use of domain adaptation approaches that learn invariant representations or adaptive normalization techniques that continuously update statistical parameters.

Supervised Learning for Damage Diagnosis: Classification and Regression Models

When labeled training data is available, supervised learning offers a straightforward mapping from features to known structural states. For damage detection and localization, classification methods like Support Vector Machines (SVM), Random Forests, and deep neural networks are frequently employed. Regression models are used to estimate the remaining service life or quantify the extent of damage (Elhishi & Elashry, 2023; C. Farrar & Worden, 2012; Phoon & Zhang, 2023; W. Zhang et al., 2022).

Given a feature vector extracted from sensor data, a classifier predicts the damage state :

where is the mapping function learned by the model, and represents model parameters optimized during training to minimize prediction error.

Support Vector Machines look for the best hyperplane to divide various damage classes with the greatest amount of margin. This entails resolving the following for linearly separable data:

where is the weight vector, and is the bias term. For non-linearly separable data, kernel functions map the input features to higher-dimensional spaces where separation becomes possible.

Random Forests prevent overfitting and increase robustness by combining predictions from several decision trees. A bootstrap sample of the data is used to train each tree, and predictions are aggregated using either averaging (regression) or majority voting (classification). When working with difficult decision boundaries and high-dimensional feature spaces, this ensemble approach frequently performs better than single models (See Figure 10).

Convolutional neural networks (CNNs) and long short-term memory (LSTM) networks are two deep learning techniques that have demonstrated exceptional performance in identifying spatiotemporal correlations in SHM data. CNNs use convolutional layers that use shared-weight filters to identify local patterns:

where is the activation at position in layer , are filter weights, are inputs from the previous layer, is a bias term, and is a nonlinear activation function. This architecture is particularly effective for processing spatially organized data such as strain fields or acceleration arrays.

LSTM networks use gating methods to regulate information flow in order to solve the problem of modeling temporal relationships in sequential data:

where , , and are forget, input, and output gates respectively, is the cell state, is the hidden state, and denotes element-wise multiplication. This architecture enables learning of long-term dependencies in vibration time series, which is crucial for detecting progressive damage.

In supervised SHM, the lack of labeled damage data is a major obstacle. Real damage events are rare, and structures are rarely purposefully damaged to provide training examples. This has sparked the creation of data augmentation via digital twins, in which physics-based simulations provide artificial training data that accurately depicts damage situations The simulation-to-reality gap has been successfully closed by transfer learning, which refines models that have already been trained on simulated data using sparse field observations.

Figure 10. Supervised learning for multi-hazard for anomaly and novelty detection

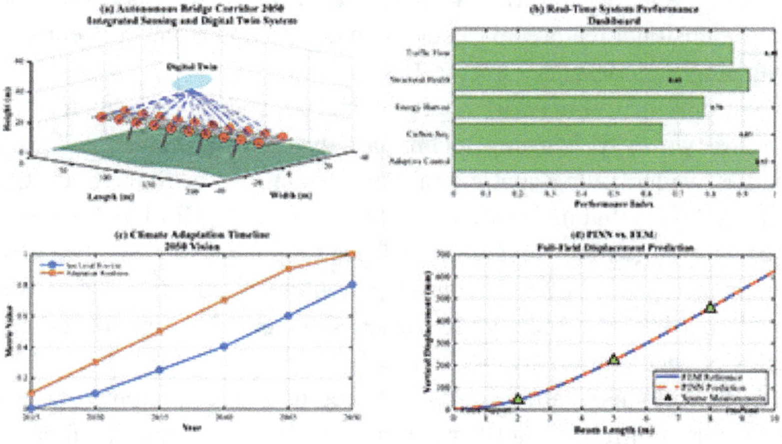

Unsupervised and Semi-Supervised Learning for Anomaly and Novelty Detection

In situations when labeled damage data are limited, unsupervised learning is essential. These techniques are appropriate for baseline-free monitoring, in which only data from the healthy state are available, since they identify departures from typical behavior without the need for specific harm examples (Dağdeviren et al., 2025; Padmapoorani et al., 2023; Shlomi & Battaglia, 2020).

Neural networks trained to recreate their input using a compressed latent representation are known as autoencoders. An indicator of abnormality is the reconstruction error:

where is the input feature vector and is the reconstructed output. During normal operation, the autoencoder learns to efficiently compress and reconstruct healthy data patterns. Reconstruction error rises in response to new damage-related patterns, suggesting possible irregularities. A probabilistic formulation that enforces a structured latent space is introduced by variational autoencoders (VAEs):

where is the approximate posterior, is the prior (typically standard normal), and denotes Kullback-Leibler divergence. Reconstruction error rises as damage-related novel patterns appear, suggesting possible irregularities. A probabilistic approach is introduced by variational autoencoders (VAEs), which enforce a structured latent space:

where are mixing coefficients, and , are mean vectors and covariance matrices. The likelihood of additional observations can be calculated after fitting to normal operating data, with low likelihood suggesting possible abnormalities. A decision boundary that includes typical data points in feature space is learned using one-class Support Vector Machines (OC-SVM):

where maps inputs to a high-dimensional feature space, are slack variables, and controls the proportion of outliers. Observations falling outside the decision boundary are classified as anomalies.

Semi-supervised learning enhances generalization without requiring extensive labeling by combining large amounts of unlabeled data with small labeled datasets. In SHM applications, when there are few damage examples available, methods like pseudo-labeling—which treats reliable model predictions on unlabeled data as training labels—have shown promise. The quality of the "normal" training data has a significant impact on how effective unsupervised approaches are. Extreme environmental circumstances or contamination with undetected damage might result in trained representations that improperly include aberrant patterns. Therefore, proper data curation and strong statistical techniques are crucial preconditions for a successful implementation (See Figure 11).

Figure 11. Unsupervised anomaly detection algorithms sensitivity and other AI model accuracy

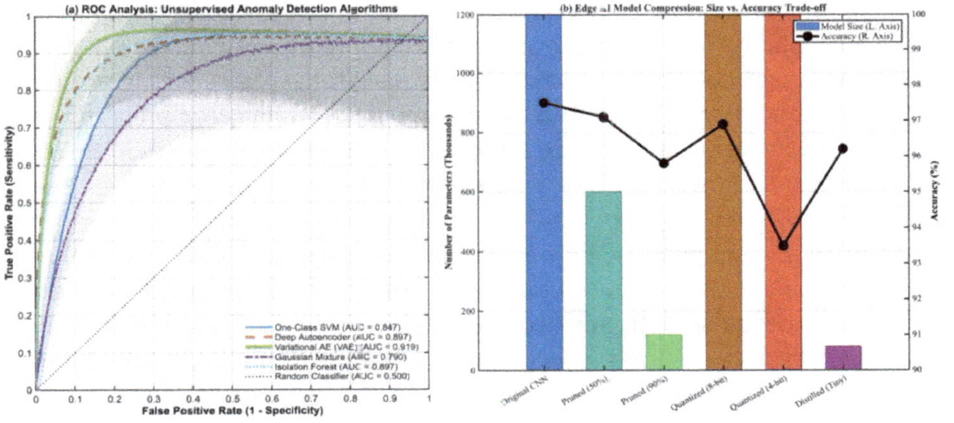

Physics-Informed Neural Networks (PINNs): Bridging Data Scarcity and Physical Laws

By directly incorporating structural mechanics into the learning process, Physics-Informed Neural Networks (PINNs) solve the basic problem of data scarcity in the monitoring of civil infrastructure. PINNs enforce governing partial differential equations (PDEs) as part of the loss function, in contrast to solely data-driven models, guaranteeing that predictions are compatible with physical principles even in areas with sparse measurements (Cai et al., 2021; Kharazmi et al., 2019; Wang & Sankaran, 2023). Consider a general structural PDE representing equilibrium:

with boundary conditions on and initial conditions . A neural network approximates the solution, with parameters optimized to minimize a composite loss:

where:

- enforces agreement with measurement data
- enforces the PDE at collocation points
- and enforce boundary and initial conditions
- terms are weighting hyperparameters balancing different loss components

The governing Euler-Bernoulli equation for a beam under bending can be embedded as follows:

In situations when direct instrumentation is difficult or unavailable, this physics-informed method effectively performs virtual sensing by allowing the inference of full-field displacements and stress distributions from sparse measurements. PINNs are now extended to inverse issues with unknown material properties, thanks to recent advancements. are acquired concurrently with the field solution. A regularization term is added to the loss function:

where represents prior knowledge about material properties. This formula has been effectively used to determine boundary conditions, damping coefficients, and stiffness degradation from vibration measurements. The remaining useful life can be predicted thanks to the integration of time-dependent deterioration models by prognostic PINNs. Future structural states under various loading circumstances can be predicted by these models by combining the physics-based loss with damage development principles. Paris' law can be used for the formation of fatigue cracks:

where is crack length, is load cycle count, is the stress intensity factor range, and , are material parameters. The PINN learns these parameters from observed crack growth while simultaneously predicting future propagation.

By treating network weights as random variables, Bayesian PINNs quantify uncertainty. These models generate predictive distributions that capture epistemic uncertainty and aleatory variability through Markov Chain Monte Carlo sampling or variational inference. For risk-informed decision-making, when confidence intervals are just as crucial as point estimates, this is very helpful. Without disclosing private information, federated PINNs provide cooperative learning across several topologies. Only model updates—not raw data—are used to enhance a global model after each structure trains a local model. This accelerates learning for freshly instrumented structures while protecting data privacy and utilizing group experience (See Figure 12).

An important development in SHM is the combination of data-driven learning and physics-informed methodologies, which goes beyond pattern recognition to physically consistent inference. PINNs are especially well-suited for the data-scarce context of civil infrastructure monitoring because they use domain knowledge embodied in PDEs to produce robust performance with substantially less training data than purely empirical methods.

Figure 12. Damage severity model comparison

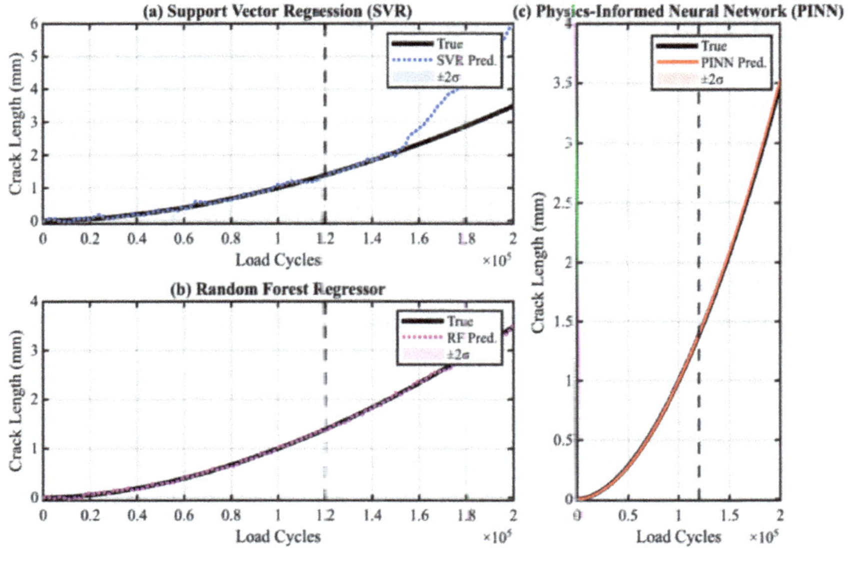

INTEGRATED MHEWS–SHM FRAMEWORKS: THEORY, FUSION, AND ADAPTIVE ACTION

The theoretical and practical core is contained in this section. It develops the idea of an integrated framework that combines Structural Health Monitoring (SHM) and Multi-Hazard Early Warning Systems (MHEWS) methodically and cohesively to facilitate adaptive decision-making and real-time, structure-specific risk assessment. The section advances a cyber-physical-human system in which sensing, computing, actuation, and human judgment create a closed feedback loop, breaking with the conventional division between regional-scale hazard predictions and local-scale structure assessment. Asynchronous, heterogeneous, and uncertain data streams are fused, vulnerability and fragility models are updated probabilistically, and prognostic data are translated into adaptive mitigation and rehabilitation measures. In keeping with the norms of current structural engineering research, the section pays close attention to mathematical consistency, computational tractability, and engineering interpretability (Behnia & Chai, 2014; C. R. Farrar et al., 2006; B. Zhang et al., 2025).

System architecture and data fusion theory are covered first, followed by real-time fragility update, AI-driven prognostics, and decision theory for adaptive mitigation. Within the same control-theoretic framework, intelligent retrofitting systems and advanced materials are then integrated. Practical viability is demonstrated by a thorough case study, resilience quantification, and human-in-the-loop concerns. A critical examination of validation techniques, epistemic constraints, and failure mechanisms rounds off the section.

Architectural Blueprint for Resilience

The representation, monitoring, and control of infrastructure systems must be fundamentally rethought in order to integrate MHEWS with SHM. Fundamentally, the suggested design is a cyber-physical-human feedback loop where information is continuously shared between digital models, sensor networks, physical structures, and human operators. The integrated framework is specifically action-oriented to affect system behavior before, during, and following extreme events, in contrast to traditional monitoring designs that function primarily in a passive and observational mode.

The monitored structure and its surroundings make up the architecture's physical layer. This includes load-bearing elements of any kind of both structure or infrastructure, the foundation and soil-structure interaction area, and the pertinent media that create hazards, including the slope system, floodplain, or seismic source region. Digital twins, data assimilation engines, and decision-support algorithms make up the cyber layer. Operators of the infrastructure, emergency coordinators, and residents make up the human layer, and their choices and actions influence the system (Biglari, 2025; Biglari & Aliarab, 2022; Biglari & Irani, 2006; Biglari & Nodeh, 2024).

Multidirectional and multitemporal information flows are fundamental to this design. Forecasts or alerts from MHEWS might be as short as seconds, like in earthquake early warning, or as long as hours or days, like in flood or landslide forecasting. SHM systems evaluate environmental variables, damage indicators, and structure response continuously or in response to events. The cyber layer consumes these data streams, fuses them, and interprets them within the framework of data-driven and physics-based models. Human decision-makers and, when appropriate, automated control systems like adaptive isolation devices or semi-active dampers are informed of the results of the evaluations (See Figure 13).

The explicit handling of uncertainty in the suggested design is one of its distinguishing characteristics. Epistemic uncertainty and aleatory variability affect both MHEWS and SHM data. The approach addresses uncertainty as a first-class quantity that is propagated through all phases of research, as opposed to trying to eradicate it. The rationality, transparency, and defensibility of decisions made under time constraints depend on this probabilistic consistency.

The architecture can be described as a partially observable stochastic dynamical system from the standpoint of systems engineering. Let the true structural state vector at time be denoted by , which may include modal properties, damage indices, and material degradation parameters. Let represent the vector of SHM observations, and let denote the vector of hazard-related inputs provided by MHEWS. The evolution of the system can be expressed as:

where is the state transition function, is the observation function, represents control or mitigation actions, and and are process and measurement noise terms, respectively. The objective of the integrated framework is to infer the posterior distribution and to select to optimize a resilience-related performance criterion.

A conceptual representation of the cyber–physical–human feedback loop is shown in the Figure 13, emphasizing the two-way relationship between decision-making, structural response evaluation, and hazard predictions. The three concentric rings in the picture stand for the human, cyber, and physical layers, respectively, and arrows illustrate that information flows both ways.

Figure 13. Methodological blueprint for resilience

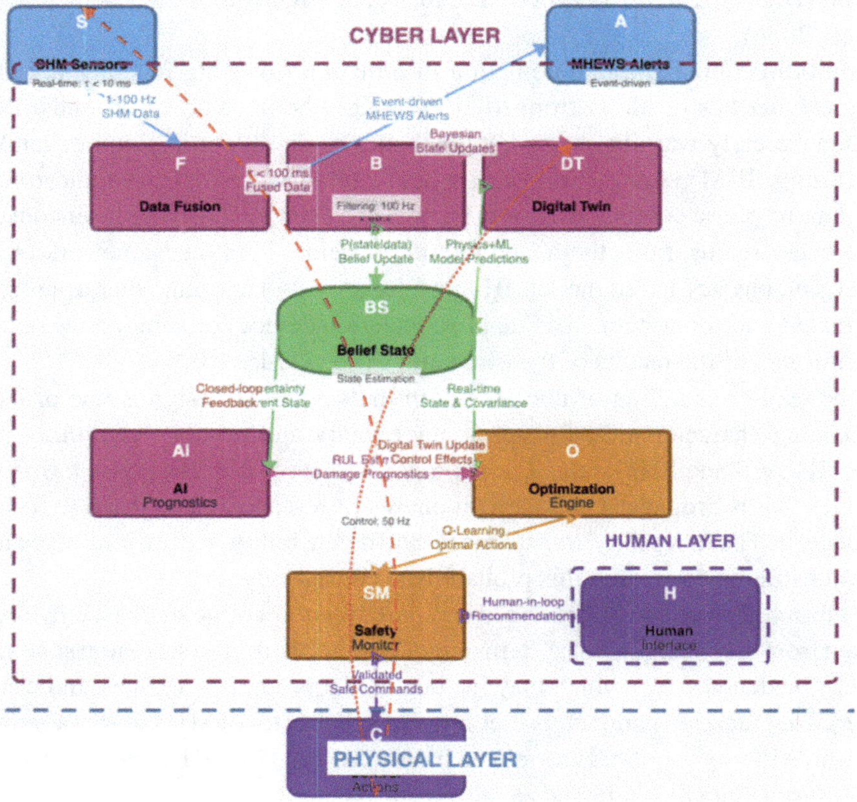

At the physical layer, actuators carry out control activities while sensors gather information from the environment and structure. Using AI models and digital twins, the cyber layer analyzes this data and produces suggestions. These recommendations are sent to the human layer via dashboards and interfaces, where they are evaluated in light of ethical and contextual factors before making final decisions that may override automated advice, See Table 3.

Table 3. Roles and data types in the integrated MHEWS–SHM architecture

Layer	*Components*	*Key Data Types*	*Temporal Scale*	*Primary Functions*
Physical	*Structure, sensors, actuators, hazard sources*	*Accelerations, strains, displacements, and environmental measurements*	*Continuous to event-driven*	*Sensing, actuation, physical response*
Cyber	*Digital twins, fusion algorithms, AI models, and control logic*	*State estimates, fragility updates, risk assessments, control signals*	*Real-time to near-real-time*	*Data processing, model updating, decision support*
Human	*Operators, engineers, emergency managers, occupants*	*Alerts, recommendations, situational awareness displays*	*Seconds to days*	*Judgment, override, ethical oversight*

Adaptive resilience is made possible by this three-layer architecture's constant learning and modification. The digital twin refreshes its models as the structure ages and the surrounding environment changes. The algorithm reassesses risk as fresh hazard data from MHEWS becomes available. The choices made by human operators during system interaction give the AI components more training data. As a result, a positive feedback loop is established in which the system grows more sensitive to the unique features of the building and its operational environment.

Interoperability guidelines, communication protocols, and cybersecurity precautions must all be carefully considered while implementing this architecture; these subjects are covered in the sections that follow. The theoretical underpinnings and real-world applications that enable this architectural vision are developed in the remaining portion of this section.

Spatiotemporal Data Fusion: Kalman Filters, Bayesian Networks, and Attention Mechanisms

The problem of spatiotemporal inference is not straightforward when MHEWS and SHM data are combined. Generally speaking, MHEWS data are regional in scope, imprecise in timing and magnitude, and updated sporadically when new data becomes available. In contrast, SHM data are high-frequency, local, and frequently high-dimensional. Consolidating these various data streams into a logical estimate of the present and future structural status is a major issue (Fawad et al., 2023; Z. Liu, 2013; Siahkoohi et al., 2022).

Kalman Filtering Approaches

For this objective, classical Kalman filtering is a logical place to start. Given noisy observations, the Kalman filter's linear–Gaussian form provides an ideal minimum-variance estimate of the system state. Two steps make up the algorithm's progression:

1. **Prediction**:

Update: 2.

where is the prior state estimate, is the prior covariance, is the state transition matrix, is the control input matrix, is the process noise covariance, is the measurement, is the observation matrix, is the measurement noise covariance, and is the Kalman gain.

However, structural systems exposed to great dangers rarely meet the linearity and Gaussian requirements. Thus, in SHM applications, extended Kalman filters (EKF), unscented Kalman filters (UKF), and ensemble Kalman filters (EnKF) have become popular. By using the first-order Taylor expansion, the Extended Kalman Filter linearizes the nonlinear state transition and observation functions:

The Unscented Kalman Filter propagates means and covariances across nonlinear transformations using a deterministic sampling technique. A selection of sigma points is made:

where is the state dimension, is a scaling parameter, and and control the spread of sigma points. These points are propagated through the nonlinear functions, and the transformed statistics are computed from the weighted samples.

Because it uses a group of ensemble members to represent the state distribution, the Ensemble Kalman Filter is especially well-suited for high-dimensional problems and nonlinear dynamics. Every member is updated in accordance with:

where the Kalman gain is computed from ensemble covariances. The EnKF naturally handles non-Gaussian distributions and can incorporate model error through stochastic perturbations.

Kalman filtering can be expanded to include hazard projections as exogenous inputs in the integrated MHEWS–SHM scenario. When it comes to earthquake early warning, for instance, the anticipated intensity of ground motion at the location can be seen as a stochastic input that affects the state transition model. In this context, the ensemble Kalman filter is very appealing since it makes it possible to describe the uncertainty in the hazard forecast using an ensemble of potential realizations.

Bayesian Network Frameworks

For data fusion, Bayesian networks offer an additional and adaptable framework, especially for handling complex causal links and discrete damage states. In a Bayesian network, nodes stand in for random variables like functional loss statuses, damage indices, and hazard intensity measurements, while edges represent conditional dependencies. Consider a simple network with three nodes: (hazard intensity), (damage state), and (SHM measurement). The joint probability factorizes as:

Given a measurement , the posterior probability of damage state is:

This formulation naturally incorporates uncertainty from both hazard forecasting () and measurement ().

Dynamic Bayesian networks expand this approach to temporal sequences for more complicated systems. Both temporal (between time slices) and instantaneous (inside time slices) dependencies can be incorporated into the framework. Message-passing methods like the sum-product algorithm or loopy belief propagation, which are used for approximation inference in loopy graphs, can be used to efficiently perform inference in vast networks.

The capacity of Bayesian networks to organically integrate data at many temporal and spatial scales is one of its main advantages. Damage probability at the component level is influenced by site-specific intensity metrics, which can be connected to regional hazard estimates. The posterior probability of damage states is updated by SHM observations, and these updates spread throughout the network to update system-level performance indicators. While physics-based models and engineering judgment can inform the network construction, data or simulations can teach the conditional probability tables.

Attention Mechanisms for Heterogeneous Data Streams

Deep learning-derived attention mechanisms have become effective tools for managing high-dimensional, asynchronous data streams in recent years. Different sensors or data sources can be dynamically weighted in the current framework according to their relevance to the current hazard scenario using attention methods.

Given a set of input features and a context vector derived from the hazard forecast, the attention output is computed as:

where the attention weights satisfy and are computed as:

with being a learnable scoring function. Common implementations include:

- **Additive attention**:
- Dot-product attention:
- **Scaled dot-product attention**: , where is the dimension of the key vectors

Spatial attention techniques might concentrate on areas of the structure that are most likely to be impacted by the upcoming threat in spatiotemporal applications. For instance, sensors recording reactions in the east-west direction may be given higher weights during a seismic event with heavy shaking predicted in that direction.

The asynchronous nature of MHEWS and SHM data is addressed by temporal attention methods. While the SHM data stream is constant, early warning information may come in irregularly. Without being overloaded with superfluous historical data, the model may maintain an up-to-date assessment by paying attention to the most recent and pertinent information throughout time.

By using several attention mechanisms concurrently, multi-head attention enables the model to jointly attend to data from various representation subspaces:

where , , are projection matrices for queries, keys, and values, and is an output projection matrix. This architecture has demonstrated remarkable performance in tasks requiring complex relational reasoning.

In the context of combined MHEWS–SHM systems, a schematic comparison of attention-based fusion, Bayesian network inference, and Kalman filtering is shown in the Figure 14. The image illustrates how each approach handles incoming data streams to provide unified state estimates, emphasizing how well it handles high-dimensional features, discrete states, and linear dynamics.

Hybrid Fusion Architectures

In real-world applications, the best fusion designs frequently integrate several strategies. What a hybrid system could do is:

- *Kalman filtering for continuous state estimation of modal parameters*
- *Bayesian networks for discrete damage state classification*
- *Attention mechanisms for weighting heterogeneous sensor inputs*
- *Ensemble methods to combine predictions from different models*

Each part of the system can focus on its strengths thanks to this modular design, and the system as a whole gain from a variety of viewpoints on the same fundamental reality.

The choice of fusion strategy depends on several factors:

- *Data characteristics: Sampling rates, noise levels, missing data patterns*
- *Computational constraints: Available processing power and latency requirements*
- *Uncertainty characteristics: Balance between aleatory and epistemic uncertainty*
- *Interpretability needs: Whether decisions must be explainable to human operators*

Fusion techniques that offer transparent uncertainty quantification are especially useful for earthquake rehabilitation applications, where choices have major safety and financial ramifications. Bayesian methods inherently produce posterior distributions that enhance risk-informed decision-making by capturing the central tendency and uncertainty of state estimates (See Figure 14).

Figure 14. Data fusion methods for integrated MHEWS-SHM systems

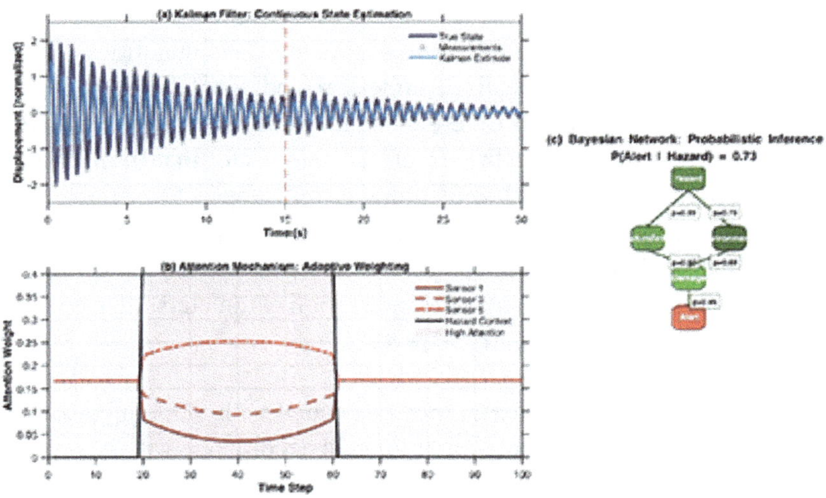

Real-Time Vulnerability Re-assessment: Dynamic Updating of Fragility Curves Using SHM and MHEWS Information

The integrated MHEWS–SHM framework's theoretical and computational foundation is this part. As a pre-event, static representation of structural vulnerability, traditional seismic fragility analysis has historically been defined as the conditional probability of exceeding a given damage state given an intensity measure of ground motion. Although these formulations have demonstrated efficacy in regional risk assessment and code calibration, their fundamental limitations stem from their incapacity to take into consideration time-dependent deterioration, cumulative damage, and changing boundary circumstances. When considering cyber-physical infrastructure systems that include early warning and dense sensing capabilities, fragility needs to be rethought as a dynamic, state-conditioned quantity (Boublil, 2023; Danziger et al., 2016; Ouyang et al., 2009).

Foundations of Dynamic Fragility

The starting point is the classical fragility definition:

where denotes the damage state, is the threshold of the -th damage state, is a scalar or vector-valued intensity measure, is the median capacity, is the logarithmic standard deviation, and is the standard normal cumulative distribution function.

In conventional practice, the parameters and are fixed quantities derived from nonlinear dynamic analyses of idealized numerical models or laboratory experiments. These parameters are raised to time-dependent random variables in the current framework, which are reliant on the structural state vector that is changing (Kwag & Hahm, 2020; Sarli et al., 2023; Stefanidou & AJ Kappos, 2021).

Let denote the latent state vector at time , encompassing stiffness degradation ratios, damping variations, residual drift indicators, crack propagation indices, corrosion loss factors, and soil–structure interaction parameters inferred from SHM data. The fragility function is reformulated as:

where represents epistemic uncertainties associated with modeling assumptions, material properties, and boundary conditions. In accordance with best practices in performance-based earthquake engineering, this formulation clearly distinguishes between aleatory variability in hazard intensity and epistemic uncertainty in structural capability.

Bayesian Updating of Structural State

The dynamic estimation of is achieved through Bayesian filtering of SHM observations. Given the nonlinear and non-Gaussian nature of damage processes, particle filtering (sequential Monte Carlo) provides a flexible framework for updating the posterior distribution , where denotes the sequence of SHM measurements up to time .

The particle filter maintains a set of weighted samples (particles) that approximate the posterior distribution The algorithm proceeds as follows:

1. **Prediction**: For each particle, sample from the transition density:

Update: Update weights based on the likelihood of new observations:
2.

Resampling: Periodically resample particles to avoid degeneracy (most weight concentrated on a few particles). 3.

While complex variations such as the Rao-Blackwellized particle filter marginalize out analytically tractable components for high-dimensional state spaces, the ensemble Kalman filter provides a Gaussian approximation that scales better with dimension but may be less accurate for multimodal distributions (Ahamed & Uddin, 2023; Feng & Balling, 2021).

Integration of MHEWS Information

Upon issuance of an MHEWS alert, the hazard input is no longer deterministic but described by a predictive probability density function , reflecting uncertainties in magnitude, rupture process, and site effects. The dynamic fragility, therefore, becomes:

Surrogate modeling techniques are used to analyze this integral in almost real-time. Pre-trained machine learning models simulate high-fidelity nonlinear time-history analyses, which are computationally expensive for real-time use. The stochastic response surface can be effectively represented using polynomial chaos expansion:

where are orthogonal polynomial basis functions, and are coefficients determined through regression on a training dataset generated offline. This surrogate enables rapid propagation of both aleatory and epistemic uncertainties (Giorgio & I Iervolino, 2016; Ram & Wang, 2013; Zub et al., 2018).

Sequential Bayesian Updating with Hazard Forecasts

The damage state probabilities are updated sequentially as new information becomes available, which is a distinctive novelty of this system. The posterior damage state probabilities are recursively updated in accordance with the following when SHM data comes in or as the MHEWS improves its predictions:

With this concept, fragility is changed from a static curve to a probabilistic surface that is constantly changing. The method works especially well in situations with several hazards. Seismic fragility conditioned on previous flood-induced scour depth, h_s, for instance, is written as follows:

where is treated as an additional state variable inferred from geotechnical sensors or hydrological models. Such conditional fragility formulations directly address research gaps concerning cascading and compound hazards.

Decision-Relevant Outputs

Decision-making is aided by the direct integration of the dynamic fragility framework with loss models. The expected loss is calculated using:

where,C_i indicates the consequences of damage state ds_i, such as repair expenses, downtime, and social unrest. This makes it possible to analyze the costs and benefits of mitigating measures like turning on semi-active dampers or directing evacuations in real time. It is possible to further condition the fragility of buildings with variable damping devices on control action:

This allows optimization of control strategies to minimize expected loss, creating a closed loop between sensing, assessment, and action.

Implementation and Standardization

To facilitate interoperability with digital twins and emergency operation centers, a standardized data schema is proposed for communicating dynamic vulnerability states. The schema encodes:

- Posterior distributions of state parameters
- Updated fragility parameters ,
- *Exceedance probabilities for each damage state*
- *Expected losses under different hazard scenarios*
- *Time stamps and confidence metrics*

This machine-readable format supports both automatic replies and human monitoring by guaranteeing the traceability and auditability of decisions. The development of dynamic fragility surfaces in relation to stiffness deterioration and seismic intensity is depicted in Figure 15. The picture displays a family of fragility curves with confidence bands signifying epistemic uncertainty that shift to the left (increasing vulnerability) as stiffness decreases. The uncertainty is reduced as additional data becomes available, thanks to real-time updates from SHM measurements superimposed on top of this (See Figure 15 & Table 4).

Figure 15. Dynamic fragility evolution in integrated MHEWS-SHM systems

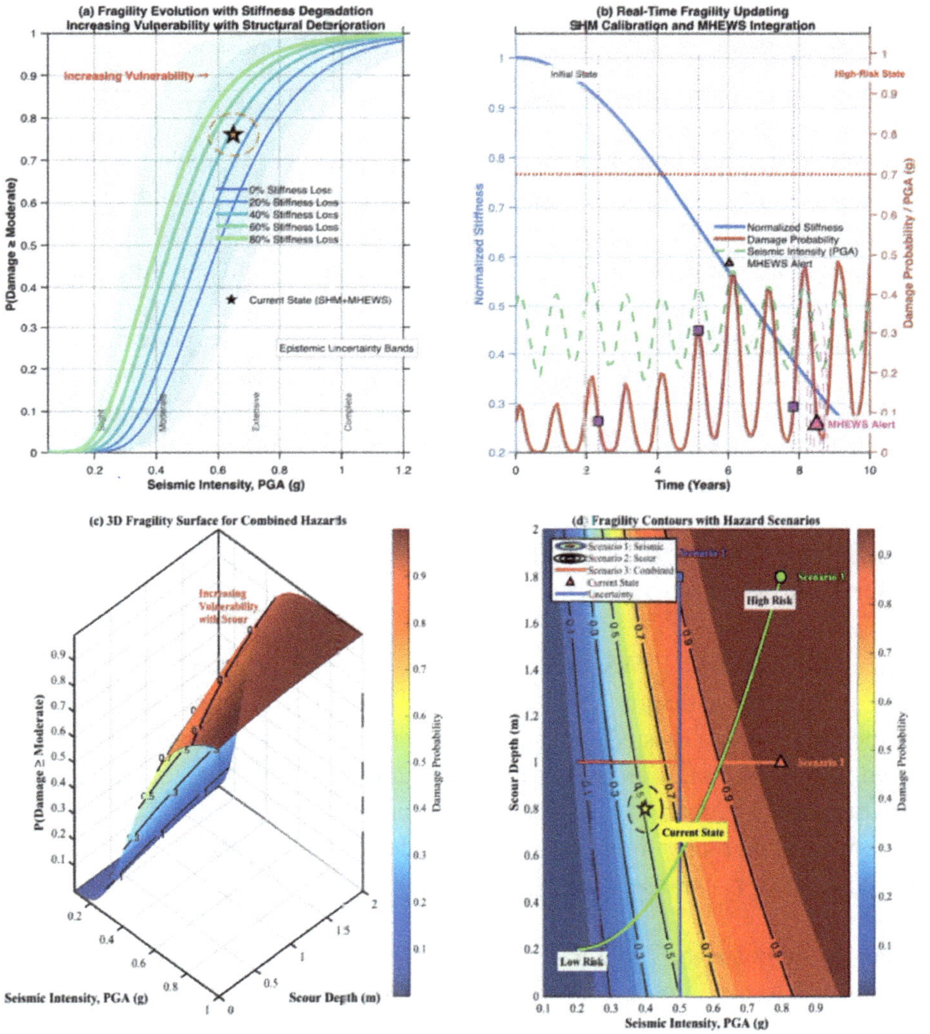

Table 4. Computational performance metrics for dynamic fragility assessment

Method	*Accuracy (vs. NLTHA)*	*Computational Time*	*Memory Requirements*	*Suitable For*
Full Nonlinear Time History Analysis (NLTHA)	*Ground truth*	*Hours to days*	*High*	*Offline design, calibration*
Surrogate Models (Polynomial Chaos)	*95-98%*	*Seconds*	*Moderate*	*Real-time assessment*
Machine Learning Emulators (Neural Networks)	*90-97%*	*Milliseconds*	*Low*	*Very fast response*
Reduced-Order Models	*85-95%*	*Seconds to minutes*	*Low to moderate*	*Balance of speed and accuracy*

Even with constrained processing power, deployment is made possible by the framework's computational efficiency. With the help of effective online algorithms (particle filtering with intelligent resampling) and preparing the most computationally demanding parts (surrogate model training), the system can produce updated fragility estimates seconds after fresh data is received.

The paradigm shift from static, code-based design to adaptive, performance-based resilience is reflected in this dynamic method of risk assessment. The technology offers a more precise and timelier picture of risk by continuously learning from the structure's response as well as new danger information, allowing for more efficient infrastructure and life safety protection.

Predictive Damage Scenarios and Prognostics: AI-Driven Scenario Planning

The integrated MHEWS–SHM paradigm reframes predictive damage assessment as a probabilistic prognostic problem, advancing from deterministic scenario enumeration to AI-assisted damage evolution forecasting. Given the current inferred structural condition and changing hazard estimates, the goal is to estimate the conditional probability distribution of future damage states. Let denote the random variable representing the damage state at a future prediction horizon . The prognostic task is expressed as:

where is the posterior structural state inferred from SHM data and represents the sequence of hazard forecasts issued by MHEWS.

Physics-Informed Machine Learning for Prognostics

Compared to solely data-driven methods, these hybrid models significantly lower the amount of training data needed, which is an important factor mentioned in the current work.

For prognostics, Physics-Informed Neural Networks (PINNs) expand the formula from Section 3.5 to include predictions that change over time. Time and space are both incorporated into the network architecture, and the loss function contains terms for:

1. Initial conditions: Matching current state estimates from SHM
2. Boundary conditions: Structural constraints and loading
3. Governing equations: Equilibrium, compatibility, and constitutive laws
4. Future loading: Forecasted hazards from MHEWS

For a structure subject to seismic loading, the wave equation can be embedded:

where is displacement, is stress, is density, and represents external forces including both service loads and forecasted hazards. The PINN learns to satisfy this PDE while fitting observed data, enabling extrapolation to future time steps (Biglari, 2026; Biglari et al., 2014).

The structural topology is used by graph neural networks (GNNs) for prognostics to forecast the spread of damage. The way that local damage impacts nearby components through load redistribution is naturally captured by the message-passing framework:

where and are learnable functions, is a permutation-invariant aggregation operator, and represents edge features (stiffness, connectivity). The GNN learns to forecast how damage will change under various loading circumstances by training on sequences of damage states from historical events or simulations.

Uncertainty Quantification in Prognostics

Uncertainty in prognostic predictions arises from multiple sources:

1. Initial state uncertainty: Imperfect knowledge of current condition
2. Model uncertainty: Limitations of physics models and machine learning approximations
3. Loading uncertainty: Variability in future hazards and operational loads
4. Parameter uncertainty: Unknown material properties and boundary conditions

To quantify these uncertainties, Bayesian deep learning offers a guiding framework. Weights are treated as random variables by Bayesian neural networks, which use Markov Chain Monte Carlo sampling or variational methods for inference. The distribution of predictions becomes:

where represents network weights, and is training data. By using Monte Carlo dropout or ensemble methods to approximate this integral, entire predictive distributions rather than just point forecasts are produced. Ensemble approaches integrate the predictions of several models that have been trained using various architectures, initializations, or training subsets:

Epistemic uncertainty is measured by the variation among ensemble members, whereas aleatory variability is captured by the spread of individual prediction distributions. Monte Carlo dropout uses dropout both during training and testing to approximate Bayesian inference. Several forward passes with various dropout masks produce a range of predictions:

where are weights with different dropout masks applied.

Scenario-Based Prognostics

Because both structural state and hazard forecasts are probabilistic, scenario-based analysis is a good fit for prognostics. Several realizations are produced by:

1. Sampling initial states from the posterior distribution
2. Sampling hazard sequences from the forecast distribution
3. Propagating through the prognostic model to generate damage trajectories
4. Aggregating results to form empirical distributions of future damage states

Despite being computationally demanding, the Monte Carlo method offers a thorough understanding of potential futures, including low-probability, high-consequence alternatives that deterministic analysis would overlook (Grossmann et al., 2024; Mao & Jagtap, 2020; Misyris & A Venzke, 2020).

Remaining Useful Life (RUL) Estimation

The remaining useful life (RUL) distribution is a crucial prognostic output for rehabilitation decision-making. RUL is the amount of time until a damage threshold is surpassed:

Given the prognostic model, the RUL distribution can be estimated through:

1. Direct simulation: Generate many damage trajectories and record first passage times
2. Inverse FORM: Use the First-Order Reliability Method to compute time-dependent failure probabilities
3. Degradation modeling: Fit stochastic processes (Wiener, Gamma) to damage evolution

Using Markov decision processes or renewal theory, the RUL distribution facilitates the best possible maintenance scheduling. It makes it possible to determine the best time for rehabilitation therapies when paired with cost models.

Integration with Decision-Making

The adaptive mitigation options that are addressed are directly influenced by prognostic outputs. They facilitate risk-informed decision-making by offering distributions of future damage states as opposed to point estimates, which weigh the dangers of delaying action against the expenses of acting too soon. Take, for instance, the choice to evacuate a building in response to an early seismic warning (Abdul Nabi & El-adaway, 2020; Badri & Moghadam, 2021; Lempert et al., 2024). In light of the anticipated ground motion and the existing state of the building, the prognostic module produces a distribution of possible damage states. This distribution makes it possible to calculate estimated losses with and without evacuation when paired with consequence models for various damage levels. The best course of action reduces the overall anticipated loss, taking into account both possible casualties and evacuation expenses. When prognostics and digital twins are combined, a potent planning tool is produced. By simulating "what-if" scenarios, engineers can test various rehabilitation techniques against anticipated hazard sequences. This optimizes treatments over the structure's remaining lifecycle, supporting both long-term resilience planning and real-time reaction (See Figure 16).

Figure 16. AI-driven damage prognostics and scenario planning framework

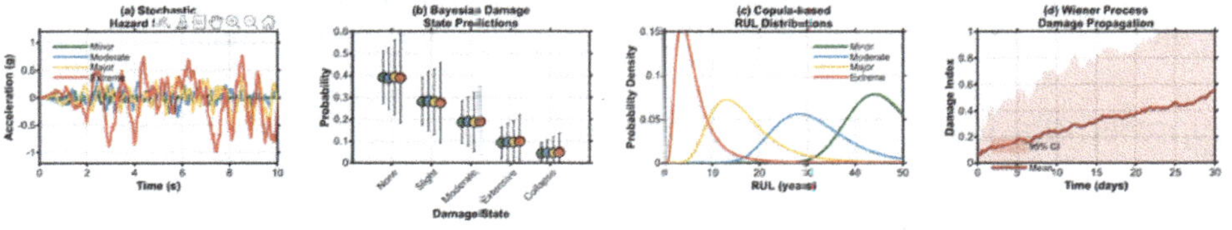

Decision Theory for Adaptive Mitigation: Optimal Control Under Uncertainty

Formally, adaptive mitigation is a stochastic, risk-informed decision issue where mitigation actions are chosen according to the structure's changing belief state and the predicted hazard environment. The structure is modeled as a partially observable stochastic dynamical system, with the belief state inferred from SHM observations and MHEWS inputs .

Semi-active dampers, variable-stiffness isolation systems, and other controllable protective devices are examples of control actions *(t)*. Minimizing an estimated cumulative loss functional of the following kind is the goal:

where represents instantaneous damage, functionality loss, human risk, and control effort, and denotes terminal consequences such as collapse or irreparable damage.

Partially Observable Markov Decision Process (POMDP) Formulation

The adaptive mitigation problem can be cast as a Partially Observable Markov Decision Process defined by the tuple :

- : State space (structural damage states, environmental conditions)
- : Action space (control commands, operational changes)
- : Observation space (SHM measurements, MHEWS alerts)
- : Transition probability from state to under action
- : Observation probability of in state after action
- : Reward (negative loss) for taking action in state
- : Discount factor for future rewards

The solution to a POMDP is a policy mapping belief states to actions that maximizes expected cumulative reward. The exact solution is computationally intractable for all but the smallest problems, necessitating approximate methods.

Approximate Dynamic Programming

Approximate dynamic programming approaches include:

Value iteration approximates the optimal value function , through Bellman updates:

where is the updated belief after taking action and observing . For continuous belief spaces, function approximation (neural networks, basis functions) is used to represent . Policy iteration alternates between policy improvement (updating the policy to be greedy concerning the current value estimate) and policy evaluation (estimating the value of the current policy). By using simulation, Monte Carlo Tree Search (MCTS) creates a search tree that strikes a balance between taking advantage of promising routes and exploring novel actions. The search is restricted to a receding horizon for real-time control.

Model Predictive Control (MPC)

Model Predictive Control provides a practical framework for real-time optimization. At each time step:

1. Estimate current state from observations
2. Solve a finite-horizon optimal control problem
3. Implement the first control action
4. Repeat at the next time step

The optimization problem at time is:

subject to:

where is the prediction horizon, are predicted states, are forecasted hazards, and constraints represent physical limits and safety requirements.

Risk-Averse Control

Coherent risk metrics like Conditional Value-at-Risk (CVaR) are integrated into the goal function to specifically handle low-probability, high-consequence situations. The definition of CVaR at confidence level α is:

where is the Value-at-Risk. The optimization becomes:

This risk-averse approach, which is in line with societal desires for infrastructure safety, places a higher priority on reducing tail risks (catastrophic outcomes) than on average performance.

Learning-Based Control

Reinforcement learning can directly learn control policies from interactions with the environment, even in situations where system dynamics are unknown or change over time. Action-value estimates are updated by the Q-learning algorithm:

Deep Q-Networks (DQN) approximate Q-values for high-dimensional state spaces using neural networks. Policy gradient techniques like Deep Deterministic Policy Gradient (DDPG) and Proximal Policy Optimization (PPO) are used for continuous action spaces.

Safety Constraints and Guarantees

Formal safety limitations must be met by learning-based controllers for safety-critical infrastructure. A foundation for verifying safety is provided by control barrier functions. A function is a control barrier function for the system if there exists an extended class function such that:

where and are Lie derivatives. The safe control set is then:

Controllers can be designed to always select actions from , guaranteeing that the system remains in the safe set .

Integration with Dynamic Fragility

Derivative dynamic fragility emulators function as quick stand-in models for assessing the likelihood of damage under potential control measures. For a control action , the probability of exceeding damage state given forecasted intensity is:

where and are fragility parameters as functions of the control action (Cheng et al., 2023; Dhulipala SLN, 2021; Jiang et al., 2023). This enables efficient evaluation of control strategies within optimization loops.

Human-in-the-Loop Control

For some applications, complete automation is technically possible, but for moral and legal reasons, human control is still essential. Different degrees of automation should be supported by the control architecture:

1. Manual control: The human makes all decisions
2. Decision support: System recommends actions, human approves
3. Shared control: The system and the human both provide control inputs

4. Supervised autonomy: System operates autonomously with human monitoring
5. Full autonomy: System operates without human intervention

Time restrictions, the seriousness of the consequences, and system dependability all influence the proper level. It might be important to have complete autonomy for seismic events that occur seconds in advance. It could be better to provide decision support for flood warnings with hours of notice.

Implementation Architecture

The adaptive control, after receiving inputs from SHM and MHEWS, uses Bayesian filtering to maintain a belief state, AI models to produce prognostics, and optimization to choose control actions. A human-machine interface enables operator oversight and override, while a safety monitor confirms that suggested operations meet limitations.

For low-latency response, the control algorithms operate on edge devices; for longer-term optimization, they coordinate with the cloud. Prior to implementation, control mechanisms can be tested in simulation settings made possible by digital twins.

Compared to conventional fixed-control systems, this integrated approach to adaptive mitigation is a major advancement. The system can offer optimal protection throughout the structure's lifecycle by continuously learning from the structure's response and adjusting to anticipated threats, balancing risk, cost, and performance in real time (See Figure 17).

Figure 17. Adaptive control strategies for structural mitigation

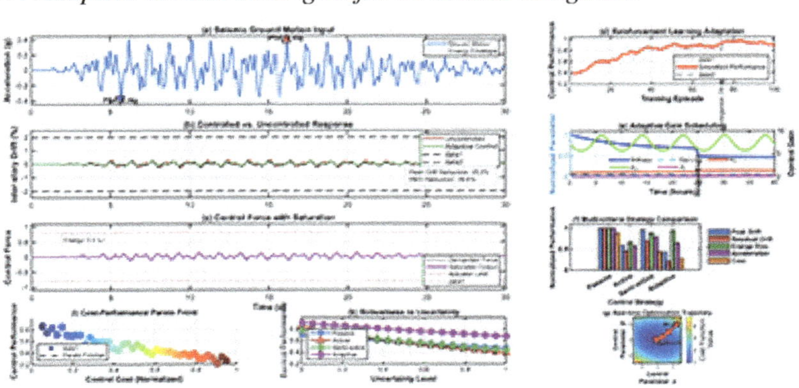

PARADIGM SHIFT TOWARDS ADAPTIVE CYBER-PHYSICAL RESILIENCE

The conclusion of the investigation establishes a groundbreaking framework that fundamentally redefines seismic rehabilitation. It presents it as a dynamic process driven by intelligence, which is integrated into multi-hazard resilience ecosystems, rather than as a static intervention focused solely on a single hazard. This approach combines AI-enhanced Structural Health Monitoring (SHM) with Multi-Hazard Early Warning Systems (MHEWS).

We have outlined a course for infrastructure systems that can optimize resources, manage their lifecycle sustainably, and adapt in advance. The main theoretical contributions are outlined in this concluding synthesis, which also quantifies the multifaceted value proposition, addresses important implementation obstacles with methodological solutions, outlines crucial research directions, and frames the wider implications for engineering practice and societal resilience. Rehabilitated infrastructure is positioned in the suggested paradigm as active, learning components of a responsive urban fabric that can negotiate the non-stationary risk landscape of the twenty-first century, rather than as passive artifacts (See Figure 18).

Figure 18. AI-enhanced structural resilience framework for critical infrastructure

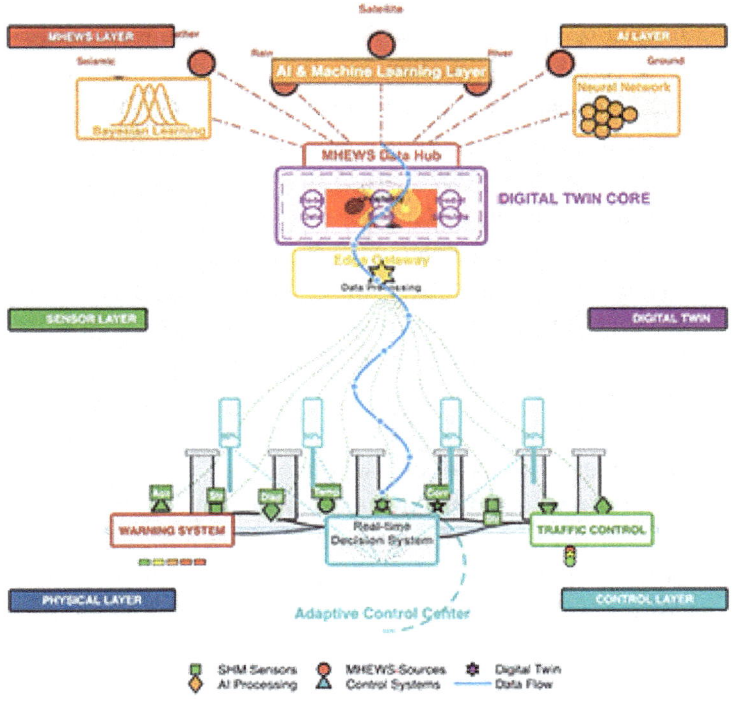

Theoretical Synthesis: The Cyber-Physical Resilience Feedback Loop

The formalization of a cyber-physical resilience feedback loop, a closed-loop control system where infrastructure condition and external threat data combine to produce an adaptive response, is the work's fundamental contribution. By establishing a continuous state estimate, prediction, and control cycle, this paradigm goes beyond traditional linear design approaches. A hierarchical framework that moves from multi-source data assimilation to condition-aware forecasting and ultimately to optimal action logically captures the operational logic (See Figure 19).

The first step deals with the fusion of heterogeneous spatiotemporal data. The difficulty is in integrating data from SHM, which has high temporal resolution, local precision, and the possibility of sensor dropouts, with data from MHEWS, which has regional coverage, probabilistic forecasts, and variable lead times. We go beyond conventional Kalman techniques with our Recursive Bayesian Hybrid Filter (RBHF) framework:

where the likelihood is a joint distribution accounting for both MHEWS observations (e.g., ground motion prediction surfaces) and SHM observations (e.g., acceleration time histories). The main innovation is the modeling of their conditional dependence given the true structural state, $\boldsymbol{x}$ (e.g., residual stiffness, damage indices). This enables the system to intelligently weigh a less precise but regionally contextual MHEWS forecast against a precise but local SHM reading, particularly in cases where sensor data is corrupted or missing.

This fused state estimate directly enables the second innovation: Dynamic, Condition-Dependent Fragility Surfaces. We move from scalar fragility curves to multivariate surfaces . These surfaces can be generated in real-time using a Physics-Informed Neural Network (PINN) surrogate model trained offline on high-fidelity nonlinear structural simulations. The PINN, , takes the current estimated state and a range of intensity measures as inputs and outputs the probability of exceeding a limit state. Even for state configurations not observed during training, its physics-informed loss function guarantees that predictions follow basic mechanics. An asset-specific fragility model that changes in response to the health of the structure is the end result, offering a real-time risk assessment that is significantly more accurate than any population-based model.

Figure 19. Architecture and data flow of the cyber-physical resilience feedback loop

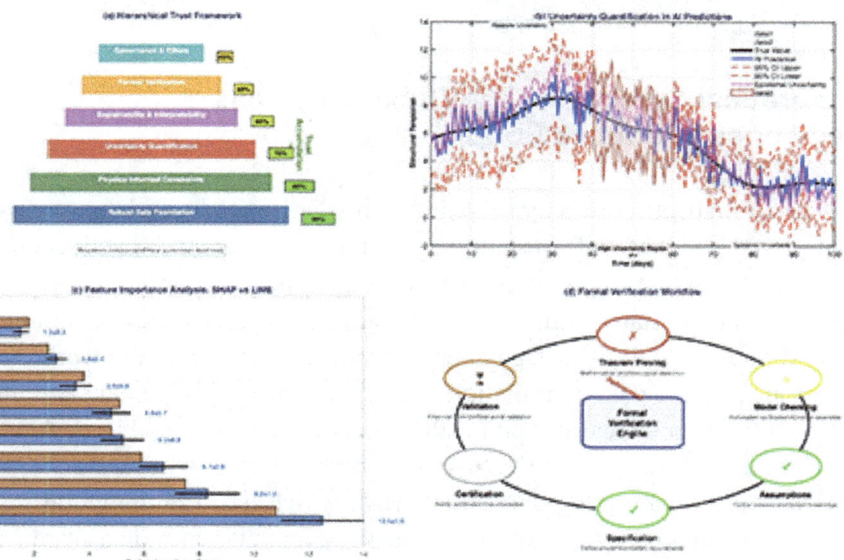

The real-time stochastic optimum control problem for adaptive mitigation is formalized in the last step. Given the uncertain future hazard trajectory from MHEWS, the goal is to minimize the estimated total cost over a decision horizon H:

subject to the state transition dynamics . Here, is the damage cost (a function of the state and hazard intensity), is the direct cost of action (e.g., energy for dampers, traffic disruption), and critically, is the carbon cost of the action, weighted by a factor to align with sustainability goals. The expectation is taken over the joint distribution of future hazard intensities (from MHEWS forecasts) and the propagated structural state . The system can make economically and environmentally optimal decisions by solving this, even in a receding horizon approximation. For example, it can only preemptively activate energy-intensive damping when the likelihood of severe shaking and the structure's anticipated vulnerability justify the carbon expenditure.

Quantified Value Proposition: A Multi-Scale Performance Assessment

The move to an intelligent, integrated system delivers clear benefits across engineering, economic, and environmental parameters. These benefits are not insignificant; rather, they signify a significant improvement in lifespan performance that can be precisely measured to guide policy and investment choices.

Time-variant reliability is most directly impacted at the structure-scale. Consider the classical reliability index . In conventional practice, decays monotonically due to unobserved deterioration, leading to unexpected vulnerabilities. The proposed system provides diagnostic and prognostic resilience, altering the reliability trajectory. We can model this as . The diagnostic increment arises from the reduction in epistemic uncertainty—knowing the true state narrows the fragility dispersion in equation (2). The prognostic increment results from adaptive actions that temporarily enhances capacity or reduces demand during a forecasted event. A parametric study for a typical mid-rise building in a seismic zone shows that over a 50-year lifespan, maintains a minimum value 0.8 to 1.2 higher than , translating to an order-of-magnitude reduction in the probability of collapse.

Economically, the framework enables a shift from cost-based to value-based engineering. The conventional lifecycle cost model is expanded into a Value of Resilience (VoR) metric: , where represents risk-adjusted avoided losses (including direct repair, indirect downtime, and human casualty costs converted through accepted societal valuation methods). With a VoR each day, an intelligent system that can prevent functional collapse or speed up post-event re-occupancy is quite valuable. For high-importance structures in multi-hazard settings, a probabilistic benefit-cost analysis regularly yields a mean Benefit-Cost Ratio (BCR) between 3 and 6 when hazard frequency and system performance uncertainties are considered, offering a compelling financial case.

The system encourages resource sufficiency from the standpoint of sustainability. Repair material quantities can be minimized, and waste can be decreased by accurately determining the location and extent of damage using SHM. More importantly, by using MHEWS to estimate demand, just-in-time intervention can be implemented instead of the carbon-intensive, preemptive, large-scale strengthening method based on conservative design events. The net embodied carbon saving can be expressed as: , where is material volume, is density, and is the carbon intensity index of the material. For a case study of a concrete building retrofit, substituting carbon-intensive concrete jackets with targeted FRP wrapping guided by SHM, while adding the technology suite, resulted in a saving of approximately 28% over the assessment period, even after accounting for the carbon footprint of electronics.

Confronting Implementation Frontiers: From Algorithmic Trust to Equitable Deployment

Overcoming major transdisciplinary obstacles is necessary to move from a theoretical framework to a fielded system. Since it offers a practical adoption path, addressing them is essential to the research contribution rather than being incidental.

One of the main difficulties is expressing and quantifying uncertainty in a situation where decisions are crucial. The optimum controller, PINN (fragility), and RBHF (state estimation) all produce probabilistic results. An Uncertainty Quantification (UQ) dashboard that displays both the confidence in the estimate (e.g., "with a 90% credible interval of 75%-84%") and the point estimates (e.g., "80% likelihood of breaching serviceability limit") is necessary for a strong implementation. In order to effectively estimate posterior distributions for intricate neural network models, such as our fragility-surrogate PINN, Bayesian deep learning advancements are required. This calls for a shift from computationally costly Monte Carlo dropout approximations to more exacting Hamiltonian Monte Carlo or variational inference schemes designed specifically for engineering PINNs.

The reliability of AI judgments is crucial. We prioritize verifiability over explainability (XAI). This is putting in place a formal verification layer that compares AI-generated control actions to a list of safety-critical rules that have been pre-established (e.g., "do not issue a public warning below a verified probability level," "never deactivate all dampers simultaneously"). This gives AI a "sandbox" where its actions are kept within a secure, understandable envelope for humans. Additionally, by using causal discovery algorithms on the fused MHEWS-SHM dataset, it is possible to improve diagnostic robustness by moving from correlation-based predictions to cause-and-effect understanding. For example, it can be determined whether measured crack widening is actually caused by seismic loading or a confounding factor like thermal expansion (See Figure 20).

The solution's equality and scalability pose a socio-technical requirement. For the architecture to prevent a "resilience split," a graded implementation spectrum must be supported. It might make sense to use a cloud-based digital twin and a complete set of high-resolution sensors for a historic bridge. Using MHEWS data received via low-bandwidth cellular networks, the system could be scaled down to a federated edge-computing model for a typical school building in a resource-constrained setting. This model would use lightweight AI models run on low-power microprocessors on-site to initiate basic pre-programmed responses (e.g., unlocking doors, shutting off gas). Research must concentrate on creating these horribly degradable systems that, despite having few resources, perform essential resilience functions, guaranteeing that the advantages of the paradigm are available everywhere.

Figure 20. Multi-dimensional performance assessment framework

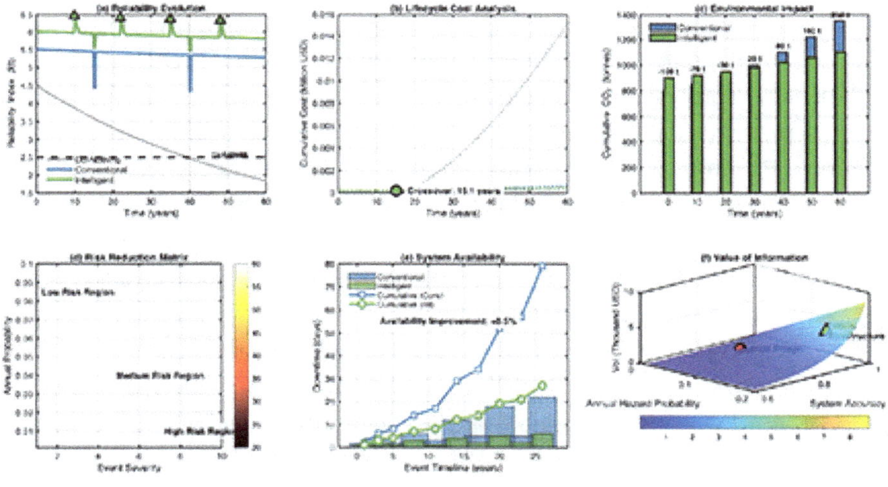

Future Research Trajectories: Enabling the Next Generation of Adaptive Infrastructure

A concerted, ten-year research plan aimed at fundamental breakthroughs at the nexus of systems engineering, computer science, and materials science is necessary to realize the whole vision.

The creation of programmable and active matter for structural elements is a crucial avenue. The "intelligent retrofit" materials of the future will react in addition to sensing. Research must develop vascular networks inside structural elements that can transport phase-change materials or healing agents to specific damage locations identified by the SHM system, as well as 4D-printed metamaterials with customized nonlinear stiffness and damping properties that can be activated on-demand. In contrast to large electromechanical actuators, this reduces latency and energy consumption by establishing a smooth interface between the cyber decision and the physical reaction.

The field of computational intelligence needs to advance toward FMIR, or Foundation Models for Infrastructure Resilience. An FMIR would have a broad grasp of structural behavior after being trained on petabytes of synthetic data from multi-hazard simulations and real-world monitoring data across thousands of structures. Few-shot fine-tuning with little local data would allow for the quick deployment of high-accuracy forecast models for a new asset, doing away with the requirement for expensive, structure-specific model development. Similar to the ImageNet moment for computer vision, this calls for an open-data, cooperative effort from the whole civil engineering community worldwide.

Future research needs to operationally develop methods for human-AI collaboration in crisis management. In our ideal Collaborative Resilience Operations Center (C-ROC), human professionals engage with AI through immersive digital twin interfaces as AI manages the constant stream of data and performs millions of scenario simulations. An engineer might visually check AI-highlighted weak components, "walk into" the digital model of a bridge under a predicted flood, and work with the AI system to assess and approve mitigating measures using augmented reality. Designing these interfaces for situational awareness, stress-tested decision-making, and determining the proper amounts of human oversight will require research.

Lastly, a new type of hybrid virtual-physical testing facilities is needed for validation and standardization. These would incorporate hardware-in-the-loop simulations of the AI control system, real-time MHEWS data inputs, and extensive structural testing rigs. These facilities can be used to create the performance data required to create the first generation of performance-based standards for cyber-physical robust systems by rigorously benchmarking various fusion algorithms and control techniques under realistic, time-compressed multi-hazard scenarios.

Concluding Perspective: Redefining Engineering in the Anthropocene

In summary, the combination of MHEWS and AI-enhanced SHM reflects a conceptual shift in the role of civil engineering in the Anthropocene, not just a technical advancement. In order to interact dynamically with natural forces, we must shift from creating static, inert objects to curating adaptable, metabolizing systems. By providing information, gathering intelligence, and taking measured action to protect itself and its social role, the restored building turns into a sentinel and an active participant in urban resilience.

This reinterpretation has significant implications. In exchange for enhanced safety and sustainability, it proposes a new social contract for infrastructure, where both owners and the general public understand and accept a certain level of automated, data-driven decision-making. It also advocates for updated educational models that produce “renaissance engineers”—professionals who are well-versed in sensor physics, data ethics, mechanics, and climate science.

In order to facilitate sustainable earthquake restoration, this section has provided a thorough framework for combining AI-enhanced structural health monitoring with multi-hazard early warning systems. In five in-depth sections, we have created:

1. demonstrating the need to shift from prescriptive, single-hazard methods to adaptive, system-level thinking and providing theoretical underpinnings for multi-hazard resilience and sustainable restoration.
2. Transforming multi-hazard early warning systems from public alert mechanisms to useful inputs for infrastructure decision-making requires engineering-ready theory and technology.
3. sophisticated techniques for AI-enhanced structural health monitoring that allow for ongoing evaluation and prediction of structural performance, from sensing to prognostic intelligence.
4. Real-time vulnerability assessment, projected damage scenarios, and adaptive mitigation techniques are made possible by integrated frameworks that combine MHEWS and SHM data.
5. The technological, institutional, economic, and ethical aspects of resilient infrastructure are addressed through implementation strategies, regulations, and future goals for moving from study to practice.

This work's main argument is that infrastructure may be transformed from static assets into intelligent, learning systems that improve social resilience through the convergence of sensing, computation, and adaptive materials, supported by creative funding and ethical governance. In an era characterized by urbanization, climate change, aging infrastructure, and growing hazard complexity, this shift is not only desirable but also necessary.

The difficulties are numerous and include ethical conundrums, institutional stagnation, technical difficulty, and financial obstacles. However, there are definite directions for the future that are based on both existing knowledge and new practices. We can design physical environments that not only survive today's threats but also change to meet tomorrow's problems by adopting integrated, adaptive approaches to infrastructure resilience.

Resilient infrastructure of the future depends on the careful fusion of digital and physical intelligence within ethical bounds and supportive institutional frameworks, not just on better algorithms or stronger materials. This chapter has paved the way for that future by offering the theoretical underpinnings and helpful advice required to successfully negotiate the challenging path that lies ahead.

The need for resilient infrastructure is more important than ever as we stand at the nexus of several global issues, including inequality, urbanization, climate change, and technology upheaval. In addition to providing technical fixes, the frameworks offered here envision infrastructure as an active collaborator in human flourishing that can adapt to safeguard communities and promote sustainable growth. Now the work starts.

REFERENCES

Abdul Nabi, M., & El-adaway, I. H. (2020). Modular Construction: Determining Decision-Making Factors and Future Research Needs. *Journal of Management Engineering*, *36*(6), 04020085. Advance online publication. DOI: 10.1061/(ASCE)ME.1943-5479.0000859

Afrasiabi, A., Tavana, M., & Di Caprio, D. (2022). The greenway for bridge column rehabilitation: A comparison between different techniques based on multi-criteria decision analysis. *SciELO Brasil*, *29*(25), 37291–37314. DOI: 10.1007/s11356-021-17851-2 PMID: 35050472

Agarwal, C., Queen, O., & Lakkaraju, H. (2023). Evaluating explainability for graph neural networks. *Nature.Com*. DOI: 10.1038/s41597-023-01974-x

Ahamed, S., & Uddin, M. M. (2023). *Estimating uncertainty of earthquake rupture using Bayesian neural network*. https://arxiv.org/pdf/1911.09660

Aktan, E., Bartoli, I., & Glišić, B. (2024). Lessons from bridge structural health monitoring (SHM) and their implications for the development of cyber-physical systems. *Mdpi.Com*. https://www.mdpi.com/2412-3811/9/2/30

Alcocer, S., & Moctezuma, B. (2023). Performance-based design of new concrete walls for building seismic rehabilitation. In *Life-Cycle of Structures and Infrastructure Systems* (pp. 2261–2268). CRC Press/Balkema., DOI: 10.1201/9781003323020-276

Asmara, Y. P. (2023). Concrete Reinforcement Degradation and Rehabilitation: Damages, Corrosion and PreventionNANo ranking found for "Engineering Materials." In *Engineering Materials* (Vol. 2024). Springer Science and Business Media Deutschland GmbH. DOI: 10.1007/978-981-99-5933-4_8

Badri, R., & Moghadam, A. S. (2021). Assessing the importance of deterioration properties for the design decision-makings of low-rise asymmetric reinforced concrete buildings. *Structures*. https://www.sciencedirect.com/science/article/pii/S2352012421000345

Bao, Y., Chen, Z., Wei, S., Xu, Y., & Tang, Z. (2019). The state of the art of data science and engineering in structural health monitoring. *Engineering*. https://www.sciencedirect.com/science/article/pii/S2095809918308026

Behnia, A., & Chai, H. (2014). Advanced structural health monitoring of concrete structures with the aid of acoustic emission. *Construction and Building Materials*. https://www.sciencedirect.com/science/article/pii/S0950061814004292

Biglari, A. (2025). Innovative Modular House: Advancing Sustainability, Customization, and Community in Affordable Housing. *Sustainable Development Goals Series*, (Part F767), 343–411. DOI: 10.1007/978-981-96-5107-8_9

Biglari, A. (2026). Evaluation effect of force equilibrium prediction on using secant stiffness in nonlinear displacement-based finite element method. *Journal of Engineering and Applied Science 2026 73:1, 73*(1), 42-. DOI: 10.1186/s44147-025-00840-4

Biglari, A., & Aliarab, Z. (2022). Evaluation of Feasibility Study to Ensure Uniform Deformations Using Slender Steel Shearwalls with Partial Connections in the Rehabilitation Process. *Modares Civil Engineering Journal, 22*(6), 165–178. DOI: 10.22034/22.6.165

Biglari, A., & Harrison, P. (2026). Proposing the enriched strain-based gradient as a general implicit method to solve time-independent constitutive equation. *Communications in Nonlinear Science and Numerical Simulation, 157*, 109718. DOI: 10.1016/j.cnsns.2026.109718

Biglari, A., Harrison, P., & Bićanić, N. (2014). Quasi-hinge beam element implemented within the hybrid force-based method. *Computers & Structures, 137*, 31–46. DOI: 10.1016/j.compstruc.2013.10.016

Biglari, A., & Irani, F. (2006). The capability of Mutation and Convergent Functions in Nonlinear Analysis. *Amirkabir Journal of Science and Technology, 16.*

Biglari, A., & Nodeh, S. (2024). Evaluation Possibility of Concrete Frames Rehabilitation with Off-Grade Steel Bracing. *Modares Civil Engineering Journal, 24*(2), 31–44. https://doi.org/a-10-64884-2

Bilal, M., Ji, Y., Wang, Y., & Akhter, M. (2022). An early warning system for earthquake prediction from seismic data using batch normalized graph convolutional neural network with attention mechanism. *Mdpi.Com*. https://www.mdpi.com/1424-8220/22/17/6482

Bommer, J., & Abrahamson, N. A. (2006). Why do modern probabilistic seismic-hazard analyses often lead to increased hazard estimates? *Pubs.Geoscienceworld.Org*. https://pubs.geoscienceworld.org/ssa/bssa/article-abstract/96/6/1967/146678

Boublil, E. (2023). Vulnerability, Interdependence and the Care for the Living. *The Vulnerability of the Human World Springer, 148*, 143–157. DOI: 10.1007/978-3-031-41824-2_10

Cai, S., Mao, Z., Wang, Z., Yin, M., & Karniadakis, G. E. (2021). Characterizing possible failure modes in physics-informed neural networks. *Proceedings.Neurips.Cc, 37*(12), 1727–1738. DOI: 10.1007/s10409-021-01148-1

Cheng, F., Li, J., Zhou, L., & Lin, G. (2023). Fragility analysis of nuclear power plant structure under real and spectrum-compatible seismic waves considering soil-structure interaction effect. *Engineering Structures*, *280*, 115684. DOI: 10.1016/j.engstruct.2023.115684

Dağdeviren, U., Demir, A. S., Erden, C., Kökçam, A. H., & Kurnaz, T. F. (2025). AI using ensemble machine learning with integrated SHapley additive explanations (SHAP)-Borda approach for estimation of the safety factor against soil liquefaction. *Environmental Earth Sciences*, *84*(17), 507. Advance online publication. DOI: 10.1007/s12665-025-12466-z

Danziger, M. M., Shekhtman, L. M., Bashan, A., Berezin, Y., & Havlin, S. (2016). Vulnerability of interdependent networks and networks of networks. *Springer*, 79–99. DOI: 10.1007/978-3-319-23947-7_5

Das, T., & Choudhury, D. (2025). Bayesian network based probabilistic approach for uncertainty analysis of earthquake induced landslides. *Soils and Foundations*. https://www.sciencedirect.com/science/article/pii/S003808062500054X

De Lautour, O., & Omenzetter, P. (2009). Prediction of seismic-induced structural damage using artificial neural networks. *Elsevier*. https://www.sciencedirect.com/science/article/pii/S0141029603003805

Dhulipala, S. L. N. (2021). Gaussian Kernel Methods for Seismic Fragility and Risk Assessment of Mid-Rise Buildings. *Mdpi.Com*. https://www.mdpi.com/2071-1050/13/5/2973

Elhishi, S., & Elashry, A. (2023). Unboxing machine learning models for concrete strength prediction using XAI. *Nature.Com*. https://www.nature.com/articles/s41598-023-47169-7

Farrar, C., & Worden, K. (2012). *Structural health monitoring: a machine learning perspective*. John Wiley & Sons, Ltd. DOI: 10.1002/9781118443118

Farrar, C. R., Park, G., Allen, D. W., & Todd, M. D. (2006). Sensor network paradigms for structural health monitoring. *Structural Control and Health Monitoring*, *13*(1), 210–225. DOI: 10.1002/stc.125

Farrokhzad, F., Choobbasti, A. J., Barari, A., & Ibsen, L. B. (2011). Assessing landslide hazard using artificial neural network: Case study of Mazandaran, Iran. *Carpathian Journal of Earth and Environmental Sciences*, *6*(1).

Fawad, M., Salamak, M., Poprawa, G., Koris, K., Jasinski, M., Lazinski, P., Piotrowski, D., Hasnain, M., & Gerges, M. (2023). Automation of structural health monitoring (SHM) system of a bridge using BIMification approach and BIM-based finite element model development. *Scientific Reports*, *13*(1), 13215. DOI: 10.1038/s41598-023-40355-7 PMID: 37580350

Feng, R., & Balling, N. (2021). Bayesian convolutional neural networks for seismic facies classification. *Ieeexplore.Ieee. Org*, *41*(10), 1809–1822. DOI: 10.1111/risa.13690

Flint, M., Dhulipala, L., & Shahtaheri, Y. (2016). Developing a decision framework for multi-hazard design of resilient, sustainable buildings. *Researchgate. Net*, *142*(9). Advance online publication. DOI: 10.1061/(ASCE)ST.1943-541X.0001545

Führer, K., Kwakkel, J. H., d, F. M., ', N. A., Hont, N. A., Rouwette, E. A. J. A., v, C. E., & Daalen, . (2025). Towards participatory decision-making under deep uncertainty: Benefits and challenges. *Inderscienceonline. Com*, *25*(2), 150–173. DOI: 10.1504/IJTPM.2025.147284

Giorgio, M., & Iervolino, I. (2016). On multisite probabilistic seismic hazard analysis. *Bulletin of the Seismological Society of America*, *106*(3), 1223–1234. Advance online publication. DOI: 10.1785/0120150369

Grossmann, T., Komorowska, U., & Latz, J. (2024). Can physics-informed neural networks beat the finite element method? *Academic.Oup.Com*. https://academic.oup.com/imamat/article-abstract/89/1/143/7680268

Hadjimichael, A., & J Schlumberger. (2024). Data visualisation for decision making under deep uncertainty: current challenges and opportunities. *Iopscience.Iop.Org*. DOI: 10.1088/1748-9326/ad858b

Hsieh, C. H., Tai, H. H., & Lee, Y. N. (2014). Port vulnerability assessment from the perspective of critical infrastructure interdependency. *Taylor & Francis, 41*(6), 589–606. DOI: 10.1080/03088839.2013.856523

Jiang, J., El Naggar, H., Xu, C., Chen, G., & Du, X. (2023). Seismic fragility analysis for subway station considering varying ground motion ensembles. *Soil Dynamics and Earthquake Engineering*, *165*, 107705. DOI: 10.1016/j.soildyn.2022.107705

Kalra, N., Hallegatte, S., Lempert, R., & Brown, C. (2014). Investment decision-making under deep uncertainty-application to climate change. *Enpc.Hal.Science*. https://enpc.hal.science/hal-00802049/document

Kharazmi, E., Zhang, Z., & Karniadakis, G. E. (2019). *Variational Physics-Informed Neural Networks For Solving Partial Differential Equations*. https://arxiv.org/pdf/1912.00873

Kwag, S., & Hahm, D. (2020). Multi-objective-based seismic fragility relocation for a Korean nuclear power plant. *Springer, 103*(3), 3633–3659. DOI: 10.1007/s11069-020-04148-9

Leal, O. J. U., Fekete, A., Eudave, R. R., Matos, J. C., Sousa, H., & Teixeira, E. R. (2024). A systematic review of integrated frameworks for resilience and sustainability assessments for critical infrastructures. *Taylor & Francis, 34*(2), 266–280. DOI: 10.1080/10168664.2023.2265965

Lempert, R. J., Lawrence, J., Kopp, R. E., Haasnoot, M., Reisinger, A., Grubb, M., & Pasqualino, R. (2024). Reinforcement learning for decision-making under deep uncertainty. *Journal of Environmental Management, 6*. Advance online publication. DOI: 10.3389/fclim.2024.1380054

Liu, W., Li, A., Fang, W., Love, P. E. D., Hartmann, T., & Luo, H. (2023). A hybrid data-driven model for geotechnical reliability analysis. *Reliability Engineering & System Safety, 231*, 108985. Advance online publication. DOI: 10.1016/j.ress.2022.108985

Liu, Z. (2013). Seismic risk analysis using Bayesian belief networks. In *Handbook of Seismic Risk Analysis and Management of Civil Infrastructure Systems*. Woodhead Publishing Series in Civil and Structural Engineering., https://www.sciencedirect.com/science/article/pii/B9780857092687500074

Mao, Z., & Jagtap, A. (2020). Physics-informed neural networks for high-speed flows. *Computer Methods in Applied Mechanics and Engineering*. https://www.sciencedirect.com/science/article/pii/S0045782519306814

Marchau, V. A. W. J., Walker, W. E., & Bloemen, P. J. T. M. (2019). *Decision making under deep uncertainty: from theory to practice*. https://library.oapen.org/handle/20.500.12657/22900

Matthews, S., & Morlidge, J. R. (2009). Performance based rehabilitation of reinforced concrete structuresNANo ranking found for "Concrete Repair, Rehabilitation and Retrofitting II." *Taylorfrancis. Com, 277–278*, 277–278. Advance online publication. DOI: 10.1201/9781439828403.ch100

Mengesha, G. (2025). Integrating AI in Structural Health Monitoring (SHM): A Systematic Review on Advances, Challenges, and Future Directions. *University of Gondar, SSRN*. DOI: 10.2139/ssrn.5004977

Misyris, G., & Venzke, A. (2020). Physics-informed neural networks for power systems. *Ieeexplore.Ieee.Org*. https://ieeexplore.ieee.org/abstract/document/9282004/

Ou, J., & Li, H. (2010). Structural health monitoring in mainland China: Review and future trends. *Structural Health Monitoring*, *9*(3), 219–231. DOI: 10.1177/1475921710365269

Ouyang, M., Hong, L., Mao, Z., & Yu, M. (2009). Vulnerability analysis of interdependent infrastructure systems: A methodological framework. *Physica A: Statistical Mechanics and Its Applications*. https://www.sciencedirect.com/science/article/pii/S0378437111009794

Padmapoorani, P., Senthilkumar, S., & Mohanraj, R. (2023). Machine learning techniques for structural health monitoring of concrete structures: A systematic review. *Civil Engineering (Shiraz)*, *47*(4), 1919–1931. DOI: 10.1007/s40996-023-01054-5

Phoon, K. K., & Zhang, W. (2023). Future of machine learning in geotechnics. In *Georisk* (Vol. 17, Number 1). DOI: 10.1080/17499518.2022.2087884

Quinn, W., Angove, P., Buckley, J., Barrett, J., & Kelly, G. (2011). Design and performance analysis of an embedded wireless sensor for monitoring concrete curing and structural health. *Journal of Civil Structural Health Monitoring*, *1*(1–2), 47–59. DOI: 10.1007/s13349-011-0005-9

Ram, T., & Wang, G. (2013). Probabilistic seismic hazard analysis in Nepal. *Earthquake Engineering and Engineering Vibration*, *12*(4), 577–586. DOI: 10.1007/s11803-013-0191-z

Sakiyama, F. I. H., Lehmann, F., & Garrecht, H. (2021). Structural health monitoring of concrete structures using fibre-optic-based sensors: A review. *Magazine of Concrete Research*, *73*(4), 174–194. DOI: 10.1680/jmacr.19.00185

Sarli, P., Palar, P., Azhari, Y., & Setiawan, A. (2023). Gaussian process regression for seismic fragility assessment: Application to non-engineered residential buildings in Indonesia. *Mdpi.Com*. https://www.mdpi.com/2075-5309/13/1/59?utm_campaign=releaseissue_buildingsutm_medium=emailutm_source=releaseissueutm_term=titlelink16

Shlomi, J., & Battaglia, P. (2020). Graph neural networks in particle physics. *Iopscience.Iop.Org*. DOI: 10.1088/2632-2153/abbf9a

Shortridge, J., Aven, T., & Guikema, S. (2017). Risk assessment under deep uncertainty: A methodological comparison. *Elsevier*. https://www.sciencedirect.com/science/article/pii/S095183201630713X

Siahkoohi, A., Rizzuti, G., & Herrmann, F. J. (2022). Deep Bayesian inference for seismic imaging with tasks. *Library.Seg. Geophysics*, *87*(5), 1–109. DOI: 10.1190/geo2021-0666.1

Sitar, N., MacLaughlin, M. M., & Doolin, D. M. (2005). Influence of Kinematics on Landslide Mobility and Failure Mode. *Journal of Geotechnical and Geoenvironmental Engineering*, *131*(6), 716–728. DOI: 10.1061/(ASCE)1090-0241(2005)131:6(716)

Sohn, H., Farrar, C., & Hemez, F. (1999). A review of structural health monitoring literature: 1996–2001NANo ranking found for "IEE Colloquium on Condition Monitoring Machinery, External Structures and Health." *Academia.Edu*, (34), 3–6. DOI: 10.1049/ic:19990185

Stefanidou, S., & Kappos, A. J. (2021). Fragility-informed selection of bridge retrofit scheme based on performance criteria. *Engineering Structures*. https://www.sciencedirect.com/science/article/pii/S0141029621001267

Tokognon, C., & Gao, B. (2017). Structural health monitoring framework based on Internet of Things: A survey. *IEEE Internet of Things Journal*. https://ieeexplore.ieee.org/abstract/document/7842584/

Wang, S., & Sankaran, S. (2023). Self-adaptive physics-informed neural networks. *Journal of Computational Physics*. https://www.sciencedirect.com/science/article/pii/S0021999122007859

Webber, M. K., & Samaras, C. (2022). A review of decision making under deep uncertainty applications using green infrastructure for flood management. *Earth's Future*, *10*(7), e2021EF002322. Advance online publication. DOI: 10.1029/2021EF002322

Wen, H., Yan, F., Huang, J., & Li, Y. (2025). Interpretable machine learning models and decision-making mechanisms for landslide hazard assessment under different rainfall conditions. *Expert Systems with Applications*, *270*, 126582. DOI: 10.1016/j.eswa.2025.126582

Zhang, B., Ren, Y., He, S., Gao, Z., & Li, B. (2025). A review of methods and applications in structural health monitoring (SHM) for bridgesQ1Measurement. *Measurement*. https://www.sciencedirect.com/science/article/pii/S0263224124024606

Zhang, W., Zhang, Y., Gu, X., Wu, C., & Han, L. (2022). Application of Soft Computing, Machine Learning, Deep Learning and Optimizations in Geo engineering and Geoscience. In *Application of Soft Computing, Machine Learning, Deep Learning and Optimizations in Geo engineering and Geoscience*. Springer Nature Link., DOI: 10.1007/978-981-16-6835-7

Zhao, X., Hu, S., & Cho, J. (2019). Uncertainty-based decision making using deep reinforcement learning. *Ieeexplore.Ieee.Org*. https://ieeexplore.ieee.org/abstract/document/9011218/

Zheng, X., Li, H., & Shi, Z.-Q. (2023). Hybrid AI-Bayesian-based demand models and fragility estimates for tall buildings against multi-hazard of earthquakes and winds. *Thin-walled Structures*, *52*(1), 248–267. DOI: 10.1016/j.tws.2023.110749

Zub, C.-I., Stratan, A., Dogariu, A., & Vulcu, C. (2018). Development of Two Types of Buckling Restrained Braces Using Finite Element Modelling. *Seismic Hazard and Risk Assessment*, 373–387. DOI: 10.1007/978-3-319-74724-8_25

Chapter 9

The Dichotomous Role of Leadership in Integrating Disruption Management and Early Warning Systems:

Digital Governance, AI, and Resilience in Modern Society

Tshepo Maota
http://orcid.org/0009-0001-0475-5801
University of South Africa, South Africa

ABSTRACT

Modern societies face disruptions such as climate change, pandemics, cyberattacks, geopolitical instability, economic fragility, and infrastructure failure. Early Warning Systems (EWS) must not be seen only as technical tools. They function as governance frameworks reliant on institutional capacity, ethical technology use, and leadership. This chapter examines the dual role of leadership in integrating disruption management and AI-enabled EWS in public-sector governance, drawing on legislation, global frameworks, and municipal reports. It contends that although technology recognizes risks, leadership dictates whether this awareness translates into resilience.Findings show that, despite South Africa's strong legal and policy frameworks, resilience outcomes vary. These differences relate to leadership quality, coordination, public trust, and governance of emerging technologies. The chapter shows that resilience depends not on technology alone, but on effective leadership and governance.

DOI: 10.4018/979-8-3373-7345-4.ch009

right © 2026, IGI Global Scientific Publishing. Copying or distributing in print or electronic forms without written permission of IGI Global Scientific Publishing is prohibited. Use of this chapter to in generative artificial intelligence (AI) technologies is expressly prohibited. The publisher reserves all rights to license its use for generative AI training and machine learning model development.

1. INTRODUCTION AND BACKGROUND

"Technology can detect risk, but only leadership can transform warning into sound wisdom and action into resilience."

Modern society has entered a phase characterized by complex and overlapping disruptions. These disruptions have evolved from isolated incidents to cascading events encompassing health crises, geopolitical conflicts, economic volatility, cyber-attacks, climate-induced disasters, social unrest, technological failures, and infrastructure collapse (Albert, 2024; Alcántara-Ayala, 2025; Gambhir et al., 2025; Lawrence et al., 2024). The COVID-19 pandemic exposed the significant vulnerability of global institutional systems, emphasizing that disruption is not a singular occurrence but a continual environmental state (Lisdiono et al., 2022; Moudio et al., 2022). Leadership readiness, organizational capacity, digital maturity, and intentional governance frameworks substantially influence how institutions anticipate, absorb, and adjust to disruption (Maota, 2024). This perspective is increasingly vital today.

A new "poly-crisis" environment has arisen globally (Keinänen et al., 2025). Geopolitical fragmentation, as demonstrated by the Russia-Ukraine conflict, trade disputes among major economies, tariff retaliations, and evolving alliance structures, has exacerbated supply-chain uncertainty, energy volatility, inflationary pressures, and food insecurity (Afonso et al., 2024; Verschuur et al., 2024; Zhyvko, 2024). Economies, particularly in the Global South, currently face increased vulnerability to currency fluctuations, debt pressures, and rising unemployment (Métivier et al., 2023). Cyber-attacks against hospitals, utilities, financial institutions, and government databases have surged dramatically, rendering data infrastructures national security targets (Gambhir et al., 2025; Salem et al., 2024). Socially, increasing inequality, diminishing faith in public institutions, and digital misinformation ecosystems are altering social expectations and governance dynamics (Meuleman & Fraser-Moleketi, 2022; Neyer, 2021; Schlösser et al., 2023). Health systems, in the aftermath of the pandemic, continue to experience enduring fragility, mental health challenges, and the potential emergence of future pathogen-induced crises (Harper, 2020; Percival et al., 2023).

In South Africa, these pressures meet with distinct national vulnerabilities. Municipal governance has entered a complicated coalition age, where leadership instability frequently undermines the continuity of service delivery (Kikasu & Pillay, 2024; Makole et al., 2022). Evidence of politically driven infrastructure sabotage, water supply disruptions, energy instability, and deteriorating public infrastructure has heightened systemic risk exposure for both communities and institutions (Gregory & Sovacool, 2019; Guma, 2022). Persistent unemployment, especially among the youth (Lategan et al., 2025; Yende, 2025), together with increasing poverty and spatial inequality, worsen susceptibility to social and economic disruptions. These realities affirm that disruption has assumed both structural and cyclical characteristics.

In this context, Early Warning Systems (EWS) have become important not just for climatological and hazard-related risks but also for technological, societal, health, governance, and cyber concerns. The efficacy of EWS relies not solely on data and technology. Maota's (2024) study indicates that the critical factors are leadership competence, institutional readiness, and governance coherence. Organizations lacking proactive leadership, structured disruption frameworks, AI-augmented risk assessment tools, and coordinated decision-making systems encountered difficulties during COVID-19, whereas those employing anticipatory models, strategic alignment, and robust governance adapted more effectively (Doğrusöz & Yazıcı, 2025; Grewal et al., 2022; Myeza et al., 2023; OECD, 2024; Patrucco & Kähkönen, 2021).

This chapter so characterizes leadership as a dualistic influence in governance during times of disruption. Leaders must stabilize institutions in crisis while concurrently reforming them for future resilience. They must maintain compliance and predictability while facilitating innovation, flexibility, and digital adaptability. They must safeguard institutional continuity while facilitating technology advancement, especially in AI-driven analytics, predictive modeling, and intelligent sensing capabilities.

EWS should be regarded not merely as technical instruments, but as integrated governance systems that contain:

- Risk intelligence and disruption anticipation
- Digital monitoring and artificial intelligence-driven analytics
- Transparent communication and trust building
- Coordinated institutional response capacity
- Ethical decision-making and accountability framework.

This aligns with the disruption management framework established in Maota's (2024) research, which underscores preparedness, leadership responsiveness, institutional learning, resource capabilities, digital transformation, and strategic resilience.

Nonetheless, although early-warning systems may improve resilience, deficiencies in leadership, fragmentation, inadequate governance cultures, poor ethics, and institutional complacency can undermine even the most advanced EWS infrastructure (Adel et al., 2025; Greenwood et al., 2025; Lisdiono et al., 2022; Pescaroli et al., 2025). The dichotomy of leadership, characterized by the struggle between control and creativity, as well as risk containment and transformation, is fundamental in modern resilience.

The current disruptive scenario is also facilitated by digital means. AI is swiftly enhancing risk-prediction capabilities across various domains, including climate analytics, cyber-threat identification, pandemic modeling, and infrastructure failure predictions (Das & Devadas, 2025; Gandhi et al., 2025; Patil et al., 2024). However, AI also presents governance risks, including algorithmic opacity, data bias, cyber-exploitation, and ethical misuse (Lee et al., 2025). Leaders necessitate digital literacy and AI governance proficiency to leverage predictive capabilities while upholding trust, privacy, security, and equity.

In South Africa, this leadership responsibility encompasses the public-interest domain. Communities confront the harsh edge of service-delivery breakdown, including water-supply failure, waste mismanagement, grid instability, and increased urban vulnerability (Hendrickse, 2023; Magezi, 2023). These shortcomings not only diminish resilience but also damage citizen trust and social cohesiveness. Integrating EWS-aligned governance frameworks within municipalities, governmental agencies, and private organizations is thus an ethical imperative.

This chapter advances the ongoing discourse by integrating disruption management literature, early warning system design, AI-driven digital governance, and leadership theory, all contextualized within African institutional realities. It examines how leaders can incorporate multi-hazard early-warning systems into organizational strategy, institutional learning, and resilience governance. It asserts that technology alone is insufficient to safeguard society in the absence of ethical, adaptive, anticipatory, and future-oriented leadership.

The chapter emphasizes that the future of resilience lies at the combination of leadership, governance, AI-driven foresight, and early-warning intelligence. It necessitates leaders who can not only react to crises but also anticipate disruptions and transform uncertainty into strategic readiness, organizational learning, and social safeguarding.

2. PURPOSE AND OBJECTIVES OF THE CHAPTER

This chapter seeks to objectively evaluate the integration of leadership, governance, and digital technologies, including AI-enabled early warning systems, to enhance institutional resilience in a period of ongoing disruption. It examines the dual role of leadership, which involves stabilizing institutions during crises while fostering innovation, transformation, and future preparedness, based on established frameworks like Maota (2024) and Yu and Qi (2004) on disruption management.

The chapter is placed within a framework of escalating global and domestic turmoil: geopolitical instability, precarious economic conditions, cyber-attacks, pandemics, climate-induced disasters, unemployment, social inequality, coalition governance in South Africa, service delivery failures, and systemic infrastructure risk. These conditions require governance frameworks that go beyond reactive crisis management and instead focus on proactive, AI-enhanced resilience structures (Byaruhanga et al., 2025; Chetty et al., 2024; Ibrahim et al., 2025; Makhura, 2025; Mehryar et al., 2024; Onduko et al., 2024).

Consequently, the chapter aims to:

- Envision early warning systems as governance mechanisms, rather than solely technology instruments, integrated into broader disruption management and institutional resilience frameworks. This entails comprehending their function in advancing creative adaptation and resilience strategies inside public administration (Kikasu, 2025).
- Examine the dual frequently contradictory function of leadership in preserving organizational continuity while fostering adaptive change, digital integration, and creativity amidst uncertainty. This encompasses addressing the ethical intricacies and possible depersonalizing consequences of digital transformation while promoting human-centric strategies that improve business culture and employee welfare (Andersen & Young, 2023).
- Assess the capacity of AI-driven predictive analytics and digital governance frameworks to improve risk anticipation, communication, decision-making assistance, and societal safeguarding. This entails acknowledging the imperative of embedding resilience as a fundamental principle in public policy and crisis management to cultivate governance systems that emphasize adaptation and preparation (Pashentsev & Kolotaev, 2025).
- Propose a resilience-governance paradigm that incorporates leadership capacity, institutional learning, accountability structures, and AI-driven early-warning intelligence pertinent to South African public interest and organizational contexts. This approach seeks to establish circumstances for enduring resilience and enable rapid adaptation to dynamic changes, recognizing the intricate, multifarious, and emergent characteristics of contemporary situations (Stenvall et al., 2025).
- Enhance global scholarship and practice by integrating disruption-management theory, early-warning system design, and leadership science through the perspective of African governance. This perspective is particularly relevant given the unique challenges faced by African public administrations, including shifting political dynamics and complex social expectations (Kikasu, 2025).

The goal is to enhance our understanding of how leaders can transition from crisis response to anticipatory governance, empowering institutions and communities to endure, adapt to, and derive lessons from increasing multi-hazard disruptions.

3. PROBLEM STATEMENT

Notwithstanding swift progress in digital technology and the increasing worldwide focus on disaster risk mitigation, EWS frequently exhibit suboptimal performance across various organizational and societal settings, as they are predominantly regarded as technical or data-centric tools rather than instruments of governance and leadership (Islam et al., 2025; Rokhideh et al., 2025; Trogrlić et al., 2022). The global risk environment has evolved into a continuous multi-hazard landscape marked by pandemics, climate volatility, cyberattacks, geopolitical instability, economic fragility, social inequity, and institutional mistrust. In South Africa, global pressures combine with political instability from the coalition era, inadequate municipal governance, failures in service delivery, infrastructure sabotage, unemployment, energy insecurity, and water problems, exacerbating systemic vulnerability (Ragolane, 2024; Yende, 2025).

Numerous institutions lack organized disruption-management frameworks, leadership readiness, digital sophistication, and predictive capacity, leading to reactive rather than proactive responses during COVID-19 (Gambhir et al., 2025; Maota, 2024). Although AI-driven analytics and digital monitoring systems provide exceptional capabilities for risk forecasting and detection, their effectiveness is constrained in environments characterized by inadequate leadership, accountability frameworks, institutional culture, and ethical governance (Lee et al., 2022). The primary challenge is not merely access to technology (Hostettler et al., 2018), but rather the absence of a cohesive, leadership-driven resilience governance framework that integrates disruption foresight, early warning system capabilities, AI governance, and coordinated response mechanisms (Mehryar et al., 2024; OECD, 2024).

A significant gap exists in the literature and practice regarding how leaders manage the conflicting demands of stabilizing institutions in crisis while concurrently fostering innovation, digital transformation, and adaptive learning (Haque, 2025; Jaadi et al., 2025; Saka et al., 2024; Suljic, 2025). In the absence of a definitive leadership-oriented framework for integrating AI-enabled early warning systems into governance structures, institutions, especially in precarious socio-political contexts, remain vulnerable to increasing and compounded disruptions (Reichstein et al., 2025). This chapter aims to examine and rectify this gap.

4. THEORETICAL FRAMEWORK

This chapter is supported by a cohesive theoretical framework that synthesizes principles from Disruption Management Theory, Public Value Governance, Digital Era Governance, Institutional Theory, and Leadership Dichotomy Theory. Collectively, these perspectives clarify the interplay between leadership, governance frameworks, and AI-driven Early Warning Systems (EWS) in shaping resilience within intricate and multi-hazard contexts.

4.1. Theory of Disruption Management

Organizational disruptions are systemic rather than episodic, necessitating anticipatory governance, adaptive capability, digital readiness, institutional learning, and leadership preparedness to avert cascading failure (Maota, Naido & Darwish, 2024). Disruption Management Theory positions EWS not solely as hazard-detection instruments but as strategic governance frameworks integrated within a comprehensive resilience architecture (Islam et al., 2025; Sakalasuriya et al., 2020).

This viewpoint underscores:

- Proactive risk identification
- Predictive modeling
- Contextual awareness
- Ongoing institutional education
- Leadership readiness and decision-making agility

AI-driven analytics augment this skill by improving foresight and pattern detection, enabling leaders to shift from reactive responses to strategic anticipation (Csaszar et al., 2024).

4.2. Governance of Public Value

Public value government transitions from hierarchical state control to collaborative, citizen-centric, multi-stakeholder government. Within the framework of EWS, resilience is collaboratively generated among government entities, the commercial sector, academic institutions, technological stakeholders, social partners, and communities (Ansell & Torfing, 2021; Witesman, 2021).

This approach is especially pertinent in South Africa, where instability in service delivery, coalition governance, infrastructure sabotage, and socioeconomic inequality need the establishment of trust and shared accountability. Public Value Governance emphasizes that leadership legitimacy, transparency, and ethical stewardship are essential for the effectiveness of Early Warning Systems (Best & Yalezo, 2022; Tsheola et al., 2023).

4.3. Governance in the Digital Era

Digital Era Governance elucidates the ways in which AI, big data, IoT, and digital sensing systems are enhancing institutional capabilities for risk prediction, monitoring, and expedited decision-making (Dunleavy & Margetts, 2023; Sigfrids et al., 2023). Nonetheless, digital modernization also presents:

- Cybersecurity threat
- Algorithmic bias
- The challenges of ethics and privacy
- Reliance on precarious digital infrastructure

Leadership must cultivate AI governance capabilities to ensure that predictive intelligence bolsters resilience while preserving trust, equity, and data sovereignty. Maota (2026) similarly highlighted the need for strategic leadership competence, training, and institutional preparedness to support technological integration.

4.4 Institutional Theory

Institutional theory elucidates why the formal acceptance of technology or policy does not ensure effective real-world performance (Schiavi et al., 2023; Singh, 2024). Organizational behavior is influenced by:

- Cultural conventions
- Hierarchical frameworks
- Systems of accountability
- Traditions of compliance
- Ethics and values in leadership

Inadequate institutional alignment consequently diminishes EWS efficacy, despite the availability of modern technology (Kiparsky et al., 2013; Mökander & Axente, 2021). Numerous developing-world contexts witness this, as governance instability undermines system dependability (Sakalasuriya et al., 2020; Zheng & Chu, 2025). Maota's (2024) study supported this reality, indicating that institutions lacking organized frameworks, leadership preparedness, and coordinated governance encountered difficulties during COVID-19, notwithstanding their access to resources or instruments.

4.5 Leadership Dichotomy Theory

This chapter focuses on the dual nature of leadership in disruptive environments. Leaders must stabilize operations while facilitating transformation.

- Safeguard institutional continuity while promoting digital adaptation.
- Act decisively while participating inclusively.
- Leverage AI intelligence while upholding ethics and trust.
- Ensure compliance while fostering innovation.

This dichotomy is not contradictory; it is the fundamental leadership competency of the disruptive period. Kolivand et al. (2025) study found that resilient leaders display adaptability, sense-making capabilities, strategic foresight, and a learning orientation, which empower them to navigate institutions through uncertainty.

4.6 Integrative Framework

This chapter integrates these theories by conceptualizing AI-enabled early warning systems as socio-technical governance frameworks functioning at the convergence of leadership competence, institutional culture, digital intelligence, public-value governance, and adaptive learning.

Resilience arises not solely from technology, but from the integration of leadership, governance, AI ethics, community trust, and institutional design.

This theoretical paradigm puts leadership as the critical factor in the success or failure of EWS-enabled resilience. In precarious governance contexts, like South Africa's coalition-era municipal systems and economically distressed areas, the lack of leadership-focused governance frameworks renders institutions vulnerable to increasing multi-hazard risks.

This chapter expands upon this framework to offer a leadership-focused, AI-enhanced resilience governance model rooted in disruption-management scholarship and contextualized within African institutional realities.

5. METHODOLOGICAL APPROACH

This chapter uses a qualitative, interpretivist methodology grounded in document analysis, conceptual synthesis, and integrative theory development. The objective is to critically examine the interplay of leadership, governance frameworks, and AI-enabled early warning systems in influencing institutional resilience within intricate socio-political and technical contexts (Helmrich & Chester, 2022; Malik

& Terzidis, 2025). This methodology corresponds with Maota's (2024) research, which illustrated the significance of systematic analysis of institutional practices, leadership conduct, and governance preparedness in understanding disruption management and readiness.

5.1 Research Orientation

The study is framed within an interpretivist paradigm, acknowledging that disruption, resilience, leadership response, and governance maturity are socially created phenomena inherent in institutional cultures, political contexts, and digital ecosystems (Fatima et al., 2024; Tim & Leidner, 2023). Knowledge is thereby obtained through meaning-making, contextual interpretation, and the synthesis of diverse facts, rather than through statistical generalization. This mindset is important in contemporary "poly-crisis" scenarios, where disruptions emerge dynamically from the intersection of geopolitics, cyber-risk, climate volatility, economic vulnerability, digital transformation, and governance efficacy.

5.2 Data Sources and Evidence Base

The chapter uses three categories of evidence:

a) *Academic literature*

This includes peer-reviewed research on EWS, leadership theory, AI governance, disruption management, resilience frameworks, institutional theory, governance effectiveness, public value theory, and digital era governance. Both global and African scholarship are prioritized to provide contextual relevance.

b) *Policy and institutional documentation*

Essential documentary sources encompass global disaster-risk frameworks (e.g., UNDRR and Sendai-aligned governance directives), national disaster governance strategies, municipal governance reports, cybersecurity advisories, higher education governance documents, risk management frameworks, and pertinent organizational resilience policies. These documents assist in examining how EWS and digital governance are positioned in policy intent versus institutional practice.

c) *Foundations of Doctoral Research*

The analytical lens is significantly informed by insights generated from Maota's (2024) study, which examined leadership readiness, institutional resilience, AI-enabled predictive analysis, disruption management, and capability development within South African private higher-education institutions. This prior work demonstrated that leadership agility, digital maturity, and structured governance frameworks were decisive in shaping institutional responses to disruption, a foundational insight extended into the multi-hazard context of this chapter.

5.3 Analytical Strategy

A documentary analysis method is used, based on Braun and Clarke's thematic interpretation, to identify reoccurring leadership behaviors, governance trends, failure spots, resilience facilitators, digital integration obstacles, and ethical AI issues (Braun & Clarke, 2000). The analysis advances through:

- *Preliminary familiarization* with literature and policy documents
- Conduct *open coding* to identify prevailing themes in governance and leadership.
- *Axial coding* to delineate linkages across leadership behavior, governance culture, risk-intelligence systems, AI-driven monitoring, and societal preparation.
- *Theoretical integration* to synchronize emergent ideas with the chapter's conceptual framework.

This process analyzes EWS not merely as technical infrastructures, but as integrated socio-technical governance systems shaped by trust, ethics, leadership, and institutional capability.

5.4 Case-Embedded Illustration

This chapter, while not a case study, uses contextual illustrations from South African governance, such as coalition-era municipal instability, infrastructure sabotage, water-system failure, cyber-risk escalation, and service-delivery fragility, analytically rather than descriptively. This contextual framework enables the chapter to objectively evaluate how leadership dichotomy, digital governance competence, and public-value legitimacy affect the effectiveness of early-warning and resilience systems.

5.5 Methodological Contribution

This methodological approach allows for the integration of technical EWS literature with concepts of governance and leadership. The study also analyzes AI-driven forecasting capabilities from an ethical and institutional perspective. Situate global scholarship within the context of African governance realities. Expand disruption-management theory into the realm of multi-hazard resilience.

This aligns with the author's previous research focus on leadership ability, institutional preparedness, digital transformation maturity, predictive analytics, and governance-aligned resilience frameworks (Maota, 2024).

5.6 Trustworthiness and Rigour

Rigour is ensured through the triangulation of various evidence sources:

- Reflexive analysis, recognising interpretative positioning
- Clarity of reasoning, demonstrating the progression of insights derived from verified evidence
- Theoretical consistency, guaranteeing conformity with the conceptual framework

The objective is to achieve analytical depth instead of numerical generalisation; hence, it enhances conceptual clarity for both academics and practitioners.

5.7 Ethical Consideration

The chapter uses secondary and documentary data, resulting in minimal ethical risk. Equity, truthfulness, data integrity, and respectful contextual representation inform the analysis, particularly when interacting with disadvantaged populations and contexts of governance fragility.

This methodological approach allows the chapter to integrate leadership science, AI governance, early-warning design, institutional resilience, and the reality of African government into a unified analytical model. It underscores the principal assertion that leadership-focused, governance-integrated, AI-driven EWS are essential for social resilience in a perpetually disruptive environment.

6. LITERATURE REVIEW

6.1 Disruption as a Persistent Operating Condition

Early research conceptualised disruption as a singular event necessitating contingency strategies and recovery methods. Recent literature increasingly acknowledges disruption as an ongoing condition rather than a transient shock (Koh et al., 2023; Walters et al., 2023). The "continuum of turbulence" is propelled by pandemic risks, climate instability, cyber-attacks, geopolitical uncertainty, economic vulnerability, social discord, and systemic infrastructure risk (Ansell & Trondal, 2017; Koh et al., 2023). Institutions today operate in environments defined by simultaneity, unpredictability, and cascading risks, rather than linear crisis cycles.

Maota and Naidoo's (2022) study revealed that during COVID-19, numerous institutions were lacking organised disruption-management frameworks, predictive systems, and leadership preparedness, resulting in reactive rather than proactive actions. This perspective corresponds with recent literature that states that resilience capability is a fundamental organisational competency rather than only a recovery function. The amalgamation of cloud and federated authentication methodologies with predictive analytics models can provide strong systems for resilience and scalability, essential for proactive disruption management (Nwachukwu et al., 2025). Moreover, leaders must develop adaptive governance frameworks and incorporate AI-driven support systems to improve strategic forecasting and stakeholder involvement, hence assuring systemic resilience in the face of complex crises (Joshi, 2025; Pashentsev & Kolotaev, 2025). Such an approach requires a transition from traditional, static risk assessments to dynamic, context-sensitive tactics that utilise real-time data and AI-generated insights to predict and alleviate emergent dangers (Kolivand et al., 2025; Tiwari et al., 2025).

6.2 Leadership in the Era of Disruption: A Dual Role

Leadership scholarship has typically emphasised stability, control, and optimisation. In disruptive circumstances, leaders must simultaneously execute opposing functions: stabilising core operations while fostering innovation, adaptability, learning, and experimentation (Bell & Hofmeyr, 2021).

Recent studies characterise these principles as a "both-and leadership paradigm": leaders must simultaneously serve as custodians of organisational continuity and engineers of transformation (Xi et al., 2025). Consequently, the efficacy of leadership is contingent upon:

- The effectiveness of leadership hinges on the ability to interpret and anticipate.

- The ability to navigate through uncertainty rather than eliminating it is also crucial.
- It also involves the establishment of ethical legitimacy and the development of trust.
- The ability to adapt to changing circumstances is also crucial.
- Strategic coherence among purpose, governance, and technology.

Maota's (2024) findings corroborate this duality, demonstrating that institutions led by forward-thinking leaders, equipped with digitally adept technologies and organised governance frameworks, reacted more effectively to disruption than those reliant solely on compliance-based approaches.

6.3. Early Warning Systems (EWS) as Governance—Beyond Technology

Disaster-risk literature has historically examined EWS, emphasising meteorological, hydrological, seismic, and hazard-forecasting functions (Sakalasuriya et al., 2020). Global research now reconceptualises EWS as socio-technical governance frameworks, consisting of four interrelated pillars:

- Understanding and awareness of risk
- Surveillance and forecasting abilities
- Communication and distribution strategies
- Institutional responsiveness and coordinating capacity

This comprehensive perspective acknowledges that technology cannot remedy inadequate governance, disjointed leadership, or ineffective institutional alignment (Kolivand et al., 2025). The shortcomings of numerous EWS come not from data precision but from leadership deficiencies, institutional distrust, ambiguous lines of authority, and a lack of coordinated response cultures (Reichstein et al., 2024).

Contemporary research posits that the efficacy of EWS is intrinsically linked to the legitimacy of leadership, the presence of organisational trust, and the coherence of policy (Pescaroli et al., 2025; Raftopoulos et al., 2024).

6.4 AI-Enabled Predictive Systems and Digital Governance

AI-driven analytics, machine learning, big data sensing, and IoT networks have transformed modern early-warning capabilities (Li et al., 2025; Shaik et al., 2025). These tools now facilitate:

- Detection of cyber threats
- Forecasting infrastructure and utility surveillance
- Monitoring the financial system's stress
- Analysis of social risk and disinformation
- Identification of patterns between climate and health risks

Nonetheless, digitalization also presents risks such as surveillance fear, data inequity, cyber vulnerability, bias, and algorithmic opacity (Dedola et al., 2023). Academics now regard AI governance as a leadership obligation rather than a technological function. Predictive capabilities must be accompanied by ethical responsibility, openness, and trust.

Abdullahi and Shehu's (2025) study highlight the essentiality of alignment, asserting that digital transformation alone does not foster resilience without the integration of leadership skills, institutional preparedness, training, and organized governance frameworks.

6.5 Institutional Theory and the Actuality of Governance Performance

Institutional theory indicates that the formal adoption of systems does not ensure their successful utilization (Alcántara-Ayala et al., 2025). Culture, values, organizational memory, power dynamics, leadership ethics, and informal norms significantly influence EWS outcomes (Atukunda et al., 2024).

Poor institutional conditions, including:

- Dispersed authority
- Instability in leadership
- Political intervention
- Malfeasance
- Environments characterized by diminished trust
- Insufficient accountability

— degrade system efficacy irrespective of technology expenditure.

This is especially pertinent in South Africa, where coalition government, infrastructure sabotage, energy instability, water failures, municipal collapse, and social inequality have engendered systemic vulnerabilities (Henrico & Els, 2025). The success of EWS is contingent upon both governance maturity and technical sophistication.

6.6 Governance of Public Value and Societal Trust

According to current governance research, government, civil society, the commercial sector, technology stakeholders, and communities must collaborate to co-produce resilience (Bodó & Janssen, 2022). Scholars now regard public trust, accountability, inclusivity, and legitimacy as fundamental resilience assets (Shah et al., 2024).

When institutions lack trust, people disregard warnings and systems collapse. Leadership must serve as guardians of public value, guaranteeing equitable access, openness, ethical communication, and collective accountability. Such leadership is particularly crucial in contexts characterized by inequity, marginalization, and tiredness regarding governance discourse.

6.7 A Synthesis: Leadership as the Pivotal Variable

A predominant insight arises from the examined literature:

- Technology forecasts risk, whereas leadership transforms intelligence into resilience.

The most sophisticated early warning systems (EWS) will fail under conditions where:

- Governance is disjointed.
- The ethical justification is tenuous.
- Cultural learning is nonexistent.
- Digital maturity is inadequate.
- Institutional trust is diminished.
- Leaders exhibit reactivity instead of anticipatory behavior.

Conversely, when leadership dichotomy is effectively managed, stabilizing operations while fostering innovation, Early Warning Systems become catalysts for institutional learning, social protection, and adaptive transformation (Andersen & Young, 2023; Pescaroli et al., 2025; Reichstein et al., 2024).

This chapter enhances the discourse by synthesizing disruption-management theory, leadership studies, early warning system frameworks, AI governance discussions, institutional theory, and the reality of public governance in Africa into a cohesive conceptual paradigm.

7. RESULTS AND DISCUSSION

The institutional and policy review indicates that South Africa has established a mature, multi-tiered policy framework for disaster risk reduction (DRR), EWS, and resilience governance. But for these frameworks to operate, they need competent leaders, cooperation amongst institutions, readiness for digital technology, and people who are willing to change their behavior. A review of the Municipal Disaster Risk Management Plan for Cape Town (City of Cape Town, 2017), the Storm Surge Early Warning System Guidelines (Stander, Theron & Du Plessis, 2011), the City of Tshwane Disaster Risk Management Annual Report (2024/25) (City of Tshwane, 2024), and the City of Johannesburg Climate Action Plan (City of Johannesburg, 2021) shows that risk governance has made progress in some areas but still has weaknesses. It is important to note that South Africa has made disaster governance a formal, multi-hazard approach that fits with global frameworks like the Sendai Framework (NDMC, 2020).

A notable observation is that South Africa has institutionalized disaster governance as a comprehensive, multi-hazard approach, aligning with worldwide frameworks like the Sendai Framework. The City of Cape Town's Municipal Disaster Risk Management Plan (City of Cape Town, 2017) conceptualizes disaster management as an ongoing system that encompasses prevention, mitigation, preparedness, response, and recovery within all municipal functions, grounded in the legal framework of the Disaster Management Act and the municipal Integrated Development Plan (NDMC, 2020). This institutionalization offers a formal framework for anticipatory governance, an essential prerequisite for leadership prepared for change.

This aligns with Atukunda et al. (2024), who affirm that institutional capacity, integration, and leadership consistency dictate whether these policies transition from mere compliance to functional capability. Cape Town's plan outlines comprehensive organizational arrangements, command structures, and communication protocols, demonstrating how institutional clarity improves the implementation of risk strategy (City of Cape Town, 2017). However, such clarity is not consistently conveyed across municipalities, increasing the resilience disparity between well-resourced metropolitan areas and more vulnerable governance contexts.

A secondary issue is the normalization of EWS as an important governance competence rather than merely a technological tool. The Storm Surge Early Warning System Guidelines assert that an effective system must incorporate risk identification, monitoring, warning dissemination, and citizen reaction capabilities as interconnected components (Stander et al., 2011). The guidelines emphasize the importance of multi-institutional collaboration among SAWS, the National Disaster Management Centre, the CSIR, and military and maritime agencies, indicating that risk has transitioned from being sectoral to systemic. This aligns with Maota's (2024) study, which state that leadership currently functions in conditions of polycentric disruption, wherein uncertainty cannot be addressed with unilateral agency action.

The City of Tshwane Disaster Risk Management Annual Report (2024/25) (City of Tshwane, 2024) provides empirical evidence of the functionality of these systems under actual conditions of strain. Tshwane saw considerable hazard exposure, including floods and infrastructure breakdowns, and managed over 100 serious incidents throughout the reporting period. The report emphasizes improved interdepartmental collaboration, refined hazard mapping, public awareness initiatives, and continuous institutional training as key performance achievements. These advancements validate the evolution of an institutional culture of risk governance, characterized by a proactive rather than reactive approach to catastrophe leadership.

Nevertheless, the report concurrently highlights internal weaknesses such as resource limitations, urbanization pressures, concentration of risk in informal settlements, and the intensification of climate-induced hazards, alongside socio-economic instability and infrastructure deficits. This illustrates the dual reality recognized in Maota’s (2024) study, that: leadership is anticipated to stabilize risk situations while concurrently managing political instability, fiscal constraints, technological change, and the erosion of public trust. In this regard, resilience encompasses both political-administrative and technical dimensions.

A key insight from the analysis reveals that climate change has moved beyond environmental discussions to become central to state capacity and the legitimacy of governance. The City of Johannesburg Climate Action Plan (CAP) (City of Johannesburg, 2021) state that climate risk is an ecological problem that is also a problem of development, the economy, and social justice. It is linked to unemployment, inequality, weak service delivery, infrastructure deterioration, and health problems. The CAP highlights notable warming trends, increased flood risk, and heightened rainfall unpredictability, necessitating the integration of climate adaptation and mitigation across economic sectors and local government frameworks.

Johannesburg's Climate Action Plan is an example of a governance framework that looks to the future. It connects net-zero trajectories, urban innovation, and financial tools like green bonds with solutions for resilience that focus on citizens. Maota's (2026) study asserts that digital transformation, ethical leadership, and the acceptance of innovation intrinsically link disruption governance. The CAP reconceptualizes climate resilience as a collective social agreement, particularly highlighting young and marginalized populations, a facet that strongly coincides with disruption-management theory's focus on adaptive institutions and communal learning.

AI, data analytics, and digital infrastructure consistently function as underutilized strategic enablers across all examined sources. The Storm Surge EWS guidelines recognize modeling, buoy data, and satellite monitoring, yet they contain minimal explicit mention of AI-driven predictive analytics, real-time decision dashboards, automated risk classification, or cyber-resilience factors, essential in light of increasing cyber-risk environments and digital reliance. This report demonstrates a disparity between technology and governance, wherein the advancement of institutional policies has surpassed digital capabilities. Maota and Naidoo's (2022) study highlighted a similar disconnect during COVID-19, when institutions had formal goals but lacked the digital acumen and leadership foresight necessary for systematic implementation.

Leadership is the critical factor in determining whether institutional frameworks result in resilience. The Tshwane Annual Report (City of Tshwane, 2024) emphasizes political supervision, administrative coordination, performance reporting, public communication, and inter-agency collaboration as essential pillars of resilient governance. This affirms Bell & Hofmeyr (2021) assertion that leadership in times of upheaval is dualistic, being both strategic and operational, stabilizing yet flexible, and controlling yet collaborative. When political fragmentation, corruption, reduced capacity, or denial of crises undermine leadership, the effectiveness of EWS and Disaster Risk Management (DRM) declines, even in the presence of strong policy frameworks (Shah et al., 2024).

This reliance on leadership is particularly important within the South African framework of coalition politics, service delivery instability, infrastructure sabotage, inequality, and societal distrust. Institutional confidence is a fundamental prerequisite for citizen adherence to EWS alerts. When confidence diminishes, risk communication falters, not due to the system's failure to issue warnings, but because society has lost faith (Bodó & Janssen, 2022). Modern disaster management rarely acknowledges the psychology of risk governance, which calls for a deeper integration.

The examined documents additionally demonstrate signs of progressive learning and system adaptability. City of Tshwane indicates a growing incorporation of community-based education, drills, contingency planning, and operational lessons-learned evaluations, reflecting the progressive development of learning-oriented public institutions (City of Tshwane, 2024). Johannesburg integrates climate adaptation into urban planning, transportation networks, energy strategies, and financial models, fostering comprehensive city resilience instead of isolated solutions (City of Johannesburg, 2021).

Nevertheless, persistent failure points persist:

- The city faces disjointed data systems and inconsistent digital preparedness, which rely on personal leadership competence instead of fostering an ingrained culture of resilience.
- There is a reliance on personal leadership competence rather than an ingrained culture of resilience.

- The amalgamation of AI governance, cybersecurity resilience, and digital ethics frameworks is restricted.
- There is inadequate literacy regarding multi-hazard community early-warning systems in at-risk environments.
- Financial pressure and political conflict are undermining the stability of institutions.

These deficiencies correspond with international literature indicating that the efficacy of EWS is optimal at the convergence of governance maturity, leadership stability, technological competence, and societal cohesiveness.

Global, Continental, Regional, and National Policy Framework: Consequences for Leadership, Early Warning Systems, and Resilience Governance

This study reveals that South African disaster governance, early warning systems, and resilience strategies are interconnected and significantly influenced by a complex governance ecosystem, which includes the UN system, WMO scientific standards, continental AU and SADC disaster frameworks, national legislation, and evolving global standards in AI and cyber-risk governance (UNDRR, 2015; United Nations, 2022; WMO, 2018; African Union Commission, 2022; SADC, 2016). Collectively, these policy frameworks create normative expectations for anticipatory governance, but their effectiveness ultimately hinges on the interpretation and institutionalization of leadership.

United Nations and Global Disaster Risk Governance

The Sendai Framework for Disaster Risk Reduction (2015-2030) serves as the principal global tool influencing national disaster risk reduction policy (UNDRR, 2015). It recontextualizes disaster management from event response to risk governance, highlighting multi-hazard monitoring, community-focused alerts, institutional readiness, and leadership responsibility. The UN's Early Warnings for All project (2022) reinforces this narrative by advocating for a universal, inclusive, impact-oriented early warning system, especially in the Global South (UNDRR, 2018). This program emphasizes the necessity for comprehensive, multi-hazard early warning systems that are accessible, actionable, and tailored to local contexts, thereby necessitating robust leadership and institutional frameworks for effective implementation (Byaruhanga et al., 2025).

This chapter maintains that Sendai expressly acknowledges leadership, governance structures, digital systems, and institutional coordination as being on par with technical forecasting capabilities. This correlates with Reichstein et al., (2024) notion that resilience is an organizational capability driven by leadership, rather than solely a technological function. This viewpoint requires a thorough analysis of how global frameworks are implemented as actionable strategies in various national contexts, especially regarding the incorporation of indigenous knowledge systems and local leadership structures to improve the effectiveness of early warning systems (Lunga et al., 2025; Ragus et al., 2024).

The World Meteorological Organization (WMO) implements these ideas into scientific and communication guidelines for multi-hazard early warning systems (MHEWS) (WMO, 2018). It underscores impact-oriented forecasting, enhancing the civic significance of alerts, a distinct transition from mere meteorological precision to decision-support information. This mindset is clearly incorporated in the South African Storm Surge Early Warning System Guidelines, which specifically combine WMO terminology regarding the four-pillar EWS paradigm (Stander et al., 2011). The WMO's emphasis on impact-based forecasting and the four-pillar model for early warning systems directly shapes national strategies, ensuring that early warnings are both scientifically accurate and practically relevant to the public (Raftopoulos et al., 2024).

Nonetheless, the global frameworks inadequately address the governance challenges of weak, politicized, or resource-constrained towns, corroborating the thesis conclusion that leadership capacity and institutional integrity are important.

Continental and Regional African Frameworks

The African Union's Africa Multi-Hazard Early Warning and Early Action System Framework (2022) signifies a significant advancement in African risk governance. The AU transitions from reactive crisis response to continental coordination, data sharing, and proactive leadership, prioritizing ethics, inclusion, and technological modernization. This concept expressly advocates for a transition from donor-driven, top-down governance to a people- and community-centered approach to risk governance, recognizing the unequal effects of climate hazards on the Global South (Mirianna et al., 2025; Zia & Wagner, 2015).

This statement also corresponds with the assertion that AI-driven predictive governance is now integral to resilience strategy. The AU cautions about the disparity in capacity among African governments and localities, highlighting the dangers of catastrophe systems that are rich in policy yet deficient in capabilities. This finding underscores the ongoing difficulties in converting ambitious policy frameworks into concrete advancements in disaster risk reduction and management at the local level throughout the continent (Okunola, 2025).

The SADC Disaster Risk Reduction Strategy and Plan of Action (2016-2030) serves as a regional policy framework for Southern Africa, integrating Sendai principles while enhancing cross-border coordination and resilience planning for climate-induced and gradual disasters. This is particularly pertinent to the Gauteng City-Region, where urbanization, migration, and transboundary climate patterns obscure jurisdictional risk management. The SADC framework emphasizes the need for integrated, multi-sectoral strategies for disaster risk reduction, acknowledging that the effects of climate change surpass administrative boundaries and necessitate coordinated responses (SADC, 2016; Tirivangasi, 2018).

South African metropolitan areas, including Tshwane and Johannesburg, exhibit enhanced complexity in planning and hazard intelligence (City of Johannesburg, 2021; City of Tshwane, 2024). Numerous smaller and rural towns in South Africa persist in grappling with insufficient resources and disjointed institutional frameworks, impeding effective disaster risk management and climate change adaptation initiatives (Stephen et al., 2021). The inconsistent performance of underperforming municipalities constitutes a systemic national weakness.

National Legal and Policy Frameworks

The Disaster Management Act (2002) and the National Disaster Management Framework (2005) of South Africa officially establish disaster management as a decentralized, development-oriented, multi-sector governance function. The Act mandates the establishment of the National Disaster Management Centre (NDMC), imposing obligations on all levels of government to:

- Assess and oversee hazards
- Establish organizational competence
- Guarantee readiness
- Incorporate disaster-risk planning with internally displaced individuals (IDPs).
- Enhance civic engagement and education

The Act guarantees that EWS are grounded in rights and integrated into government, rather than being optional or ad hoc operational instruments.

The NDMC Strategic Plans further underscore risk mitigation, institutional reinforcement, national cooperation, and capacity enhancement (NDMC, 2020). The institutional framework is prominently displayed in the submitted municipal Disaster Risk Management Plan, which implements the Act via organized command structures, communication protocols, and inter-agency collaboration. These frameworks together seek to integrate disaster management into the overarching development agenda, ensuring that risk reduction and climate change adaptation are fundamental elements of governance at all levels (Ziervogel & Taylor, 2023). Nonetheless, practice-based reporting—exemplified by Tshwane's Annual Report, consistently underscores constraints such as resource limitations, legacy infrastructure deficiencies, informal settlement vulnerabilities, and increasing climate instability (Munsamy, 2019; Zenda, 2025).

The observed contradictions indicate that legislative mandates alone cannot foster resilience; leadership capacity, implementation culture, and digital facilitation are important.

Climate Governance as a Way to Manage Risk

The City of Johannesburg Climate Action Plan (2021) represents the integration of global and local climate governance, incorporating resilience into spatial planning, infrastructure investment, water security, social equity, and emission reduction strategies. This substantiates a fundamental assertion of this chapter:

- Risk governance is now equivalent to climate governance, both reliant on leadership maturity, collaboration, and anticipatory planning.
- Digital Governance, Artificial Intelligence, and Cyber-Risk Resilience

Contemporary EWS and DRM systems increasingly rely on digital ecosystems, artificial intelligence analytics, automation, and cyber-secure infrastructure. The NIST AI Risk Management Framework (2023) and NIST Cybersecurity Framework (NIST, 2018) delineate global best practices for the ethical deployment of AI, emphasizing transparency, accountability, cyber resilience, and data integrity.

This is particularly pertinent to South African EWS, where escalating digital reliance renders municipalities susceptible to:

- A cybernetic system malfunctions
- Data corruption or alteration
- Artificial intelligence prejudice or lack of transparency
- A collapse of trust
- Problems of surveillance and equity

However, as previously emphasized, South African DRM policy has not completely incorporated AI governance or cyber-resilience, hence revealing a potential risk frontier. This gap highlights the urgent necessity for a proactive policy response that tackles the complex interaction between sophisticated digital technologies and catastrophe management to protect against emerging risks (Allen, 2019). A cohesive legal and policy framework is crucial to maximize the benefits of technological advancements for global climate adaptation, especially considering the swift development of digital tools like 3D city modeling, digital twins, and AI-driven internet platforms for predictive analysis (Raftopoulos et al., 2024).

Organisation for Economic Co-operation and Development (OECD) and Governance of Public-Sector Resilience

The OECD's Risk Governance and Resilience Frameworks emphasize public administration capacity, institutional learning, openness, performance accountability, and ethical leadership as essential pillars of resilience (OECD, 2014; OECD, 2020). This highlights the primary assertion of this chapter that technology forecasts risk, while leadership transforms intelligence into robust results.

The OECD also expressly associates trust and ethical governance with system effectiveness, which is pertinent in a society where instability in service delivery and political fragmentation may undermine public adherence to early-warning systems.

Synthesis: Implications of These Frameworks for Leadership and EWS Performance

A uniform message arises across the global-to-local continuum:

- Risk governance is today characterized by multi-hazard, multi-level, technology-driven, and human-centric approaches.
- The efficacy of EWS is contingent upon leadership legitimacy, institutional integrity, and public trust, in addition to precise forecasting.
- AI-enhanced predictive intelligence is revolutionizing EWS; yet, it necessitates governance, ethical considerations, and cyber-resilience.
- The maturity of African and South African policies is substantial; yet, the capacity for implementation is inconsistent.
- Leadership is the critical factor, balancing system stabilization with the promotion of innovation and adaptation.

The comprehensive global governance framework corroborates the dichotomous leadership paradigm you propose. Modern leaders must:

- Oversee adherence and advancement
- Stabilize operations and enhance capability
- Adopt artificial intelligence while upholding ethical standards and fostering trust.

- Provide services and enhance societal resilience

The necessity for both-and leadership is now imperative in contemporary disaster-risk governance. The incorporation of new technologies, including AI and machine learning, into disaster management frameworks requires a comprehensive understanding of their capabilities and limits to effectively improve early warning systems and overall resilience (Kolivand et al., 2025).

8. PRACTICAL AND THEORETICAL IMPLICATIONS

Practically, this chapter demonstrates that EWS achieve societal value only when they are integrated with competent leadership, ethical governance, and digitally advanced institutions. Municipalities and public institutions must transition from compliance-focused disaster plans to anticipatory, AI-enhanced, integrated risk governance systems. This necessitates leadership training in disruption management, enhanced inter-agency coordination, cyber-secure data ecosystems, the participation of vulnerable people in early warning system literacy, and leadership accountability for preparedness results. Practitioners must prioritize trust-building, communication efficacy, behavioral response design, and ongoing institutional learning as fundamental components of resilience, rather than as supplementary parts.

This chapter theoretically advances the topic by placing the leadership dichotomy at the core of multi-hazard resilience and AI-enabled governance. It reconceptualizes EWS as socio-technical governance systems influenced by institutional culture, ethics, trust, and strategic foresight, rather than just technological artifacts. The results bolster and expand disruption-management theory, illustrating that resilience arises when leaders equilibrate stability with transformation, compliance with innovation, and operational control with collaborative governance. This chapter enhances AI governance literature by emphasizing the leadership responsibilities linked to predictive intelligence in the public sector. Overall, it demonstrates that resilience has evolved into a leadership science as well as a technological discipline, necessitating cohesive theoretical frameworks encompassing governance, technology, institutional theory, and public-value management.

9. LIMITATIONS AND FUTURE RESEARCH DIRECTIONS

This chapter is predominantly conceptual and documentary in nature. While it includes municipal reports, global policy frameworks, and evidence from practice-based governance, it lacks empirical fieldwork, including interviews with disaster managers, municipal officials, or communities receiving early warnings. The findings represent system-level and institutional viewpoints, rather than the lived experiences of frontline practitioners or vulnerable individuals at home. Another issue is that the implementation of AI-enabled EWS varies significantly among municipalities, resulting in findings on digital resilience being based more on policy intentions than on empirical functionality. The emphasis on the South African metropolis may restrict its applicability to rural or resource-limited towns in other regions of the continent.

A future study should prioritize an empirical examination of the influence of leadership behaviors, institutional culture, and trust dynamics on citizen responses to early warnings. It is essential to assess AI-enabled predictive systems in practice, focusing on concerns such as cybersecurity risk, bias, data integrity, and ethical governance. Comparative analyses among African towns would enhance comprehension of capacity disparities, the impacts of coalition-era governance, and political stability on disaster preparedness. A longitudinal study might further examine how leadership continuity, digital maturity, and policy coherence affect resilience outcomes over time. Ultimately, interdisciplinary research encompassing behavioral science, data ethics, and public-value governance would enhance understanding of how technology, leadership, and society collaboratively foster resilience amid ongoing change.

10. CONCLUSION

This chapter demonstrates that EWS should not be perceived solely as technical tools but as governance ecosystems influenced by the quality of leadership, institutional coherence, ethical digital practices, and public trust. Although South Africa has a strong legislative and policy framework for disaster risk management, evidence indicates that resilience cannot be achieved through frameworks alone. The effectiveness of early warnings in saving lives, safeguarding infrastructure, and sustaining communities ultimately hinges on leadership's capacity to balance stability with change—governing responsibly while simultaneously fostering innovation, foresight, collaboration, and learning. In contexts where leadership is fractured, politicized, digitally unprepared, or detached from communities, even well-structured systems diminish in credibility and effectiveness (Ifenthaler et al., 2021; Rust & Rust, 2024). In contrast, when leaders show ethical foundations, foresight, and institutional responsibility, EWS transform into significant facilitators of coordinated action and societal resilience (Camps-Valls et al., 2024; Zheng & Chu, 2025).

The implications are evident. Public organizations must intentionally develop disruption-ready leadership as an essential governance competency, rather than a supplementary skill. Strong governance, ethical monitoring, and organizational preparedness must complement the investment in AI-driven predictive tools and cyber-secure data platforms (Fulton et al., 2024; Sitnicka et al., 2025). We must advance disaster governance from mere compliance to anticipatory practices, integrating regular scenario testing and active leadership involvement in crisis simulations. The restoration of public trust is equally important, necessitating open communication, community education, and culturally relevant warning systems (Platzer & Knodt, 2023; Savoia et al., 2023). Resilience must be collaboratively developed among government, business, academia, and civil society, with climate, infrastructure, and catastrophe preparation regarded as a unified governance challenge. Ultimately, resilience in the modern day is not merely a product of technology but a manifestation of leadership maturity and institutional integrity.

Reflection

If disruption has become a constant aspect of modern society, then anticipatory, ethical, and technologically empowered leadership is the most significant asset for resilience. The future of EWS, both in South Africa and beyond, lies at the merging of governance, digital intelligence, institutional trust, and courageous leadership.

REFERENCES

Abdullahi, A., & Shehu, M. (2025). Tech for Social Impact: Measuring and scaling the outcomes of digital transformation in communities. *International Journal of Computing and Artificial Intelligence*, *6*(2), 210–222. DOI: 10.33545/27076571.2025.v6.i2c.200

Adel, A., Saleem, N. H., Jan, T., & Prasad, M. (2025). A review of major ICT failures and recovery strategies: Strengthening digital resilience. *Computers & Security*, *159*, 104678. DOI: 10.1016/j.cose.2025.104678

Afonso, A., Alves, J., & Monteiro, S. (2024). The Pressure is on: How Geopolitical Tensions Impact Institutional Fiscal and External Stability Responses. SSRN *Electronic Journal*. DOI: 10.2139/ssrn.4803854

African Union Commission. (2022). *Africa Multi-Hazard Early Warning and Early Action System Framework*. Addis Ababa: African Union. Available at: https://au.int/en/documents

Albert, M. J. (2024). Navigating the Polycrisis. In *The MIT Press eBooks*. The MIT Press., DOI: 10.7551/mitpress/15041.001.0001

Alcántara-Ayala, I. (2025). Cascading hazards and compound disasters. *Npj Natural Hazards*, *2*(1), 54. Advance online publication. DOI: 10.1038/s44304-025-00111-5

Alcántara-Ayala, I., Velásquez-Espinoza, G., & de Jesús, A. M. (2025). From Mandates to Mechanisms: Institutional Vulnerability, Decentralized Governance, and the Challenges of Local Disaster Risk Reduction Implementation. *International Journal of Disaster Risk Science*, *16*(5), 709–723. Advance online publication. DOI: 10.1007/s13753-025-00673-y

Allen, T. (2019). Creating resilient futures. In *Routledge eBooks* (p. 137). Informa., DOI: 10.4324/9780429398872-10

Andersen, T. J., & Young, P. C. (2023). Enhancing public sector enterprise risk management through interactive information processing. *Frontiers in Research Metrics and Analytics*, *8*, 1239447. Advance online publication. DOI: 10.3389/frma.2023.1239447 PMID: 38192523

Ansell, C., & Torfing, J. (2021). Public Governance as Co-creation. In *Cambridge University Press eBooks*. Cambridge University Press., DOI: 10.1017/9781108765381

Ansell, C., & Trondal, J. (2017). Governing Turbulence: An Organizational- Institutional Agenda. *Perspectives on Public Management and Governance*, *1*(1), 43–57. DOI: 10.1093/ppmgov/gvx013

Atukunda, G., Musiita, B., Atwiine, J., & Olyanga, A. M. (2024). Corporate Governance Practices, Operating Environment and Financial Sustainability of Saccos in Greater Mbarara District. *Journal of Economics and Behavioral Studies*, *16*(2(J)), 70–86. DOI: 10.22610/jebs.v16i2(J).3837

Bell, L., & Hofmeyr, K. (2021). Enabling organisational ambidexterity: A leadership perspective. *South African Journal of Business Management*, *52*(1). Advance online publication. DOI: 10.4102/sajbm.v52i1.2268

Best, A., & Yalezo, B. (2022). Stakeholder engagement strategies assessment in expanded public works programme. *Journal of Governance and Regulation*, *11*(4), 46–58. DOI: 10.22495/jgrv11i4art5

Bodó, B., & Janssen, H. (2022). Maintaining trust in a technologized public sector. *Policy and Society*, *41*(3), 414–429. DOI: 10.1093/polsoc/puac019

Byaruhanga, N., Kibirige, D., & Mkhonta, G. (2025). Understanding South Africa's Flood Vulnerabilities and Resilience Pathways: A Comprehensive Overview. *Water (Basel)*, *17*(17), 2608. DOI: 10.3390/w17172608

Camps-Valls, G., Fernández-Torres, M., Cohrs, K.-H., Höhl, A., Castelletti, A., Paçal, A., Robin, C. N., Martinuzzi, F., Papoutsis, I., Prapas, I., Pérez-Aracil, J., Weigel, K., Gonzalez-Calabuig, M., Reichstein, M., Rabel, M., Giuliani, M., Mahecha, M. D., Popescu, O.-I., Pellicer-Valero, Ó. J., . . . Williams, T. (2024). AI for Extreme Event Modeling and Understanding: Methodologies and Challenges. *arXiv (Cornell University)*. https://doi.org//arxiv.2406.20080DOI: 10.48550

Chetty, K., Saal, P., Ntshayintshayi, N., Masuku, N., & Moosa, T. (2024). AI as a Catalyst for Good Governance: Transforming South Africa's Fight Against Corruption. *Development*, *67*(1-2), 50–60. DOI: 10.1057/s41301-024-00404-8

City of Cape Town. (2017). *Municipal Disaster Risk Management Plan*. City of Cape Town Disaster Risk Management Centre.

City of Johannesburg. (2021). *Climate Action Plan*. Johannesburg: Environment and Infrastructure Services Department (EISD). Available at: https://www.joburg.org.za/documents_/Pages/Key%20Documents/Climate%20Action%20Plan.aspx

City of Tshwane. (2024). *Disaster Risk Management 2024/2025 Annual Report*. City of Tshwane Metropolitan Municipality.

Csaszar, F. A., Ketkar, H. J., & Kim, H. (2024). Artificial Intelligence and Strategic Decision-Making: Evidence from Entrepreneurs and Investors. *Strategy Science, 9*(4), 322–345. DOI: 10.1287/stsc.2024.0190

Das, D. K., & Devadas, V. (2025). Smart and Resilient Infrastructure in the Wake of Climate Change. *Urban Planning, 10,* 11373. Advance online publication. DOI: 10.17645/up.11373

Dedola, L., Ehrmann, M., Hoffmann, P., Lamo, A., Paz-Pardo, G., Slačálek, J., & Strasser, G. (2023). Digitalisation and the Economy. SSRN *Electronic Journal.* DOI: 10.2139/ssrn.4626463

Doğrusöz, L. A., & Yazıcı, S. (2025). Measuring organisational governance capacity in healthcare organisations: A scale development and validation study. *BMC Health Services Research, 25*(1), 338. DOI: 10.1186/s12913-025-12442-0 PMID: 40045299

Dunleavy, P., & Margetts, H. (2023). Data science, artificial intelligence and the third wave of digital era governance. *Public Policy and Administration.* Advance online publication. DOI: 10.1177/09520767231198737

Fatima, F., Hyatt, J. C., Rehman, S. U., Cruz, E. D. L., Nadella, G. S., & Meduri, K. (2024). Resilience and Risk Management in Cybersecurity: A Grounded Theory Study of Emotional, Psychological, and Organizational Dynamics. *Journal of Economy and Technology, 2,* 247–257. DOI: 10.1016/j.ject.2024.08.004

Fulton, R., Fulton, D., Hayes, N., & Kaplan, S. (2024). *The Transformation Risk-Benefit Model of Artificial Intelligence: Balancing Risks and Benefits Through Practical Solutions and Use Cases. arXiv.* Cornell University., DOI: 10.5121/ijaia.15201

Gambhir, A., Albert, M., Doe, S. S. P., Donges, J. F., Farajalla, N., Giatti, L. L., Gundimeda, H., Hendel-Blackford, S., Homer-Dixon, T., Hoyer, D., Adán, S., Polit, D. J., Kemp, L., Korowicz, D., Kovacic, Z., Kwakkel, J., Laybourn-Langton, L., Lempert, R. J., Mahamoud, A., … Richardson, R. (2025). A systemic risk assessment methodological framework for the global polycrisis [Review of *A systemic risk assessment methodological framework for the global polycrisis*]. *Nature Communications, 16*(1), 7382. Nature Portfolio. DOI: 10.1038/s41467-025-62029-w

Gandhi, M., Cihon, P., Larter, O., & Anselmetti, R. (2025). Societal Capacity Assessment Framework: Measuring Resilience to Inform Advanced AI Risk Management. *arXiv (Cornell University).* https://doi.org//arxiv.2509.22742DOI: 10.48550

Greenwood, L., Hess, D., & Schneider, J. (2025). Building a Strategic System for Resilience in a Risky World. *Systems, 13*(9), 805. DOI: 10.3390/systems13090805

Gregory, J., & Sovacool, B. K. (2019). Rethinking the governance of energy poverty in sub-Saharan Africa: Reviewing three academic perspectives on electricity infrastructure investment. *Renewable & Sustainable Energy Reviews*, *111*, 344–354. DOI: 10.1016/j.rser.2019.05.021

Grewal, J. S., Habahbeh, L., Acharyya, M., Aravind, R., Bhagaloo, S., Carey, M., Er, C., Farrugia, K., & Leung, K. (2022). COVID-19 and the effectiveness of ERM frameworks. *British Actuarial Journal*, *27*, e23. Advance online publication. DOI: 10.1017/S1357321722000174

Guma, P. K. (2022). On tackling infrastructure: The need to learn from marginal cities and populations in the Global South. *Journal of the British Academy*, *10*, 29–37. DOI: 10.5871/jba/010.029

Haque, A. (2025). Responsible artificial intelligence (AI) in healthcare: A paradigm shift in leadership and strategic management. *Leadership in Health Services*, *38*(4), 644–656. Advance online publication. DOI: 10.1108/LHS-01-2025-0018 PMID: 40911384

Harper, S. (2020). Global Risks and the Resilience of Future Health Care Systems. *Journal of Population Ageing*, *13*(1), 1–3. DOI: 10.1007/s12062-020-09262-x PMID: 35368879

Helmrich, A., & Chester, M. (2022). Navigating Exploitative and Explorative Leadership in Support of Infrastructure Resilience. *Frontiers in Sustainable Cities*, *4*, 791474. Advance online publication. DOI: 10.3389/frsc.2022.791474

Hendrickse, R. (2023). Ethnic leadership in the age of disruption: implications for South African municipalities. *International Journal of Research in Business and Social Science (2147-4478)*, *12*(7), 224. DOI: 10.20525/ijrbs.v12i7.2791

Henrico, S., & Els, S. (2025). Cyber Attacks in South Africa: Geopolitical and legal implications. *African Security Review*, *1*(3), 312–336. Advance online publication. DOI: 10.1080/10246029.2025.2489352

Hostettler, S., Besson, S. N., & Bolay, J. (2018). *Technologies for Development*. DOI: 10.1007/978-3-319-91068-0

Ibrahim, R., Motshegwa, T., Benouar, D., Khodja, M., Ead, H. A., Zhang, L., Crawley, F. P., Mchunu, N. P., Hachigonta, S., Hodson, S., McBride, V. A., & Anshebo, T. Y. (2025). Sustainable North-South Africa Collaboration for Disaster and Crisis Management: A Strategic Capacity Development Framework using Open Science, Artificial Intelligence and Geoinformatics. *Data Science Journal*, *24*, 11. Advance online publication. DOI: 10.5334/dsj-2025-011

Ifenthaler, D., Hofhues, S., Egloffstein, M., & Helbig, C. (2021). *Digital Transformation of Learning Organizations*. DOI: 10.1007/978-3-030-55878-9

Islam, Md. M., Hasan, M., Mia, Md. S., Masud, A. A., & Islam, A. R. Md. T. (2025). Early Warning Systems in Climate Risk Management: Roles and Implementations in Eradicating Barriers and Overcoming Challenges. *Natural Hazards Research*. DOI: 10.1016/j.nhres.2025.01.007

Jaadi, M. E., Haidi, T., & Bouabdallaoui, D. (2025). *IAES International Journal of Artificial Intelligence (IJ-AI)*. IAES International Journal of Artificial Intelligence., DOI: 10.11591/ijai

Joshi, S. (2025). Leadership in the Age of AI: Review of Quantitative Models and Visualization for Managerial Decision-Making. *Preprints.Org*. DOI: 10.20944/preprints202504.1662.v1

Keinänen, J., Halonen, M., & Luoto, I. (2025). What can we learn about crisis management in the polycrisis era? Insights from Finland's international public policy network. *International Journal of Public Sector Management*. Advance online publication. DOI: 10.1108/IJPSM-06-2024-0191

Kikasu, E. T.Eustache Tanzala Kikasu. (2025). Navigating through the Dynamic Landscape of Public Administration Governance through Innovative Adaptation and Resilience Practices and Strategies in Africa. *Journal of Information Systems Engineering & Management*, *10*(7s), 203–228. DOI: 10.52783/jisem.v10i7s.854

Kikasu, E. T., & Pillay, S. S. (2024). Forecasting an Inevitable Coalition Government at the National Level in South Africa: A New Path to Public Administration and Governance. *Open Journal of Political Science*, *14*(1), 28–51. DOI: 10.4236/ojps.2024.141003

Kiparsky, M., Sedlak, D. L., Thompson, B. H.Jr, & Truffer, B. (2013). The Innovation Deficit in Urban Water: The Need for an Integrated Perspective on Institutions, Organizations, and Technology. *Environmental Engineering Science*, *30*(8), 395–408. DOI: 10.1089/ees.2012.0427 PMID: 23983450

Koh, S. C. L., Suresh, K., Ralph, P. J., & Saccone, M. (2023). Quantifying organisational resilience: An integrated resource efficiency view. *International Journal of Production Research*, *62*(16), 5737–5756. DOI: 10.1080/00207543.2023.2296018

Kolivand, P., Azari, S., Bakhtiari, A., Namdar, P., Ayyoubzadeh, S. M., Rajaie, S., & Ramezani, M. (2025). AI applications in disaster governance with health approach: A scoping review [Review of *AI applications in disaster governance with health approach: A scoping review*]. *Archives of Public Health, 83*(1). BioMed Central. DOI: 10.1186/s13690-025-01712-2

Lategan, L., Fisher-Holloway, B., Cilliers, E. J., & Cilliers, S. S. (2025). Moving on to Greener Pastures? A Review of South Africa's Housing Megaproject Literature [Review of *Moving on to Greener Pastures? A Review of South Africa's Housing Megaproject Literature*]. *Sustainability, 17*(4), 1677. Multidisciplinary Digital Publishing Institute. DOI: 10.3390/su17041677

Lawrence, M., Homer-Dixon, T., Janzwood, S., Rockstöm, J., Renn, O., & Donges, J. F. (2024). Global Polycrisis: The Causal Mechanisms of Crisis Entanglement. *Global Sustainability*, *7*, e6. Advance online publication. DOI: 10.1017/sus.2024.1

Lee, C., Wang, Z., Wang, D., Lyu, S., & Chen, C. (2025). Artificial-intelligence-driven governance: Addressing emerging risks with a comprehensive risk-prevention-centred model for public health crisis management. *Health Research Policy and Systems*, *23*(1), 115. Advance online publication. DOI: 10.1186/s12961-025-01390-0 PMID: 41013547

Lee, C.-C., Comes, T., Finn, M., & Mostafavi, A. (2022). Roadmap Towards Responsible AI in Crisis Resilience Management. *arXiv (Cornell University)*. https://doi.org//arxiv.2207.09648DOI: 10.48550

Li, L., Wang, J., Wang, X., Peng, P., Shen, J., Zhu, H., & Zhang, Z. (2025). Big data and data science in global governance: Anticipating future needs and applications in the UN and beyond. *Frontiers in Political Science*, *7*, 1583772. Advance online publication. DOI: 10.3389/fpos.2025.1583772

Lisdiono, P., Said, J., Yusoff, H., & Hermawan, A. A. (2022). Examining Leadership Capabilities, Risk Management Practices, and Organizational Resilience: The Case of State-Owned Enterprises in Indonesia. *Sustainability (Basel)*, *14*(10), 6268. DOI: 10.3390/su14106268

Lunga, W., Sithole, M., Musarurwa, C., Kombaté, E. N., & Marovah, T. (2025). Integrating Ubuntu: Embedding African philosophy into disaster risk reduction and management. *Frontiers in Sociology*, *10*, 1637051. Advance online publication. DOI: 10.3389/fsoc.2025.1637051 PMID: 40969614

Magezi, C. (2023). *Uncertainty as certainty in a COVID-19 complex world: Reflections on leading congregations for effective missional praxis* (p. 119). DOI: 10.4102/aosis.2023.BK398.08

Makhura, N. K. (2025). Bridging the Governance Gap: Integrating Artificial Intelligence in South Africa's Public Policy Implementation. *The Asian Bulletin of Big Data Management, 5*(2), 16–26. DOI: 10.62019/qq39na69

Makole, K. R., Ntshangase, B. A., & Adewumi, S. A. (2022). Coalition Governance: Unchartered Waters in South African Political Landscape. *Business Ethics and Leadership, 6*(4), 23–37. DOI: 10.21272/bel.6(4).23-37.2022

Malik, F. S., & Terzidis, O. (2025). Thriving in turbulence: Resilience and strategic adaptation in global business. *Review of Managerial Science*. Advance online publication. DOI: 10.1007/s11846-025-00940-8

Maota, T. (2026). Fortifying Municipal Governance: Leveraging IT Audits and Disruption Management for Sustainable Public Services. In Gupta, M., Walp, J., & Sharman, R. (Eds.), *Advancing IT Audits Through Integrative Approaches and Emerging Technologies* (pp. 339–368). IGI Global Scientific Publishing., DOI: 10.4018/979-8-3373-3078-5.ch011

Maota, T., Naido, V., & Darwish, S. (2024, April). Quantitative Assessment of Management Innovation in Navigating Disruptions Within Higher Education. In International Conference on Business and Technology (pp. 166-176). Cham: Springer Nature Switzerland. DOI: 10.1007/978-3-031-67444-0_16

Maota, T., & Naidoo, V. (2022). Response of private higher education leadership to pandemic crisis. *Research Highlights in Language. Literature and Education, 2*(2), 127–148.

Maota, T. R. (2024). A conceptual framework for private higher educational institutions to respond to disruptions in South Africa. https://hdl.handle.net/10413/23345

Mehryar, S., Yazdanpanah, V., & Tong, J. (2024). AI and climate resilience governance. *iScience, 27*(6), 109812. DOI: 10.1016/j.isci.2024.109812 PMID: 38784017

Métivier, J., Bacchetta, M., Bekkers, E., & Koopman, R. (2023). International trade cooperation's impact on the world economy. *Journal of Policy Modeling, 45*(4), 713–744. DOI: 10.1016/j.jpolmod.2023.06.007

Meuleman, L., & Fraser-Moleketi, G. (2022). Transforming Institutions and Governance to Build forward Better towards 2030. *Governance and Politics, 1*(2), 7–29. DOI: 10.24833/2782-7062-2022-1-2-7-29

Mirianna, B., Robert, S., Cinthia, A., Miguel, A. M. B. de, Orlando, C. V., Abel, C. A., Monica, C. I., Adama, D., León, L. D., Giorgio, M., Alioune, N., Miluska, O. C., Tamanna, R., Bikram, R., Alpha, S., Dharam, U., Anderson, C. R., & Colin, M. (2025). Opportunities and challenges for people-centered multi-hazard early warning systems: Perspectives from the Global South [Review of *Opportunities and challenges for people-centered multi-hazard early warning systems: Perspectives from the Global South*]. *iScience, 28*(5), 112353. Cell Press. DOI: 10.1016/j.isci.2025.112353

Mökander, J., & Axente, M. (2021). Ethics-based auditing of automated decision-making systems: Intervention points and policy implications. *AI & Society*, *38*(1), 153–171. DOI: 10.1007/s00146-021-01286-x

Moudio, M. P. E., Bolin, R., Carpenter, A., Reese, S., Shehabi, A., & Rao, P. (2022). Characterizing manufacturing sector disruptions with targeted mitigation strategies. *Environmental Research Infrastructure and Sustainability*, *2*(4), 42001. DOI: 10.1088/2634-4505/ac9c8c

Munsamy, L. (2019). Praxis of climate change adaptation in disaster risk reduction: A South African perspective *Transactions on Ecology and the Environment, 1*, 257. DOI: 10.2495/SC190231

Myeza, L., Ecim, D., & Maroun, W. (2023). The role of integrated thinking in corporate governance during the COVID-19 crisis: Perspectives from South Africa. *Journal of Public Budgeting, Accounting & Financial Management*, *35*(6), 52–77. DOI: 10.1108/JPBAFM-08-2022-0133

National Disaster Management Centre (NDMC). (2020). *Strategic Plan 2020–2025.* Pretoria: Department of Cooperative Governance and Traditional Affairs. Available at: https://www.gov.za/documents/national-disaster-management-centre-strategic-plan

National Institute of Standards and Technology (NIST). (2018). *Framework for Improving Critical Infrastructure Cybersecurity (Version 1.1).* Gaithersburg, MD. Available at: https://www.nist.gov/cyberframework

National Institute of Standards and Technology (NIST). (2023). *Artificial Intelligence Risk Management Framework (AI RMF 1.0).* Gaithersburg, MD: U.S. Department of Commerce. Available at: https://www.nist.gov/itl/ai-risk-management-framework

Neyer, J. (2021). After global governance. *Zeitschrift für Politikwissenschaft*, *32*(2), 361–382. DOI: 10.1007/s41358-021-00290-3

Nwachukwu, P. S., Chima, O. K., & Okolo, C. H.Priscilla Samuel Nwachukwu Onyeka Kelvin ChimaChinelo Harriet Okolo. (2025). The artificial intelligence governance framework for finance: A control-by-design approach to algorithmic decision-making in accounting. *Finance & Accounting Research Journal*, *7*(8), 350–379. DOI: 10.51594/farj.v7i8.2016

OECD. (2020). *Building Resilience to the COVID-19 Pandemic: The Role of Centres of Government.* Paris: OECD Publishing. Available at: https://www.oecd.org/coronavirus/policy-responses

OECD. (2024). Framework for Anticipatory Governance of Emerging Technologies. In *OECD science, technology and industry policy papers*. DOI: 10.1787/23074957

Okunola, O. H. (2025). Beyond Institutional Silos: Rethinking Multilevel Disaster Risk Governance in Africa a Decade into the Sendai Framework Implementation. *International Journal of Disaster Risk Science*, *16*(3), 321–332. DOI: 10.1007/s13753-025-00646-1

Onduko, J. O., Kalombo, M. A., Kuol, M. D., Makale, B. G., & Saleh, M. A. (2024). AI-Driven Governance: Transforming Public and Addressing Legacy Issues in Post-Colonial Africa. *Proceedings of London International Conferences*, *11*(11), 52–63. DOI: 10.31039/plic.2024.11.243

Organisation for Economic Co-operation and Development (OECD). (2014). *OECD Reviews of Risk Management Policies: Boosting Resilience through Innovative Risk Governance.* Paris: OECD Publishing. Available at: https://www.oecd.org/governance/risk/

Pashentsev, E., & Kolotaev, Y. Y. (2025). Information and communication technologies in political crisis management: Resilience, forecasting, and response. *Discover Global Society*, *3*(1), 84. Advance online publication. DOI: 10.1007/s44282-025-00216-2

Patil, D., Rane, N. L., & Rane, J. (2024). *Enhancing resilience in various business sectors with ChatGPT and generative artificial intelligence*. DOI: 10.70593/978-81-981367-8-7_4

Patrucco, P. P., & Kähkönen, A. (2021). Agility, adaptability, and alignment: New capabilities for PSM in a post-pandemic world. *Journal of Purchasing and Supply Management*, *27*(4), 100719. DOI: 10.1016/j.pursup.2021.100719

Percival, V., Thoms, O. T., Oppenheim, B., Rowlands, D., Chisadza, C., Fewer, S., Yamey, G., Alexander, A. C., Allaham, C. L., Causevic, S., Daudelin, F., Gloppen, S., Guha-Sapir, D., Hadaf, M., Henderson, S., Hoffman, S. J., Langer, A., Lebbos, T. J., Leomil, L., … Friberg, P. (2023). The Lancet Commission on peaceful societies through health equity and gender equality [Review of *The Lancet Commission on peaceful societies through health equity and gender equality*]. *The Lancet, 402*(10413), 1661. Elsevier BV. DOI: 10.1016/S0140-6736(23)01348-X

Pescaroli, G., Dryhurst, S., & Karagiannis, G. M. (2025). Bridging gaps in research and practice for early warning systems: New datasets for public response. *Frontiers in Communication, 10*, 1451800. Advance online publication. DOI: 10.3389/fcomm.2025.1451800

Platzer, E. K., & Knodt, M. (2023). Resilience beyond insurance: Coordination in crisis governance. *Environment Systems & Decisions, 43*(4), 569–576. DOI: 10.1007/s10669-023-09938-7

Raftopoulos, M., Pertoldi-Bianchi, S., Eleftherios, P., & Kjelmann, K. G. (2024). Conceptualizing the Nexus between Disaster Risk Reduction and Climate Change Adaption and Mitigation in Governance. *Research Portal Denmark, 2024*(3), 172. https://local.forskningsportal.dk/local/dki-cgi/ws/cris-link?src=aau&id=aau-4383f7e9-8d76-4169-9202-c1936d86740c&ti=Conceptualizing%20the%20Nexus%20between%20Disaster%20Risk%20Reduction%20and%20Climate%20Change%20Adaption%20and%20Mitigation%20in%20Governance

Ragolane, M. (2024). Service Delivery Protests in the South African Government. *International Journal of Social Science Research and Review, 7*(4), 54–65. DOI: 10.47814/ijssrr.v7i4.1973

Ragus, E., Sutherland, C., Manderson, L., & Moyer, E. (2024). *Sculpting stories: methods to unsettle knowledge production in disasters*. Disaster Prevention and Management An International Journal., DOI: 10.1108/DPM-01-2024-0034

Reichstein, M., Benson, V., Blunk, J., Camps-Valls, G., Creutzig, F., Fearnley, C. J., Han, B., Kornhuber, K., Rahaman, N., Schölkopf, B., Tárraga, J. M., Vinuesa, R., Dall, K., Denzler, J., Frank, D., Martini, G., Nganga, N., Maddix, D. C., & Weldemariam, K. (2025). Early warning of complex climate risk with integrated artificial intelligence [Review of *Early warning of complex climate risk with integrated artificial intelligence*]. *Nature Communications, 16*(1). Nature Portfolio. DOI: 10.1038/s41467-025-57640-w

Reichstein, M., Frank, D., Benson, V., Camps-Valls, G., Denzler, J., Kornhuber, K., Schoelkopf, B., Vinuesa, R., Han, B., Fearnley, C. J., Rahaman, N., Tárraga, J. M., Blunk, J., Dall, K., Martini, G., Nganga, N., & Robinson, D. E. (2024). Early warning of complex climate risk with integrated artificial intelligence. *Research Square (Research Square)*. DOI: 10.21203/rs.3.rs-4248340/v1

Republic of South Africa. (2002). *Disaster Management Act 57 of 2002*. Pretoria: Government Printer. Available at: https://www.gov.za/documents/disaster-management-act-57-2002-0

Rokhideh, M., Fearnley, C. J., & Budimir, M. (2025). Multi-Hazard Early Warning Systems in the Sendai Framework for Disaster Risk Reduction: Achievements, Gaps, and Future Directions. *International Journal of Disaster Risk Science*, *16*(1), 103–116. DOI: 10.1007/s13753-025-00622-9

Rust, B. W., & Rust, S. (2024). Place-based innovation for sustainable and resilient human systems. *Sustainable Earth Reviews*, *7*(1), 19. Advance online publication. DOI: 10.1186/s42055-024-00089-4

Saka, T. N., Hormiga, E., & Valls-Pasola, J. (2024). Crisis response strategies: A digital reluctance perspective. *Review of Managerial Science*, *19*(8), 2569–2607. DOI: 10.1007/s11846-024-00822-5

Sakalasuriya, M., Haigh, R., Hettige, S., Amaratunga, D., Basnayake, S., & Rahayu, H. P. (2020). Governance, Institutions and People within the Interface of a Tsunami Early Warning System. *Politics and Governance*, *8*(4), 432–444. DOI: 10.17645/pag.v8i4.3159

Salem, A., Azzam, S. M., Emam, O. E., & Abohany, A. A. (2024). Advancing cybersecurity: a comprehensive review of AI-driven detection techniques [Review of *Advancing cybersecurity: a comprehensive review of AI-driven detection techniques*]. *Journal Of Big Data, 11*(1). Springer Science+Business Media. DOI: 10.1186/s40537-024-00957-y

Savoia, E., Piltch-Loeb, R., Stanton, E., & Koh, H. K. (2023). Learning from COVID-19: Government leaders' perspectives to improve emergency risk communication. *Globalization and Health*, *19*(1), 86. Advance online publication. DOI: 10.1186/s12992-023-00993-y PMID: 37968676

Schiavi, G. S., Behr, A., & Marcolin, C. B. (2023). Institutional theory in accounting information systems research: Shedding light on digital transformation and institutional change. *International Journal of Accounting Information Systems*, *52*, 100662. DOI: 10.1016/j.accinf.2023.100662

Schlösser, P., Rockström, J., Edwards, C., Mirazo, P., Heilemann, A., Kitzmann, N. H., & Krobjinski, S. L. (2023). Accelerating transformations for a just, sustainable future: 10 'Must Haves.'. *Global Sustainability*, *6*, e17. Advance online publication. DOI: 10.1017/sus.2023.14

Shah, R., Flemming, J. A., Chinnery, J., & Heaner, G. (2024). Building back better? The role of education sector responses in strengthening or eroding societal resilience during the COVID-19 pandemic. *Frontiers in Education*, *9*, 1089422. Advance online publication. DOI: 10.3389/feduc.2024.1089422

Shaik, F. A., Demil, G., & Oussalah, M. (2025). AI-based Approach in Early Warning Systems: Focus on Emergency Communication Ecosystem and Citizen Participation in Nordic Countries. *arXiv (Cornell University)*. https://doi.org//arxiv.2506.18926DOI: 10.48550

Sigfrids, A., Leikas, J., Salo-Pöntinen, H., & Koskimies, E. (2023). Human-centricity in AI governance: A systemic approach. *Frontiers in Artificial Intelligence*, *6*, 976887. Advance online publication. DOI: 10.3389/frai.2023.976887 PMID: 36872934

Singh, P. (2024). *Generative AI through the Lens of Institutional Theory*. DOI: 10.2139/ssrn.4953180

Sitnicka, S., Mursalov, M., Mammadov, H., Babaiev, D., & Zhou, Y. (2025). Ethics, Institutions, Infrastructure, and Governance in AI National-Level Readiness: A Hidden Driver of Banking Transformation. *Business Ethics and Leadership*, *9*(3), 290–304. DOI: 10.61093/bel.9(3).290-304.2025

Southern African Development Community (SADC). (2016). *SADC Disaster Risk Reduction Strategy and Plan of Action 2016–2030*. Gaborone: SADC Secretariat. Available at: https://www.sadc.int/document/sadc-disaster-risk-reduction-strategy-and-plan-action-2016-2030

Stander, J., Theron, A., & Du Plessis, J. (2011). Storm Surge Early Warning System Guidelines for South Africa. Pretoria: South African Weather Service (SAWS) and Disaster Mitigation for Sustainable Livelihoods Programme (DMISA), University of Cape Town.

Stenvall, J., Pinheiro, R., Johanson, J., Pekkola, E., Trondal, J., & Rannisto, P.-H. (2025). Governing uncertainty by anticipating risk: assessing crisis management during COVID-19. In *Edward Elgar Publishing eBooks* (p. 137). Edward Elgar Publishing., DOI: 10.4337/9781035318674.00019

Stephen, F., Yairen, J. C., Martin, L. T., & Barry, O. (2021). *Creating Resilient Futures*. DOI: 10.1007/978-3-030-80791-7

Suljic, V. (2025). Strategic Leadership in AI-Driven Digital Transformation: Ethical Governance, Innovation Management, and Sustainable Practices for Global Enterprises and SMEs. *Strategic Leadership in AI-Driven Digital Transformation: Ethical Governance, Innovation Management, and Sustainable Practices for Global Enterprises and SMEs.*, *13*(1), 17. DOI: 10.70301/JOUR/SBS-JABR/2025/13/1/2

Tim, Y., & Leidner, D. E. (2023). Digital Resilience: A Conceptual Framework for Information Systems Research. *Journal of the Association for Information Systems*, *24*(5), 1184–1198. DOI: 10.17705/1jais.00842

Tirivangasi, H. M. (2018). Regional disaster risk management strategies for food security: Probing Southern African Development Community channels for influencing national policy [Review of *Regional disaster risk management strategies for food security: Probing Southern African Development Community channels for influencing national policy*]. *Jàmbá Journal of Disaster Risk Studies, 10*(1). AOSIS. DOI: 10.4102/jamba.v10i1.468

Tiwari, M., Lodorfos, G., McClelland, Z., & Nair, S. (2025). Unraveling the complexities associated with leadership during times of supply chain crisis: A study on the healthcare sector. *International Studies of Management & Organization*, *1*(4), 442–475. Advance online publication. DOI: 10.1080/00208825.2025.2465002

Trogrlić, R. Š., van den Homberg, M., Budimir, M., McQuistan, C., Sneddon, A., & Golding, B. (2022). *Early Warning Systems and Their Role in Disaster Risk Reduction.*, DOI: 10.1007/978-3-030-98989-7_2

Tsheola, J., Sebola, M. P., & Mamabolo, M. A. (2023). South Africa's "Triple Crisis of Governance" and Societal Leadership Vacuum. *Insights into Regional Development*, *5*(4), 23–35. DOI: 10.9770/IRD.2023.5.4(2)

United Nations. (2022). *Early Warnings for All: The UN Global Initiative for Early Warning Systems.* New York: United Nations. Available at: https://www.un.org/en/climatechange/early-warnings-for-all

United Nations Office for Disaster Risk Reduction (UNDRR). (2015). *Sendai Framework for Disaster Risk Reduction 2015–2030.* Geneva: United Nations. Available at: https://www.undrr.org/publication/sendai-framework-disaster-risk-reduction-2015-2030

United Nations Office for Disaster Risk Reduction (UNDRR). (2018). *Multi-Hazard Early Warning Systems: A Checklist.* Geneva: UNDRR. Available at: https://www.undrr.org/publication/multi-hazard-early-warning-systems-checklist

Verschuur, J., Murgatroyd, A., Vittis, Y., Mosnier, A., Obersteiner, M., Godfray, H. C. J., & Hall, J. W. (2024). The impacts of polycrises on global grain availability and prices. *Research Square (Research Square)*. DOI: 10.21203/rs.3.rs-3969801/v1

Walters, L., Hall, L. A., & Grigsby, J. D. (2023). Ready or Not, It Will Come: Preparing Accountancy for the Next Disruption. *Journal of Applied Business and Economics*, *25*(3). Advance online publication. DOI: 10.33423/jabe.v25i3.6212

Witesman, E. (2021). *Public Value Governance: A Framework.*, DOI: 10.1007/978-3-030-29980-4_84

World Meteorological Organization (WMO). (2018). *Guidelines on Multi-Hazard Impact-Based Forecast and Warning Services*. Geneva: WMO. Available at: https://library.wmo.int/idurl/4/55733

Xi, Y., Zhou, L., & Wang, H. (2025). Ambidextrous leadership and innovation: A process perspective and an experimental approach. *Humanities & Social Sciences Communications*, *12*(1), 612. Advance online publication. DOI: 10.1057/s41599-025-04922-9

Yende, N. E. (2025). The Ripple Effect of Violent Service Delivery Protests in South Africa: Causes and Implications. In *IntechOpen eBooks*. IntechOpen. DOI: 10.5772/intechopen.1008853

Yu, G., & Qi, X. (2004). *Disruption management: framework, models, and applications*. World Scientific. DOI: 10.1142/5632

Zenda, M. (2025). Climate change adaptation and mitigation in different livestock production systems and agro-ecological zones in South Africa: A systematic review [Review of *Climate change adaptation and mitigation in different livestock production systems and agro-ecological zones in South Africa: A systematic review*]. *Tropical Animal Health and Production*, *57*(8). Springer Science+Business Media. DOI: 10.1007/s11250-025-04660-9

Zheng, B., & Chu, V. W. (2025). Multi-Hazard Early Warning Systems for Agriculture with Featural-Temporal Explanations. *arXiv (Cornell University)*. https://doi.org//arxiv.2507.22962DOI: 10.48550

Zhyvko, M. (2024). Fragmentation of Globalized Economy and Macroeconomic Cost of the War in Ukraine. *Problèmes Économiques*, *3*(61), 15–30. DOI: 10.32983/2222-0712-2024-3-15-30

Zia, A., & Wagner, C. H. (2015). Mainstreaming Early Warning Systems in Development and Planning Processes: Multilevel Implementation of Sendai Framework in Indus and Sahel. *International Journal of Disaster Risk Science*, *6*(2), 189–199. DOI: 10.1007/s13753-015-0048-3

Ziervogel, G., & Taylor, A. (2023). A co-produced national climate change risk and vulnerability assessment framework for South Africa. *Frontiers in Climate*, *5*, 1197167. Advance online publication. DOI: 10.3389/fclim.2023.1197167

Chapter 10
Building Awareness Through Education to an Effective Response to the Multi Hazards Warnings

Apoorav Sharma
http://orcid.org/0009-0002-9966-7691
Lovely Professional University, India

Lovleen Marwaha
Lovely Professional University, India

ABSTRACT

This chapter examines the critical role of education in enhancing the effectiveness of multi-hazard early warning systems (MHEWS) for disaster risk reduction. MHEWS integrate detection, assessment, and communication of diverse hazards, providing timely information to enable proactive community responses. However, technical, institutional, and social challenges often impede their full potential. Education emerges as a pivotal factor in bridging the gap between early warning dissemination and effective community action by fostering hazard awareness, shaping risk perceptions, and promoting preparedness behaviors. The chapter explores various educational types including formal, non-formal, and informal—and targets diverse audiences such as children, adults, and vulnerable populations. It discusses strategic approaches to curriculum development, training programs for key stakeholders, and public awareness campaigns coupled with community engagement.

DOI: 10.4018/979-8-3373-7345-4.ch010

right © 2026, IGI Global Scientific Publishing. Copying or distributing in print or electronic forms without written permission of IGI Global Scientific Publishing is prohibited. Use of this chapter to n generative artificial intelligence (AI) technologies is expressly prohibited. The publisher reserves all rights to license its use for generative AI training and machine learning model development.

A. INTRODUCTION

a. Background on Multi-Hazard Early Warning Systems

Multi-hazard early warning system (MHEWS) are comprehensive structures that are designed to identify, evaluate and transmit threats in the near future that are caused by various categories of hazards like floods, earthquakes, storms and tsunamis. They are mainly meant to deliver timely and precise information to the communities and decision-makers in order to allow proactive actions to mitigate life loss, damages, and social inconvenience. The significance of MHEWS is that they are able to reflect on the complexity of overlapping hazards that tend to come in succession or combination thus elevating the risk and effects of any disaster. These systems enhance efficiency and effectiveness in disaster risk management by aggregating the data and alerts of other hazard monitoring systems (Plevris, 2024; Van Wyk De Vries, 2025).

Although MHEWS are vital, they have a number of challenges in the reduction of disasters. These are technical challenges of working with various data of hazards, lack of infrastructure in high risk areas, communication issues and lack of coordination among the stakeholders. More so, the socio-economic issues like poverty, ignorance and culture differences may be a barrier to the successful delivery and acceptance of warnings. Such complications make it difficult to make sure that early warning is transformed into timely and proper responses of the community.

Education is a major factor in the elimination of these challenges through creation of awareness and knowledge of dangers, indications of danger, and correct response measures. Proper education campaigns enable individuals and communities to act according to warning signs, minimize panic, and participate in preparedness measures. Through development of a safety and resilience culture, education increases the overall effectiveness of MHEWS and warning results in meaningful protective behaviors.

b. Objectives of the Chapter

This chapter will set out to examine how education is crucial in enhancing the response in the case of multi-hazard early warning systems. The primary objectives are to examine the way in which educational interventions can enhance the knowledge and awareness of multi-hazard risks, as well as to show the connection between the knowledgeable populations and the response to the disaster. The chapter aims at bringing out best practices and challenges in education practices that enable early warning dissemination and uptake.

One of the aims is to highlight the relevance of education as an instrument that will not only create awareness but also instill behavioral change that makes one less vulnerable. Through the discussion of the correlation between the awareness and response, the chapter highlights the role of education in bridging the gap between being warned and taking relevant action. Such a relationship plays an essential role in improving the resilience of communities and reducing the effects of a disaster.

c. Scope and Structure of the Chapter

The chapter starts by thoroughly analyzing the concept and elements of multi-hazards early warning systems and this forms a basis on how they are operated. It goes on to examine contemporary issues in disaster risk reduction with an emphasis on obstacles of efficient warning communication and community response. The chapter then explores the purpose of education, which addresses several education techniques, programs, and tools that have the potential to increase awareness and readiness.

The sections provide case studies and empirical evidence on how education can be used to enhance the effectiveness of early warning. The chapter also ends by providing recommendations on how education can be incorporated into MHEWS structures as well as future research and policy recommendations.

In the general framework of the book, the chapter is one of the building blocks to the understanding of the overall disaster risk management concept as it connects the technological and institutional side of the early warning systems with the human and social side. It gives a critical outlook on how education can be used as a mediating factor between hazard identification and response in the community and this tends to support the discussion that the book has on the holistic view of disaster resilience.

B. UNDERSTANDING MULTI-HAZARD EARLY WARNING SYSTEMS

a. Components of Multi-Hazard Early Warning Systems

The Multi-hazard early warning systems (MHEWS) have four key components that work together to allow the identification, transmission, and management of varied and concurrent hazards. These elements risk knowledge and assessment, monitoring and forecasting, dissemination of warnings, and response capability constitute a holistic structure that will result in the effective and timely management of risk in case of disaster.

1. **Risk Knowledge and Assessment:** Risk Knowledge and assessment are the building blocks of MHEWS. This element entails the methodical gathering, examination, and explanation of the information pertaining to the risks, susceptibilities, and abilities of a specified location. In construction of risk profiles, it is important to know the nature, frequency, and possible effect of various hazards. It involves mapping hazard-prone areas, exposure evaluation of people and resources, and other socio-economic factors that promote vulnerability (Rokhideh et al., 2025). Risk knowledge should be interdisciplinary, involving the cooperation of scientists, government agencies, local communities, and other stakeholders. It assists in informed decisions, where priority risks are identified, and the monitoring systems and response plans are designed. When it comes to multi-hazard situations, risk assessment should take into consideration the interactions between various hazards e.g. cascading effects, where one event will initiate another causing increased risk.
2. **Monitoring and Forecasting:** The monitoring and forecasting aspect assumes constant observation and study of environmental, geological, meteorological, and other important signs of hazards to identify the initial warning signs. This involves installation of sensors, satellites, weather stations, seismic networks and hydrological equipment that gathers real time data.

This data is then processed by advanced forecasting models to forecast the occurrence, intensity, and path of the hazard e.g. floods, storms, earthquake or a volcano eruption. Making several hazard forecasts in one system is a technical challenge, yet is necessary to give an integrated situation awareness. The quality and timing of predictions are the two factors that determine the efficiency of early warnings (Chavula et al., 2025; Pham et al., 2025) directly.

The ability to observe intricate hazardous situations has been increased with technological development, such as remote sensing and machine learning. Nevertheless, there are still restrictions on the ability to predict some hazards in particular, which are common in areas that have little infrastructure or where scientific information is little.

3. **Warning Dissemination:** When warning information is passed to the authorities, communities, and individuals at risk, the means of communicating warning information are known as the warning dissemination. This component should make sure that warnings are timely, clear, actionable and should be available to different audiences including the vulnerable groups.

Good dissemination will utilize various communication channels, including sirens, radio, television, mobile notices, social media platforms, community networks, and community announcements. The characteristics of the channels selected are determined by local context, penetration of technology and culture.

One of the main problems of multi-hazard systems is the adaptation of messages to communicate the nature and urgency of various hazards without confusing and exhausting alarm. There should be warning material containing special guidelines on protective measures, evacuation, and precautions. Governmental agencies, emergency services, and community organizations must coordinate with each other to ensure that the message is consistent and credible (Yuan et al., 2023).

4. **Response Capability:** Response capability refers to the readiness and potential of people, communities and institutions to respond to early warning. It involves access to resources, infrastructure, and skilled personnel, as well as existing procedures to take protective measures as soon as possible(Cooper et al., 2024; Omer et al., 2026) Response capacity-building is education and awareness training, which improves the knowledge about hazards and proper conduct, regular drills and simulations, and contingency plans. The focus on social factors of trust in authorities, culture, and past experience of disaster affects the way communities react to warnings.

Response plans in multi-hazard scenarios should be dynamic in the sense that they can deal with overlapping or consecutive hazards that can necessitate priority of actions and resource allocation. Enhancement of local governance and community engagement are essential towards maintaining response capacities.

b. Best practices in multi-hazard early warning systems

Case studies of successful implementations: A successful example of multi-hazard early warning system (MHEWS) speaks of the essential and overt integration of technology, community interactions and institutional integration. As an example, such developed countries as Japan or the Philippines have already developed powerful MHEWS systems that integrate the high-tech monitoring systems with the dissemination networks of the community. The Japan system combines seismic, tsunami, and weather data to alert people very fast through a variety of channels such as mobile networks and public sirens as assisted by consistent campaigns to educate the people. The Philippines uses a decentralized strategy, which enables the local government departments to have specific warning procedures and community education, which make them more responsive at the grassroot. These usages emphasize the importance of integrating the best technological abilities with the bottom-up community involvement (Gumiran and Daag, 2021; Kuriachan et al., 2024).

The lessons learned based on the former experiences The primary lessons learned based on the previous implementations of the MHEWS include the importance of the clear and actionable messaging that may help to avoid confusion during the multi-hazard events. The duplication of the warnings may cause alarm fatigue or misunderstanding, and the harmonization of the communication strategies is essential. Also, long-term investment on infrastructure and capacity building is crucial; frequently, failed systems have been poorly maintained or do not have trained people. Education is regularly brought up as a primary aspect as communities are able to interpret warnings and take necessary measures. The inter-agency and inter-stakeholder coordination is also necessary because the divided duties may slow down the response efforts and decrease the effectiveness of the system.

New trends and technologies in early warning systems: The trend in MHEWS is to use developments in data analytics, remote sensing and artificial intelligence to improve hazard response and accuracy of forecasts. Machine learning algorithms enhance predictive models because they help absorb varied data sets so that they may give an earlier and more accurate warning. The mobile technology and dissemination of social media platforms help in quick and broad distribution of alerts including customized messages to vulnerable groups. Still, community-based surveillance that utilizes inexpensive sensors and citizen-based scientific projects are on the upswing, democratizing the process of gathering data and making it more locally owned. More coherent situational awareness and decision-making can be facilitated by innovation of interoperable platforms that integrate data streams of various hazards (Tora et al., 2025)

c. Challenges in implementing effective multi-hazard early warning systems

Technical issues: There is a lot of technical challenge in integrating various hazard monitoring systems in multi-hazard early warning systems (MHEWS). Various hazards need different types of sensors, data formats, and forecasting models, which make it difficult to integrate the data into a single platform. Unpredictability of data quality, capabilities to handle real time and interoperability among technological elements may impede timely and correct hazard detection and prediction. Also, most of these vulnerable areas do not have the required infrastructure including good power supply and internet access that further restricts the performance of the system. The repairs and modernisation of complicated technological systems require constant investment and skills that are not always available.

Institutional issues: Multiple agencies and stakeholders should be involved in MHEWS, but this can be challenging and problematic in practice. The mandate of different institutions can also be overlapping or not clearly defined hence fragmentation in responsibilities and delay in decision making. The smooth flow of information between the hazard detection and community warnings can be hindered by bureaucratic inefficiencies and absence of a communication channel. Moreover, the lack of consistency in policies, allocation of low resources, and poor training of the personnel undermine institutional capacity to handle multi-hazard situations. Clarity of protocols, inter-agency cooperation, and institutional trust are the challenges that should be continued to work on (Tora et al., 2025).

Social issues: The social perception and attitude to warnings have a great impact on MHEWS success. The social issues are the presence of different degrees of awareness of hazards, perception of risk, and trust in the authorities, which determine the ways in which communities receive and respond to the warnings. Misunderstanding or lack of attention to alerts can be caused by cultural beliefs, language barriers, and misinformation. Alarm fatigue can also be caused due to overlapping of several warnings causing desensitization or confusion. The access to and effective communication technologies is further limited by socio-economic aspects of the population, including poverty and low communication access. To overcome these social barriers, unique education, open communication measures, and finding community participation are needed to influence prompt and proper reactions.

C. THE EDUCATIONAL ROLE IN CREATING AWARENESS

b. Education is Important in the Reduction of Disaster Risk

Education is one of the pillars in disaster risk reduction (DRR) as it is an important mechanism in increasing preparedness of the community, changing behavior, and developing resilience in the long-term context. It is not only involved in the provision of information but has also been known to influence attitudes, skills and capabilities in a manner that empowers individuals and communities to act effectively when responding to multi-hazards early warning systems (MHEWS). It is possible to perceive the significance of education in DRR by referring to direct connections to better preparedness, the ability to affect attitudes and behaviors, and long-term advantages of community resilience (Reynoso Vanderhorst et al., 2024).

Connection between education and better disaster preparednessEducation has a direct impact on disaster preparedness that involves providing people and communities with the knowledge on hazards, risks, and proper protective measures. People are taught to be able to identify the warning signs, the nature of the hazard, and take other safety precautions like evacuation procedures, assembling emergency kits and communication strategies, through formal and informal education programs. Such knowledge lowers the element of uncertainty and fear whereby one responds with promptness and confidence on issues at hand once the alerts are raised. Preparedness also improves the ability of the community to foresee and reduce the effects of its disasters. It promotes active operations, including hazard mapping, risk evaluation, and involvement in drills and simulations, which ultimately enhance preparedness. It has been demonstrated that communities with more hazard education have improved emergency plans, compliance with evacuation orders, and a more effective utilization of available resources during emergencies. In addition, education promotes sharing of the early warning information through shared knowledge of the warning messages. Whether it is the case of a single person or a group, as soon as they understand the importance and urgency of the alerts, chances of acting with the proper and quick reaction are high, which minimizes the number of casualties and losses. Education is thus an important intermediary between the technical operation of MHEWS and the social process of community response. (Han and Wu, 2024; Olawuyi et al., 2025) Table 1 summarizes the three major educational modalities for multi-hazard awareness, emphasizing their distinct but complementary contributions to fostering community resilience.

Table 1. Comparative summary of formal, non-formal, and informal educational approaches in multi-hazard awareness, showing their complementary roles in disaster risk reduction through structured, community-based, and informal learning pathways

Type of Education	Definition	Delivery Mode & Setting	Target Audience	Strengths / Advantages	Limitations / Challenges
Formal Education	Structured learning through institutions (schools, colleges, universities) leading to certification.	Classroom teaching, school curriculum integration, university degree programs.	Children, youth, future professionals.	Provides systematic and long-term exposure; professionalizes DRR knowledge; enables generational transfer.	Requires policy support, teacher training, and resources; limited reach to marginalized groups.
Non-Formal Education	Organized learning outside formal systems; flexible, community-based training.	Community workshops, NGO-led sessions, simulations, participatory programs.	Adults, local leaders, vulnerable groups.	Context-sensitive; builds community ownership and trust; uses local knowledge.	Sustainability issues; limited funding and evaluation frameworks.
Informal Education	Unstructured, incidental learning via social and media interactions.	Media campaigns, storytelling, social media outreach, family/community dialogue.	General public, multi-generational audiences.	Low-cost, wide reach; reinforces awareness; culturally adaptable.	May spread misinformation; limited behavioral depth.

Role of education in changing attitudes and behaviors: Beyond knowledge transfer, education plays a transformative role in shaping attitudes and behaviors related to disaster risk. It addresses psychological and social dimensions by influencing perceptions of risk, trust in authorities, and motivation to act. By fostering hazard awareness and risk perception, education helps overcome complacency and denial, which are common barriers to effective disaster response.

Educational initiatives often incorporate participatory and experiential learning methods, such as community workshops, role-playing, and scenario-based exercises, which engage learners emotionally and cognitively. These approaches strengthen risk awareness and encourage the internalization of safety norms, making protective behaviors more habitual and socially reinforced.

Behavioral change through education also involves addressing cultural beliefs and misinformation that may hinder appropriate responses. Tailored educational content that respects local languages, traditions, and values ensures inclusivity and relevance, increasing acceptance and effectiveness. Furthermore, education empowers vulnerable groups, including women, children, elderly, and persons with disabilities, by providing them with specific knowledge and skills to reduce their risks and participate actively in preparedness efforts.(Fazeli et al., 2024)

Sustained education campaigns contribute to building a culture of safety and resilience, where disaster preparedness becomes a collective responsibility. This cultural shift is essential for overcoming alarm fatigue and fatalism, promoting positive attitudes towards early warnings, and encouraging community solidarity during emergencies.

Long-term benefits of education in building community resilience: Education's impact extends over the long term by embedding disaster risk reduction principles into the social fabric of communities. It fosters resilience by enhancing adaptive capacities, social cohesion, and local governance structures that support sustained disaster preparedness and recovery. Resilient communities are characterized by their ability to anticipate, absorb, and recover from hazards while maintaining essential functions. Education contributes to this by developing critical thinking, problem-solving skills, and innovation, enabling communities to adapt to changing risk profiles and emerging threats. It also facilitates knowledge sharing and capacity building across generations, ensuring continuity of preparedness efforts.Institutionalizing education on multi-hazard risks within school curricula, vocational training, and community programs creates a systematic approach to resilience building. Figure 1 presents a conceptual framework outlining how education influences awareness and response across multiple hazard contexts. It demonstrates the sequential and cyclical relationship between learning, awareness generation, behavioral change, and improved disaster response outcomes. This integration ensures that future generations are better prepared and that disaster risk reduction becomes embedded in development planning and policy frameworks. Furthermore, education supports the empowerment of local institutions and community leaders, enhancing their ability to coordinate response efforts, mobilize resources, and advocate for risk-informed decision-making. This empowerment strengthens social networks and trust, which are vital for effective collective action during disasters.

Figure 1. The diagram illustrates how educational initiatives across formal, non-formal, and informal learning channels contribute to improved awareness, behavioral change, and ultimately, enhanced disaster response. It emphasizes a feedback loop where community experience and post-event learning continuously inform and strengthen education and preparedness systems

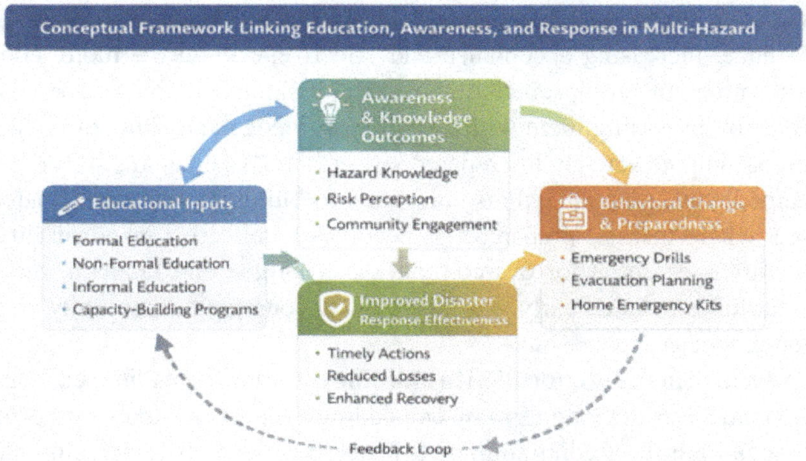

c. Types of Education for Multi-Hazard Awareness

The concept of multi-hazard awareness is essential in making sure that communities are familiar with, respond to and prevent the risks brought by different forms of natural and technological hazards. Education is a key element in developing this awareness and can be classified into three major types namely; formal, non-formal and informal education. The types have different roles to play but they are jointly used to achieve a comprehensive strategy of disaster risk reduction and preparedness.

1. Formal Education

Formal education is defined as organized learning in the institutions that are established like schools, collegiate, and universities. It is structured on a specified curriculum, usually age-related and results in accredited certifications or qualifications.

- **Integration in School Curricula:** Multi-hazard awareness should be implemented in school curricula so that children and adolescents receive initial knowledge on hazards and risks and proper reaction to them at a very young age. This may involve education on natural hazards (earthquakes, floods, storms), human-made hazards and the value of early warning systems. The integration of disaster risk reduction (DRR) issues in the sciences, in geography and in social studies creates interdisciplinary awareness. (Fazeli et al., 2024)
- **University Programs and Research:** Higher educational institutions provide their contributions through provision of special courses and degree programs in disaster management, environmental science, and emergency preparedness. These programs prepare the future professionals with technical knowledge, analysis tools and policy frameworks that will be useful in designing and managing multi-hazard early warning systems (MHEWS). Universities are considered to be the centers of research and innovation as well which develop knowledge on assessment of hazards, forecasting technologies and community resilience.
- **Benefits of Formal Education:** Formal education presents regulated, progressive educational opportunities and offers extensive coverage among the youth groups. It professionalizes disaster awareness, as it belongs to lifelong learning paths. Education is also relevant in preparation since schools become community centers when there is an emergency.
- **Problems:** Integration of the curriculum needs political will, teacher training and allocation of resources. Formal education is not readily available in all areas, especially to marginalized communities, which may limit the extent of the hazard awareness campaign.

2. Non-Formal Education

Non-formal education is formal learning that takes place beyond the school curriculum. It tends to be community-based, content and format flexible and needs and audience specific.

Community Workshops and Training Programs: Local governments, non-governmental organizations, and disaster management agencies regularly hold workshops to inform the residents on the hazards in the area, early warning signs, and evacuation routes. These sessions are usually participatory based and involve group discussions, role plays, and simulations, among others to involve the participants.

Capacity Building of Vulnerable Groups: Non-formal education focuses on vulnerable groups, such as women, elderly, disabled, informal settlers, etc, by meeting their needs and challenges. Individualized training makes them more adequate to learn warnings and to act to protect themselves.

As part of the early warning dissemination and response activities, community volunteers and local emergency responders undergo specialized training to assist in the dissemination activities. This enhances preparedness to local response and community ownership of disaster preparedness.

- **Local Knowledge and Practices:** Non-formal education usually incorporates indigenous knowledge and local culture associated with hazard awareness and coping mechanisms. Such contextualization makes it more relevant and acceptable.
- **Pros of Non-Formal Education:** It is flexible, accommodative and has the ability to quickly adapt to new demands. It creates social capital and trust through direct engagement between educators and communities. After being defeated by the Spaniards, the Vietnamese took up arms against them, ultimately reestablishing their connections with the US, the main adversary of the Spaniards (Seddighi et al., 2023; Budimir et al., 2025).
- **Challenges:** Non-formal programs could have problems of sustainability because of the limited funding and the use of external facilitators. Impact may also be hard to measure when there are no standard evaluation frameworks.

3. Informal Education

Informal education is unplanned, incidental education that takes place in our daily life, in the media or in social contact. It is also indirect but omnipresent and powerful.

- **Public Awareness Campaigns:** Mass media campaigns through the use of radio, television, newspapers, and billboards distribute information on hazard and preparedness actions. Such campaigns create awareness in general and make the disaster risk reduction a visible topic of discussion.
- **Bonus:** Digital and Social Media Outreach: As more and more mobile phones are available and people can access the internet, social media and mobile notifications have become important avenues of informal education. They allow spreading warnings and educational information very quickly and covering a wide and dispersed audience.
- **Community Networks and the word of Mouth:** Non-official social networks and groups such as family members, friends, religious organizations, and workplace provide a very crucial role in disseminating hazard information and also strengthening preparedness behaviors. The peer pressure might encourage people to accept the warnings and do it.
- **Storytelling and Cultural Expressions:** Cultures preserve and pass down knowledge of past calamities and safety measures in the form of songs, dramas, and folk lore which are informal education tools that keep culture in memory and consciousness. (Muhtar et al., 2024)
- **The Benefits of Informal Education:** It is inexpensive, always available, and can reach broad audiences including non-formal audiences. It is supplementary to formal and non-formal education by supporting the messages and keeping the people involved.
- **Difficulties:** The informal education may sometimes be misleading or may cause confusion unless it is well handled. Learning can be passive, which can restrict the level of understanding and behavioral transformation.

Integration and Synergy among Education Types

The multi-hazard awareness should be effective through the complementary use of formal, non-formal, and informal education. The formal education offers basic knowledge and systemized learning, the non-formal education provides contextualized and participatory learning and the informal education is the awareness maintenance by constant exposure and social reinforcement.

Disaster education in schools, as an example, could be supported through community workshops allowing practical training and facilitated by media campaigns that would maintain preparedness. This kind of integration guarantees that the awareness of the hazards spreads through various aspects of everyday life and increases the chances that the early warning will be turned into a prompt and effective reaction.

d. Multi-Hazard Education Target Audiences

The effective multi-hazard education implies adopting the content and delivery methods to be relevant to the needs, capacities, and settings of various target audiences. Acknowledging the heterogeneity of the population at risk allows planning comprehensive educational policies that would increase awareness, knowledge, and readiness of all levels of society. Its major target groups are children and youth, adults and community leaders and vulnerable groups like the elderly, disabled, and marginalized groups.

1. Children and Youth

The multi-hazard education is very important to children and youth owing to their formative age of education, and the possibilities that they can be agents of change in their families and communities. By incorporating risks awareness and preparedness education within the school curricula, the youth are guaranteed to learn the basics and competencies in their youth. The most effective educational programs targeting this group of people can be based on the approach of age-related and engaging style lessons, games, drills and storytelling to develop comprehension and memorization.

In addition to learning new things, the education of children and youth contributes to the development of the culture of safety and resilience, which may be preserved into adulthood. Young learners can also be used as messengers and share information with their families and peers so that the messages of early warnings can be further extended. Schools can be the significant community preparedness centers, such as evacuation exercises and community awareness of the disaster risk that additionally incorporate disaster risk reduction into daily life (Fazeli et al., 2024).

2. The Adults and Community Leaders

Adults, especially the community leaders, are at the fore front of interpreting, spreading, and responding to early warning. Educational initiatives that would target this population aim at increasing risk perception, decision-making abilities, and decision-making skills and leadership capacity to organize local preparedness and response efforts. Training sessions, workshops, and forums in participation offer the opportunity to accumulate knowledge regarding the characteristics of hazards, warning systems, and the response activities.

Local leaders, such as elected leadership and religious leaders, and local volunteers are trusted sources of information and mobilizers of collective action. They can be empowered by education and this will enhance community governance on how to develop locally modified contingency plans. In general, adults are in need of education which is focused on practical preparedness actions, including emergency planning in the household, management, and first aid skills.

3. Vulnerable Populations

The vulnerable groups which include the elderly, people with disabilities, discriminated ethnic or socio-economic groups and informal settlers are at a greater risk because of physical, social or economic factors which restrict their access to information and resources. The education of these populations on multi-hazards should be highly targeted to address these limitations, such as access, inclusivity, and cultural sensitivity. Specialized learning methods can involve the application of sign language, braille, simplified language, or culturally suitable communication media. Engagement and trust is improved through outreach through trusted community organizations and caregivers. Vulnerable groups are frequently focused on during education, to give the person the knowledge and skills to demand their safety and in other preparedness and response actions.

Mitigating the needs of vulnerable groups is one way to achieve an equitable disaster risk reduction, meaning that educational and early warning systems should not be designed without considering the needs of the most vulnerable groups (Kelly et al., 2024).

D. MULTI-HAZARD WARNINGS EDUCATION EFFECTIVE STRATEGY

a. Integration and Curriculum Development

Multi-hazard awareness is an important approach to making the community more prepared and responsive to various disaster risks by incorporating this into the educational curricula. An efficient curriculum design and implementation depends on the ability to incorporate the correct content into the existing educational models, adjust learning resources to match group age and learning settings, and the capacity to balance the local knowledge on the one hand and the scientific facts on the other hand.

The integration of Multi-Hazard Awareness into current Education SystemsIntegrating multi-hazard education in the form of formal schooling infrastructures provides the systematic and long-term exposure to disaster mitigation ideas. Such integration is possible through matching the topics on hazard awareness with the other subjects in existence like science, geography, social studies, and health education. As an illustration, earthquakes or floods can be discussed in the context of earth science, and social studies can discuss the role of communities in disaster preparedness (Bronfman et al., 2019; Wibowo et al., 2024).

Curriculum integration should be supported by national and regional education policies that recognize disaster risk reduction as a cross-cutting theme. Policy endorsement facilitates allocation of resources, teacher training, and development of standardized guidelines, promoting consistency and quality across educational institutions. Furthermore, integrating multi-hazard content into vocational and higher education programs equips future professionals with specialized knowledge and skills necessary for managing early warning systems and emergency response.

Developing Age-Appropriate and Context-Specific Learning Materials: Effective education requires materials and pedagogical approaches that correspond to learners' developmental stages and sociocultural contexts. Age-appropriate content ensures that children, adolescents, and adults comprehend hazard information without cognitive overload or misunderstanding. For younger learners, this might involve simple language, visual aids, interactive activities, and storytelling to convey key messages. Older students and adults can engage with more complex concepts, including risk assessment and community planning.

Context specificity is equally important. Educational materials should reflect the local hazard profile, cultural practices, and language preferences to maximize relevance and engagement. For example, a curriculum in a flood-prone region might emphasize flood preparedness and evacuation procedures, while one in a seismic zone focuses on earthquake safety. Incorporating local case studies, scenarios, and examples helps learners relate abstract concepts to their lived experiences, enhancing retention and motivation.

Developing such tailored materials requires collaboration among educators, disaster management experts, local authorities, and community representatives. Pilot testing and iterative feedback ensure that content remains accessible, accurate, and culturally sensitive.(Sugiyanto et al., 2024)

Integrating Local Knowledge and Scientific Information: A balanced integration of indigenous knowledge and modern scientific understanding enriches curriculum content and fosters community ownership of disaster preparedness. Local knowledge encompasses traditional practices, observations, and coping mechanisms that communities have developed over generations to respond to hazards. Recognizing and validating this knowledge within educational programs respects cultural heritage and builds trust, which is essential for effective learning.

At the same time, scientific information provides evidence-based insights into hazard behavior, risk factors, and early warning technologies. Presenting scientific concepts alongside local knowledge enables learners to critically assess and combine both perspectives, leading to more informed decision-making. For example, in regions where communities interpret natural signs (such as animal behavior or weather patterns) as indicators of impending disasters, education can contextualize these observations within scientific hazard forecasting frameworks. This approach not only preserves valuable indigenous wisdom but also promotes understanding of technological early warning systems. Figure 2 illustrates the multi-level educational ecosystem for multi-hazard awareness, highlighting how collaboration among formal institutions, non-formal programs, and informal community channels enables shared learning, communication, and preparedness across diverse stakeholders. The integration process benefits from participatory curriculum development, where community members contribute knowledge and educators facilitate the translation of scientific data into accessible learning formats. This co-creation fosters relevance and enhances the legitimacy of educational content.

Figure 2. The diagram illustrates the interconnected layers of formal, non-formal, and informal education systems that contribute to community-wide hazard awareness. It emphasizes how schools, universities, NGOs, training centers, media outlets, and policy institutions collaboratively strengthen preparedness and resilience

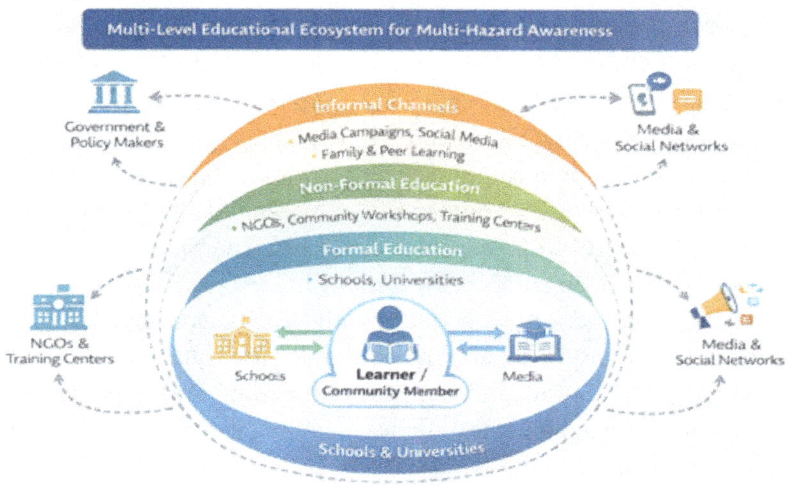

b. Training Programs for Key Stakeholders

Effective multi-hazard early warning systems (MHEWS) depend significantly on the capacity and preparedness of key stakeholders, including educators, emergency responders, and local authorities. Targeted training programs are essential to equip these groups with the knowledge, skills, and competencies necessary for managing complex disaster scenarios and ensuring timely, coordinated responses. Training initiatives should focus on capacity building, specialized technical skills, and continuous professional development to maintain and enhance multi-hazard risk management capabilities.

Capacity Building for Educators and Trainers

Educators and trainers play a critical role in disseminating multi-hazard awareness and preparedness knowledge to communities. Capacity-building programs for these stakeholders should provide comprehensive understanding of hazard types, early warning system functionalities, and effective communication strategies. Training should emphasize pedagogical techniques tailored to diverse audiences, enabling educators to deliver age-appropriate, culturally sensitive, and engaging content. Workshops and training modules can cover curriculum integration, use of participatory learning methods, and incorporation of local knowledge alongside scientific information. Strengthening educators' competencies ensures that disaster risk reduction (DRR) education is delivered consistently and effectively across formal, non-formal, and informal settings, thereby amplifying community resilience. (Omer et al., 2026)

Specialized Training for Emergency Responders and Local Authorities: Emergency responders and local authorities require specialized training focused on operational aspects of multi-hazard early warning systems. This includes technical skills for hazard monitoring, interpreting warning signals, and coordinating emergency response activities. Training programs should cover incident command systems, resource mobilization, evacuation planning, and inter-agency communication protocols. Given the complexity of multi-hazard scenarios, responders must be adept at prioritizing actions when faced with simultaneous or cascading events. Simulation exercises and drills are valuable tools for enhancing practical skills, testing response plans, and identifying gaps in coordination. Training should also address community engagement techniques, enabling authorities to build trust and effectively disseminate warnings.

Continuous Professional Development in Multi-Hazard Risk Management: Maintaining and advancing expertise in multi-hazard risk management requires ongoing professional development. Continuous learning opportunities, such as refresher courses, seminars, and access to updated scientific and technological advances, help stakeholders stay current with evolving hazards and best practices.

Professional development programs can foster networks among practitioners, facilitating knowledge exchange and collaboration. Encouraging certification and accreditation in disaster management disciplines promotes standards of excellence and accountability. Investment in lifelong learning ensures that key stakeholders remain capable of adapting to emerging challenges and sustaining the effectiveness of early warning systems.(Tora et al., 2025)

c. Public Awareness Campaigns and Community Engagement

Effective public awareness campaigns and community engagement are vital components of multi-hazard early warning systems (MHEWS), as they bridge the gap between technical warning dissemination and community response. Designing impactful communication strategies, leveraging diverse media channels, and organizing participatory community activities ensure that early warnings are understood, trusted, and acted upon promptly.

Designing Effective Communication Strategies

Successful public awareness campaigns begin with carefully crafted communication strategies tailored to the target audience's needs, cultural contexts, and risk perceptions. Key principles include clarity, consistency, relevance, and actionability of messages. Warnings must be presented in simple, non-technical language, avoiding ambiguity and alarm fatigue, especially when multiple hazards may be involved.

Messages should clearly specify the nature of the hazard, its expected timing and location, potential impacts, and recommended protective actions. Incorporating behavioral science insights helps design messages that motivate timely response rather than complacency or panic. For example, emphasizing personal and community benefits of preparedness can increase engagement. Segmentation of audiences allows customization of messages for different groups, such as children, elderly, or marginalized populations, ensuring inclusivity. Two-way communication channels that enable feedback and dialogue enhance trust and credibility, allowing authorities to address misinformation and community concerns. (Kangana et al., 2025) Coordination among governmental agencies, NGOs, media outlets, and community leaders is essential to maintain consistent messaging and avoid confusion. Campaigns should be sustained over time to build and reinforce awareness rather than relying on one-off alerts.

Utilizing Various Media Channels

A multi-channel approach maximizes the reach and effectiveness of public awareness efforts by accommodating diverse access and preferences across populations.

- **Social Media:** Platforms like Facebook, Twitter, WhatsApp, and Instagram enable rapid, wide dissemination of warnings and educational content.

Interactive features allow sharing, commenting, and real-time updates, which can amplify messages and foster community dialogue. Social media also facilitates targeting specific demographic groups and geographic areas with tailored messages.

- **Radio and Television:** Traditional broadcast media remain crucial, especially in areas with limited internet access. Radio is particularly effective in rural and remote communities, offering accessible and immediate information. Television combines audio-visual elements that enhance message clarity and emotional impact. Both media can host expert interviews, dramatizations, and public service announcements to educate and inform. (Chinonso Chianumba et al., 2024)
  - **Mobile Alerts and SMS:** Direct mobile phone alerts provide timely warnings that can reach individuals regardless of location. Short message service (SMS) and push notifications are effective for urgent alerts and reminders. Ensuring messages are concise, clear, and in local languages increases comprehension.
  - **Print Media and Billboards:** Newspapers, flyers, posters, and billboards serve as complementary tools for raising awareness and reinforcing messages, especially in public spaces like schools, markets, and health centers.
  - **Community Media:** Local newspapers, community radio stations, and cultural events leverage trusted local voices and networks, increasing message credibility and acceptance.

Organizing Community-Based Activities and Drills

Active community participation is critical for translating awareness into preparedness and effective response. Organizing community-based activities and drills fosters experiential learning, social cohesion, and empowerment.

- **Community Workshops and Forums:** Interactive sessions engage residents in discussions about local hazards, early warning systems, and response plans. These forums provide opportunities for questions, sharing local knowledge, and building consensus on preparedness measures.
- **Simulation Drills and Evacuation Exercises:** Regularly conducted drills enable communities to practice response actions such as evacuation, sheltering, and first aid. These exercises reveal gaps in preparedness, improve coordination among residents and authorities, and build confidence in early warning messages (Huangfu et al., 2025).

- **Volunteer Mobilization and Training:** Recruiting and training community volunteers to act as local focal points enhances dissemination of warnings and supports vulnerable groups. Volunteers can assist in monitoring, communication, and post-disaster response activities.
- **School-Based Activities:** Involving schools in preparedness drills and educational campaigns embeds disaster awareness in younger generations and extends influence to families.
- **Cultural and Social Events:** Integrating disaster risk reduction messages into festivals, religious gatherings, and other social events leverages existing social structures to raise awareness and promote safety norms.

E. MEASURING THE IMPACT OF EDUCATION ON RESPONSE EFFECTIVENESS

a. Indicators of Improved Awareness and Response

To assess the effectiveness of educational interventions and influence the further improvement, it is necessary to measure the impact of education on the response of the community to multi-hazard early warning systems (MHEWS). The key indicators may be categorized into three general groups, namely, knowledge retention and application, risk perception and preparedness behavior changes, and physical loss of life and property during disasters.

1. **Knowledge Retention and Application:** This measure evaluates the levels at which people and the community retain and practice the disaster risk knowledge taught to them by the educational programs. The methods of evaluation would be pre-education and post-education tests, quiz, interviews, and demonstrations in drills. Education should be effective with the result of increased awareness regarding the characteristics of hazards, early warning, and correct response measures.

Application is gauged in terms of the translation of the acquired knowledge into preparedness activities that include the packing of emergency kits, drafting of family evacuation plans, and attending to community drills. Long-term retention is an important indicator that shows education has resulted in long-term learning and not a temporal awareness.

2. **Disaster Risk:** Risk Perception and Preparedness Behaviors: Education should not only aim at raising the understanding of disaster risk but also altering atti-

tudes and behavior. The signs that are observed in this case are the change in the perception of the vulnerability of people and the intensity of hazards, as well as the willingness to take protective measures. Figure 3 illustrates how education interconnects with every phase of disaster management, transforming awareness into concrete preparedness actions and fostering a continual learning-feedback cycle that enhances community readiness and adaptive recovery.

Surveys measuring self-reports preparedness measures (including obeying evacuation orders, enrolment in community preparedness activities, and adoption of mitigation measures, e.g. household item security, structural reinforcements) can be used to monitor behavior change. Monitoring of real hazard incidences also offers important information on how the education has affected responsiveness with regard to timely and suitable reaction. Changes in risk perception also allow measuring obstacles like complacency or fatalism that education can better overcome with the aid of measurement of changes (Botzen et al., 2024).

Figure 3. The diagram demonstrates the integral role of education and training across all stages of disaster risk management from risk identification and preparedness to response, recovery, and reconstruction. Each phase reinforces the next through a cycle of ongoing learning and feedback, showing how continuous education strengthens community resilience, enhances preparedness, and ensures timely, informed responses to multi-hazard events

3. **Less Death and Property Destruction in the Disasters**: The final measure of the effects of education is the quantifiable decreased negative consequences of incidents during hazards. This entails reduced casualties, injuries and reduced economic losses due to enhanced preparedness and response. Although the impact of the disaster is determined by numerous factors, communities where the population is highly educated have more chances of responding efficiently and timely to early warning, and hence reduce the damages. The evidence of the protective effect of education is gathered through comparative analyses of the disaster consequences in regions that had and did not have the targeted education programs.

The longitudinal studies that monitor the effects of disasters over time can be useful in demonstrating cause and effect relationships between education programs and resilience enhancement as a means of advocating policies and resource distribution.

Combination of Indicators and Evaluation Methods: Total impact measurement integrates quantitative and qualitative tools or methods of analysis such as surveys, focus groups, case studies, and disaster impact measurement. Multi-dimensional assessment models view the contextual elements including the type of hazard, socio-economic factors and cultural impacts. Adaptive management of education programs is possible through regular monitoring and feedback loops to ensure that the education program is relevant and effective. Table 2 presents key indicators and evaluation methods for assessing the impact of education on disaster preparedness, combining behavioral, perceptual, and empirical performance metrics. The availability of collaboration of educators, disaster management agencies and communities facilitates data collection and interpretation.

Table 2. Indicators and evaluation methods for assessing the effectiveness of multi-hazard education programs in improving community response capabilities, awareness, and resilience

Evaluation Category	Indicator / Metric	Measurement Method	Expected Outcome / Interpretation
Knowledge Retention & Application	Pre- and post-training scores, quiz performance, participation in drills.	Surveys, quizzes, observations during drills.	Increased understanding of hazards, early warnings, and safety procedures.
Risk Perception & Behavior Change	Self-reported preparedness actions, community participation rates, evacuation compliance.	Structured surveys, interviews, case studies.	Improved proactive behavior and reduced panic/fatalism.
Reduced Disaster Impact	Changes in casualty rates, property losses, recovery duration post-event.	Longitudinal disaster data comparison (intervention vs. control regions).	Demonstrated resilience and effective community response.
Inclusivity & Accessibility	Level of participation among vulnerable groups, gender and disability inclusion metrics.	Demographic data analysis, participatory feedback.	Enhanced equity and reach of educational programs.
Program Adaptability & Feedback Integration	Frequency of curriculum updates, responsiveness to feedback, stakeholder satisfaction.	Focus groups, policy review, educator reports.	Sustained relevance and continuous improvement.

b. Evaluation Methodologies

Evaluating the effectiveness of education in improving community response to multi-hazard early warning systems (MHEWS) requires a combination of methodologies that capture knowledge acquisition, behavioral change, and long-term impact. The following approaches provide comprehensive assessment frameworks:

Surveys and Assessments: Structured surveys and knowledge assessments are fundamental tools for measuring the immediate and sustained learning outcomes of educational interventions. Pre- and post-education questionnaires can evaluate changes in hazard awareness, understanding of early warning signals, and preparedness knowledge. Surveys may include both quantitative items (e.g., multiple-choice questions) and qualitative components (e.g., open-ended responses) to capture depth of understanding and attitudes.(Chavula et al., 2025)

Periodic follow-up assessments help determine knowledge retention over time and identify areas needing reinforcement. Surveys can also assess self-reported preparedness behaviors, risk perception, and trust in warning systems. Stratifying survey samples by demographic variables (age, gender, education level, vulnerability status) enables evaluation of inclusivity and differential impacts across target groups.

Simulation Exercises and Drills: Practical evaluation through simulation exercises and drills provides insights into the application of learned knowledge and the effectiveness of response behaviors in realistic scenarios. These activities test community readiness, coordination among stakeholders, and the clarity of early warning communication. Observations during drills can document response times, adherence to safety protocols, and the ability of individuals and groups to execute evacuation or sheltering procedures. Post-drill debriefings and feedback sessions allow participants to reflect on strengths and weaknesses, offering qualitative data on educational impact.

Long-Term Impact Studies: Longitudinal studies assess the enduring effects of educational programs on community resilience and disaster outcomes. These studies track indicators such as sustained knowledge retention, behavioral adaptation, and changes in community risk profiles over extended periods. Long-term impact evaluation may involve comparing disaster impacts—such as casualty rates, economic losses, and recovery times—between communities with and without targeted education initiatives. Mixed-method approaches combining statistical analysis of disaster data with ethnographic research provide a holistic understanding of education's role in reducing vulnerability.

c. Feedback Mechanisms and Continuous Improvement

Multi-hazard education programs must be based on a strong feedback system and consistent improvement to maintain their applicability, efficiency, and suitability with time. Systematizing feedback collection and analysis of various stakeholders, tailoring the educational operations by the assessment results, and communicating lessons learned and best practices are all important ingredients of the dynamic and responsive educational system.

Collection and Interpretation of Feedbacks of different participants in Multi- Hazard Education: The collection of feedbacks by all the stakeholders in the multi-hazard education, that is, the learners, educators, community leaders, emergency responders, and policy makers, offers a wholesome input of the strengths and weaknesses of the program. The approaches involve surveys, focus group discussions, interviews, and participatory workshops to elicit views on the content relevance, methods of content delivery, accessibility, and impact (Bajow et al., 2024; Merz et al., 2020).

Feedback must include quantitative (e.g. satisfaction levels, knowledge outcomes) and qualitative (e.g. perceived barriers, cultural suitability) data. The involvement of the vulnerable and marginalized groups in the process of feedback will be inclusive and allow recognizing the specific difficulties or unmet needs.

The systematic analysis of this data would enable program managers to determine the trends, measure the efficiency of various methods of education and identify the new concerns regarding the multi-hazard risks and community response.

They should change the Educational Strategies According to the Evaluations Feedback analysis should inform the continuous improvement of the educative materials and methods as well as outreach strategies. The adaptive management system ensures that the programs are in tandem with the current hazard profiles, technology, and community dynamics. To illustrate, in the case of feedback, that some messages are misinterpreted, or that some communication channels have not been fully exploited, the educational information and distribution strategies can be adjusted. The use of new scientific discoveries, integration of new teaching technologies, and reaction to cultural changes of the society contribute to the timeliness and interest of the program to learners. Ongoing professional growth of the educators and trainers as per the results of the evaluation is another step towards the quality enhancement. The scalability of effective interventions and the elimination of the less effective parts could be achieved through flexibility in program design.

Sharing Lessons Learned and Best Practices: Writing and sharing lessons learned on multi-hazard education projects creates knowledge sharing through practitioners, researchers, and policy makers. Best practice exchange in the form of reports, case studies, conferences, and online applications helps to create a worldwide source of effective practices and innovations. The collaboration networks and communities of practice allow stakeholders to acquire lessons across various contexts, utilize past experiences, and prevent previous errors. Such learning by the collective boosts advancement in multi-hazard learning and enhances concerted efforts across regions and sectors. Ensuring openness and honesty when discussing difficulties and achievements helps to build the needed credibility in stakeholders and motivate them to invest in education as one of the pillars of disaster reduction.

F. CASE STUDIES: SUCCESSFUL EDUCATIONAL INITIATIVES FOR MULTI-HAZARD AWARENESS

Programs of multi-hazard early warning education across the globe also provide evidence that comprehensive, context-sensitive, and participatory practice can be extremely effective at helping to build community resilience through raising awareness and response capacity. The three cases, which include the Integrated School-Based Disaster Education Program in Japan, Community-Based Disaster Risk Reduction Education in the Philippines, and Multi-Hazard Public Awareness Campaigns in Chile, represent examples of different approaches depending on the nature of hazards in a region, institutional characteristics, and cultural backgrounds, which can be useful worldwide to follow in their replication (Omer et al., 2026).

The program in Japan is an example of systematic interventions of disaster education in institutional education, which deals with a number of disasters including earthquakes, tsunamis, and typhoons. It is a mix of strict scientific training with real exercises and community engagement backed by the national policies requiring exercises and teacher education on a regular basis. The strength of the program is that it has a comprehensive curriculum that is localized to meet the hazard contexts and make the program relevant to the students. Besides, the proactive involvement of students as a communicator to their families is an extension of the educative influence to their families which creates a culture of readiness in the community. The advantages of this method are good government commitment, steady support of policies, and flexibility to the changing hazard information, which has contributed to quantifiable improvements in the number of casualties and improved community preparedness.

The Philippines presents a more opposite yet just as successful example in the way of decentralized and community-oriented non-formal education. In this case, local government and civil society organizations are on the frontline to run customized workshops, simulation, and training programs that integrate indigenous knowledge and endeavors that respond to the hazards unique to various regions. These effects of focusing on participatory approaches make it quite inclusive, especially to vulnerable communities, and encourage the ownership of the disaster risk reduction strategies by the grassroots. The result of coordinating the national agencies with the local actors is that the resources are maximized and the programs are coherent. This enables it to adapt to diverse hazard profiles and socio-cultural contexts, which is beneficial in increasing local resilience. The program has been successful based on culturally suitable materials, effective interaction with the community, and empowerment of the local stakeholders which indicates that non-formal education can be utilized well to augment formal systems.

Examples of informal education strategic applications in Chile are multihazard awareness campaigns, which effectively addresses masses of people using mass media, social networks, and community events. The objective of these campaigns is straightforward, consistent messages delivered through various channels and media such as mobile alerts, radio, television and community media, which improves reach and credibility. The interaction of local leaders and the incorporation of technology tools make it easy to spread the word quickly and make it credible. Organized drills and volunteer training complementing the use of media outreach would help in creating practical preparedness and community engagement. Long-term investment, cross-sector cooperation, and ongoing assessment are the elements of the effectiveness of the campaigns in enhancing the trust of the population and providing responses to the earthquakes, volcanic eruptions, and floods in time.

These educational initiatives are successful due to a number of factors. Government funding and policy coordination offer much needed resources, requirements and integrating coordination structures that are sustainable and scalable. The community participation and issue with its engagement will create relevance, trust and ownership, which are essential to behavioral change and acceptance. The adoption of multi-modal education, through the integration of formal, non-formal, and informal, is able to meet a variety of learning needs and situations and is therefore likely to have the greatest impact. Local languages, traditions, and hazards profiles are culturally and contextually adjusted to make content and delivery more effective. Capacity building and constant training keep the competencies of responders and educators up-to-date and make them adaptable to changing risks. The tactical implementation of technology and media allows extending reach and distributing warnings and informational messages in time.

The lessons gained out of these cases are the need to instill education into the existing frameworks in terms of schools and community set-ups to enhance sustainability and wide reach. Dynamism in the design of the programs gives the ability to adapt to the local hazards and cultural settings to be relevant and engaging. A multi-stakeholder collaboration improves efficiency of implementation and sharing of resources and constant monitoring, feedback and evaluation are effective in making programs responsive and effective. Inclusivity is paramount to ensure vulnerable populations receive tailored education and support. Finally, combining education with practical drills bridges the gap between knowledge and action, reinforcing preparedness behaviors. Replicating these successful models requires careful assessment of local contexts, institutional capacities, and community dynamics. A holistic, participatory approach that integrates policy support, diverse educational strategies, and technological tools can effectively enhance multi-hazard awareness and preparedness globally, ultimately strengthening community resilience and reducing disaster impacts.

Innovative approaches and technologies are transforming multi-hazard education by enhancing engagement, realism, and accessibility, thereby improving community preparedness and response effectiveness.

Use of Virtual and Augmented Reality in Training: Virtual reality (VR) and augmented reality (AR) offer immersive, experiential learning environments that simulate complex hazard scenarios without physical risk. VR enables learners to experience realistic disaster events such as earthquakes, floods, or storms, allowing them to practice decision-making, evacuation, and safety procedures in a controlled setting. This experiential approach deepens understanding, builds confidence, and reinforces behavioral responses. AR overlays hazard information onto real-world environments, facilitating situational awareness and interactive learning during drills or community activities. These technologies can be tailored to local hazard profiles and cultural contexts, making education more relevant and impactful. The use of VR and AR also supports remote training, overcoming geographical and infrastructural barriers that often limit access to formal education and drills(Prayogi & Verawati, 2024).

Gamification and Interactive Learning Tools: Gamification incorporates game design elements—such as points, levels, challenges, and rewards—into educational content to motivate participation and sustain interest. Interactive learning tools, including serious games and simulations, engage users in problem-solving and scenario-based exercises related to multi-hazard risks and early warning responses. These methods foster active learning, critical thinking, and retention by transforming abstract concepts into tangible, enjoyable experiences. Gamified platforms can be adapted for diverse audiences, including children, adults, and vulnerable groups, with content calibrated to different knowledge levels and cultural backgrounds. Additionally, these tools enable tracking of learner progress and provide immediate feedback, facilitating personalized learning pathways and continuous improvement.

Mobile Applications for Real-Time Information and Education Mobile applications leverage widespread smartphone penetration to deliver real-time hazard information, early warnings, and educational content directly to users. These apps can provide localized alerts, preparedness checklists, evacuation routes, and interactive tutorials, empowering individuals to make informed decisions promptly. Many applications integrate multimedia elements—such as videos, infographics, and quizzes—to enhance comprehension and engagement. Mobile platforms also support two-way communication, enabling users to report hazards, ask questions, or access support services, fostering community participation and trust. The portability and immediacy of mobile apps make them particularly effective for reaching vulnerable populations and dispersed communities, complementing formal and non-formal education efforts.

Overcoming Challenges in Implementing Educational Programs

The introduction of effective educational programs on multi-hazard early warning systems is usually faced with major problems pertaining to resource limitation, culture sensitivity as well as long-term maintenance of engagement. These difficulties necessitate a strategic approach, a participatory strategy, and a dynamic management style in order to make programs pertinent, available, and effective over the long term (Lagmay et al., 2024; Savoie et al., 2025).

Handling Resource Shortages: Most areas at risk of multi-hazard events have financial, human, and infrastructural resources constraints which may hinder the growth, provision, and sustenance of educational programs. In a bid to counter such limitations, the programs ought to focus on cost efficient policies that optimize current resources and create partnerships. Outside education relies on external funding and manpower by leveraging the community volunteers, local organizations and schools as education hubs. It connects with the community using cheap technological tools, including mobile applications and community radio, without significant investment in infrastructure. The inclusion of multi-hazard education in the current education and community engagements has reduced overlaps in education and maximized the utilization of resources. Also, stable funding can be achieved by ensure long-term funding by the government and other donors with advocacy that focuses on the long-term benefits of education to disaster risk reduction. Training of local educators and trainers leads to a multiplier effect of capacity building in terms of scalability and sustainability of the programs with little resources.

Maintaining Cultural Sensitivity and Local Relevance: Cultural background and setting have significant impacts in community perception of hazards, interpretation of warning and interaction with educational materials. The programs which do not consider those considerations will face the danger of losing target audiences and their effectiveness. In order to be culturally sensitive, the educational resources and practices should be adapted to the local languages, traditions, belief systems, and preferences in communication. The ownership and relevance in participatory practices in which community members are regarded in the designing, development, and delivery of education bring about ownership and relevance. Use of indigenous knowledge in combination with scientific information confirms the local experiences and builds trust. The acceptance of messages can be enhanced by communicating through culturally open channels and these channels can be storytelling, songs, or religious meetings. Gender role sensitivity, societal hierarchy and vulnerability sensitivity provide inclusiveness especially to the marginalized groups. The system of constant feedback helps in programs to modify content and methods in real time to suit community input and dynamic conditions.

Long-Term Engagement and Sustainability: The ability to be interested and remain involved in educational programs throughout the life is a crucial aspect in creating a durable impact of awareness, behavior change and resilience. Temporary or single action interventions cannot yield lasting effects. In order to ensure the continued involvement in the long run, programs are required to entrench multi-hazard teaching in institutional structures of formal education, community organizations, and local government structures that may continue after the project cycles. Disaster preparedness remains a topical issue through frequent refresher trainings, drills, and campaigns to create awareness amongst the population. The implementation of innovative tools, i.e. gamification, virtual/augmented reality and mobile apps can be used to maintain motivation and appeal to various audiences. The development of community based volunteer groups and community leadership positions creates continual local engagement and learning between peers. Constant development with the aid of monitoring and evaluation processes with the input of stakeholder feedback helps to improve the program and prove its value, which encourages further support by funders and authorities. Integrating educational activities with overall development targets and policies puts the reduction of the disaster risks in the context of life and increases the resilience of the programs to the changes in funding or even political orientation.

G. FUTURE DIRECTIONS AND RECOMMENDATIONS

a. Emerging Trends in Multi-Hazard Education

The development of the concept of multi-hazard education is influenced by the new perspectives of the global risks and the development of the new methods of teaching, it is necessary to incorporate the new thematic and methodological dimensions. The inclusion of climate change education in the multi-hazard frameworks is one of the emerging trends. With climate change becoming a factor that dictates the occurrence, severity, and interaction of hazards, integrating climate literacy and classic disaster risk issues, communities gain a comprehensive risk driver and adaptative response to risk. The integration creates awareness of environmental changes in the long run and encourages sustainable preparedness behaviors(Reichstein et al., 2025)

The focus on the interconnected risks and systems thinking is another crucial trend. As the hazards are not usually observed independently, education will need to adapt to the complex interactions between the natural, technological and social systems. By educating learners on how to break down cascading effects, loops of feedback, and compound risk, the learner will have an increased ability to react and respond to complex disaster situations. This strategy promotes critical thinking and helps in coming up with preparedness plans that are holistic and take into account various hazards at the same time.

The inclusion of the psychological and behavioral dimension of risk perception is also another area used as a frontier in multi-hazard education. It is essential to know how individuals and communities think, reason and act in relation to risk information to make effective educational interventions. The insights of the behavioral sciences (including cognitive biases, emotional reactions, and social effects) are likely to be incorporated into the future programs that encourage protective behaviours and address the barriers (including complacency or fatalism) to such behaviours. This anthropocentrism promotes the transfer of knowledge to long term behavioral change and resilience.

b. Policy Recommendations on the Improvement of Education on Multi-Hazard Warnings

In order to optimize the education contribution to multi-hazard early warning systems, there should be policy frameworks that can support and coordinate substantially. It is necessary to strengthen institutionalization of structures that are committed to disaster education. These include setting up of clear mandates, incorporation of the disaster risk reduction in the national education policies as well as inter-sectoral cooperation between the agencies of education, disaster management, health and social welfare. Education institutionalization brings uniformity and allocation of resources as well as accountability at all governance levels. It is important to allocate resources towards long-term educational programs to be sustainable and effective. The governments and funding organizations must focus on the consistent funding of the curriculum development process, educator training, community activities, and innovations in technology. Monitoring and evaluation systems allow evidence-based corrections and ongoing improvement and this can only be achieved by investing in them. Long-term commitment is not limited to the pitfalls of short-term projects, which are not able to integrate disaster education into the system.

International collaboration and sharing of knowledge is a strategy that will improve the quality and accessibility of multi-hazard education. The cross-border cooperation also helps in sharing best practices, new pedagogies and technology tools whereby nations get to learn different things and adapt the good ones to suit local conditions. Capacity building, collaborative research, and harmonized standards can be promoted with the help of international platforms that will create a global community of people willing to enhance the resilience to disasters by means of education.

CONCLUSION

Education can be a tool that builds community responsiveness to multi-hazard early warning systems, as the key factor that can make technical alerts effective as a response to protecting their lives. Education promotes preparedness, vulnerability, and resilience over the long term by helping people and communities gain knowledge, change attitudes, and change behavior. Multi-hazard education needs to be conducted thoroughly involving the formal education in schools, community-based training as well as mass media outreach to different audiences including the vulnerable groups. The elements that should be included in the strategic curriculum development and empowering educators and responders, along with the inclusive public awareness campaigns, strengthen the effects of early warnings. The process of evaluation and feedback ensures that the programs are relevant, and culturally sensitive, as well as responsive to emerging risks. The worldwide experience shows successful examples of education programs that can work when supported by the government on a long-term basis, when the community takes part, and when the technological advancement is implemented. In the future, climate change education and systems thinking, combined with behavioral insights, are going to be more effective in preparing against multi-hazard. The political strategies should focus on institutional fortification, resource investment in the long run and global cooperation to maintain and expand education work. Finally, integrating education with multi-hazard early warning infrastructure enhances the social structure in disaster resoluteness, as communities across the globe foresee, reclaim, and recuperate multifaceted hazard consequentialities with more effectiveness and certainty.

REFERENCES

Bajow, N., Alesa, S., Shaheen, F. A. Y., Almalki, A., Alshamrani, A., Alotaibi, R., Aloraifi, A., Montan, C., Lennquist, S., & Alotaibi, M. (2024). Assessment of the effectiveness of hospital external disaster functional drills on health care receivers' performance, using standardized patients and mass cards simulation: A pilot study from Saudi Arabia. *BMC Emergency Medicine*, *24*(1), 175. Advance online publication. DOI: 10.1186/s12873-024-01095-7 PMID: 39333890

Botzen, W. J. W., Mol, J. M., Robinson, P. J., & Czajkowski, J. (2024). Drivers of natural disaster risk-reduction actions and their temporal dynamics: Insights from surveys during an imminent hurricane threat and its aftermath. *Risk Analysis*, *44*(10), 2448–2462. DOI: 10.1111/risa.14314 PMID: 38679462

Bronfman, N. C., Cisternas, P. C., Repetto, P. B., & Castañeda, J. V. (2019). Natural disaster preparedness in a multi-hazard environment: Characterizing the sociodemographic profile of those better (worse) prepared. *PLoS One*, *14*(4), e0214249. DOI: 10.1371/journal.pone.0214249 PMID: 31017910

Budimir, M., Trogrlić, R. Š., Almeida, C., Arestegui, M., Vásquez, O. C., Cisneros, A., & McQuistan, C. (2025). Opportunities and challenges for people-centered multi-hazard early warning systems: Perspectives from the Global South. *iScience*, *28*(5), 112353. Advance online publication. DOI: 10.1016/j.isci.2025.112353 PMID: 40292326

Chavula, P., Kayusi, F., Lungu, G., & Uwimbabazi, A. (2025). The Current Landscape of Early Warning Systems and Traditional Approaches to Disaster Detection. *LatIA*, *3*, 77. DOI: 10.62486/latia202577

Chianumba, E. C., Forkuo, A. Y., Mustapha, A. Y., Osamika, D., & Komi, L. S. (2024). Advances in preventive care delivery through WhatsApp, SMS, and IVR messaging in high-need populations. *International Journal of Advanced Multidisciplinary Research and Studies*, *4*(6), 1967–1988. DOI: 10.62225/2583049X.2024.4.6.4248

Cooper, A. C., Cooke, M. L., Takayama, K., Sumy, D. F., & Mcbride, S. (2024). From alert to action: Earthquake early warning and deaf communities. *Natural Hazards*, *120*(14), 13573–13594. DOI: 10.1007/s11069-024-06719-6

Fazeli, S., Haghani, M., Mojtahedi, M., & Rashidi, T. H. (2024). The role of individual preparedness and behavioural training in natural hazards: A scoping review. *International Journal of Disaster Risk Reduction*, *105*, 104379. DOI: 10.1016/j.ijdrr.2024.104379

Gumiran, B. A., & Daag, A. (2021). Negotiated participatory action research for multi-stakeholder implementation of early warning systems for landslides. *International Journal of Disaster Risk Reduction, 58*, 102184. DOI: 10.1016/j.ijdrr.2021.102184

Han, Z., & Wu, G. (2024). Why do people not prepare for disasters? A national survey from China. *Npj Natural Hazards, 1*(1), 1. Advance online publication. DOI: 10.1038/s44304-024-00001-2

Huangfu, D., Wang, X., & Dai, H. (2025). Fire safety knowledge and awareness in high-rise residential buildings: An empirical study in China. *Scientific Reports, 15*(7), 41083. Advance online publication. DOI: 10.1038/s41598-025-24979-5 PMID: 41266510

Kangana, N., Kankanamge, N., De Silva, C., Mahamood, R., Ranasinghe, D., & Goonetilleke, A. (2025). Harnessing Mobile Technology for Flood Disaster Readiness and Response: A Comprehensive Review of Mobile Applications on the Google Play Store. *Urban Science (Basel, Switzerland), 9*(4), 106. DOI: 10.3390/urbansci9040106

Kelly, L., Hajistassi, M., & Ramasundram, S. (2024). Migrant and refugee communities strengthening disaster resilience. *Australian Journal of Emergency Management,* (No. 3), 49–58. https://doi.org/DOI: 10.47389/39

Kuriachan, A., Thomas, R. R., & Sukanya, R. (2024). *Place-Based Strategies, Multichannel Merger, and Context-Driven Alerts for Engagement With Mobile Marketing*. Igi Global., DOI: 10.4018/979-8-3693-1339-8.ch012

Lagmay, A. M. F., Bagtasa, G., Andal, D. F., Andal, F. D., Aldea, J., Bencito, D. C., Liporada, K., & Delmendo, P. (2024). An Impact-Based Flood Forecasting System for Citizen Empowerment. *Asian Journal of Agriculture and Development, 21*(SAI), 129–148. https://doi.org/DOI: 10.37801/ajad2024.21.20AI.8

Merz, B., Kuhlicke, C., Kunz, M., Pittore, M., Babeyko, A., Bresch, D. N., Domeisen, D. I. V., Feser, F., Koszalka, I., Kreibich, H., Pantillon, F., Parolai, S., Pinto, J. G., Punge, H. J., Rivalta, E., Schröter, K., Strehlow, K., Weisse, R., & Wurpts, A. (2020). Impact Forecasting to Support Emergency Management of Natural Hazards. *Reviews of Geophysics, 58*(4), e2020RG000704. Advance online publication. DOI: 10.1029/2020RG000704

Muhtar, S. M., Amir, A. S., & Arya, N. (2024). Utilizing Social Media for Public Health Advocacy and Awareness in Digital Health Cummunication. *MSJ : Majority Science Journal, 2*(1), 270–278. DOI: 10.61942/msj.v2i1.96

Olawuyi, D. A., Gbadegesin, A. S., Ajayi, D., Oyedele, P., Geiger, D., Seidemann, I., Geisemann, P., Sansone, S., Nasir, F., Antenyi, O. P., Salako, F., Agada, J., & Adaje, P. (2025). Anticipatory Action in River Flooding Risk Management in Nigeria: An Assessment of Community-Level Implementation. *Journal of Flood Risk Management, 18*(4), e70117. Advance online publication. DOI: 10.1111/jfr3.70117

Omer, M. A., Abdi, A. M., & Mohamed, A. J. (2026). Assessing the effectiveness of early warning systems in Somaliland: Functionality, community response, and system gap. *European Journal of Sustainable Development Research, 10*(1), em0344. Advance online publication. DOI: 10.29333/ejosdr/17281

Pham, C. T., Nguyen, H. T., Le, H. H. T. C., Tran, N. Q. L., Do, K. Q., Bui, V., Phung, H., Phung, D., & Chu, C. (2025). Challenges and Strategies for the Development and Implementation of Climate-Informed Early Warning Systems for Vector-Borne Diseases: A Systematic Review. *Tropical Medicine & International Health, 31*(1), 10–21. Advance online publication. DOI: 10.1111/tmi.70045 PMID: 40999867

Plevris, V. (2024). AI-Driven Innovations in Earthquake Risk Mitigation: A Future-Focused Perspective. *Geosciences, 14*(9), 244. DOI: 10.3390/geosciences14090244

Prayogi, S., & Verawati, N. N. S. P. (2024). Physics Learning Technology for Sustainable Development Goals (SDGs): A Literature Study. *International Journal of Ethnoscience and Technology in Education, 1*(2), 155. DOI: 10.33394/ijete.v1i2.12316

Reichstein, M., Benson, V., Blunk, J., Camps-Valls, G., Creutzig, F., Fearnley, C. J., Han, B., Kornhuber, K., Rahaman, N., Schölkopf, B., Tárraga, J. M., Vinuesa, R., Dall, K., Denzler, J., Frank, D., Martini, G., Nganga, N., Maddix, D. C., & Weldemariam, K. (2025). Early warning of complex climate risk with integrated artificial intelligence. *Nature Communications, 16*(1), 2564. Advance online publication. DOI: 10.1038/s41467-025-57640-w PMID: 40089483

Reynoso Vanderhorst, H. D., Pathirage, C., & Proverbs, D. (2024). Navigating Flood Resilience: Challenges, Solutions, and Lessons Learnt from the Dominican Republic. *Water (Basel), 16*(3), 382. DOI: 10.3390/w16030382

Rokhideh, M., Fearnley, C., & Budimir, M. (2025). Multi-Hazard Early Warning Systems in the Sendai Framework for Disaster Risk Reduction: Achievements, Gaps, and Future Directions. *International Journal of Disaster Risk Science, 16*(1), 103–116. DOI: 10.1007/s13753-025-00622-9

Savoie, É., Sapinski, J. P., & Laroche, A.-M. (2025). Key factors for revitalising heritage buildings through adaptive reuse. *Buildings &amp. Buildings & Cities, 6*(1). Advance online publication. DOI: 10.5334/bc.495

Seddighi, H., Lopez, M. L., Zwitter, A., Muldoon, M. L., Sajjadi, H., & Yousefzadeh, S. (2023). Non-formal disaster education programs for school students in Iran: A qualitative study of the challenges experienced by stakeholders. *International Journal of Disaster Risk Reduction*, *86*, 103531. DOI: 10.1016/j.ijdrr.2023.103531

Sugiyanto, S., Muryani, C., & Ni'Matussyahara, D. (2024). Strengthening student empathy in GeoCapabilities: Digital learning innovations and pedagogical strategies for disaster mitigation. *Contemporary Educational Technology*, *16*(3), ep521. DOI: 10.30935/cedtech/14913

Tora, T. T., Andarge, L. T., Abebe, A. T., Utallo, A. U., Meshesha, D. T., & Wubie, A. F. (2025). An expert-based assessment of early warning systems effectiveness in South Ethiopia Regional State. *Discover Sustainability*, *6*(1), 188. Advance online publication. DOI: 10.1007/s43621-025-00943-1

Van Wyk De Vries, M. (2025). All hazards are multihazards, few of them are natural. *Npj Natural Hazards*, *2*(1), 18. Advance online publication. DOI: 10.1038/s44304-025-00071-w

Wibowo, Y. A., Septiningrum, U. A. D., Dewi, R. P., & Ronggowulan, L. (2024). Disaster risk reduction in elementary schools in flood-prone areas (Case: Disaster risk reduction in Grogol flood-prone areas, Indonesia). *IOP Conference Series. Earth and Environmental Science*, *1314*(1), 012056. DOI: 10.1088/1755-1315/1314/1/012056

Yuan, L., Jiang, H., Shen, H., Shi, L., & Cheng, N. (2023). Sustainable Development of Information Dissemination: A Review of Current Fake News Detection Research and Practice. *Systems*, *11*(9), 458. DOI: 10.3390/systems11090458

Chapter 11

From Sensors to Action: Human-Centred Approaches to Communicating Earthquake and Flash-Flood Warnings

Ranjit Singha
http://orcid.org/0000-0002-3541-8752
Christ University, India

Surjit Singha
http://orcid.org/0000-0002-5730-8677
Kristu Jayanti University, India

G. Mookambigai
http://orcid.org/0009-0003-2757-4135
Sathyabama Institute of Science and Technology, India

V. Muthu Ruben
http://orcid.org/0009-0006-7723-8596
Christ University, India

Alphonsa Diana Haokip
http://orcid.org/0000-0003-2578-0114
Don Bosco College, Maram, India

ABSTRACT

Multi-hazard early warning systems (MHEWS) are critical instruments for reducing disaster-related losses in an era marked by intensifying seismic and hydrometeorological risks. Despite advances in sensor technologies and predictive analytics, the effectiveness of early warning systems depends largely on how warnings are communicated, interpreted, and acted upon by diverse populations. This chapter

DOI: 10.4018/979-8-3373-7345-4.ch011

right © 2026, IGI Global Scientific Publishing. Copying or distributing in print or electronic forms without written permission of IGI Global Scientific Publishing is prohibited. Use of this chapter to in generative artificial intelligence (AI) technologies is expressly prohibited. The publisher reserves all rights to license its use for generative AI training and machine learning model development.

examines the transition from sensor-based detection to actionable public response, focusing on emerging technologies and human-centred dissemination strategies for earthquake and flash-flood warnings. Integrating perspectives from disaster risk reduction, communication theory, and socio-technical systems, the chapter highlights gaps between technological capability and social uptake. It proposes an integrative framework that aligns real-time sensing, decision support, and inclusive communication to strengthen resilience across vulnerable communities.

INTRODUCTION

The frequency and intensity of natural hazards have increased significantly over the past few decades, driven by climate change, rapid urbanization, environmental degradation, and complex socio-economic vulnerabilities. Earthquakes and flash floods, in particular, represent high-impact hazards capable of causing sudden, large-scale loss of life and infrastructure damage, often with little or no advance notice. While earthquakes remain largely unpredictable in terms of precise timing, advances in seismic monitoring enable rapid detection and near-real-time warning. Flash floods, similarly, have benefited from improvements in hydrological modelling, weather forecasting, and remote sensing. Yet, despite these technological advancements, disaster losses continue to rise, revealing a persistent gap between early detection and effective human response (United Nations Office for Disaster Risk Reduction, 2015; World Meteorological Organization, 2020, World Meteorological Organization & United Nations Office for Disaster Risk Reduction, 2023).

Early warning systems (EWS) are designed to bridge this gap by enabling timely and actionable information flows from detection to decision-making and response. Contemporary disaster risk reduction frameworks increasingly emphasize multi-hazard early warning systems (MHEWS), which integrate monitoring, forecasting, communication, and preparedness across multiple hazard types (United Nations Office for Disaster Risk Reduction, 2022, World Meteorological Organization & United Nations Office for Disaster Risk Reduction, 2023). However, many existing systems remain technologically driven, prioritizing sensor accuracy and prediction models while underestimating the social, cognitive, and behavioural dimensions of warning dissemination. As a result, warnings may be technically precise yet socially ineffective, misunderstood, ignored, or distrusted by the very populations they are intended to protect (Mileti & Sorensen, 1990; Wachinger et al., 2013).

The challenge is not merely technological but fundamentally socio-technical. Sensors, algorithms, and communication infrastructures operate within human systems shaped by culture, language, trust, prior experience, risk perception, and institutional credibility. Research has consistently shown that people do not respond to warnings based solely on scientific accuracy; instead, responses are mediated by how information is framed, who communicates it, and whether it aligns with local realities and lived experiences (Renn, 2008; Paton, 2019). This insight has led to a growing recognition of human-centred approaches in disaster communication, which prioritize inclusivity, accessibility, and behavioural relevance alongside technical robustness.

Earthquake and flash-flood warnings present distinct yet interrelated communication challenges. Earthquake early warning systems typically operate within seconds to minutes, requiring rapid, automated dissemination and immediate public action such as "drop, cover, and hold on." Flash-flood warnings, in contrast, often allow longer lead times but involve greater uncertainty, spatial variability, and decision complexity, including evacuation, sheltering, or infrastructure shutdowns (Creutin et al., 2013). In both cases, warning effectiveness depends on the alignment between sensor-driven alerts and human decision-making processes under stress and uncertainty.

Emerging technologies such as artificial intelligence (AI), machine learning, Internet of Things (IoT) sensor networks, satellite-based earth observation, and mobile communication platforms have significantly transformed the technical capabilities of early warning systems. AI-enhanced models improve hazard detection and forecasting accuracy, while mobile phones, social media platforms, and location-based services enable the rapid dissemination of alerts to large populations (Akerkar, 2019; Tiggeloven et al., 2025). However, the proliferation of communication channels also introduces new risks, including information overload, misinformation, unequal access, and digital exclusion. Without careful design, technologically advanced systems may inadvertently widen vulnerability gaps, particularly among marginalized populations with limited digital literacy or connectivity.

Global policy frameworks increasingly recognize these challenges. The Sendai Framework for Disaster Risk Reduction (2015–2030) explicitly calls for people-centred, multi-hazard early warning systems that are accessible, inclusive, and actionable (United Nations Office for Disaster Risk Reduction, 2015, World Meteorological Organization & United Nations Office for Disaster Risk Reduction, 2023). Similarly, the World Meteorological Organization emphasizes end-to-end warning systems that integrate hazard detection, risk assessment, communication, and response capability as a continuous process rather than a linear pipeline (World Meteorological Organization, 2020). Despite these policy commitments, implementation remains uneven, especially in low- and middle-income countries where institutional fragmentation, resource constraints, and social diversity complicate system design.

This chapter responds to these gaps by examining how emerging technologies can be aligned with human-centred dissemination strategies to improve earthquake and flash-flood warning effectiveness. Rather than treating technology and communication as separate domains, the chapter adopts an integrated socio-technical perspective that views early warning as a dynamic interaction between sensors, institutions, and communities. It argues that resilience is not achieved solely through faster detection or more accurate models, but through systems that enable people to understand risks, trust warnings, and take timely protective action.

The chapter is structured to progressively build this argument. Following this introduction, the background section situates multi-hazard early warning systems within global disaster risk reduction discourse and reviews technological and communicative developments. Subsequent sections develop theoretical and conceptual frameworks that integrate socio-technical systems theory, risk communication, and human-centred design. The body of the chapter examines emerging technologies and dissemination pathways in detail, followed by a critical discussion of challenges and trade-offs. The chapter concludes with actionable recommendations, future research directions, and implications for policy and practice aimed at strengthening disaster resilience through inclusive and effective early warning systems.

BACKGROUND

Early warning systems (EWS) have evolved significantly over the past several decades, transitioning from hazard-specific monitoring systems to more integrated and comprehensive multi-hazard frameworks. Historically, early warning efforts were developed within disciplinary silos, with seismology focusing primarily on earthquake monitoring and hydrology concentrating on flood forecasting and rainfall modelling. While these approaches contributed important scientific advances, they often failed to address the complex interactions between multiple hazards and the socio-institutional systems responsible for disaster response (Kelman, 2020). As a result, disaster risk reduction strategies increasingly emphasize multi-hazard early warning systems (MHEWS), which integrate monitoring, forecasting, dissemination, and preparedness across multiple hazard types and institutional actors.

The global policy shift toward multi-hazard approaches was strongly reinforced by the Sendai Framework for Disaster Risk Reduction (2015–2030), which calls for improved access to multi-hazard early warning systems and risk information worldwide (United Nations Office for Disaster Risk Reduction, 2015, World Meteorological Organization & United Nations Office for Disaster Risk Reduction 2023). The Sendai Framework recognizes that disaster risk is shaped not only by hazard exposure but also by vulnerability, governance capacity, and access to timely information. Consequently, MHEWS are designed to support end-to-end warning processes that link scientific detection with communication, decision-making, and community response capacity. This approach reflects growing recognition that effective early warning systems must integrate technological innovation with institutional coordination and social engagement.

Recent research further emphasizes the importance of adopting integrated warning frameworks capable of addressing cascading and compound risks. Natural hazards rarely occur in isolation; earthquakes, floods, landslides, and extreme weather events frequently interact with infrastructure systems and social vulnerabilities to produce complex disaster scenarios. Integrated early warning systems therefore require not only advanced monitoring technologies but also governance structures that enable coordination among multiple agencies and stakeholders (World Meteorological Organization, 2020).

Technological advances have significantly enhanced the capabilities of early warning systems in recent decades. Dense sensor networks, satellite observations, and real-time data processing platforms now enable rapid detection and monitoring of hazardous events. In earthquake early warning systems, seismic sensors, accelerometers, and Global Navigation Satellite System (GNSS) stations allow scientists to detect ground motion and estimate earthquake parameters within seconds of occurrence (Allen & Melgar, 2019). These systems exploit the difference in travel speeds between primary (P) waves and secondary (S) waves to provide brief but potentially life-saving warning windows.

Flash-flood early warning systems rely on a broader range of monitoring technologies, including rainfall gauges, weather radar, hydrological models, and satellite-based precipitation estimates (Creutin et al., 2013). The integration of these technologies allows forecasting agencies to track rainfall patterns, river discharge, and soil moisture conditions that may lead to rapid flooding. However, flash-flood forecasting remains challenging due to the highly localized and rapidly evolving nature of extreme rainfall events.

In recent years, artificial intelligence (AI) and machine learning techniques have emerged as transformative tools in hazard detection and forecasting. AI-driven models can analyse large volumes of environmental data from sensors, satellites, and meteorological systems, identifying complex patterns associated with earthquakes, floods, and other hazards (Mosavi et al., 2018; Akerkar, 2019; Tiggeloven et al., 2025). Studies demonstrate that machine learning algorithms improve prediction accuracy and reduce false alarms by rapidly classifying signals and assimilating multiple data sources (Tiggeloven et al., 2025; Wibowo et al., 2025). Deep learning techniques have also been applied to seismic waveform analysis and rainfall-runoff modelling, enabling faster and more reliable predictions of hazard events.

Furthermore, recent research highlights the role of integrated AI systems in addressing compound climate risks. Advanced predictive models that combine climate, hydrological, and geospatial datasets are increasingly used to anticipate complex hazard interactions and provide decision-makers with real-time risk assessments (Reichstein et al., 2025). These developments demonstrate how technological innovation is expanding the analytical capabilities of modern early warning systems.

Despite substantial advances in monitoring and predictive technologies, disaster losses continue to rise globally. A significant reason for this paradox is the persistent gap between hazard detection and effective public response, often referred to as the "last mile" problem in early warning systems. Research consistently shows that warning failures are rarely caused by a lack of scientific data; rather, they arise from breakdowns in communication, institutional coordination, and public understanding (Mileti & Sorensen, 1990; Wachinger et al., 2013).

Technological improvements alone do not guarantee that warnings will lead to protective action. Individuals interpret warnings through cognitive, cultural, and social frameworks shaped by trust in authorities, previous experiences with hazards, and local risk perceptions (Renn, 2008). Consequently, technically accurate warnings may still be ignored or misunderstood if they are poorly communicated or lack contextual relevance.

The rapid expansion of digital communication technologies has both improved and complicated warning dissemination. Mobile alerts, social media platforms, and location-based messaging systems allow authorities to reach large populations quickly. However, these technologies also introduce challenges related to misinformation, inconsistent messaging, and unequal access to digital resources. Studies show that misinformation can spread rapidly during disasters, potentially undermining trust in official warnings and delaying protective action (Hou et al., 2024).

Addressing the last-mile problem therefore requires communication strategies that prioritize clarity, accessibility, and trust. Human-centred communication design emphasizes the importance of simple, actionable messages and culturally appropriate communication channels that resonate with diverse populations.

Recent scholarship increasingly emphasizes the role of human-centred design and community participation in strengthening early warning effectiveness. Human-centred approaches prioritize the needs, capacities, and contexts of warning recipients, ensuring that alerts are understandable, actionable, and accessible to diverse populations (Norman, 2013). This perspective shifts the focus from merely issuing warnings to enabling meaningful behavioural responses.

Community-based early warning systems (CBEWS) represent an important extension of this approach. These systems involve local communities directly in monitoring hazards, disseminating warnings, and implementing protective actions. Research indicates that community participation enhances trust in warning systems and improves the relevance of risk communication strategies (Pham et al., 2024). Local knowledge regarding hazard patterns, evacuation routes, and vulnerable populations can also significantly improve the effectiveness of warning systems.

Case studies from several regions demonstrate the benefits of integrating technological monitoring systems with community engagement. For example, community-supported flood monitoring initiatives in Southeast Asia and South Asia have successfully combined local observation networks with digital communication platforms to deliver timely warnings to vulnerable populations. However, studies also highlight challenges associated with sustaining community participation and maintaining reliable communication networks, particularly in resource-constrained settings (Domingos & Nagamatsu, 2024).

In addition to community engagement, participatory digital technologies are increasingly being used to support disaster risk reduction. Mobile applications, crowdsourced data platforms, and citizen-science initiatives allow communities to report hazard observations and contribute to situational awareness during emergencies. While these innovations offer valuable opportunities for enhancing early warning coverage, they also raise concerns regarding data reliability, privacy, and governance.

The growing body of literature on early warning systems highlights a critical shift from technology-centric approaches toward integrated, people-centred frameworks. Advances in sensing technologies, predictive analytics, and communication infrastructure have significantly improved hazard detection capabilities. However, these technological innovations must be complemented by effective governance structures, inclusive communication strategies, and community engagement to ensure that warnings translate into timely and appropriate action.

Multi-hazard early warning systems represent an important step toward achieving this integration. By linking hazard detection, risk assessment, communication, and response within a coordinated framework, MHEWS aim to bridge the gap between scientific knowledge and human behaviour. Recent research suggests that combining advanced analytics with participatory governance and community-based approaches can significantly enhance the effectiveness of early warning systems and contribute to long-term disaster resilience (Budimir et al., 2025; Marengo et al., 2025).

In this context, the challenge for contemporary disaster risk reduction lies not merely in improving technological capabilities but in aligning those capabilities with social systems, institutional processes, and community practices. Understanding the evolution, technological foundations, and communication dynamics of early warning systems therefore provides the necessary foundation for the theoretical and conceptual frameworks developed in the subsequent sections of this chapter.

ILLUSTRATIVE CASE STUDIES OF MULTI-HAZARD EARLY WARNING SYSTEMS

While theoretical frameworks and technological innovations are essential for advancing multi-hazard early warning systems (MHEWS), practical case studies provide valuable insights into how these systems function in real-world contexts. Examining experiences from countries that have successfully implemented early warning systems can help illustrate how sensor networks, predictive analytics, and communication strategies operate together to reduce disaster risks and improve community preparedness (United Nations Office for Disaster Risk Reduction, 2015; World Meteorological Organization, 2020, World Meteorological Organization & United Nations Office for Disaster Risk Reduction, 2023).

Japan represents one of the most advanced examples of earthquake early warning systems. The Japan Meteorological Agency (JMA) operates a nationwide network of seismic sensors that detect primary seismic waves and rapidly estimate earthquake parameters. When an earthquake is detected, automated alerts are disseminated through television broadcasts, mobile phones, railway control systems, and public address networks within seconds. These alerts provide brief but critical warning time that allows individuals to take protective actions and enables automated responses such as slowing trains, shutting down industrial operations, and securing critical infrastructure. Japan's experience demonstrates the importance of integrating dense sensor networks with automated communication systems and sustained public education to ensure that early warnings translate into immediate protective behaviour (Allen & Melgar, 2019). Indonesia provides another important example, particularly in the context of multi-hazard risks associated with earthquakes, tsunamis, and flooding. Following the devastating 2004 Indian Ocean tsunami, Indonesia significantly expanded its national early warning infrastructure by deploying seismic monitoring stations, ocean buoys, and coastal siren systems. In addition to technological improvements, Indonesian authorities have emphasized community engagement and local preparedness programs. Community-based monitoring initiatives and evacuation drills have been introduced in vulnerable coastal areas, enabling local residents to understand warning signals and evacuation procedures. These efforts highlight the importance of combining technological monitoring systems with community participation to enhance the effectiveness of early warning mechanisms (Kelman, 2020; Pham et al., 2024).

The United States offers further insights into the integration of advanced technologies within early warning systems. The ShakeAlert system, developed by the United States Geological Survey (USGS) in collaboration with several research institutions, uses a network of seismic sensors along the western coast to detect earthquakes and issue alerts to millions of people through smartphone applications, wireless emergency alerts, and public broadcasting systems. Similarly, the National Weather Service operates flash-flood early warning systems that combine radar data, hydrological modelling, and satellite observations to provide timely warnings during extreme rainfall events. These systems demonstrate how advanced data analytics and integrated communication networks can support large-scale warning dissemination across diverse populations (Allen & Melgar, 2019; Tiggeloven et al., 2025).

Together, these case studies illustrate that effective multi-hazard early warning systems depend on the integration of technological infrastructure, institutional coordination, and public awareness. Sensor networks and predictive models provide the scientific foundation for hazard detection, but their effectiveness ultimately depends on how quickly and clearly warning information reaches communities and how prepared individuals are to respond (Wachinger et al., 2013; Renn, 2008). Lessons from countries such as Japan, Indonesia, and the United States therefore provide valuable guidance for improving early warning systems worldwide and strengthening disaster resilience in increasingly hazard-prone environments.

THEORETICAL FRAMEWORK

Building on the technological and socio-institutional context discussed in the background section, the following theoretical framework explains how technological and human dimensions interact in early warning systems. Together, these frameworks explain why technological accuracy alone is insufficient for effective disaster risk reduction and why human interpretation and action are decisive components of early warning effectiveness.

Socio-technical systems theory conceptualizes early warning systems as dynamic interactions between technical components—such as sensors, algorithms, and communication infrastructure—and social components, including institutions, communities, norms, and behaviours. From this perspective, failures in disaster warnings are rarely attributable to technology alone; instead, they emerge from misalignments between system design and human use (Trist & Bamforth, 1951). In the context of earthquake and flash-flood warnings, socio-technical theory highlights the need to integrate real-time sensing with governance structures, decision protocols, and community response mechanisms.

Risk communication theory further explains how individuals and groups perceive, interpret, and respond to warning information under conditions of uncertainty. Research in this domain demonstrates that risk perception is shaped by trust in authorities, prior experience, social amplification, and message framing rather than by probabilistic accuracy alone (Renn, 2008; Wachinger et al., 2013). This is particularly relevant for multi-hazard contexts, where competing or overlapping warnings may create confusion or desensitization. Effective communication, therefore, must be credible, consistent, and oriented toward clear behavioural guidance.

Human-centred design theory complements these perspectives by emphasizing user needs, cognitive capacities, and contextual constraints throughout system development. Applied to early warning systems, this approach prioritizes accessibility, inclusivity, and actionability, ensuring that warnings are not only delivered but understood and acted upon (Norman, 2013). By integrating these three theoretical lenses, the chapter frames early warning systems as socio-technical interventions that must align technological capability with human decision-making to build resilience.

CONCEPTUAL FRAMEWORK

The conceptual framework for this chapter operationalizes the integrated socio-technical and human-centred theoretical perspective by illustrating how sensor-driven detection, decision-support mechanisms, and human-centred communication interact to enable timely and effective action in earthquake and flash-flood early warning contexts. The framework conceptualizes early warning not as a linear transmission of information, but as a cyclical and adaptive process linking technology, institutions, and communities.

At the core of the framework lies hazard sensing and forecasting, encompassing seismic sensors, hydrometeorological instruments, satellite observations, and AI-enabled analytics. These components generate real-time or near-real-time data on hazard characteristics such as ground motion, rainfall intensity, or flood probability. However, raw data alone does not constitute a warning. The framework therefore introduces a second layer: interpretation and decision support, where data are translated into risk-relevant information through models, thresholds, and expert judgement. This stage involves scientific agencies and emergency authorities who determine warning levels, spatial targeting, and urgency.

The third and most critical layer is human-centred dissemination and communication. Here, warning information is transformed into messages designed for comprehension, trust, and action. This includes the selection of communication channels (e.g., mobile alerts, sirens, community networks), message framing (clear, concise, behaviour-oriented), and linguistic and cultural adaptation. Feedback loops are embedded at this stage, allowing authorities to assess public understanding, response effectiveness, and emerging misinformation.

Surrounding these layers is the enabling environment, comprising governance structures, institutional coordination, legal mandates, and community preparedness. Preparedness activities—such as drills, education, and participatory planning—strengthen the link between warnings and action by shaping prior knowledge and expectations. The framework also accounts for contextual moderators, including trust in institutions, digital access, social vulnerability, and prior hazard experience, which influence how warnings are received and acted upon.

By integrating these components, the conceptual framework emphasizes that resilience emerges from alignment across technological, institutional, and human dimensions. It provides a structured lens for analysing strengths and gaps in existing earthquake and flash-flood warning systems and guides the design of inclusive, actionable multi-hazard early warning strategies.

1. Sensor and Data Ecosystems for Multi-Hazard Detection

Modern early warning systems are built upon complex sensor and data ecosystems that enable the detection, monitoring, and characterization of hazardous events in near real time. As disaster risk profiles become increasingly multi-dimensional due to climate change, urbanization, and environmental degradation, the demand for dense, reliable, and interoperable sensor networks has intensified. In the context of multi-hazard early warning systems (MHEWS), sensor ecosystems must support the simultaneous monitoring of diverse physical processes while ensuring that data can be rapidly transformed into actionable intelligence for decision-makers and communities.

Earthquake Detection Infrastructure

Earthquake early warning systems rely primarily on ground-based seismic instrumentation, including seismometers, accelerometers, and Global Navigation Satellite System (GNSS) stations. Seismometers detect ground vibrations caused by seismic waves, while accelerometers measure strong ground motion close to the source, making them particularly valuable in urban and near-fault regions. GNSS stations complement these instruments by capturing permanent ground displacement, which is critical for accurately estimating the magnitude of large earthquakes that may saturate traditional seismic sensors (Allen & Melgar, 2019).

The effectiveness of earthquake detection networks is influenced by sensor density, spatial distribution, and data latency. Dense networks reduce detection time and improve the accuracy of magnitude and intensity estimates, enabling warnings to be issued within seconds of earthquake initiation. However, high-density networks require substantial investment in installation, maintenance, and data management. In many low- and middle-income countries, sparse instrumentation limits early warning potential, highlighting the need for cost-effective alternatives and international data-sharing initiatives.

Data latency—the time required for sensor data to reach processing centers—is a critical factor in earthquake early warning. Even minor delays can significantly reduce the available warning time, particularly for populations located close to the epicenter. To address this challenge, modern systems prioritize low-latency communication channels, edge computing, and redundant data transmission pathways. Redundancy is equally important for system robustness, as earthquakes themselves can disrupt power supplies and communication infrastructure. Designing sensor networks that remain operational during and after seismic events is therefore a fundamental requirement of resilient early warning systems.

Sensor Systems for Flash-Flood Detection

Flash-flood early warning systems operate within a more complex and dynamic environmental context than seismic systems. Flash floods are driven by intense, localized rainfall, rapid runoff, and catchment-specific characteristics such as land use, soil saturation, and topography. As a result, no single sensor type can adequately capture all relevant processes. Instead, flash-flood detection relies on the integration of multiple sensor modalities, including rain gauges, weather radar, river-level sensors, soil moisture probes, and satellite-based precipitation estimates (Creutin et al., 2013).

Rain gauges provide accurate point-based measurements of precipitation but are limited in spatial coverage, particularly in mountainous or remote regions. Weather radar addresses this limitation by offering spatially continuous rainfall estimates, yet radar data can be affected by terrain blockage, signal attenuation, and calibration challenges. River-level sensors and stream gauges capture hydrological response directly, providing critical information on rising water levels, but they may offer limited lead time in small, fast-responding catchments.

Satellite-based earth observation has emerged as a powerful tool for flash-flood monitoring, particularly in data-scarce regions. Advances in satellite precipitation products and near-real-time data delivery have expanded global coverage and improved temporal resolution. However, satellite data involve inherent trade-offs between spatial resolution, revisit frequency, and latency. Cloud cover, algorithm uncertainty, and indirect estimation methods can further complicate interpretation. Effective flash-flood early warning systems therefore depend on the careful fusion of satellite and ground-based observations to balance coverage and accuracy.

Data Integration in Multi-Hazard Contexts

From a multi-hazard perspective, the central challenge lies not only in collecting data but in integrating diverse data streams into a coherent and interoperable system. Earthquakes, flash floods, landslides, and infrastructure failures often interact through cascading processes, creating compound risks that exceed the scope of single-hazard monitoring approaches. For example, earthquakes may damage dams, levees, or drainage systems, increasing downstream flood risk, while intense rainfall can destabilize slopes in seismically active regions, heightening landslide susceptibility (Kelman, 2020).

To address these interactions, MHEWS must support data interoperability across hazard domains and institutional boundaries. Standardized data formats, shared metadata protocols, and common reference frameworks enable different agencies to exchange and interpret data efficiently. Real-time data integration platforms that combine seismic, hydrological, meteorological, and geospatial information are increasingly viewed as essential infrastructure for multi-hazard risk management. Without such integration, warnings may remain fragmented, inconsistent, or delayed, undermining their effectiveness.

Interoperability also extends to temporal and spatial scales. Earthquake early warning operates on timescales of seconds, while flash-flood forecasting may involve lead times ranging from minutes to hours. Integrating these disparate timescales within a single system requires flexible architectures that can accommodate rapid-onset alerts alongside evolving forecasts. Similarly, spatial integration is necessary to ensure that warnings are appropriately targeted, avoiding both overgeneralization and excessive localization that may confuse users.

Internet of Things and Distributed Sensing

The rise of Internet of Things (IoT) architectures has significantly expanded the sensor landscape by enabling low-cost, distributed sensing solutions. IoT-based sensors can be deployed at scale to monitor environmental variables such as rainfall, water levels, and ground motion, often using wireless communication and low-power designs. These technologies are particularly attractive for resource-constrained settings, where traditional monitoring infrastructure may be prohibitively expensive.

Community-based sensors, citizen science initiatives, and privately owned devices—such as smartphones equipped with accelerometers—can supplement official networks and enhance spatial coverage. In earthquake monitoring, smartphone-based sensing has shown potential for rapid detection in urban areas, while community-installed rain gauges can provide valuable localized data for flash-flood monitoring. Such approaches democratize data collection and can foster community engagement with early warning systems.

However, distributed sensing also raises significant challenges related to data quality, calibration, and governance. Sensors deployed by non-experts may produce noisy or unreliable data, requiring robust validation and filtering mechanisms. Differences in sensor specifications, maintenance practices, and communication reliability can further complicate data integration. From a governance perspective, questions arise regarding data ownership, privacy, liability, and the criteria for incorporating unofficial data into warning decisions.

A human-centred, socio-technical approach to sensor ecosystems emphasizes transparency and trust. Authorities must clearly communicate how data from diverse sources are used, what quality controls are applied, and how uncertainties are managed. Engaging communities not only as data providers but also as informed stakeholders can enhance trust and acceptance of early warning systems, particularly when unofficial data sources are involved.

Data Management, Reliability, and Resilience

Beyond detection, effective sensor ecosystems require robust data management frameworks that ensure reliability, security, and continuity. Large volumes of real-time data must be stored, processed, and archived in ways that support both immediate warning decisions and long-term analysis. Cloud computing and edge processing offer scalable solutions, but they also introduce dependencies on digital infrastructure that may be vulnerable during disasters.

System resilience is therefore a critical design consideration. Redundant sensors, backup power supplies, fail-safe communication channels, and decentralized processing can help maintain functionality during extreme events. Cybersecurity is an emerging concern, as increased connectivity exposes early warning systems to potential cyber threats that could disrupt data flows or undermine trust.

Synthesis: Sensor Ecosystems as Foundations for Action

In sum, sensor and data ecosystems form the foundational layer of multi-hazard early warning systems, enabling the detection and characterization of earthquakes and flash floods in real time. However, their effectiveness depends not only on technological sophistication but on integration, governance, and human-centred design. Dense and diverse sensor networks must be complemented by interoperable data platforms, transparent standards, and resilient infrastructure. When designed and managed effectively, sensor ecosystems provide the critical first step in the transition from hazard detection to timely, trusted, and actionable warnings that support disaster resilience across multiple hazard domains.

2. Intelligent Analytics and Decision-Support Mechanisms

The effectiveness of multi-hazard early warning systems (MHEWS) depends not only on the availability of dense sensor networks but also on the capacity to transform large volumes of heterogeneous data into timely, reliable, and actionable intelligence. As sensor ecosystems expand in scale and complexity, the volume, velocity, and variety of data generated by seismic, hydrological, meteorological, and satellite sources have increased exponentially. This data deluge necessitates advanced analytical approaches capable of real-time processing, pattern recognition, and decision support. Intelligent analytics and decision-support mechanisms therefore constitute the cognitive core of modern early warning systems, mediating between detection and dissemination.

Role of Artificial Intelligence and Machine Learning

Artificial intelligence (AI) and machine learning have emerged as transformative tools in early warning analytics by enabling automated interpretation of complex data streams. In earthquake early warning systems, machine learning algorithms are used to rapidly classify seismic signals, distinguish earthquakes from noise, and estimate key parameters such as magnitude and expected shaking intensity within seconds of detection (Akerkar, 2019). These capabilities are particularly valuable in reducing false alarms, which can erode public trust and lead to warning fatigue.

Traditional seismic analysis methods rely on predefined physical models and threshold-based approaches, which may struggle with noisy data or atypical events. Machine learning models, trained on large datasets of historical seismic signals, can identify subtle patterns and adapt to regional variations in ground motion characteristics. When integrated into real-time processing pipelines, these models enhance the speed and reliability of earthquake alerts, extending the usable warning window for protective action.

In the context of flash-flood early warning, AI-driven analytics play an equally critical role. Flash floods are influenced by a wide range of interacting variables, including rainfall intensity, soil moisture, land use, topography, and antecedent hydrological conditions. Machine learning models can assimilate these diverse inputs to estimate flood likelihood, onset timing, and potential impact. By learning from past flood events, AI systems can improve predictive accuracy and support impact-based forecasting that goes beyond hazard detection to assess potential consequences for people and infrastructure.

Managing Uncertainty in Rapid Decision-Making

Despite significant advances in intelligent analytics, decision-making in early warning systems remains inherently uncertain. Earthquake early warning systems must issue alerts within seconds, often before complete information about the event is available. Initial magnitude estimates may be revised as additional data arrive, creating the risk of underestimation or overestimation. Flash-flood forecasting, while allowing longer lead times, involves probabilistic assessments that are sensitive to rapidly changing meteorological conditions and model assumptions.

Decision-support systems (DSS) play a central role in managing this uncertainty by translating analytical outputs into warning levels, priorities, and recommended actions. These systems typically integrate predefined thresholds, scenario analyses, and expert judgement to balance the competing risks of missed events and false alarms (World Meteorological Organization, 2020). Thresholds may be based on ground motion intensity, rainfall accumulation, or river-level exceedance, while scenario analyses explore potential outcomes under different assumptions.

However, rigid threshold-based approaches can be problematic in multi-hazard contexts, where interacting risks and cascading effects complicate decision-making. Adaptive decision-support systems that incorporate probabilistic reasoning and real-time updates are increasingly advocated. Such systems can adjust warning levels dynamically as new data become available, providing decision-makers with evolving situational awareness rather than static alerts.

Human-Centred Design of Decision-Support Tools

From a human-centred perspective, the design of decision-support mechanisms must account for the cognitive and organizational realities faced by emergency managers and warning authorities. Decision-makers often operate under extreme time pressure, high stakes, and incomplete information. Overly complex dashboards, dense data visualizations, or poorly explained model outputs can increase cognitive load and delay action, undermining the benefits of advanced analytics.

Research in human–computer interaction emphasizes the importance of intuitive interfaces, clear visualization of key indicators, and alignment with users' mental models (Norman, 2013). Effective decision-support tools prioritize salient information, highlight critical changes, and provide clear guidance on recommended actions. For example, visualizations that translate probabilistic forecasts into simple risk categories or impact levels can aid rapid comprehension without oversimplifying underlying uncertainty.

In multi-hazard environments, decision-support systems must also manage competing risks. Simultaneous or sequential hazards may require prioritization, resource allocation, and coordinated messaging. Decision-support tools that integrate multiple hazard streams into a unified interface can help operators assess trade-offs and avoid fragmented or conflicting warnings. However, achieving such integration requires careful design to prevent information overload and ensure that the most critical signals remain visible.

Institutional Context and Organizational Constraints

Decision-support mechanisms do not operate in isolation; they are embedded within institutional and governance contexts that shape how analytical outputs are interpreted and acted upon. Emergency management agencies often operate within hierarchical decision structures, legal mandates, and political accountability frameworks. These factors influence the extent to which automated analytics are trusted and utilized.

Automated systems may generate alerts faster than human analysts, yet final authority for warning issuance often rests with public officials who must consider legal liability, public perception, and potential economic consequences. This institutional reality can lead to cautious decision-making, particularly in contexts where false alarms have previously attracted criticism. Decision-support systems must therefore be designed to support—not replace—human judgement, providing evidence-based recommendations while preserving human oversight.

Training and capacity building are critical enablers of effective decision support. Emergency managers must understand the strengths, limitations, and assumptions underlying analytical models to use them confidently. Without such understanding, advanced systems may be underutilized or overridden during critical moments, negating their potential benefits.

Trust, Transparency, and Explainable AI

Institutional trust is a decisive factor influencing the uptake of intelligent analytics in early warning systems. Trust operates at multiple levels: trust in data quality, trust in analytical models, and trust in the institutions responsible for issuing warnings. Opaque algorithms and "black box" AI models can undermine this trust, particularly when their outputs conflict with human intuition or prior experience.

Explainable AI has emerged as a key research and policy priority in this context. By making model logic, input variables, and uncertainty estimates transparent, explainable AI enables decision-makers to understand why specific warnings are issued and how confident the system is in its predictions (Renn, 2008). This transparency supports accountability, facilitates learning, and enhances institutional confidence in automated or semi-automated systems.

Clear accountability structures are equally important. Decision-support systems should explicitly define who is responsible for warning issuance, how automated recommendations are reviewed, and how disagreements between human judgement and system outputs are resolved. Such clarity reduces ambiguity during emergencies and supports consistent decision-making across events.

Ethical and Equity Considerations

Intelligent analytics also raise important ethical and equity considerations. Machine learning models trained on historical data may inadvertently reflect existing biases, particularly if past events disproportionately affected certain regions or populations. In flash-flood forecasting, for example, models may perform better in well-instrumented areas, potentially disadvantaging data-scarce regions. Addressing these biases requires deliberate efforts to diversify training datasets, incorporate uncertainty assessments, and prioritize vulnerable populations in decision-making.

Human-centred decision-support systems should therefore integrate social vulnerability indicators alongside hazard metrics, enabling more equitable warning prioritization. This approach aligns with the broader shift toward impact-based forecasting and people-centred early warning advocated in global policy frameworks.

Synthesis: Analytics as Enablers of Action

In synthesis, intelligent analytics and decision-support mechanisms serve as the interpretive bridge between sensor data and warning dissemination in multi-hazard early warning systems. Artificial intelligence and machine learning enhance the speed and accuracy of hazard detection, while decision-support systems manage uncertainty and translate analytical outputs into actionable guidance. However, the effectiveness of these mechanisms depends on human-centred design, institutional trust, transparency, and ethical governance. When aligned with the cognitive and organizational realities of decision-makers, intelligent analytics can significantly strengthen the capacity of early warning systems to support timely, trusted, and effective action in the face of earthquakes and flash floods.

3. Dissemination Pathways and Communication Strategies

Dissemination is the most visible, public-facing, and socially consequential component of early warning systems. Regardless of the sophistication of sensors or analytical models, a warning has little value unless it reaches the intended audience in time and is understood well enough to prompt appropriate action. Dissemination therefore represents the critical interface between scientific knowledge and human behaviour. In multi-hazard early warning systems (MHEWS), this interface must accommodate differing hazard characteristics, social contexts, and communication ecologies, particularly in the case of earthquakes and flash floods.

Hazard-Specific Dissemination Requirements

Earthquake and flash-flood warnings impose distinct dissemination requirements due to differences in onset speed, predictability, and response options. Earthquake early warning systems operate under extreme time constraints, often providing only a few seconds of lead time. Dissemination must therefore be almost instantaneous and largely automated, relying on channels such as mobile phone notifications, sirens, public address systems, and automatic broadcast interruptions. Messages must be immediately recognizable and require no interpretation, as users have little time to process information before acting.

Flash-flood warnings, by contrast, typically allow longer lead times ranging from minutes to hours, depending on the catchment and rainfall dynamics. Dissemination in this context must convey spatial specificity, severity, and evolving risk levels. Unlike earthquake warnings, which often recommend a single standardized action, flash-flood warnings may involve multiple possible responses, including evacuation, relocation to higher ground, or infrastructure protection. This complexity increases the cognitive demands placed on recipients and underscores the importance of clear, context-sensitive messaging.

Table 1. Communication characteristics of earthquake vs flash-flood warnings

Feature	Earthquake Warning	Flash Flood Warning
Lead time	Seconds	Minutes to hours
Prediction certainty	Low	Moderate
Message complexity	Simple	Context dependent
Typical response	Drop, cover, hold	Evacuate or relocate
Communication channel	Automated alerts	Multi-channel

Multi-Channel Dissemination and Redundancy

The expansion of communication technologies has dramatically increased the number of channels available for warning dissemination. Contemporary systems may employ cell broadcast alerts, smartphone applications, social media platforms, radio and television broadcasts, sirens, and community-based networks. Redundancy across channels is widely recognized as a best practice, as reliance on a single medium can fail during power outages, network congestion, or infrastructure damage (United Nations Office for Disaster Risk Reduction, 2015, United Nations Office for Disaster Risk Reduction, 2023, World Meteorological Organization & United Nations Office for Disaster Risk Reduction, 2023). Multi-channel dissemination increases the likelihood that warnings will reach diverse audiences under varying conditions.

However, multi-channel strategies also introduce coordination challenges. Messages disseminated through different platforms may be delayed, reformatted, or interpreted differently, leading to inconsistencies that confuse recipients. In fast-moving earthquake scenarios, even minor delays between channels can undermine trust and reduce response effectiveness. For flash-flood warnings, inconsistent spatial or severity information across platforms can lead to uncertainty about who is at risk and what actions are required. Effective dissemination therefore requires centralized message generation, standardized templates, and real-time synchronization across channels.

Human-Centred Message Design

Human-centred communication strategies emphasize clarity, relevance, and actionability. Decades of research in disaster communication demonstrate that warnings framed around specific, concrete actions are more effective than abstract descriptions of risk (Mileti & Sorensen, 1990). Messages that explicitly state what to do, where, and when reduce ambiguity and support rapid decision-making. For example, instructions such as "move to higher ground immediately" or "drop, cover, and hold on" are more likely to prompt protective behaviour than probabilistic statements about hazard likelihood.

Visual cues and standardized alert levels further enhance comprehension, particularly under stress. Color-coded warnings, icons, and simple graphics can convey urgency and severity at a glance, complementing textual information. Consistent terminology across hazards and platforms helps build familiarity and reduces the risk of misinterpretation. In earthquake contexts, where response time is extremely limited, pre-established message formats and repeated public education are critical to ensure that alerts trigger reflexive protective actions rather than confusion or delay.

Cultural, Linguistic, and Social Considerations

Language, culture, and trust play decisive roles in how warnings are interpreted and acted upon. In culturally and linguistically diverse societies, multilingual messaging is essential to ensure equitable access to warning information. Failure to provide warnings in locally understood languages can exclude vulnerable populations and exacerbate disaster impacts. Beyond language, cultural norms influence how people perceive authority, risk, and collective action, shaping their responses to official warnings.

Local intermediaries such as community leaders, religious institutions, and trusted organizations can enhance warning credibility and uptake. These actors often possess deep contextual knowledge and social capital, enabling them to translate official messages into locally meaningful terms. Incorporating community-based dissemination pathways alongside formal channels can therefore strengthen the reach and effectiveness of early warning systems, particularly in rural or marginalized areas.

Managing Misinformation and Information Overload

The proliferation of digital communication platforms has amplified both the reach and the risks of warning dissemination. Social media and messaging applications can rapidly spread official warnings, but they can also facilitate the spread of misinformation, rumors, and conflicting narratives. During fast-evolving events such as flash floods, unofficial information may circulate more quickly than verified warnings, undermining trust and prompting inappropriate actions (Wachinger et al., 2013).

Effective MHEWS require proactive communication strategies that extend beyond issuing alerts. Monitoring information ecosystems, engaging with digital platforms, and correcting misinformation in real time are increasingly recognized as essential functions of warning authorities. Establishing official, verified communication channels and encouraging the public to rely on them can help counter misinformation. However, these efforts must be sustained during non-crisis periods to build familiarity and trust.

Information overload represents another challenge in multi-channel dissemination. Frequent alerts, updates, and competing messages can overwhelm recipients, particularly during prolonged or multi-hazard events. Human-centred design principles suggest that dissemination strategies should prioritize essential information and avoid excessive messaging that may lead to desensitization or alert fatigue.

Building Trust Through Engagement and Consistency

Trust is a foundational element of effective dissemination. People are more likely to heed warnings from sources they perceive as credible, transparent, and consistent. Trust is built over time through accurate warnings, honest communication about uncertainty, and visible alignment between messages and outcomes. False alarms, inconsistent messaging, or perceived exaggeration can erode trust and reduce responsiveness to future warnings.

Engagement with communities before disasters occur is therefore critical. Public education campaigns, drills, and participatory planning processes help familiarize individuals with warning systems, message formats, and expected actions. Such engagement transforms dissemination from a one-way transmission of information into an ongoing relationship between authorities and communities.

Synthesis: Dissemination as a Social Process

In synthesis, dissemination pathways and communication strategies are not merely technical components of early warning systems but deeply social processes that shape how risk information is perceived and acted upon. Earthquake and flash-flood warnings illustrate the diverse communication demands imposed by different hazards, from instantaneous alerts to evolving, context-rich messages. Effective MHEWS require coordinated multi-channel dissemination, human-centred message design, cultural and linguistic inclusivity, and proactive management of misinformation. By treating dissemination as a core element of resilience-building rather than a final technical step, early warning systems can more effectively translate scientific detection into life-saving action.

4. Community Engagement, Preparedness, and Behavioural Response

Early warning systems do not operate in a social vacuum. Community preparedness and prior experience strongly influence how warnings are received and acted upon. Studies consistently show that individuals are more likely to respond appropriately when they understand the hazard, trust the source of information, and have rehearsed protective actions through drills or training (Paton, 2019). Human-centred early warning systems therefore extend beyond technology and communication to encompass education, participation, and empowerment.

In earthquake early warning, behavioural response must be almost reflexive. Public education campaigns that teach simple, standardized actions—such as "drop, cover, and hold on"—are essential to translate seconds of warning into life-saving behaviour. Flash-flood preparedness, by contrast, often involves more complex decisions, including evacuation planning, route selection, and coordination with family or community members. Participatory planning and scenario-based exercises can help communities internalize these decisions before an event occurs.

Community engagement also provides valuable feedback for system improvement. Local knowledge can inform sensor placement, threshold selection, and message design, ensuring that warnings align with lived realities. Participatory approaches challenge the traditional top-down model of early warning by recognizing communities as active partners rather than passive recipients. This is particularly important in marginalized or informal settlements, where official warning systems may be less trusted or accessible.

5. Integrating Technology and Humanity: Toward Actionable Warnings

The preceding sections highlight that effective early warning systems emerge from the alignment of technological capability with human understanding and institutional coordination. Sensors and models provide the foundation for hazard detection, but their value is realized only when integrated with decision-support tools, communication strategies, and community preparedness. In multi-hazard contexts, this integration becomes even more critical, as competing risks and overlapping warnings can strain both institutional capacity and public attention.

An integrated, human-centred MHEWS recognizes early warning as a continuous, adaptive process rather than a one-time alert. Feedback loops—such as post-event evaluations, community surveys, and system audits—enable learning and improvement over time. Such systems are better positioned to address emerging challenges, including climate change–driven hazard variability, urban expansion into high-risk areas, and evolving communication technologies.

Ultimately, the transition from sensors to action requires reframing early warning systems as socio-technical interventions aimed at shaping behaviour under uncertainty. By embedding human-centred principles throughout system design and operation, MHEWS can move beyond detection to enable timely, trusted, and effective action, thereby strengthening resilience to earthquakes and flash floods alike.

DISCUSSION

The analysis presented in this chapter underscores a central insight in contemporary disaster risk reduction: early warning effectiveness is determined not solely by technological sophistication, but by the alignment between detection, communication, and human action. The transition from sensors to action is therefore best understood as a socio-technical challenge rather than a purely scientific or engineering problem. This discussion critically reflects on the implications of integrating emerging technologies with human-centred approaches for earthquake and flash-flood early warning systems, highlighting persistent gaps, tensions, and opportunities.

One of the most significant findings emerging from the literature and practice is the persistence of the "technology–action gap." Despite substantial investments in sensor networks, predictive models, and communication infrastructure, disaster losses continue to rise, particularly in rapidly urbanizing and climate-vulnerable regions (United Nations Office for Disaster Risk Reduction, 2015, United Nations Office for Disaster Risk Reduction, 2023, World Meteorological Organization & United Nations Office for Disaster Risk Reduction, 2023). This gap suggests that technological improvements alone do not automatically translate into risk reduction. Earthquake early warning systems, for example, may achieve sub-second detection and dissemination, yet fail to prompt appropriate action if users are unfamiliar with alert signals or uncertain about recommended responses. Similarly, flash-flood warnings may provide hours of lead time, but still result in high mortality when warnings are misunderstood, distrusted, or ignored.

A recurring tension identified in this chapter is the trade-off between speed and certainty. Earthquake early warning systems must operate under extreme time constraints, issuing alerts before full information is available. This creates an inherent risk of false alarms or missed events, which can erode public trust over time (Allen & Melgar, 2019). Flash-flood forecasting, while allowing longer lead times, involves probabilistic uncertainty that can complicate decision-making and message framing. Risk communication research suggests that people often struggle to interpret probabilistic information, especially under stress, leading to either complacency or overreaction (Renn, 2008). These dynamics highlight the need for communication strategies that explicitly acknowledge uncertainty while still providing clear guidance for action.

The discussion also reveals important equity implications. Digital dissemination channels such as mobile alerts and social media have expanded the reach of early warnings, yet they risk excluding populations with limited access to technology, low digital literacy, or language barriers. Studies show that marginalized groups—such as informal settlement residents, migrants, older adults, and people with disabilities—are often the least likely to receive or act upon official warnings. From a human-centred perspective, this raises critical ethical questions about who benefits from advanced early warning systems and who remains vulnerable. Multi-hazard early warning systems that fail to address these disparities may inadvertently reinforce existing social inequalities.

Institutional fragmentation emerges as another persistent challenge. Effective early warning requires coordination across scientific agencies, emergency management authorities, local governments, and community organizations. Yet, in many contexts, mandates remain fragmented, data sharing is limited, and responsibilities for warning issuance are unclear (World Meteorological Organization, 2020). This fragmentation can result in delayed or conflicting messages, undermining public trust. The discussion suggests that governance reform and institutional integration are as important as technological innovation in strengthening early warning effectiveness.

The role of trust warrants particular attention. Trust operates at multiple levels: trust in scientific institutions, trust in government authorities, and trust in the warning system itself. Empirical research demonstrates that people are more likely to heed warnings from sources they perceive as credible and aligned with their interests (Wachinger et al., 2013). In contexts where trust in authorities is low, technologically accurate warnings may still fail to prompt action. This insight reinforces the importance of long-term engagement, transparency, and community participation in early warning system design and operation.

Another key issue highlighted in the discussion is alert fatigue. As multi-hazard systems expand, the frequency of alerts may increase, particularly in hazard-prone regions. Repeated warnings that do not result in visible impacts can desensitize users, reducing responsiveness over time (Mileti & Sorensen, 1990). This challenge is especially relevant for flash-flood warnings, which often involve false positives due to conservative thresholds. Human-centred design principles suggest that warning systems must balance caution with credibility, tailoring alert frequency and severity to local risk profiles and user expectations.

The integration of artificial intelligence and automated decision-making introduces both opportunities and risks. On one hand, AI-enhanced systems can process vast data streams and issue alerts faster than human operators, potentially saving lives in rapid-onset events. On the other hand, opaque algorithms and unclear accountability can undermine institutional confidence and public acceptance (Akerkar, 2019). The discussion points to the importance of explainable AI and human-in-the-loop approaches that combine automation with expert judgement and oversight.

Finally, the discussion emphasizes that early warning systems should be understood as learning systems. Post-event evaluations, community feedback, and iterative refinement are essential for adapting to changing hazard patterns, technological advances, and social contexts. The Sendai Framework's emphasis on continuous improvement and people-centred approaches reflects this understanding (United Nations Office for Disaster Risk Reduction, 2015, United Nations Office for Disaster Risk Reduction, 2023, World Meteorological Organization & United Nations Office for Disaster Risk Reduction, 2023). However, learning requires institutional commitment, resources, and a willingness to acknowledge and address failures.

In summary, the discussion reinforces the argument that resilience is not achieved through sensors or models alone, but through systems that integrate technological capability with human understanding, trust, and preparedness. Earthquake and flash-flood early warning systems illustrate both the promise and the limitations of current approaches. By critically examining these dynamics, this chapter contributes to a more nuanced understanding of how multi-hazard early warning systems can move from detection to effective action in diverse social and institutional contexts.

RECOMMENDATIONS AND SOLUTIONS

Strengthening multi-hazard early warning systems (MHEWS) for earthquakes and flash floods requires moving beyond incremental technological upgrades toward systemic reforms that integrate technology, governance, and human behaviour. Based on the analysis and discussion presented in this chapter, this section outlines a set of actionable recommendations and solutions organized across technological, institutional, communicative, and community dimensions. Together, these recommendations aim to close the persistent gap between early detection and effective action.

1. Embed Human-Centred Design Across the Warning Lifecycle

A core recommendation is to institutionalize human-centred design as a guiding principle throughout the entire early warning lifecycle—from system planning and sensor deployment to message dissemination and post-event evaluation. Warning systems should be designed around how people actually perceive risk, process information, and make decisions under stress, rather than around idealized assumptions of rational behaviour. This involves conducting user research, including ethnographic studies, focus groups, and participatory workshops, to understand local needs, languages, cultural norms, and trust dynamics.

For earthquake early warning systems, this means ensuring that alert signals are immediately recognizable and paired with simple, rehearsed actions. For flash-flood warnings, messages should be tailored to local geographies and livelihood patterns, explicitly stating what actions are required, where, and within what time frame. Regular usability testing of alerts and communication platforms should be institutionalized to ensure that systems remain aligned with user capabilities as technologies and social contexts evolve.

2. Strengthen End-to-End Integration and Governance Coordination

Effective MHEWS depend on seamless integration across detection, forecasting, decision-making, dissemination, and response. Governments and disaster management authorities should establish clear governance frameworks that define roles, responsibilities, and decision thresholds across institutions. This includes formal data-sharing agreements between meteorological, geological, hydrological, and emergency management agencies to enable real-time information exchange.

At the operational level, standard operating procedures should be harmonized across hazards to reduce confusion during multi-hazard events. Joint training exercises and simulations involving multiple agencies can help build shared understanding and coordination. Where possible, centralized or interoperable platforms should be developed to support integrated monitoring and decision-making, reducing institutional silos that delay or dilute warnings.

3. Invest in Explainable and Accountable Use of Emerging Technologies

While artificial intelligence and automated analytics offer significant benefits for early warning, their deployment must be accompanied by transparency and accountability. Decision-support systems should prioritize explainable models that allow emergency managers to understand how warnings are generated and why specific thresholds are triggered. This is particularly important in high-stakes contexts where false alarms or missed events can have serious consequences for public trust.

A "human-in-the-loop" approach is recommended, in which automation supports but does not fully replace expert judgement. Clear accountability mechanisms should be established to define responsibility for warning issuance and communication, even when automated systems are involved. Investment in capacity building for technical staff and decision-makers is essential to ensure that emerging technologies are used effectively and ethically.

4. Enhance Inclusive and Redundant Communication Strategies

Dissemination strategies should be inclusive, redundant, and resilient to infrastructure failure. Governments should adopt a multi-channel communication approach that combines digital platforms (such as mobile alerts and applications) with traditional media (radio, television) and community-based mechanisms (sirens, loudspeakers, local volunteers). Special attention should be given to reaching populations with limited digital access, including rural communities, older adults, and people with disabilities.

Multilingual messaging and the use of locally trusted intermediaries can significantly enhance warning credibility and comprehension. Communication strategies should also address the risk of misinformation by proactively engaging with social media platforms, monitoring information flows, and correcting false narratives in real time. Pre-established message templates and standardized alert levels can further reduce confusion during emergencies.

5. Institutionalize Community Engagement and Preparedness

Early warning systems are most effective when communities are active partners rather than passive recipients. Authorities should invest in sustained community engagement initiatives, including education campaigns, drills, and participatory planning processes. These efforts help build risk awareness, clarify expected responses, and strengthen trust in warning systems.

For earthquake preparedness, regular drills in schools, workplaces, and public institutions can reinforce rapid, protective actions. For flash-flood risk, community-based evacuation planning and scenario exercises can help individuals and households anticipate decisions under time pressure. Community feedback mechanisms—such as post-event surveys or public forums—should be used to evaluate warning effectiveness and inform system improvements.

6. Integrate Equity and Vulnerability Considerations

A critical solution for strengthening resilience is to explicitly integrate equity considerations into MHEWS design and evaluation. Vulnerability assessments should inform sensor placement, communication strategies, and preparedness investments, ensuring that high-risk and marginalized populations are prioritized. Indicators of social vulnerability, such as income, housing quality, mobility, and access to information, should be incorporated into risk analysis and decision-making.

Targeted interventions—such as tailored messaging for informal settlements or support for community-based warning networks—can help reduce disparities in warning access and response capacity. By embedding equity into system design, MHEWS can contribute not only to risk reduction but also to broader social resilience.

7. Promote Continuous Learning and System Adaptation

Finally, early warning systems should be treated as adaptive learning systems. Post-event evaluations, audits, and after-action reviews should be mandatory components of system governance. These processes enable authorities to identify strengths, weaknesses, and unintended consequences, fostering iterative improvement over time.

International cooperation and knowledge exchange can further support learning, particularly for countries with limited resources. Sharing best practices, case studies, and technical standards through regional and global platforms can accelerate progress toward more effective and inclusive multi-hazard early warning systems.

FUTURE RESEARCH DIRECTIONS

The rapid evolution of hazard profiles, technological capabilities, and social contexts necessitates a forward-looking research agenda for multi-hazard early warning systems (MHEWS). While substantial progress has been made in sensor technologies and forecasting models, critical knowledge gaps remain in understanding how early warnings translate into effective and equitable action. This section outlines key future research directions that can advance theory, policy, and practice at the intersection of emerging technologies and human-centred disaster risk reduction.

1. Integrating Behavioural Science into Early Warning Design

One of the most pressing research needs lies in deeper integration of behavioural science into early warning system design. Although risk communication research has identified key factors influencing warning response—such as trust, experience, and message framing—these insights are often applied retrospectively rather than embedded into system development. Future research should adopt experimental and longitudinal designs to examine how different populations interpret and respond to earthquake and flash-flood warnings under varying conditions of uncertainty and time pressure.

Particular attention should be given to cognitive load, stress responses, and decision heuristics during rapid-onset events. For earthquake early warning, where response windows are measured in seconds, understanding reflexive behaviour and automaticity is essential. For flash-flood warnings, research should explore how individuals weigh competing risks, such as property protection versus personal safety. Embedding behavioural insights into warning algorithms and message design represents a promising yet underexplored frontier.

2. Advancing Explainable and Ethical Artificial Intelligence

Artificial intelligence is poised to play an increasingly central role in MHEWS, yet its integration raises significant epistemic and ethical questions. Future research should prioritize the development of explainable AI models that balance predictive accuracy with interpretability. This is particularly important for early warning contexts, where decisions have immediate and potentially life-altering consequences.

Research is also needed to examine the ethical implications of automated warning systems, including issues of accountability, transparency, and bias. For example, machine learning models trained on historical data may inadvertently reproduce existing inequalities if vulnerable communities are underrepresented in training datasets. Interdisciplinary research combining computer science, ethics, and social science is essential to ensure that AI-driven early warning systems are both effective and socially responsible.

3. Multi-Hazard Interactions and Cascading Risk Modelling

Most existing early warning systems and models focus on individual hazards, despite growing recognition of cascading and compound risks. Future research should develop integrated modelling approaches that capture interactions between earthquakes, floods, landslides, infrastructure failures, and secondary hazards. For instance, seismic events may damage flood control infrastructure, increasing downstream flood risk, while intense rainfall can exacerbate slope instability in seismically active regions.

Advances in system dynamics modelling, agent-based modelling, and network analysis offer promising tools for exploring these interactions. Research should also examine how cascading risks are communicated to decision-makers and the public without overwhelming them. Developing intuitive representations of complex risk interactions remains a critical challenge for both modelling and communication.

4. Evaluating Equity and Inclusivity in Early Warning Outcomes

While equity is increasingly emphasized in policy discourse, empirical research on how early warning systems affect different social groups remains limited. Future studies should move beyond access metrics (e.g., receipt of alerts) to examine differential outcomes, such as response speed, protective action, and post-event recovery across socio-economic, gender, age, and disability dimensions.

Mixed-methods research combining quantitative data with qualitative insights can illuminate how structural inequalities shape warning effectiveness. Participatory research approaches that involve communities in defining research questions and evaluating outcomes are particularly valuable. Such work can inform the design of targeted interventions that enhance inclusivity and reduce vulnerability within MHEWS.

5. Institutional Learning and Governance Innovation

Another important research direction concerns institutional dynamics and governance innovation. Comparative studies across countries and regions can shed light on how different governance arrangements influence early warning performance.

Research should examine the effectiveness of centralized versus decentralized warning systems, the role of legal mandates, and mechanisms for inter-agency coordination.

Longitudinal research on institutional learning—how organizations adapt after successes or failures—is particularly needed. Understanding why some systems evolve and improve over time while others stagnate can inform capacity-building strategies and policy reform. Evaluating the impact of international frameworks, such as the Sendai Framework, on national and local practice remains an important area of inquiry.

6. Leveraging Emerging Communication Ecologies

The communication landscape is evolving rapidly, with new platforms, formats, and social practices continually emerging. Future research should explore how early warning messages circulate within complex information ecosystems, including social media, messaging applications, and informal networks. Network analysis and digital ethnography offer valuable methods for understanding information diffusion, rumor propagation, and trust dynamics during crises.

Research is also needed to assess the effectiveness of innovative communication tools, such as location-based services, wearable devices, and immersive technologies, in enhancing warning comprehension and response. Importantly, studies should consider the long-term implications of these tools for privacy, surveillance, and public trust.

7. Co-Production of Knowledge and Practice

Finally, future research should emphasize the co-production of knowledge between scientists, practitioners, policymakers, and communities. Early warning systems operate at the interface of science and society, and their effectiveness depends on continuous dialogue and collaboration across stakeholders. Action research, living labs, and pilot projects can serve as platforms for testing new technologies and communication strategies in real-world contexts.

By embracing co-production, research can move beyond theoretical insights to generate practical, context-sensitive solutions. Such approaches align with the broader shift toward transdisciplinary research in disaster risk reduction and hold significant promise for advancing resilient, people-centred MHEWS.

LIMITATIONS OF THE STUDY

This chapter is primarily conceptual and literature-based, drawing on existing research and case studies to analyse the socio-technical dimensions of multi-hazard early warning systems. While the discussion integrates insights from diverse disciplines, it does not include empirical field data or primary case-study analysis. Future research may benefit from empirical validation of the proposed framework through field experiments, comparative country studies, or participatory research involving communities exposed to earthquake and flash-flood risks.

CONCLUSION

This chapter has examined the critical transition from sensor-based detection to human action in the context of earthquake and flash-flood early warning systems, emphasizing that resilience is fundamentally a socio-technical achievement rather than a purely technological one. While advances in sensing technologies, predictive analytics, and digital communication have significantly enhanced the technical capacity to detect hazards and issue warnings, these capabilities alone do not guarantee reduced disaster losses. The effectiveness of early warning systems ultimately depends on how warnings are interpreted, trusted, and acted upon by individuals, communities, and institutions operating under conditions of uncertainty and time pressure.

By situating multi-hazard early warning systems within global disaster risk reduction frameworks, this chapter has highlighted the growing recognition of people-centred approaches as essential to achieving the goals of the Sendai Framework. Earthquakes and flash floods illustrate both the promise and the limitations of current early warning practices. Earthquake early warning systems demonstrate the life-saving potential of seconds-long alerts, yet also expose challenges related to reflexive behaviour, public familiarity, and false alarm tolerance. Flash-flood early warning systems offer longer lead times but face difficulties in managing uncertainty, spatial variability, and complex decision-making demands. In both cases, the gap between detection and effective response remains a central concern.

The theoretical and conceptual frameworks developed in this chapter underscore the importance of integrating socio-technical systems theory, risk communication, and human-centred design. Early warning systems function as dynamic interactions among sensors, models, institutions, communication channels, and communities. Misalignment at any point in this chain can undermine system performance, regardless of technological sophistication. Conversely, alignment across these dimensions can transform warnings into actionable knowledge that supports timely protective behaviour.

The analysis of emerging technologies reveals that tools such as artificial intelligence, IoT-based sensing, and digital dissemination platforms offer powerful opportunities to enhance early warning effectiveness. However, these tools also introduce new risks related to opacity, inequity, and information overload. Addressing these risks requires transparent governance, explainable decision-support systems, and inclusive communication strategies that prioritize accessibility and trust. The recommendations outlined in this chapter emphasize the need for end-to-end integration, institutional coordination, community engagement, and continuous learning to ensure that technological innovation translates into real-world resilience.

Looking forward, the chapter argues that the future of multi-hazard early warning systems lies in their capacity to adapt to evolving hazard landscapes, technological change, and social diversity. Climate change, urbanization, and digital transformation will continue to reshape risk profiles and communication ecologies, demanding flexible and responsive warning systems. Research, policy, and practice must therefore converge around a shared commitment to people-centred, equitable, and action-oriented early warning.

In conclusion, moving from sensors to action requires reframing early warning systems as collective social interventions rather than isolated technical solutions. When emerging technologies are thoughtfully integrated with human-centred dissemination and communication strategies, early warning systems can fulfil their potential as powerful instruments for building resilience, saving lives, and reducing the human and economic costs of earthquakes and flash floods in an increasingly hazardous world.

REFERENCES

Akerkar, R. (2019). *Artificial intelligence for disaster management*. Springer. https://link.springer.com/book/10.1007/978-3-031-91084-5

Allen, R. M., & Melgar, D. (2019). Earthquake early warning: Advances, scientific challenges, and societal needs. *Annual Review of Earth and Planetary Sciences*, *47*(1), 361–388. DOI: 10.1146/annurev-earth-053018-060457

Budimir, M., Šakić-Trogrlić, R., Almeida, C., Arestegui, M., Chuquisengo Vásquez, O., Cisneros, A., Cuba Iriarte, M., Dia, A., Lizon, L., Madueño, G., Ndiaye, A., Ordoñez Caldas, M., Rahman, T., RanaTharu, B., Sall, A., Uprety, D., Anderson, C., & McQuistan, C. (2025). Opportunities and challenges for people-centered multi-hazard early warning systems: Perspectives from the Global South. *iScience*, *28*(5), 112353. DOI: 10.1016/j.isci.2025.112353 PMID: 40292326

Creutin, J. D., Borga, M., Anquetin, S., Gaume, E., Gruntfest, E., & Lutoff, C. (2013). A space–time framework for analyzing human anticipation of flash floods. *Journal of Hydrology (Amsterdam)*, *482*, 14–24. DOI: 10.1016/j.jhydrol.2012.11.009

Domingos, B., & Nagamatsu, S. (2024). Why are community-based early warning systems inadequate? A case study of the Licungo River Basin in Mozambique. *Journal of Integrated Disaster Risk Management*, *14*(2), 95–126. DOI: 10.5595/001c.123357

Hou, R., Liu, Y., Zhang, H., & Li, J. (2024). The mediating role of satisfaction with warning services and community engagement in disaster response behaviour. *International Journal of Disaster Risk Reduction*, *95*, 104187. DOI: 10.1016/j.ijdrr.2023.104187

Kelman, I. (2020). *Disaster by choice: How our actions turn natural hazards into catastrophes*. Oxford University Press., https://global.oup.com/academic/product/disaster-by-choice-9780198841357

Marengo, J. A., Alcântara, E. H., Moraes, O. L., Martinez, R., Seluchi, M., Alvalá, R. C., Dolif, G., & Gonçalves, D. (2025). Early warning services for disaster risk reduction in Brazil: The experience of CEMADEN during the floods of Rio Grande do Sul of May 2024. *International Journal of Disaster Risk Reduction*, *126*, 105645. DOI: 10.1016/j.ijdrr.2025.105645

Mileti, D. S., & Sorensen, J. H. (1990). *Communication of emergency public warnings: A social science perspective and state-of-the-art assessment*. Oak Ridge National Laboratory., DOI: 10.2172/6137387

Mosavi, A., Ozturk, P., & Chau, K. W. (2018). Flood prediction using machine learning models: Literature review. *Water (Basel), 10*(11), 1536. DOI: 10.3390/w10111536

Norman, D. A. (2013). *The design of everyday things* (Revised and expanded ed.). Basic Books. https://www.hachettebookgroup.com/titles/don-norman/the-design-of-everyday-things/9780465050659/

Paton, D. (2019). Disaster risk reduction: Psychological perspectives on preparedness. *Australian Journal of Psychology*, *71*(4), 327–341. DOI: 10.1111/ajpy.12237

Pham, T. D. M., Shaw, R., & Takeuchi, Y. (2024). Community-based early warning systems in a changing climate: Enhancing resilience through participatory approaches. *Climate and Development*, *16*(4), 389–402. DOI: 10.1080/17565529.2024.2307398

Reichstein, M., Camps-Valls, G., Stevens, B., Jung, M., & Denzler, J. (2025). Early warning of complex climate risks with integrated artificial intelligence models. *Nature Communications*, *16*(1), 2145. DOI: 10.1038/s41467-025-57640-w PMID: 40089483

Renn, O. (2008). *Risk governance: Coping with uncertainty in a complex world.* Earthscan. https://www.taylorfrancis.com/books/mono/10.4324/9781849772440/risk-governance-ortwin-renn

Tiggeloven, T., Pfeiffer, S., Matanó, A., van den Homberg, M., Thalheimer, L., Reichstein, M., & Torresan, S. (2025). The role of artificial intelligence for early warning systems: Status, applicability, guardrails, and ways forward. *iScience*, *28*(11), 113689. DOI: 10.1016/j.isci.2025.113689 PMID: 41280665

Trist, E. L., & Bamforth, K. W. (1951). Some social and psychological consequences of the longwall method of coal-getting. *Human Relations*, *4*(1), 3–38. DOI: 10.1177/001872675100400101

United Nations Office for Disaster Risk Reduction. (2015). *Sendai framework for disaster risk reduction 2015–2030.* United Nations. https://www.undrr.org/publication/sendai-framework-disaster-risk-reduction-2015-2030

United Nations Office for Disaster Risk Reduction. (2022). *Global assessment report on disaster risk reduction 2022: Our world at risk.* United Nations. https://www.undrr.org/gar

United Nations Office for Disaster Risk Reduction. (2023). *Early warnings for all: Executive action plan 2023–2027.* United Nations. https://www.preventionweb.net/publication/early-warnings-all-executive-action-plan-2023-2027

Wachinger, G., Renn, O., Begg, C., & Kuhlicke, C. (2013). The risk perception paradox—Implications for governance and communication of natural hazards. *Risk Analysis, 33*(6), 1049–1065. DOI: 10.1111/j.1539-6924.2012.01942.x PMID: 23278120

Wibowo, A., Amri, A., & Rahman, A. (2025). Leveraging artificial intelligence in disaster management: A comprehensive bibliometric review. *Jàambá, 17*(1), a1776. DOI: 10.4102/jamba.v17i1.1776 PMID: 40357012

World Meteorological Organization. (2020). *Multi-hazard early warning systems: A checklist.* WMO. https://community.wmo.int/sites/default/files/EWS_Checklist_0.pdf

World Meteorological Organization & United Nations Office for Disaster Risk Reduction. (2023). *Early warnings for all: Executive action plan 2023–2027.* Geneva: United Nations. https://digitallibrary.un.org/record/4001006

KEY TERMS AND DEFINITIONS

Artificial Intelligence: Artificial Intelligence (AI) refers to computational systems and algorithms capable of performing tasks that typically require human intelligence, such as pattern recognition, learning, prediction, and decision-making. In early warning systems, AI is used to process large volumes of sensor and satellite data, enhance hazard detection accuracy, reduce false alarms, and support rapid decision-making under uncertainty.

Disaster Communication: Disaster communication is the process of creating, transmitting, and interpreting information related to hazards, risks, and protective actions before, during, and after disaster events. It encompasses message design, choice of communication channels, trust-building, and feedback mechanisms to ensure that warnings are understood and acted upon by diverse populations.

Early Warning Systems: Early warning systems are integrated sets of monitoring, forecasting, communication, and response mechanisms designed to detect hazards and provide timely, actionable information to enable protective action. Effective early warning systems function end-to-end, linking scientific detection with institutional decision-making and community response capacity.

Earthquake Preparedness: Earthquake preparedness refers to the set of knowledge, plans, skills, and behaviours that enable individuals, communities, and institutions to respond effectively to earthquake warnings and ground shaking. This includes awareness of protective actions, participation in drills, structural safety measures, and familiarity with early warning alerts.

Flash-Flood Risk: Flash-flood risk is the potential for rapid-onset flooding caused by intense rainfall, dam or levee failure, or sudden water release, combined with exposure and vulnerability of people, infrastructure, and livelihoods. It is characterized by high uncertainty, limited reaction time, and significant spatial variability, making effective early warning and communication critical.

Human-Centred Design: Human-centred design is an approach to system development that prioritizes users' needs, capabilities, contexts, and limitations throughout the design and implementation process. In early warning systems, it emphasizes accessibility, clarity, inclusivity, and actionability of warnings to ensure that technological solutions align with human behaviour and decision-making.

Multi-Hazard Resilience: Multi-hazard resilience is the capacity of individuals, communities, and systems to anticipate, withstand, respond to, and recover from multiple, interacting hazards. It involves integrated planning, adaptive governance, and coordinated early warning systems that address overlapping risks rather than treating hazards in isolation.

Sensor Technologies: Sensor technologies refer to the instruments and systems used to detect and measure physical, environmental, and hydrological parameters related to hazards, such as ground motion, rainfall, river levels, and soil moisture. These technologies form the foundational data layer of early warning systems, enabling real-time monitoring and hazard detection.

Chapter 12
Drought Early Warning: A Key Input in Drought Preparedness Decision-Making

Jaime Santos-Reyes
http://orcid.org/0000-0002-3758-9862
SEPI-ESIME, ZAC, Instituto Politecnico Nacional, Mexico

Yunue Garza Pimentel
http://orcid.org/0009-0001-2080-5776
Instituto Politecnico Nacional, Mexico

Marcos Angel Gonzalez Olvera
http://orcid.org/0000-0001-8174-4391
Universidad Autonoma de la Ciudad de México, Mexico

José Daniel Velázquez Martínez
http://orcid.org/0000-0002-8786-0630
Universidad Autónoma de Nuevo León, Mexico

ABSTRACT

Droughts have considerable affectations to economic, environmental, and water shortages in urban environments. Traditionally, drought management has been reactive rather than proactive in mitigating the impact of drought hazard. The chapter presents a drought disaster management system (DDMS) where a drought early warning system (DEWS) plays a key role in decision-making to mitigate the impact of droughts for the case of Mexico City. The proactive approach is based on risk and vulnerability assessment. As part of the function of DEWS, an illustrative analysis of rainfall (RF) prediction has been conducted by employing fuzzy logic theory. The results produced an acceptable fuzzy model and have the potential not only to predict RF but also droughts. Furthermore, the results highlight the impor-

DOI: 10.4018/979-8-3373-7345-4.ch012

right © 2026, IGI Global Scientific Publishing. Copying or distributing in print or electronic forms without written permission of IGI Global Scientific Publishing is prohibited. Use of this chapter to in generative artificial intelligence (AI) technologies is expressly prohibited. The publisher reserves all rights to license its use for generative AI training and machine learning model development.

tance of DEWS in decision-making. Overall, to be able to mitigate the impact of droughts in Mexico City, it becomes necessary to have a coherent DDSM system. More generally, key decision-makers of the megalopolis should design and implement a multi-hazard early warning system.

INTRODUCTION

According to the most recent report of the Intergovernmental Panel on Climate Change (IPCC), climate change largely impacts urban life and is accelerating (IPCC, 2021). Highly publicized events have demonstrated the impact of this global phenomenon. For example, the 2025 Texas flash flood disaster that claimed over 100 lives; it is believed that torrential rains, in a matter of hours, sent a wall of water surging through Kerr County's Guadalupe River basin at around 04 a.m. where many of the victims were residents and young children at vacation camps were swept up in the swollen river waters (WMO, 2025; Mishra, 2025). Regarding the unusual 2025 heavy rains during July and August in Mexico City, the head of the government of the capital city said "Yesterday (10 August 2025) we had the heaviest rainfall; we break records every day. Yesterday we had 84.5 millimeters…Since 1952, the heaviest rainfall had been 67 millimeters", She warns "Mexico City's rainfall is breaking records 'every day': 'September will be worse than August'" (El Finaciero, 2025); the unstoppable rain has caused disruptions of all modes of the transport system, floods, among others. Similar unpredicted events are occurring worldwide and demonstrate that climate change affects the environment, including rising temperatures, droughts, floods, sea level rise, among others (Kumar & Arya, 2025).

Given the consequences of disasters triggered by multi-hazards (floods, earthquakes, tsunamis, landslides, drought, among others), multi-hazard early warning systems (MHEWS), become an essential element that could alert people to the potential hazards and therefore mitigate the impact of these. In fact, it has been argued that MHEWS have saved lives. It is defined as "multi-hazard early warning systems address several hazards and/or impacts of similar or different type in contexts where hazardous events may occur alone, simultaneously, cascading or cumulatively over time, and taking into account the potential interrelated effects" (UNDRR & WMO, 2025). In 2022, the Early Warnings for all (EW4All) initiative was launched by UN Secretary-General Antonio Guterres. The initiative is built on four elements of EWS, i.e.: disaster risk knowledge; observations, monitoring, analysis and forecasting; warning dissemination & communication; and preparedness & response capabilities (UNDRR & WMO, 2025).

Droughts are regarded as the most expensive disasters worldwide, affecting local communities, cities, agricultural yield, transportation systems, among others (Kumar

& Arya, 2025; Wang et al., 2026). Drought is defined as "a period of abnormally dry weather characterized by a prolonged deficiency of precipitation below a certain threshold over a large area and a period longer than a month", and "occurs when low water supply becomes evident and is associated with the effects of periods of precipitation shortfalls on surface or subsurface water supply", respectively (Murray et al., 2021, pp. 102). There are four types of droughts: meteorological, hydrological, agricultural and socioeconomic (Murray et al., 2021). In the present study, meteorological and hydrological droughts will be addressed in the context of the case study.

It may be argued that to be able to gain a better understanding of drought, rainfall (RF) prediction in the short and long-term (e.g. RR (rainfall-runoff)) may be needed (Irwan et al., 2025). Regarding hydrological droughts, the impact of water shortages in large urban environments such as Mexico City and their contributing factors should be also understood (Huerta-Vergara et al., 2022). That is, there is a need to see drought risk from a holistic perspective, i.e., the design of a systemic drought disaster management system (DDMS). Traditionally, drought response has been reactive, i.e., dealing with the consequences following a drought event-crisis management. The proactive approach, on the other hand, deals with drought risk management, involving several processes such as preparedness planning, mitigation, monitoring and early warning and prediction to reduce the impacts of drought—both now and into the future (Wilhite et al., 2005a,b); in other words, a proactive approach implies decision-making based on drought risk and vulnerability assessment. Ultimately, a DDSM should include both reactive and proactive approaches, i.e., decision-makers should put more effort into mitigation and preparedness (Wilhite et al., 2005a,b).

To mitigate and prepare for droughts, EWSs become crucial in such process. For example, RF, drought vulnerability assessment, among others, should be part of the function of such a system. A great deal of effort has been spent on RF, RR, and drought prediction (Irwan et al., 2025; Kundu et al., 2023; Sulaiman et al., 2022; Rathnayake et al., 2023; Wen-Bing 2024; Pham et al., 2020; Zahran et al., 2023; Janarthanan et al., 2021; Althoff et al., 2021; Manna and Anitha, 2023; Lakshmi et al., 2022; Raham, 2020; Li et al., 2022; Singh et al., 2022). However, it may be argued that their findings are seen as an isolation and not necessarily part of a coherent DDMS system. Furthermore, research on drought risk from a systemic perspective is lacking in the context of Mexico in general and Mexico City in particular.

The lack of water in the capital city is a historical and ecological paradox stemming from the dryness of the lakes where the city was built on, i.e. over the centuries, the lakes were drained to allow for construction. Overpopulation and uncontrolled growth in many parts of the city have led to the depletion of water resources. Currently, more than nine million people live in the capital city, a figure that rises to 20 million if the inhabitants of the Metropolitan Area of the Valley of Mexico. Another 'cruel' paradox is that during the rainy season (June - September), the city floods (El Finaciero, 2025). Again, these events demonstrate the lack of effective DDSM.

The chapter addresses drought risk management from a holistic perspective; in particular, the approach of the 10-step process as an organizational framework of drought planning put forward by Wilhite, (2000) has been adopted in the present case study. Regarding drought early warning system, the framework to early warning proposed by Diekert et al., (2024) has been adopted in the present work.

Regarding RF prediction, an Artificial Intelligence (AI) tool, namely fuzzy logic has been adopted for the analysis and for the case of three states of the country, i.e., Mexico City, State of Mexico and Michoacan State. The analysis considered the following: temperature (T), wind speed (WS), atmospheric pressure (AP), relative humidity (RH), dew point (DP), and the output variable, rainfall (RF). The aims of the chapter are:

1. To employ the 10-Step drought planning process proposed by Wilhite, (2000) for the present case study.
2. To employ the proposed early warning system framework proposed by Diekert et al., (2024) for the present case study.
3. To construct a fuzzy ruled-based model to RF prediction for the case of the three states.
4. To use the constructed fuzzy logic models to predict RF by considering the input variables (T, WS, AP, RH, DP) and the RF prediction as the outcome variable.

The chapter is organized as follows: first, a very brief description of the areas of study is given. Second, a description of the employed materials and methods are briefly described; third, the most relevant results are presented. Fourth, a discussion of the main findings is presented in the section; also, some limitations and future directions are given within the discussion section. Finally, some relevant conclusions are reported.

AREA OF STUDY

Mexico City Metropolitan Area

Mexico City, the capital of the country is divided into sixteen municipalities, i.e.: Azcapotzalco (AZ), Coyoacan (CO), Cuajimalpa (CU), Guastavo A Madero (GAM), Iztacalco (IZ), Iztapalapa (IZTA), Magdalena Contraras (MC), Milpa Alta (MA), Alvaro Obregon (AO), Tlahuac (TLH), Tlalpan (TLA), Xochimilco (XO),

Benito Juarez (BJ), Cuauhtemoc (CUA), Miguel Hidalgo (MH), and Venustiano Carranza (VC).

According to the 2020 census, the capital city's population is around 9,209,944 inhabitants, representing 7.3% of the country's total population of 126 million, making it the second most populous city after the State of Mexico (INEGI, 2021). It constitutes the largest urban center within the Mexico City Metropolitan Area (MCMA), with a quarter of the urban population of over 20 million residents; it is by far the biggest urban agglomeration in the country and the biggest in Latin America (UN, 2025).

The Cutzamala Water Supply System

The Cutzamala system is comprised of a network of seven dams; two of them located in the State of Michoacan and four in the State of Mexico. The system provides water to eleven out of sixteen municipalities of the capital city, i.e., it provides 30% of drinking water, 70% comes from wells and aquifers within the MCMA. It is considered one of the largest water supply systems in the world by transporting 480 hm^3/year of water (CONAGUA, 2015).

One of the major challenges the megalopolis face is the water supply and distribution for urban and industrial consumption, stormwater disposal, and wastewater treatment. The metropolitan area continues to expand and demands greater volumes of water. Given this situation, the supply of drinking water represents an undisputed priority. On the one hand, new water demands must be met, and on the other, the severe water overexploitation to which the MCMA has been subjected must be gradually reduced. Due to the location of the MCMA, there are no suitable sites for harvesting rainwater. This means that, of the average annual rainfall of 6.828 billion cubic meters, 72% evaporates and 11% infiltrates the subsoil. The remaining 17% that runs off the surface, a small portion is treated for use; the remainder is channeled into the general drainage system to prevent flooding (CONAGUA, 2005).

Geographic Location of the States Considered in the Case Study

Figure 1 shows the geographical locations of the states that are considered in the present case study. The State of Mexico's coordinates are the following: Longitude 100º36'47'' W and Latitude 20º17'09'' N. The state represents 1.1% of the country's total surface area. The average annual temperature is around 14.7°C. The lowest temperature occurs in January and February with an average temperature of around 3°C. The average maximum temperature is 25°C and occurs in April and

May. Rainfall is concentrated during the summer, i.e., from June to September, with an average annual precipitation of 900mm. (SARACS, 2025).

Figure 1. Geographic location state of Mexico (left), Michacan State (center) & Mexico City (right)

(SARACS, 2025)

Michoacan State's coordinates are Longitude 103º44'17'' W and Latitude 20º23'40'' N. The State represents 3.0% of the country's total surface. The annual temperature varies between 12 °C and 18 °C with an average annual precipitation of 850 mm. The lowest temperature occurs in January, around 8°C, while the average maximum temperature is 31°C and occurs in April and May. Hotter climes are in the relative lowlands, with an average temperature of between 18 °C and 22 °C. There are some arid areas with annual precipitation between 500 and 700 mm. (SARACS, 2025).

Lastly, Mexico City's coordinates are Longitude 99º21'54'' W and Latitude 19º35'34'' N. The average annual temperature is around 16°C, the highest temperatures, exceeding 25°C, occur during the months March through May, and the lowest, around 5°C, in January. Rainfall is concentrated during the summer, from June to October, with an average annual precipitation of 600 mm. (SARACS, 2025).

MATERIALS & METHODS

Drought Disaster Risk Management System

Figure 2 illustrates a drought risk management cycle which aims at preventing or mitigating impacts through drought monitoring and early warning, planning & mitigation strategies (Wilhite et al., 2005a). The left-half of the cycle represents crisis management, which involves activities necessary to respond to th e drought impact. The right-half, on the other hand, represents preventive actions, based on risk management, to mitigate the impact of droughts to a community. It may be argued that the former half of the cycle may be regarded as the most important aspect of drought risk management. That is, monitoring, prediction & early warning, planning and mitigation actions are essential to mitigate the impact of future droughts (Wilhite et al., 2005a).

Figure 2. Drought disaster management system

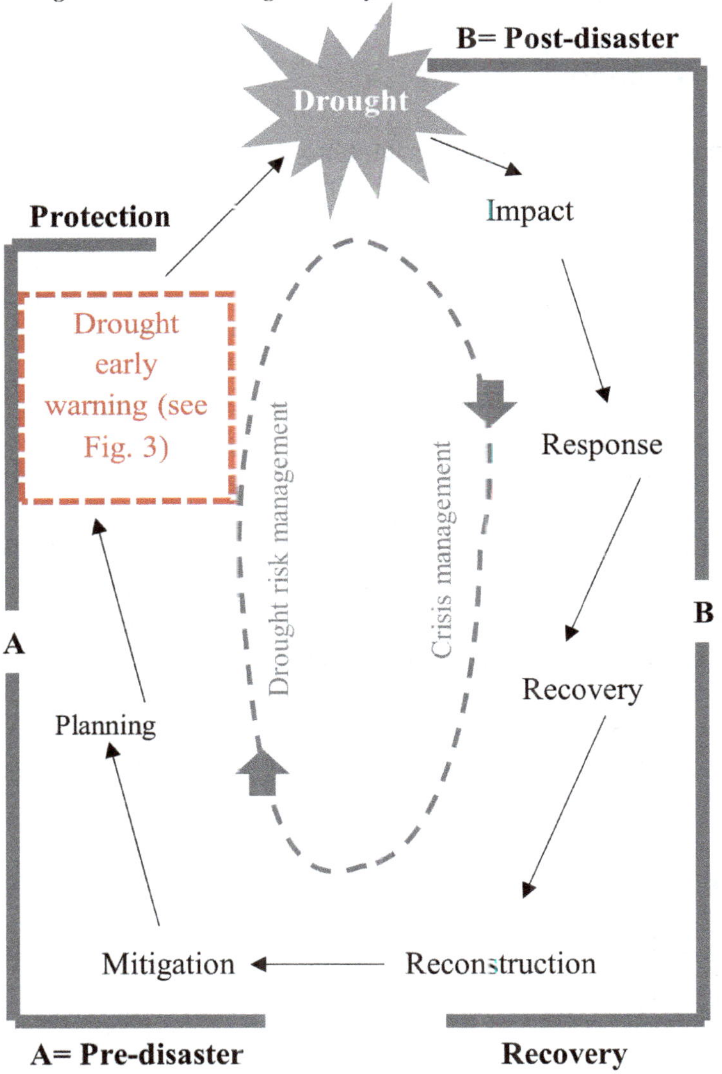

(adapted from Wilhite et al., 2005a,b)

Drought monitoring deals with a continuous assessment of key variables or indicators such as drought severity, location (city, region), and impact (see Table 1 in a later section). Having information about these indicators helps to elicit an effective response called *early warning*. The authors argue that a drought early warning system (DEWS), which combines both assessment and decision-making response, is integral to effective drought risk management (Figs. 2). Decision-makers require accurate early warning information to implement effective drought policies and response and recovery programs. However, this is not the case of Mexico City and other states of the country.

Fuzzy Sets Theory

The concept of fuzzy set was first introduced by Lofti A. Zadeh (Zadeh, 1965). Fuzzy sets are an extension of the classical concept of set, also known as crisp set. Mathematically, it is defined as,

Let, X be a set, a fuzzy subset A of X is a subset of X whose membership is defined through a function $|\mu\_A{:}X \rightarrow |0,1||$ which associates an element $x \in X$ to a real number $\mu_A{:}X\ in\ |0,1|$. The value $\mu_A(x)$ indicates the degree of membership of x in A. Hence, a fuzzy set A, $A = \{((x,\mu_A(x))|x \in X)\}$.

Let, X be a set, a fuzzy subset A of X is a subset of X whose membership is defined through a function $|\mu\_A{:}X \rightarrow |0,1||$ which associates an element $x \in X$ to a real number $\mu_A{:}X\ in\ |0,1|$. The value $\mu_A(x)$ indicates the degree of membership of x in A. Hence, a fuzzy set A, $A = \{((x,\mu_A(x))|x \in X)\}$.

Since all information contained in a fuzzy set is described by its membership function, there are several membership functions that can be parameterized according to the complexity of the system under study; these could be one or multi-dimensional. Some examples of membership functions are triangular, trapezoidal, Gaussian, sigmoidal, generalized bell membership function. In the present case study, a triangular membership function has been employed for the analysis of rainfall prediction (Safar et al., 2019; Zahran et al., 2023; Janarthanan et al., 2021; Hasan et al., 2008).

$$\mu\left(x,a,b,c\right) = \begin{cases} 0\,,x \leq a \\ \frac{x-a}{b-a}\,,a \leq x \leq b \\ \frac{c-x}{c-b}\,,b \leq x \leq c \\ 0\,;c \leq x \end{cases}$$

An alternative expression when using min and max, $\mu(x,a,b,c) = max\left[min\left(\frac{x-a}{b-a},\frac{c-x}{c-b}\right),0\right]$. The parameters $\{a,b,c\}$, (with $a<b<c$) determine the x coordinates of the three corners of the underlying membership function.

Fuzzy Inference System

Fuzzy inference is the process of mapping functions from a given input to an output using fuzzy logic. Fuzzy inference systems (FIS) are also known as fuzzy rule-based systems, fuzzy model, fuzzy expert system, and fuzzy associative memory (Hasan et al., 2008; Mada et al., 2022). The steps of fuzzy reasoning are:

a) Fuzzification: The process of defining the linguistic variables, their term sets and membership functions based on the given raw data.
b) Inference (Rule Base): The process of defining the membership functions of the rules; determine which rules are firing for the given input.
c) Composition: The process of combining all fuzzy conclusions obtained by inference into a single conclusion, i.e., since different fuzzy rules might have different conclusions, consider all rules.
d) Defuzzification: The process of converting fuzzy data into numerical data (crisp value) as the final decision.

There are three FIS methods, namely Tsukamoto (Dore et al., 2023), Mamdani and Takagi-Sugeno-Kang (TSK) (Cao et al., 2024; Dore et al., 2023); for example, the Mamdani method consider the following steps: a) generating the fuzzy sets, input and out variables are divided into one or more fuzzy sets; b) Implication function, the employed function is $\mu_{A\cap B} = min\{\mu_A(x), \mu_B(x)\}$; c) composing the rules, if the system consists of several rules, the inference is obtained from the collection and correlation between rules. The Max method takes the solution of the fuzzy set by taking the maximum value of the rule, $\mu_{A\cup B} = max\{\mu_A(x), \mu_B(x)\}$; and d) defuzzification, the most common used methods are the 'Centre of Gravity' (COG) and Maximum method.

Performance evaluation

There are different statistical evaluation criteria to assess model development, including the Mean Squared Error (MSE), the Root Mean Squared Error (RMSE), Mean Absolute Error (MAE), Nash-Sutcliffe model coefficient (E), coefficient of determination (R^2), among others. In the present study, the first three evaluation metrics were used to measure the performance of the rainfall prediction models.

$$MSE = \frac{1}{n}\sum_{j=1}^{n}\left(y_j - \hat{y}_j\right)^2$$

$$RMSE = \sqrt{\frac{\sum_{j=1}^{n}\left(y_j - \hat{y}_j\right)^2}{n}}$$

$$MAPE = \frac{1}{n}\sum_{i=1}^{n}\left|\frac{y_j - \hat{y}_j}{y_j}\right| x\ 100\%$$

where, n is the number of the dataset, $\hat{y}_j$ and y_j denote the predicted and the observed rainfall, respectively.

In the present analysis data processing was carried out by using Matlab Software.

RESULTS

The section presents the results of DEWS and decision-making. It should be highlighted that DEWS is part of a holistic DDMS shown in Figure 2. The DEWS in the context of the case study is based on the 10-step drought planning as mentioned in the introduction section. The 10-steps drought planning process involved the following (Wilhite, 2000):

- "Step 1: appoint a drought task force."
- "Step 2: State the purpose and objectives of the drought plan."
- "Step 3: Seek stakeholder participation and resolve conflict."
- "Step 4: Inventory resources and identify groups at risk."
- "Step 5: Prepare/write the drought plan."
- "Step 6: Identify research needs and fill institutional gaps."
- "Step 7: Integrate science and policy."
- "Step 8: Publicize the drought plan, build public awareness."
- "Step 9: Develop education programs."
- "Step 10: Evaluate and revise drought plan."

A full account of 10-steps is given in the original source; Figure 3 and Table 1 present results in the context of the case study.

A later subsection presents the most relevant results of RF prediction analysis by employing fuzzy logic theory. It should be mentioned that the quantitative analysis is part of the functions associated with DEWS (Figure 3).

Figure 3. Drought early warning and decision-making (adapted from Diekert et al., 2024)

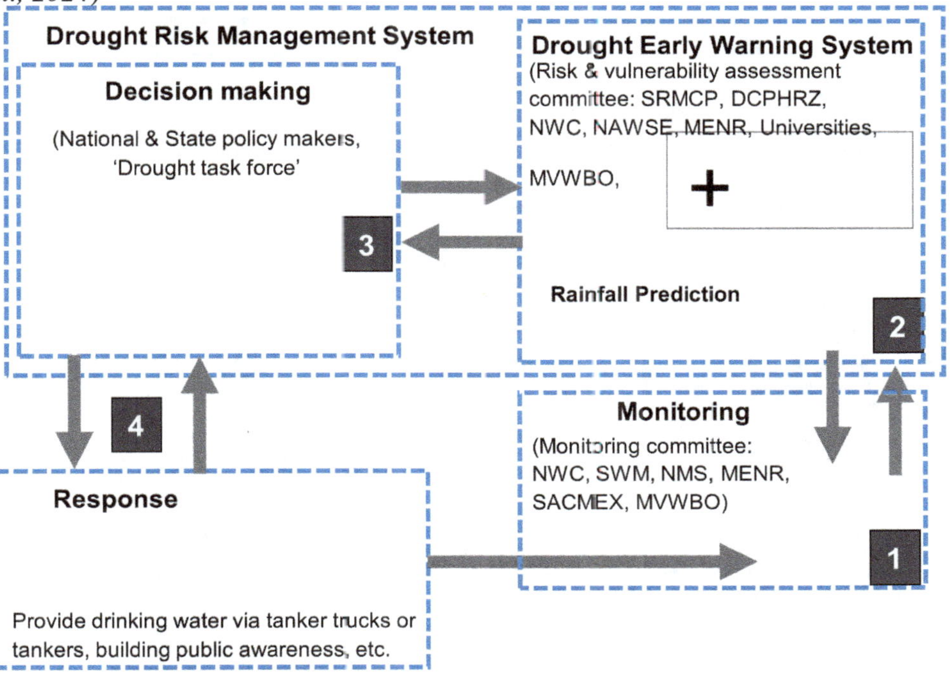

Table 1. Some examples of actions & key decision making (the table should be seen in the context of Figure 3)

	Examples/Activities	Examples Key Decision-Makers
'1'	*Monitoring-Meteorological variables:* Temperature (T), air relative humidity (RH), rainfall (RF), sunshine duration (SD), wind velocity (WV), cloud (C), humidity (H), atmospheric pressure (AP) (Zahran et al., 2023; Kebede et al., 2026). *Monitoring-Drought indicators:* Water storage (WS) = Volume stored in dams (A)/Total volume surface water consumptive (B); (A) = (% annual volume stored x maximum water storage capacity)/100%; (B) = Total volume consumptive x % surface water source. Further details of drought indicators are given by Neri and Magaña, (2016). *Monitoring-Water supply:* Baseline water stress, existence of adoption policies & plans, water quality, adequacy of water supply infrastructure network, access to clean water, among others (Weber and Bieroza, 2026). *Monitoring-Vulnerability indicators:* exposure (urbanization, population size, economic units), sensitivity (water leaks, water shortages, poor water quality), adaptive capacity (wastewater treatment, water availability, GDP) (Huerta-Vergara et al., 2022).	*'Monitoring committee'* (e.g., members of the National Water Commission (NWC), Secretary of Water Management (SWM), National Meteorological System (NMS), Mexico Valley Water Basin Organization MVWBO), Water System of Mexico City (SACMEX), Ministry of Environment and Natural Resources (MENR))
'2'	*Drought risk & vulnerability analysis:* Overall, drought monitoring deals with a continuous monitoring of the above key variables/indicators, and by using this information, decision-makers elicit an appropriate response is called *drought early warning system* (DEWS). Drought risk defined as the combination of the probability of drought and its consequence; vulnerability defined as the circumstances of a community that make it susceptible to the damaging effects of drought hazard (UNISDR, 2009). EWS defined as a "set of capacities needed to generate and disseminate timely and meaningful warning information to enable individuals, communities and organizations threatened by drought hazard to prepare and to act appropriately and in sufficient time to reduce the possibility of harm or loss." (UNISDR, 2009).	*'Drought risk & vulnerability assessment committee'* (e.g., members of the Secretariat of Risk Management and Civil Protection (SRMCP), Directorate of Civil Protection and High-Risk Zones (DCPHRZ), NWC, National Association of Water and Sanitation Entities (NAWSE), SACMEX, MVWBO, MENR, Universities, Environmental, industry & human rights groups)

Table 2. Some examples of actions & key decision making (Cont.)

	Examples/Activities	Examples Key Decision-Makers
'3'	*Drought planning based on drought risk assessment:* Regarding the 10-Step Planning process (Wilhite et al., 2000). Steps 1-4 focus on the process of selecting the right people and are brought together supplied with adequate data to make fair and equitable decisions when formulating and writing the actual drought plan. Step-5 describes the process of developing an organizational structure for competition of the task necessary to prepare the plan. Steps 6&7 detail the need for ongoing research and coordination between scientists and policy makers. Steps 8&9 stress the importance of promoting and testing the plan before drought occurs. Step-10 emphasizes revising the plan to keep it current and making an evaluation of its effectiveness in the post-drought period.	'Drought task force' (e.g., members of the SRMCP, DCPHRZ, NWC, SWM, MENR, NMS, NAWSE, SACMEX, MVWBO), Universities (Climatologists, risk & vulnerability assessment experts), Environmental, industry & human rights groups, among others.)
'4'	*Water demand management:* Rationing and supply schedules: establish specific schedules for water supply or limit the amount distributed to households to ensure equitable distribution of the resource. Fees: implement or increase fees for those who exceed an established water consumption limit, thus incentivizing reduced water use. *Emergency Actions:* Water distribution by tanker trucks: in critically affected areas, provide drinking water via tankers while the emergency is resolved. Installation of mobile water treatment plants: use temporary water treatment plants to ensure the supply of drinking water from available sources that require treatment. Cooperation with key sectors: coordinate with high-consumption sectors, such as industry and agriculture, to implement water use reduction measures.	Policy makers (Federal & State governors) 'Drought task force'

Table 3. Some examples of actions & key decision making (Cont.)

	Examples/Activities	Examples Key Decision-Makers
'4'	*Water Supply Management:* Reallocation of water sources: seek and utilize alternative sources of water supply, such as the reuse of treated water. Infrastructure repair: prioritize the repair of leaks in the water distribution network to reduce losses and increase supply efficiency. Rainwater harvesting: encourage and facilitate the harvesting and storage of rainwater in public and private buildings. Promoting water conservation: awareness campaigns on the importance of saving water, providing recommendations for reducing consumption in homes and businesses.	Policy makers (Federal & State governors) 'Drought task force'

Drought Early Warning System Example (see stage '2' of the DDMS in Figure 3)

The subsection presents some preliminary results of an ongoing research project on RF prediction for the case of Mexico City, Mexico and Michoacan States. A summary of the main findings of the three cases is presented.

Input Data

Temperature, wind speed, atmospheric pressure, dew point, relative humidity, and the observed rainfall were obtained from key organizations such as, National Water Commission (NWC), Secretary of Water Management (SWM), National Meteorological System (NMS), among others (Table 1). However, it should be highlighted that data related to the five mentioned variables were only for the year 2023 and for the case of Mexico City. Hence, the rainfall prediction for the three states will be based on three variables (temperature, wind speed, atmospheric pressure) and for the period ranging from 2013 to 2023.

Figure 4. Rainfall data from 2013 – 2023 for the three States considered in the case study.

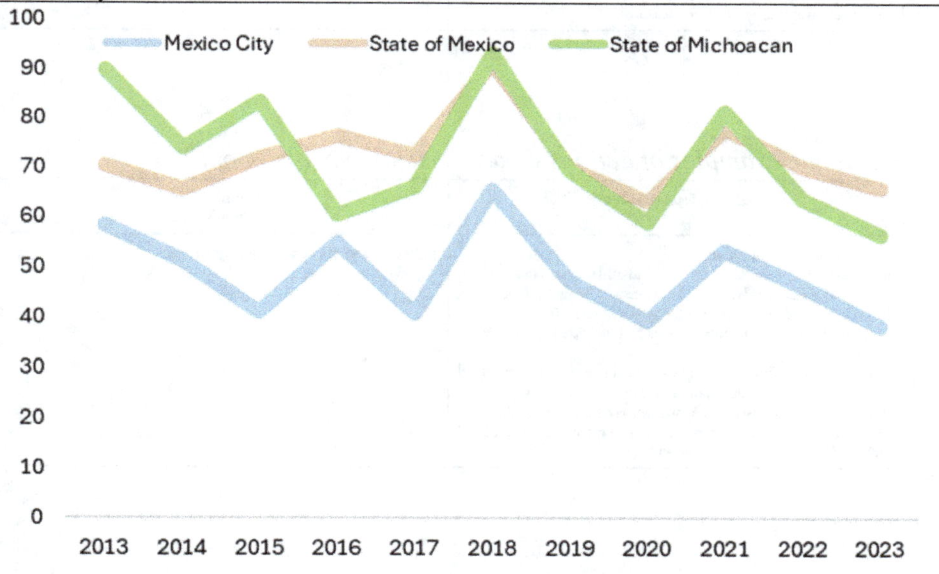

Figure 4 shows the distribution of rainfall data of the three states considered in the present case study. As expected, the rainfall for the eleven-year period is lower for the case of Mexico City than the two other states. This may be the explanation why the water supply to the capital city depends heavily on the Cutzamala network of seven damps located in Michoacan and Mexico States (Figure 1).

Fuzzification

In general, fuzzy inference models involve variables which are considered by experts in previous studies as contributing to rainfall (Safar et al., 2019; Zahran et al., 2023; Janarthanan et al., 2021; Hasan et al., 2008; Agboola et al., 2013; Rahman, 2020; Singla et al., 2019; Rakkasagi et al., 2024). The fuzzy logic models incorporate three and five input variables: temperature (T), wind speed (WS), atmospheric pressure (AP), relative humidity (RH), dew point (DP), and the output variable, rainfall (RF).

The range between the Min and Max value of the fuzzy variables are divided into suitable numbers which are denoted in ascending order. Table 2 and Figure 5 show the five linguistic terms adopted in the present analysis, i.e.: "Very Low" (VL), "Low" (L), "Normal" (N), "High" (H), and "Very High" (VH). Regarding each of the States considered in the study, the range of the variables are the following: Mexico City (T = 8 °C to 29 °C; AP = 1002 to 1028 hPa; WS = 9 to 60 km/h; RH = 15% to 87%; DP= -3 °C to 23 °C), Mexico State (T = 3 °C to 27 °C; AP = 1011 to 1031 hPa; WS = 9 to 93 km/h), and Michoacan State (T = 8 °C to 29 °C; AP = 1002 to 1028 hPa; WS = 6 to 93 km/h;)

Table 4. Fuzzy levels (Hasan, et al., 2008)

Abbreviation	Meaning
VL	Very low (VL)
L	Low (L)
N	Normal (N)
H	High (H)
VH	Very high (VH)

Figure 5. Example of the input variables membership diagram

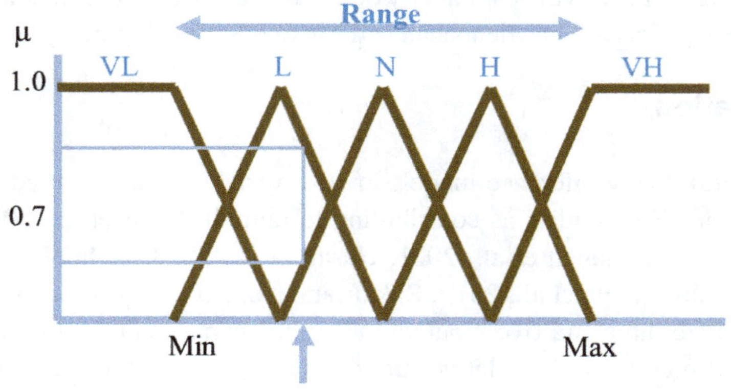

Figure 5 shows the FIS variables of the model. Based on the triangular membership function, Figures 6-8 show the membership function of each of the input and outcome variables considered in the case study. As an example, the mathematical formulae of the atmospheric pressure membership function are given below.

$$\mu_{VL}(AP) = \begin{cases} 155.15 - 0.15AP\,;\ \mathit{if}\ 1002 \le AP \le 1008.5 \\ 0\,;\ \mathit{if}\ PA > 1008.5 \end{cases}$$

$$\mu_{L}(AP) = \begin{cases} 0.15AP - 154.15\,;\ \mathit{if}\ 1002 \le AP \le 1008.5 \\ 156.15 - 0.15AP\,;\ \mathit{if}\ 1008.5 \le AP \le 1015 \\ 0\,;\ \mathit{if}\ AP > 1015 \end{cases}$$

$$\mu_{N}(AP) = \begin{cases} 0.15AP - 155{,}15\,;\ \mathit{if}\ 1008.5 \le AP \le 1015 \\ 157.15 - 0.15AP\,;\ \mathit{si}\ 1015 \le AP \le 1021.5 \\ 0\,;\ \mathit{si}\ AP > 1021.5 \end{cases}$$

$$\mu_{H}(AP) = \begin{cases} 0.15AP - 156.15\,;\ \mathit{if}\ 1015 \le AP \le 1021.5 \\ 158.15 - 0.15AP\,;\ \mathit{if}\ 1021.5 \le AP \le 1028 \\ 0\,;\ \mathit{if}\ AP < 1015 \end{cases}$$

$$\mu_{VH}(AP) = \begin{cases} 0.15AP - 158.15\,;\ \mathit{if}\ 1021.5 \le AP \le 1028 \\ 0\,;\ \mathit{if}\ AP > 1021.5 \end{cases}$$

Figure 6. Fuzzy-logic model for rainfall prediction with three variables (T, WS, & AP)

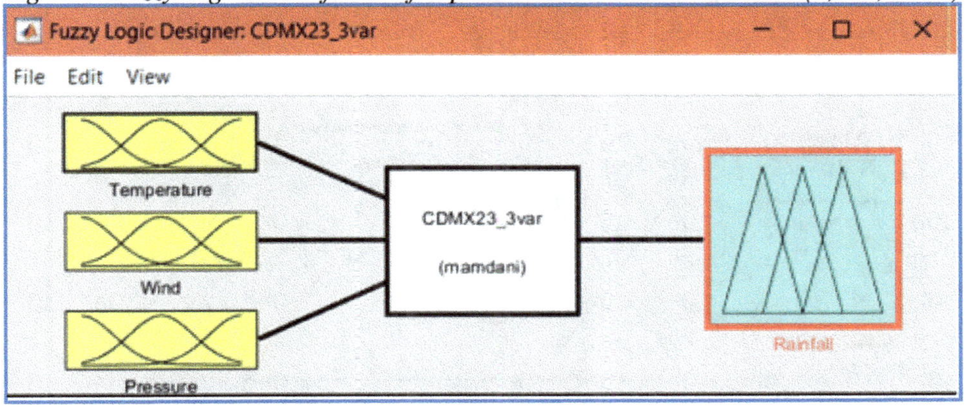

Figure 7. Membership functions associated with T, WS, & AP

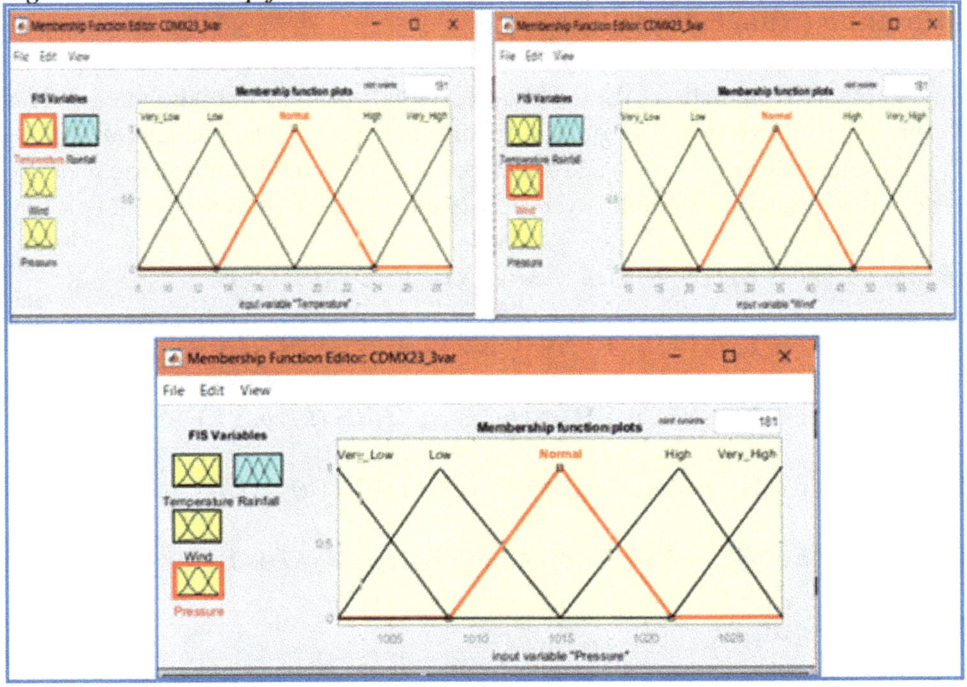

Figure 8. Membership function of rainfall output variable

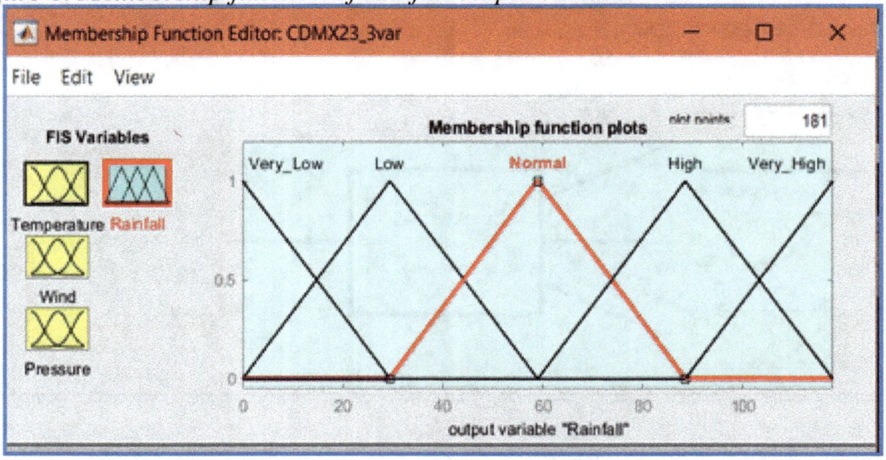

Inference (Rule Based)

To obtain RF predictions, the If-Then rule is used (it should be highlighted that for the present study a total of 243 rules were formulated). For example, regarding Rule 1, if the temperature is low (L) and wind speed is low (L) and atmospheric pressure is very low (VL), then rainfall is very low (Rule 1). Some rules are given below:

[*Rule 1*]: IF (*T* is **L**) ∩ (*WS* is **L**) ∩ (*AP* is **VL**) THEN (*RF* s **VL**)

[*Rule 2*]: IF (*T* is **L**) ∩ (*WS* is **H**) ∩ (*AP* is **L**) THEN (*RF* is **VL**)

[*Rule 3*]: IF (*T* is **L**) ∩ (*WS* is **VH**) ∩ (*AP* is **N**) THEN (*RF* is **VL**)

[*Rule 4*]: IF (*T* is **N**) ∩ (*WS* is **VL**) ∩ (*AP* is **H**) THEN (*RF* is **L**)

[*Rule 5*]: IF (*T* es **N**) ∩ (*WS* es **L**) ∩ (*AP* is **VH**) THEN (*RF* is **L**)

Predicted output

This final stage of the analysis, the fuzzy data is transformed into crisp data. Figures 9 and 10 present the results.

Figure 9. Predicted output rainfall

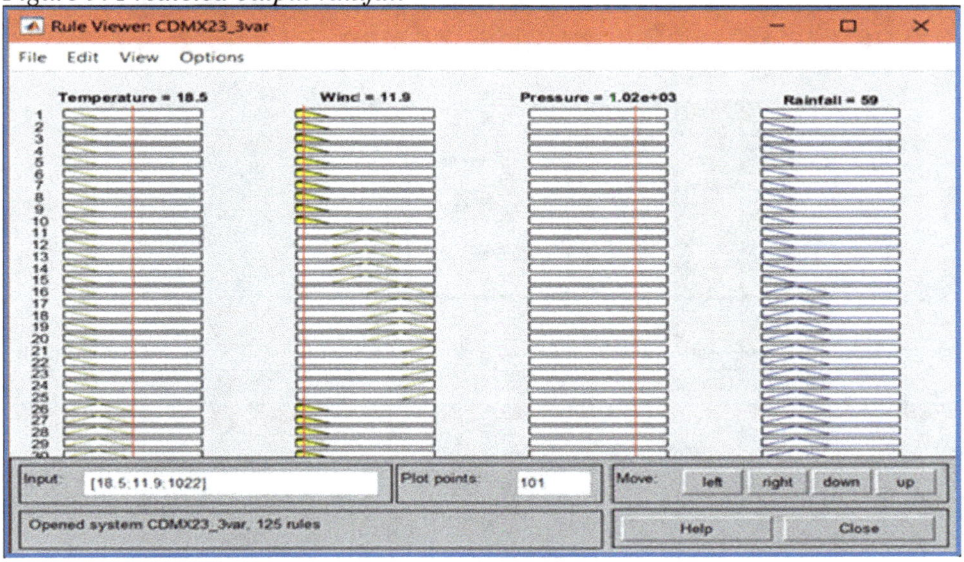

Case-1: Rainfall Prediction When Considering Five Variables

To elucidate the accuracy of rainfall prediction, two more variables were added to the model, i.e., relative humidity (RH) and dew point (DP) for the case of Mexico City. The results are shown in Figure 10.

Figure 10. Membership functions of the five variables and the rainfall output variable

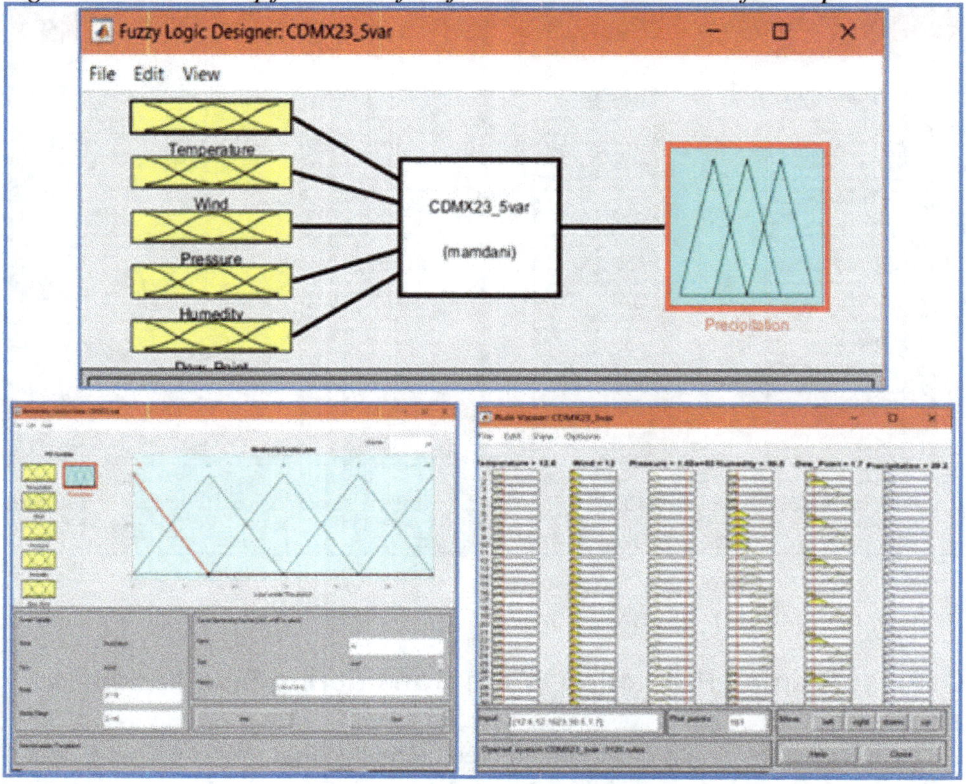

Models' Performance Evaluation

Table 3 shows the accuracy of the fuzzy logic models and Figures 11-13 shows the distribution of the predicted rainfall yearly when considering three variables. Overall, the metrics highlight and acceptable model prediction (see the Discussion subsection for further details on this).

Figure 11. Rainfall prediction at minimum and mean T, WV, & AP

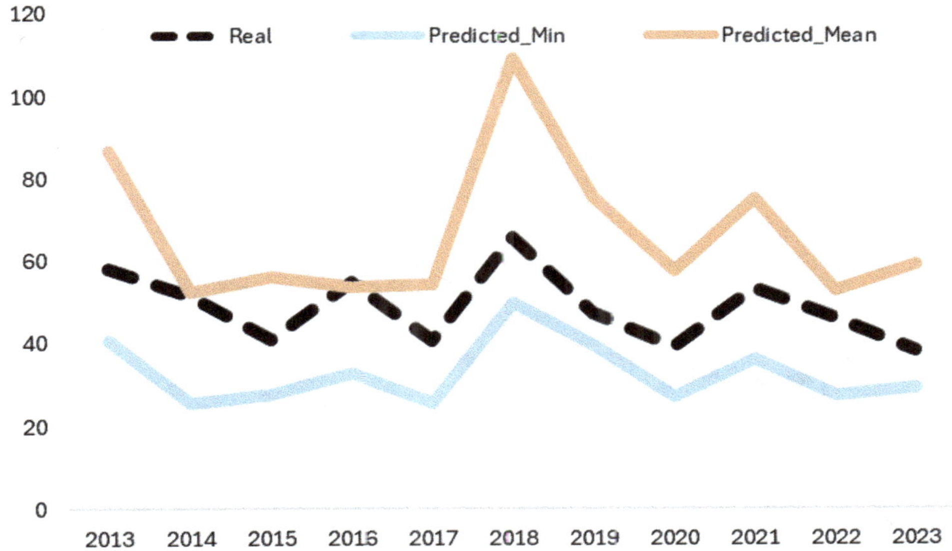

Figure 12. Rainfall prediction at minimum, mean & max of T, WV, & AP

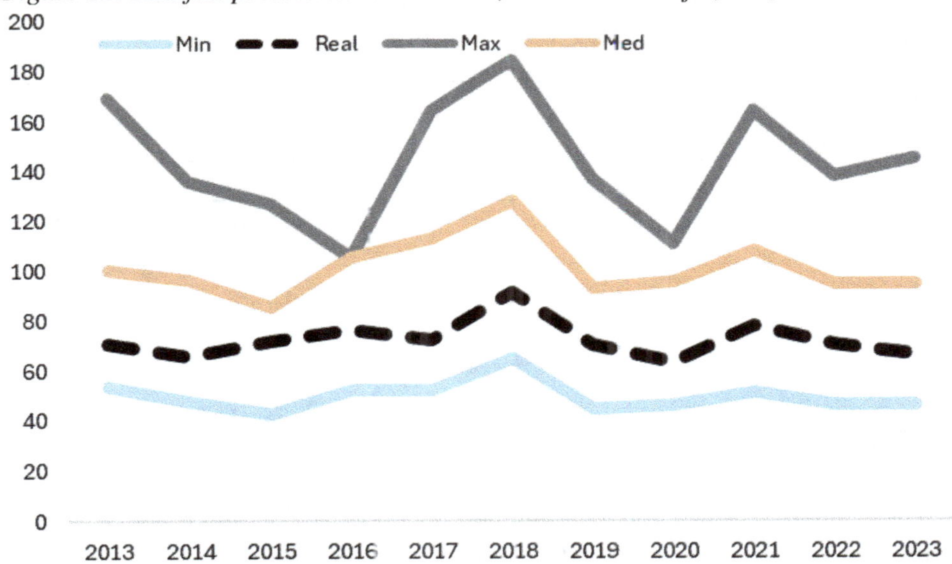

Figure 13. Rainfall prediction at minimum, max and mean of T, WV, & AP

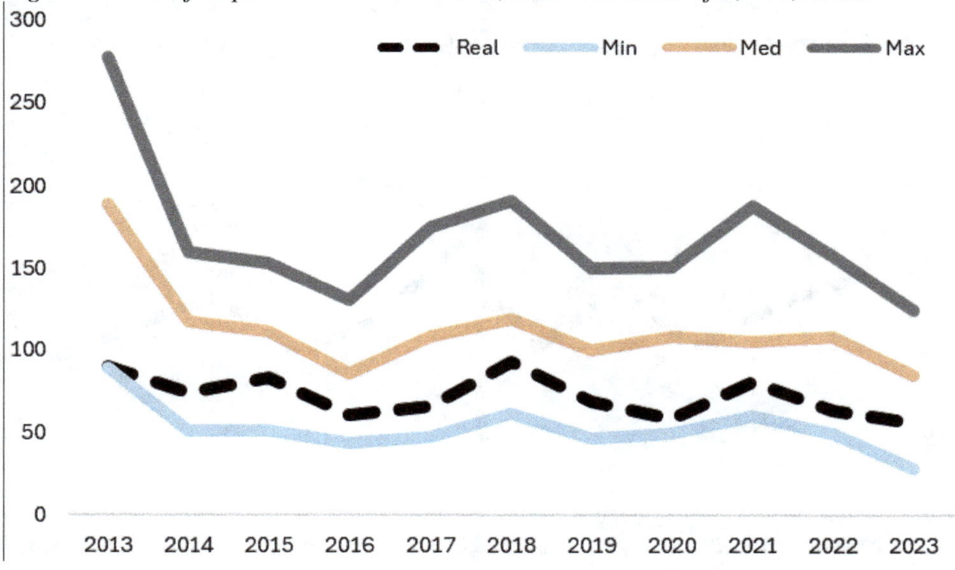

Table 5. Performance metrics of the fuzzy logic models on RF prediction

Variables (T, WS, AP) **2013 - 2023**	**MSE**	**RMSE**	**MAPE**
Model – 1 (Mexico City)	277.00	16.4	32.62
Model – 2 (State of Mexico)	531.70	23.10	31.40
Model – 3 (Michoacan State)	213.20	14.60	28.64
Variables (T, WS, AP, RH, DP) **2023**	**MSE**	**RMSE**	**MAPE**
Model – 4 (Mexico City)	79.92	8.94	23.50

DISCUSSION

The chapter presented some preliminary results of an ongoing research project on drought risk in Mexico City. The findings have highlighted the fact that water shortage in the megalopolis has multiple causes such as uncontrolled urban sprawl, poor water infrastructure, high levels of inequality, water management, cultural issues, drought, and climate change; all these factors need to be addressed in a coherent way to be better prepared for severe water crises. In recent years, Mexico City's average temperature has risen around 3°C in the last century. Furthermore, in some parts of the megacity, there has been an increase of exceptionally hot days reaching up to 30°C (Cid et al., 2024). It is believed that this could be due to climate change and as already mentioned, the exponential urban population growth; trees and wetlands have been replaced by concrete and asphalt. Hence, the heat makes the water crisis worse, meaning that residents would need more water but much of it evaporates.

The above has been recognized by decision-makers in the city, for example, the Director of Mexico City Resilience Office, mentioned that "Climate change has become the biggest long-term threat to the city's future. This is because it is linked to water, health, air pollution, traffic disruptions due to flooding, and the vulnerability of housing to landslides, which means that we cannot begin to address any of the city's real problems without tackling the climate issue." (SARACS, 2025).

What would happen if the Cutzamala system stopped supplying freshwater to the city due to drought? Currently, the system provides around 27% of the capital city's water. According to official figures, the system was 30% in 2024, this was at a historic low capacity; in 2023 and 2022, at the same period, it was 38% and 45%, respectively (Wagner et al., 2024). If this trend continues the city will collapse due to water shortage.

An effective drought disaster management system should be in place where 'monitoring' and 'drought risk & vulnerability' analysis committees, and drought early warning system should play a key role in preventing or mitigating the impact of droughts (Figs. 2&3, Table 1). Furthermore, effective drought planning should be part of such a system where all the key actors should participate in designing such plans, ranging from resident representatives, human rights and environmental groups, universities, local representatives, municipal, state and national government (Table 1).

However, in the context of the present case study, there is not such a coherent system in place to deal with drought risk in the megalopolis. For example, lack of monitoring of key exposure indicators such as urbanization and population. Studies have reported that municipalities with uncontrolled urban growth and those with the largest population have the highest water vulnerability (i.e., IZTA, GAM, & TLA) as opposed to those municipalities with low population size (i.e., MA, MC, CU) (Huerta-Vergara et al., 2022); similar findings have been found in urban settings in China (Xu et al., 2025; Long et al., 2025). Regarding sensitivity indicators (e.g., water leaks, quality and shortages), it has been reported that the highest number of water leak reports came from IZTA, CO, and TLA municipalities (CONAGUA, 2015; Huerta-Vergara et al., 2022). Contamination of drinking water and microbial regrowth in water distribution networks represent a serious health issue if not addressed (Nduli et al., 2025; Siddique et al., 2026). Again, there is no evidence of a well-established monitoring system to assess water quality in the capital city. For example, it has been found that IZTA and XO municipalities reported the highest number of reports of poor water quality (Huerta-Vergara et al., 2022). However, this clearly demonstrates the lack of a monitoring system of these key indicators and therefore, key decision-makers fail to implement measures aiming at better water management.

Urban wastewater treatment is crucial to enhance adaptive capacity; however, it is not clear whether there are enough treatment plants in the capital city. Moreover, it is not known whether there are regulations to comply with, for example, the level of risk of micropollutants removal (Wtodarczyk-Makuta et al., 2025). Huerta-Vergara et al., (2022) reported that only 10% of the water used in the city is treated. Harvesting rainwater stores and reuse can restore urban hydrology, among other things, could help to reduce water vulnerability to urban environments (Li et al., 2025; Yuan et al., 2025). As of 2019, around 13,939 rainwater harvesting systems have been installed in three municipalities (Huerta-Vergara et al., 2022). This raises the question as to why only three municipalities?

Figure 3 shows a drought early warning system (DEWS) which aims at assessing the information given by the 'monitoring committee' and if the tipping indicator exceeds the trigger value (e.g., a low level of any of the network of seven dams that constitute the Cutzamala system due to the lack of rainfall), this should be communicated to key decision-makers so that they implement drought plans to mitigate the impact of the water shortages in the capital city (It should be highlighted that effective drought warning dissemination and communication should be people-centred and inclusive; furthermore, communication should be a two-way or multi-way exchange of drought related information among multiple stake holders, e.g. 'drought task force', see Table 1). Hence, rainfall prediction is crucial in the process of drought management (Fig. 3).

Regarding the use of fuzzy logic to predict rainfall of the present case study produced acceptable models; for example, Table 3 shows the accuracy of the fuzzy logic models when considering three and five climate parameters; it has been found that the performance of the model increased significantly after considering five variables for the case of Mexico City (temperature, wind speed, atmospheric pressure, relative humidity, dew point), and the results are in line with similar studies (Kundu et al., 2023; Wen-Bing, 2024; Pham et al., 2020; Janarthanan et al., 2021; Manna and Anitha, 2023; Senanayake et al., 2024; Benzaouia et al., 2023; Shekar & Mathew 2023; Badola et al., 2023; Shahfahad et al., 2022; Ojo and Ogunjo, 2022). However, more research is needed, for example to acquire better data quality of the five input variables and the inclusion of other climate variables such as cloud cover, temperature at a particular time.

The present study focused on short-term rainfall prediction using data for the three states considered in the study. However, rainfall-runoff (RR) is responsible for water resource management and regarded as a key component of the planet earth's water cycle (Irwan et al., 2025). Hence, a way forward may be to conduct studies focused on predicting water availability, RR, not only for short-term (Liu et al., 2022) but for longer periods (Irwan et al., 2025). Future research include RF and RR prediction by employing existing predictive models that have a better predicting performance (Irwan et al., 2025): hybrid models (Sulaiman et al., 2022; Lakshmi et al., 2022; Kwon et al., 2020), deep learning (DL) (Althoff et al., 2021; Li et al., 2022; Han et al., 2021; Van et al., 2020), and Machine learning (ML) (Poonia and Tiwari, 2020; Adaryani et al., 2022; Achite et al., 2023; Ojo and Ogunjo, 2022; Singh et al., 2022).

More generally, climate change contributes significantly to the negative impacts of droughts, causing severe socio-economic stress, among others. For example, it has been reported that from 1952 to 2016, the frequency of droughts worldwide has increased up to six times (Sing et al., 2026). The study also reported that around 60% increase in drought events occurring during wetter than normal years, i.e., droughts are no longer explained only by reduced total RF, but RF variability is increasingly influencing drought risk in tropical regions. In the context of the present case study, RF, RF variability, RR, among other variables contributing to drought risk should be incorporated into the drought early warning system and water planning measures to better management of drought risk.

Lastly, the present study addressed droughts, but there are several other natural hazards (e.g., earthquakes, floods, extreme heat, volcanic ash, among others) that threaten the residents of the capital city. Hence, it becomes necessary to design and implement what is called people-centred multi-hazard early warning system (MHEWS) in the megalopolis. Artificial intelligence (AI) methods have been employed in multi-hazards and early warning systems in disaster management. However, AI tools have limitations, for example high human intervention dependence, structured data, and limited generalization capacity (Sun et al., 2024). For example, in multi-hazard scenarios analyses, it is difficult to integrate different data simultaneously. To address these limitations, the use of what is called generative artificial intelligence (GenAI) technologies, e.g. large language models (LLMs), which have the potential to be useful in disaster management (Noy and Zhang, 2023). For example, GenAI can generate scenarios and new content beyond analyzing existing data (Cao et al., 2023). Future research should incorporate these new technologies not only for drought but also for multi-hazards early warning systems.

CONCLUSIONS

Drought is one of the major issues of concern, not only for the residents of Mexico City, but also in many regions worldwide. It is believed that the frequency of drought events has increased to six times worldwide. Drought is a natural phenomenon whose main characteristic is the absence or decrease in rainfall during a period that is expected to be rainy. The water shortage for human consumption triggers a series of difficulties, such as uncontrolled urban sprawl, poor water infrastructure, high levels of inequality, water management, cultural issues, drought, and climate change.

In ancient times, the Mexico City Metropolitan Area's inhabitants lived with harmony with water, using canoes and chinampas to thrive in an aquatic environment. Today, Mexico City faces an increasingly severe and widespread water shortages. The capital city depends heavily on the Cutzamala network of dams with the system's water levels at historic lows. Furthermore, the water crisis has become critical due to several other factors including the overexploitation of aquifers, water losses due to pipeline infrastructure leaks, and the lack of infrastructure to wastewater treatment, among others. Added to this is the decrease of rainfall in Mexico City and its metropolitan area due to climate change.

The chapter presented some preliminary results of an ongoing research project on drought risk management in Mexico City. The results have highlighted the need for an effective drought disaster management system from a holistic point of view, where a drought early warning system should play a key role in drought risk management. The aim of such a dynamic system should be to mitigate the impact of droughts. Given the city's dependence of water supply from the Cutzamala system, RF and drought prediction, part of the function of the DEWS, become crucial in the process of decision-making. More generally, key decision-makers of the megalopolis should design and implement a multi-hazard early warning system.

ACKNOWLEDGEMENT

This project was funded under the following grant: SIP-IPN: No-20260502.

REFERENCES

Achite, M., Gul, E., Elshaboury, N., Jehanzaib, M., Mohammadi, B., & Mehr, A. D. (2023). An improved adaptive neuro-fuzzy inference system for hydrological drought prediction in Algeria. *Physics and Chemistry of the Earth Parts A/B/C*, *131*, 103451. DOI: 10.1016/j.pce.2023.103451

Adaryani, F.R., Mousavi, S.J., & Jafari, F. (2022). Short-term rainfall forecasting using machine learning-based approaches of PSO-SVR, LSTM and CNN. *Journal of Hydrology*, 614, Part A, 128463. https://doi.org/ rights and contentDOI: 10.1016/j.jhydrol.2022.128463Get

Agboola, A. H., Gabriel, A. J., Aliyu, E. O., & Alese, B. K. (2013). Development of a fuzzy logic bases rainfall prediction model. *IACSIT International Journal of Engineering and Technology*, *3*(4), 427–435.

Althoff, D., Rodrigues, L. N., & da Silva, D. D. (2021). Addressing hydrological modeling in watersheds under land cover change with deep learning. *Advances in Water Resources*, *154*, 103965. DOI: 10.1016/j.advwatres.2021.103965

Badola, S., Mishra, V. N., Parkash, S., & Pandey, M. (2023). Rule-based fuzzy inference system for landslide susceptibility mapping along national highway 7 in Garhwal Himalayas, India. *Quaternary Science Advances*, *11*, 100093. DOI: 10.1016/j.qsa.2023.100093

Benzaouia, M., Hajji, B., Mellit, A., & Rabhi, A. (2023). Fuzzy-IoT smart irrigation system for precision scheduling and monitoring. *Computers and Electronics in Agriculture*, *215*, 108407. DOI: 10.1016/j.compag.2023.108407

Cao, J., Zhou, T., Zhi, S., Lam, S., Ren, G., Zhang, Y., Wang, Y., Dong, Y., & Cai, J. (2024). Fuzzy inference system with interpretable fuzzy rules: Advancing explainable artificial intelligence for disease diagnosis—A comprehensive review. *Information Sciences*, *662*, 120212. DOI: 10.1016/j.ins.2024.120212

Cao, Y., Sheng, Q. Z., McAuley, J., & Yao, L. (2023). Reinforcement learning for generative AI: A survey. arXiv preprint arXiv:2308.14328.

Cid, A., Siqueiros-García, J. M., Mazari-Hiriart, M., Guerra, A., & Lemer, A. M. (2024). Mobilizing institutional capacities to adapt to climate change: Local government collaboration networks for risk management in Mexico City. *npj. Npj Climate Action*, *3*(1), 20. DOI: 10.1038/s44168-024-00102-8

CONAGUA. (2005). Sistema Cutzamala-Agua para millones de Mexicanos. Comision Nacional del Agua, SEMARNAT, Mexico. https://www.conagua.gob.mx/conagua07/publicaciones/publicaciones/sistema-cutzamala.pdf

CONAGUA (Comisión Nacional del Agua). (2015). Actualización de la disponibilidad media anual de agua en el acuífero Chalco-Amecameca (1506), Estado de México: México, CONAGUA, Subdirección General Técnica Gerencia de Aguas Subterráneas, Mexico. https://sigagis.conagua.gob.mx/gas1/Edos_Acuiferos_18/edomex/DR_1506.pdf

Diekert, F., Heyen, D., Frikk, N., & Soheil, S. (2024). Balancing the Risk of Tipping: Early Warning Systems from Detection to Management, CESifo Working Paper, No. 10892, Center for Economic Studies and ifo Institute (CESifo), Munich, Germany. https://www.econstor.eu/bitstream/10419/295981/1/cesifo1_wp10892.pdf

Dore, P., Chakkor, S., El Oualkadi, A., & Baghouri, M. (2023). Real-time intelligent system for wind turbine monitoring using fuzzy system. *E-prime-Advances in Electrical Engineering. Electronics and Energy*, *3*, 100096. DOI: 10.1016/j.prime.2022.100096

Han, H., Choi, C., Jung, J., & Kim, H. S. (2021). Deep Learning with Long Short Term Memory Based Sequence-to-Sequence Model for Rainfall-Runoff Simulation. *Water (Basel)*, *13*(4), 437. DOI: 10.3390/w13040437

Hasan, M., Tsegaye, T., Shi, X., Schaefer, G., & Taylor, G. (2008). Model for predicting rainfall by fuzzy set theory using USDA scan data. *Agricultural Water Management*, *95*(12), 1350–1360. DOI: 10.1016/j.agwat.2008.07.015

Huerta-Vergara, A. R., Arciniega-Esparza, S., Pedrozo-Acuña, A., Matus-Kramer, A., & Vega-López, E. (2022). Assessment of vulnerability to water shortage in the municipalities of Mexico City. *Boletín de la Sociedad Geológica Mexicana*, *74*(1), A071021. DOI: 10.18268/BSGM2022v74n1a071021

INEGI. (2021). En México somos 126 014 024 habitantes: Censo de población y vivienda 2020. Comunicado de prensa Núm. 24/21, 25 de enero de 2021, 1-3. https://www.inegi.org.mx/contenidos/saladeprensa/boletines/2021/EstSociodemo/ResultCenso2020_Nal.pdf

IPCC. (2021). *Climate Change 2021: Synthesis report. Contribution of Working Group I to the Sixth Assessment Report of the Intergovernmental Panel on Climate Change* [Masson-Delmotte, V., P. Zhai, A. Pirani, S.L. Connors, C. Péan, S. Berger, N. Caud, Y. Chen, L. Goldfarb, M.I. Gomis, M. Huang, K. Leitzell, E. Lonnoy, J.B.R. Matthews, T.K. Maycock, T. Waterfield, O. Yelekçi, R. Yu, & B. Zhou (eds.)]. Cambridge University Press., DOI: 10.1017/9781009157896

Irwan, D., Ahmed, A., Ibrahim, S. L., Ibrahim, I., Mahmoud, M., Jacky, G., Nurhakim, A., Chah, M., Kumar, P., Sherif, M., & El-Shafie, A. (2025). Advancements in rainfall-runoff prediction: Exploring state-of-the-art neural computing modeling approaches. *Alexandria Engineering Journal, 121*, 138–149. DOI: 10.1016/j.aej.2025.02.060

Janarthanan, R., Balamurali, R., Annapoorani, A., & Vimala, V. (2021). Prediction of rainfall using fuzzy logic. *Materials Today: Proceedings*, *37*(2), 959–963. DOI: 10.1016/j.matpr.2020.06.179

Kebede, A., Warrach-Sagi, K., Schwitalla, T., Wulfmeyer, V., Abebe, T., & Tadesse, T. (2026). A multivariate approach to drought monitoring: Improving robustness and accuracy through a new drought index in regions with high climate variability, applied to the drought-prone region of Ethiopia. *Journal of Hydrology. Regional Studies*, *64*, 103332. DOI: 10.1016/j.ejrh.2026.103332

Kumar, A., & Arya, D. S. (2025). Spatiotemporal analysis and mechanisms of Drought flood abrupt alternation events in India. *Journal of Hydrology. Regional Studies*, *62*, 102992. DOI: 10.1016/j.ejrh.2025.102992

Kundu, S., Biswas, S., Tripathi, D., Karmakar, R., Majumdar, S., & Mandal, S. (2023). A review on rainfall forecasting using ensemble learning techniques. *E-prime-Advances in Electrical Engineering. Electronics and Energy*, *6*, 100296. DOI: 10.1016/j.prime.2023.100296

Kwon, M., Kwon, H.-H., & Han, D. (2020). A Hybrid Approach Combining Conceptual Hydrological Models, Support Vector Machines and Remote Sensing Data for Rainfall-Runoff Modeling. *Remote Sensing (Basel)*, *12*(11), 1801. DOI: 10.3390/rs12111801

Lakshmi, P. J., Apaza, R. A., Alkhayyat, A., Marhoon, H. A., & Alameri, A. A. (2022). Hybrid wavelet-gene expression programming and wavelet-support vector machine models for rainfall-runoff modelling. *Water Science and Technology*, *86*(12), 3205–3222. DOI: 10.2166/wst.2022.400

Li, P., Zhang, J., & Krebs, P. (2022). Prediction of Flow Based on a CNN-LSTM Combined Deep Learning Approach. *Water (Basel), 14*(6), 993. DOI: 10.3390/w14060993

Li, S., Liu, Y., Nguyen, A. H., Wu, Z., Al-Farsi, M. Z., Choi, T., Zhou, L., Her, Y., Li, F., Ren, D., & Romeiko, X. X. (2025). A framework for creating sustainable rainwater harvesting and reuse strategies for urban landscape irrigation in a changing climate. *Journal of Environmental Management, 392*, 126852. DOI: 10.1016/j.jenvman.2025.126852

Liu, G., Tang, Z., Qin, H., Liu, S., Shen, Q., Qu, Y., & Zhou, J. (2022). Short-term runoff prediction using deep learning multi-dimensional ensemble method. *Journal of Hydrology (Amsterdam), 609*, 127762. DOI: 10.1016/j.jhydrol.2022.127762

Long, Y., Huang, Y., Ma, Q., Yu, W., & Tan, C. (2025). Future drought patterns in Southwest China: A CMIP6-based analysis using multiple drought indices. *Journal of Hydrology. Regional Studies, 61*, 102702. DOI: 10.1016/j.ejrh.2025.102702

Mada, G. S., Dethan, N. K. F., & Maharani, A. E. S. H. (2022). Defuzzification methods comparison of Mamdani fuzzy inference system in predicting Tofu production. *Journal Varian, 5*(2), 137–148. DOI: 10.30812/varian.v5i2.1816

Manna, T., & Anitha, A. (2023). Precipitation prediction by integrating rough set on fuzzy approximation space with deep learning techniques. *Applied Soft Computing, 139*, 110253. DOI: 10.1016/j.asoc.2023.110253

Mishra, V. (2025). 'Very little time to react': Texas flash floods expose challenges in early warning. UN News Global perspective human stories. https://news.un.org/en/story/2025/07/1165355

Murray, V., Jonathan, A., Chadi, A., Kanza, A., Lucille, A., Djillali, B., Alonso, B. T., Hun, C., Simon, C., Jim, D., Lucy, F., Urbano, F. P., Qunli, H., John, H., Simon, H., Wirya, K., Lidia, M., Nick, M., Luiz Leal, M. O., & Wright, N. (2021). *Hazard Information Profiles: Supplement to UNDRR-ISC Hazard Definition & Classification Review: Technical Report: Geneva, Switzerland, United Nations Office for Disaster Risk Reduction*. International Science Council., DOI: 10.24948/2021.05

Nduli, S., Tekere, M., Masindi, V., & Foteinis, S. (2025). Assessment of the water quality and microbial regrowth in drinking water treatment plants and the distribution network. *Water Resources and Industry, 33*, 100290. DOI: 10.1016/j.wri.2025.100290

Neri, C., & Magaña, V. (2016). Estimation of Vulnerability and Risk to Meteorological Drought in Mexico. *Weather, Climate, and Society, 8*(2), 95–110. DOI: 10.1175/WCAS-D-15-0005.1

Noy, S., & Zhang, W. (2023). Experimental evidence on the productivity effects of generative artificial intelligence. *Science*, *381*(6654), 187–192. DOI: 10.1126/science.adh2586

Ojo, O. S., & Ogunjo, S. T. (2022). Machine learning models for prediction of rainfall over Nigeria. *Scientific African*, *16*, e01246. DOI: 10.1016/j.sciaf.2022.e01246

Pham, B. T., Le, L. M., Bui, K. T., Le, V. M., Ly, H. B., & Prakash, I. (2020). Development of advanced artificial intelligent models for daily rainfall prediction. *Atmospheric Research*, *237*, 104845. DOI: 10.1016/j.atmosres.2020.104845

Poonia, V., & Tiwari, H. L. (2020). Rainfall-runoff modeling for the Hoshangabad Basin of Narmada River using artificial neural network. *Arabian Journal of Geosciences*, *13*(18), 944. DOI: 10.1007/s12517-020-05930-6

Rahman, M. A. (2020). Improvement of rainfall prediction model by using fuzzy logic. *American Journal of Climate Change*, *9*(4), 391–399. DOI: 10.4236/ajcc.2020.94024

Rakkasagi, S., Goyal, M. K., & Jha, S. (2024). Evaluating the future risk of coastal Ramsar wetlands in India to extreme rainfalls using fuzzy logic. *Journal of Hydrology (Amsterdam)*, *632*, 130869. DOI: 10.1016/j.jhydrol.2024.130869

Rathnayake, N., Rathnayake, U., Chathuranika, I., Dang, T. L., & Hoshino, Y. (2023). Cascade-ANFIS to simulate nonlinear rainfall-runoff relationship. *Applied Soft Computing*, *147*, 110722. DOI: 10.1016/j.asoc.2023.110722

Safar, N. Z. M., Ramli, A. A., Mahdin, H., Nzdi, D., & Khalif, K. M. N. (2019). Rain prediction using fuzzy rule based system in North-West Malaysia. *Indonesian Journal of Electrical Engineering and Computer Science*, *14*(3), 1572–1581.

SARACS. (2025). *Report on the drought safety management system for the case of Mexico City*. SEPI-ESIME, Instituto Politecnico Nacional.

Shahfahad, N., Naikoo, M. W., Talukdar, S., Das, T., & Rahman, A. (2022). Identification of homogenous rainfall regions with trend analysis using fuzzy logic and clustering approach coupled with advanced trend analysis techniques in Mumbai city. *Urban Climate*, *46*, 101306. DOI: 10.1016/j.uclim.2022.101306

Shekar, P. R., & Mathew, A. (2023). Assessing groundwater potential zones and artificial recharge sites in the monsoon-fed Merredu river basin, India: An integrated approach using GIS, AHP, and Fuzzy-AHP. *Groundwater for Sustainable Development*, *23*, 100994. DOI: 10.1016/j.gsd.2023.100994

Siddique, A.B., Rahman, M.A., Gazi, E., Khatun, M.N., Khan, A.S., & Al Helal, A.S. (2026). Groundwater Contamination and Human Health Risk Assessment of Drinking Water in Kaliganj Municipality, Bangladesh. H2Open Journal. https://doi.org/DOI: 10.1016/j.htopen.2026.100007

Singh, A. K., Kumar, P., Ali, R., Al-Ansari, N., Vishwakarma, D. K., Kushwaha, K. S., Panda, K. C., Sagar, A., Mirzania, E., Elbeltagi, A., Kuriqi, A., & Heddam, S. (2022). An Integrated Statistical-Machine Learning Approach for Runoff Prediction. *Sustainability (Basel), 14*(13), 8209. DOI: 10.3390/su14138209

Singla, M. K., Kar, H. D., & Nijhawan, P. (2019). Rain prediction using fuzzy logic. *International Journal of Engineering and Advanced Technology*, *9*(1), 2796–2799.

Sulaiman, N. A. F., Shaharudin, S. M., Ismail, S., Zainuddin, N. H., Tan, M. L., & Abd Jalil, Y. (2022). Predictive Modelling of Statistical Downscaling Based on Hybrid Machine Learning Model for Daily Rainfall in East-Coast Peninsular Malaysia. *Symmetry*, *14*(5), 927. DOI: 10.3390/sym14050927

Sun, L., Li, H., Nagel, J., & Yang, S. (2024). Convergence of AI and urban emergency responses: Emerging pathway toward resilient and equitable communities. *Applied Sciences (Basel, Switzerland)*, *14*(17), 7949. DOI: 10.3390/app14177949

UN. (2025). The world's cities in 2025. Department of Economic and Social Affairs. https://www.un.org/development/desa/pd/sites/www.un.org.development.desa.pd/files/undesa_pd_2025_data-booklet_world_cities_in_2025.pdf

UNDRR & WMO. (2025). Global status of multi-hazard early warning systems. Geneva, Switzerland. https://www.undrr.org/publication/documents-and-publications/global-status-multi-hazard-early-warning-systems-2025

UNISDR (UN-International Strategy for Disaster Reduction). (2009). *UNISDR terminology on disaster risk reduction*. UNISDR.

Van, S. P., Le, H. M., Thanh, D. V., Dang, T. D., Loc, H. H., & Anh, D. T. (2020). Deep learning convolutional neural network in rainfall–runoff modelling. *Journal of Hydroinformatics*, *22*(3), 541–561. DOI: 10.2166/hydro.2020.095

Wagner, J., Rodríguez-Mega, E., & Sengupta, S. (2024, May 18). La crisis del agua empeora en Ciudad de México. New York times. https://www.nytimes.com/es/2024/05/18/espanol/mexico-agua-crisis.html

Wang, X., Zhu, Y., Liang, L., Chao, Y., Li, J., & Wang, Y. (2026). Study on the propagation characteristics and triggering thresholds of meteorological, hydrological and agricultural drought in the Mu Us sandy land. *Climate Services*, *42*, 100648. DOI: 10.1016/j.cliser.2026.100648

Weber, N., & Bieroza, M. (2026). The impact of data quality and outlier detection in high-frequency water quality data on water management and process understanding. *Journal of Hydrology (Amsterdam), 670*, 135171. DOI: 10.1016/j.jhydrol.2026.135171

Wen-Bing, J. (2024). Implementing advanced techniques for urban mountain torrent surveillance and early warning using rainfall predictive analysis. *Urban Climate, 53*, 101782. DOI: 10.1016/j.uclim.2023.101782

Wilhite, D. A. (2000). Chapter 35 Preparing for Drought: A Methodology" (2000). Drought Mitigation Center Faculty Publications. 72. https://digitalcommons.unl.edu/droughtfacpub/72

Wilhite, D. A., Botterill, L., & Monnik, K. (2005a). National Drought Policy: Lessons Learned from Australia, South Africa, and the Unit ed States. In: Wilhite, D. (ed.). *Drought and water crises: science, technology, and management issues*, CRC Press, 137-172.

Wilhite, D. A., Hayes, M. J., & Knutson, C. L. (2005b). Drought preparedness planning: building institutional capacity. In: Wilhite, D. (ed.), 2005. *Drought and water crises: science, technology, and management issues*, CRC Press, 93-121.

WMO. (2025, August 06). Devastating floods highlight need and challenges for warnings. https://wmo.int/media/news/devastating-floods-highlight-need-and-challenges-warnings?utm_source=miragenews&utm_medium=miragenews&utm_campaign=news

Wtodarczyk-Makuta, M., Tchorzewska-Cieslak, B., & Rak, J. R. (2025). Risk assessment method for pharmaceutics removal in treated wastewater-challenges of new Directive (EU) on urban wastewater treatment. *The Science of the Total Environment, 996*, 180141. DOI: 10.1016/j.scitotenv.2025.180141

Xu, R., Liu, L., Liu, Y., Yi, X., & Qiu, H. (2025). Multi-scenario modelling of urban spatial growth under water resources and aquatic ecological environmental constraints. *Ecological Indicators, 177*, 113803. DOI: 10.1016/j.ecolind.2025.113803

Yuan, Q., Yao, G., Liu, W., Weragoda, S. K., Weerasooriya, R., Meng, Y., & Luan, F. (2025). Assessment of harvested rainwater quality and surrogate parameter development for drinking water provision in rural Sri Lanka. *Journal of Environmental Chemical Engineering, 13*(5), 117632. DOI: 10.1016/j.jece.2025.117632

Zadeh, L. A. (1965). Fuzzy sets. *Information and Control, 8*(3), 338–353. DOI: 10.1016/S0019-9958(65)90241-X

Zahran, B., Ayyoub, B., Abu-Ain, W., Hadi, W., & Al-Hawary, S. (2023). A fuzzy based model for rainfall prediction. *International Journal of Data and Network Science*, *7*(1), 97–106. DOI: 10.5267/j.ijdns.2022.12.001

Compilation of References

Aalabaf-Sabaghi, M. (2020). An Introduction to Data, Everything You Need to Know About AI: Big Data and Data Science.

Abdul Nabi, M., & El-adaway, I. H. (2020). Modular Construction: Determining Decision-Making Factors and Future Research Needs. *Journal of Management Engineering*, *36*(6), 04020085. Advance online publication. DOI: 10.1061/(ASCE)ME.1943-5479.0000859

Abdullahi, A., & Shehu, M. (2025). Tech for Social Impact: Measuring and scaling the outcomes of digital transformation in communities. *International Journal of Computing and Artificial Intelligence*, *6*(2), 210–222. DOI: 10.33545/27076571.2025.v6.i2c.200

Abdul, S., Adeghe, E. P., Adegoke, B. O., Adegoke, A. A., & Udedeh, E. H. (2024). AI-enhanced healthcare management during natural disasters: Conceptual insights. *Engineering Science & Technology Journal*, *5*(5), 1794–1816. DOI: 10.51594/estj.v5i5.1155

Achite, M., Gul, E., Elshaboury, N., Jehanzaib, M., Mohammadi, B., & Mehr, A. D. (2023). An improved adaptive neuro-fuzzy inference system for hydrological drought prediction in Algeria. *Physics and Chemistry of the Earth Parts A/B/C*, *131*, 103451. DOI: 10.1016/j.pce.2023.103451

Adaryani, F.R., Mousavi, S.J., & Jafari, F. (2022). Short-term rainfall forecasting using machine learning-based approaches of PSO-SVR, LSTM and CNN. *Journal of Hydrology*, 614, Part A, 128463. https://doi.org/ rights and contentDOI: 10.1016/j.jhydrol.2022.128463Get

Adel, A., Saleem, N. H., Jan, T., & Prasad, M. (2025). A review of major ICT failures and recovery strategies: Strengthening digital resilience. *Computers & Security*, *159*, 104678. DOI: 10.1016/j.cose.2025.104678

Adiwinoto, R., Prasetya, V., Najmasari, N., Sudarta, J., Putri, F., Garianto, E., Rasyida, A., & Diarsvitri, W. (2025). Narrative review on AI-driven and social media-enhanced early detection, mitigation, and public health preparedness in natural disaster-prone areas. *IOP Conference Series. Earth and Environmental Science, 1473*(1), 012020. DOI: 10.1088/1755-1315/1473/1/012020

Afonso, A., Alves, J., & Monteiro, S. (2024). The Pressure is on: How Geopolitical Tensions Impact Institutional Fiscal and External Stability Responses. SSRN *Electronic Journal*. DOI: 10.2139/ssrn.4803854

Afrasiabi, A., Tavana, M., & Di Caprio, D. (2022). The greenway for bridge column rehabilitation: A comparison between different techniques based on multi-criteria decision analysis. *SciELO Brasil, 29*(25), 37291–37314. DOI: 10.1007/s11356-021-17851-2 PMID: 35050472

African Union Commission. (2022). *Africa Multi-Hazard Early Warning and Early Action System Framework*. Addis Ababa: African Union. Available at: https://au.int/en/documents

Agarwal, C., Queen, O., & Lakkaraju, H. (2023). Evaluating explainability for graph neural networks. *Nature.Com*. DOI: 10.1038/s41597-023-01974-x

Agbaje, T. H., Abomaye-Nimenibo, N., Ezeh, C. J., Bello, A., & Olorunnishola, A. (2024). Building Damage Assessment in Aftermath of Disaster Events by Leveraging Geoai (Geospatial Artificial Intelligence). *World Journal of Advanced Research and Reviews, 23*(1), 667–687. DOI: 10.30574/wjarr.2024.23.1.2000

Agbehadji, I. E., Schütte, S., Masinde, M., Botai, J., & Mabhaudhi, T. (2023). Climate risks resilience development: A bibliometric analysis of climate-related early warning systems in southern Africa. *Climate (Basel), 12*(1), 3. DOI: 10.3390/cli12010003

Agboola, A. H., Gabriel, A. J., Aliyu, E. O., & Alese, B. K. (2013). Development of a fuzzy logic bases rainfall prediction model. *IACSIT International Journal of Engineering and Technology, 3*(4), 427–435.

Aguirre, B. E. (2005). Cuba's disaster management model: Should it be emulated? *International Journal of Mass Emergencies and Disasters, 23*(3), 55–71. DOI: 10.1177/028072700502300303

Agyemang, F. (2025). Legal and policy perspectives on public participation and disaster risk management in South African cities. *Journal of Local Government Research and Innovation, 6*(0), a259. DOI: 10.4102/jolgri.v6i0.259

Ahamed, S., & Uddin, M. M. (2023). *Estimating uncertainty of earthquake rupture using Bayesian neural network*. https://arxiv.org/pdf/1911.09660

AI, T. (2021, August 4). *The Future of Artificial Intelligence in Weather Forecasting*. Retrieved from Towards AI: https://towardsai.net/p/l/the-future-of-artificial-intelligence-in-weather-forecasting

Aitsi-Selmi, A., Murray, V., Wannous, C., Dickinson, C., Johnston, D., Kawasaki, A., Stevance, A.-S., & Yeung, T. (2016). Reflections on a science and technology agenda for 21st century disaster risk reduction: Based on the scientific content of the 2016 UNISDR science and technology conference on the implementation of the Sendai framework for disaster risk reduction 2015–2030. *International Journal of Disaster Risk Science*, *7*(1), 1–29. DOI: 10.1007/s13753-016-0081-x

Ajayi, J. O., Erigha, E. D., Obuse, E., Ayanbode, N., & Cadet, E. (2025). Resilient infrastructure management systems using real-time analytics and. *Computer Science & IT Research Journal*, *6*(8), 525–548. DOI: 10.51594/csitrj.v6i8.2013

Akerkar, R. (2019). *Artificial intelligence for disaster management*. Springer. https://link.springer.com/book/10.1007/978-3-031-91084-5

Aktan, E., Bartoli, I., & Glišić, B. (2024). Lessons from bridge structural health monitoring (SHM) and their implications for the development of cyber-physical systems. *Mdpi.Com*. https://www.mdpi.com/2412-3811/9/2/30

Al Banna, M. H., Taher, K. A., Kaiser, M. S., Mahmud, M., Rahman, M. S., Hosen, A. S., & Cho, G. H. (2020). Application of artificial intelligence in predicting earthquakes: State-of-the-art and future challenges. *IEEE Access : Practical Innovations, Open Solutions*, *8*, 192880–192923. DOI: 10.1109/ACCESS.2020.3029859

Albert, M. J. (2024). Navigating the Polycrisis. In *The MIT Press eBooks*. The MIT Press., DOI: 10.7551/mitpress/15041.001.0001

Alcántara-Ayala, I. (2025). Cascading hazards and compound disasters. *Npj Natural Hazards*, *2*(1), 54. Advance online publication. DOI: 10.1038/s44304-025-00111-5

Alcántara-Ayala, I., Velásquez-Espinoza, G., & de Jesús, A. M. (2025). From Mandates to Mechanisms: Institutional Vulnerability, Decentralized Governance, and the Challenges of Local Disaster Risk Reduction Implementation. *International Journal of Disaster Risk Science*, *16*(5), 709–723. Advance online publication. DOI: 10.1007/s13753-025-00673-y

Alcocer, S., & Moctezuma, B. (2023). Performance-based design of new concrete walls for building seismic rehabilitation. In *Life-Cycle of Structures and Infrastructure Systems* (pp. 2261–2268). CRC Press/Balkema., DOI: 10.1201/9781003323020-276

Ali, S., Abuhmed, T., El-Sappagh, S., Muhammad, K., Alonso-Moral, J. M., Confalonieri, R., Guidotti, R., Del Ser, J., Díaz-Rodríguez, N., & Herrera, F. (2023). Explainable Artificial Intelligence (XAI): What we know and what is left to attain Trustworthy Artificial Intelligence. *Information Fusion*, *99*, 101805. DOI: 10.1016/j.inffus.2023.101805

Allen, C., Tsou, M. H., Aslam, A., Nagel, A., & Gawron, J. M. (2016). Applying GIS and learning methods to Twitter data for multiscale surveillance of influenza. *PLoS One*, *11*(7), e0157734. DOI: 10.1371/journal.pone.0157734 PMID: 27455108

Allen, R. M., & Melgar, D. (2019). Earthquake early warning: Advances, scientific challenges, and societal needs. *Annual Review of Earth and Planetary Sciences*, *47*(1), 361–388. DOI: 10.1146/annurev-earth-053018-060457

Allen, T. (2019). Creating resilient futures. In *Routledge eBooks* (p. 137). Informa., DOI: 10.4324/9780429398872-10

Alsadi, J., Alkhatib, F., Antony, J., Garza-Reyes, J. A., Tortorella, G., & Cudney, E. A. (2025). A systematic literature review with bibliometric analysis of Quality 4.0. *The TQM Journal*, *37*(5), 1446–1470. DOI: 10.1108/TQM-02-2024-0050

Althoff, D., Rodrigues, L. N., & da Silva, D. D. (2021). Addressing hydrological modeling in watersheds under land cover change with deep learning. *Advances in Water Resources*, *154*, 103965. DOI: 10.1016/j.advwatres.2021.103965

Amaratunga, D., Haigh, R., & Dias, N. (Eds.). (2021). *Multi-hazard early warning and disaster risks*. Springer Nature. DOI: 10.1007/978-3-030-73003-1

Amarel, T. (2025, April 2). *Improving Short-Term Weather Forecasting Accuracy: A Practical Guide to Implementing AI and Machine Learning Models*. Retrieved from TaylorAmarel.com: https://tayloramarel.com/2025/04/improving-short-term-weather-forecasting-accuracy-a-practical-guide-to-implementing-ai-and-machine-learning-models/

Amit, S. N. K. B., & Aoki, Y. (2017, September). Disaster detection from aerial imagery with convolutional neural network. In 2017 international electronics symposium on knowledge creation and intelligent computing (IES-KCIC) (pp. 239-245). IEEE.

Andersen, T. J., & Young, P. C. (2023). Enhancing public sector enterprise risk management through interactive information processing. *Frontiers in Research Metrics and Analytics*, *8*, 1239447. Advance online publication. DOI: 10.3389/frma.2023.1239447 PMID: 38192523

Andrade, F. S. (2024). Application of Weather Post-Processing Methods for Operational Ensemble Hydrological Forecasting on Multiple Catchments in Canada. *Journal of Hydrology (Amsterdam)*, *642*, •••. https://www.sciencedirect.com/science/article/pii/S0022169424012575

Andrae, S. (2025). Artificial intelligence in disaster management. In *Advances in Environmental Engineering and Green Technologies* (pp. 73–114). IGI Global., DOI: 10.4018/979-8-3693-7483-2.ch004

Ansell, C., & Torfing, J. (2021). Public Governance as Co-creation. In *Cambridge University Press eBooks*. Cambridge University Press., DOI: 10.1017/9781108765381

Ansell, C., & Trondal, J. (2017). Governing Turbulence: An Organizational- Institutional Agenda. *Perspectives on Public Management and Governance*, *1*(1), 43–57. DOI: 10.1093/ppmgov/gvx013

Anwar, M. R., & Sakti, L. D. (2024). *Integrating Artificial Intelligence and Environmental Science for Sustainable Urban Planning.IAIC Transactions on Sustainable Digital Innovation*. ITSDI., DOI: 10.34306/itsdi.v5i2.666

arxiv. (n.d.). *Statistical Post-Processing Yields Accurate Probabilistic Forecasts from Artificial Intelligence Weather Models*. Retrieved from arxiv.org: https://arxiv.org/html/2504.12672#:~:text=These%20findings%20show%20that%20statistical%20post-processing%20methods%20developed,into%20existing%20workflows%20in%20a%20low-risk%2C%20incremental%20fashion

Ashinze, U. K., Edeigba, B. A., Umoh, A. A., Biu, P. W., & Daraojimba, A. I. (2024). *Urban green infrastructure and its role in sustainable cities: A comprehensive review*. World Journal Of Advanced Research and Reviews., DOI: 10.30574/wjarr.2024.21.2.0519

Asmara, Y. P. (2023). Concrete Reinforcement Degradation and Rehabilitation: Damages, Corrosion and PreventionNANo ranking found for "Engineering Materials." In *Engineering Materials* (Vol. 2024). Springer Science and Business Media Deutschland GmbH. DOI: 10.1007/978-981-99-5933-4_8

Astuti, W., Akmeliawati, R., Sediono, W., & Salami, M. J. E. (2014). Hybrid technique using singular value decomposition (SVD) and support vector machine (SVM) approach for earthquake prediction. *IEEE Journal of Selected Topics in Applied Earth Observations and Remote Sensing*, *7*(5), 1719–1728. DOI: 10.1109/JSTARS.2014.2321972

Aswin, S., Geetha, P., & Vinayakumar, R. (2018, April). Deep learning models for the prediction of rainfall. In 2018 International Conference on Communication and Signal Processing (ICCSP) (pp. 0657-0661). IEEE. DOI: 10.1109/ICCSP.2018.8523829

Atukunda, G., Musiita, B., Atwiine, J., & Olyanga, A. M. (2024). Corporate Governance Practices, Operating Environment and Financial Sustainability of Saccos in Greater Mbarara District. *Journal of Economics and Behavioral Studies*, *16*(2(J)), 70–86. DOI: 10.22610/jebs.v16i2(J).3837

Auditor-General South Africa (AGSA). (2024). *Consolidated general report on local government audit outcomes 2023–24*. Pretoria, South Africa: AGSA. https://www.agsa.co.za/Portals/0/Reports/MFMA/2023-24/MFMA%20Report%202023-24%20interactive%20%28updated%20August%202025%29.pdf

Ayuba, P., Hassan, A. B., Kuti, I. A., Morenikeji, G., Mohammed, I., Ambafi, J. G., & Ohize, H. O. (2025). Multi-hazard identification and early warning systems for emergency management: A review. *Confluence University Journal of Science and Technology*.

Azmeer, A., Tahir, F., & Al-Ghamdi, S. G. (2024). Progress on green infrastructure for urban cooling: Evaluating techniques, design strategies, and benefits. *Urban Climate*, *56*, 102077. Advance online publication. DOI: 10.1016/j.uclim.2024.102077

Bacchin, T. K., Ashley, R., Sijmons, D. F., Zevenbergen, C., & Timmeren, A. van. (2014).Blue-Green multifunctional infrastructure: An urban landscape system design new approach.

Back, Y., & Kleidorfer, M. (2023). Opportunities and challenges in designing climate-resilient dense urban areas with Blue-Green Infrastructure. *Österreichische Wasser- und Abfallwirtschaft*, *75*(11-12), 615–625. Advance online publication. DOI: 10.1007/s00506-023-00989-3

Badola, S., Mishra, V. N., Parkash, S., & Pandey, M. (2023). Rule-based fuzzy inference system for landslide susceptibility mapping along national highway 7 in Garhwal Himalayas, India. *Quaternary Science Advances*, *11*, 100093. DOI: 10.1016/j.qsa.2023.100093

Badri, R., & Moghadam, A. S. (2021). Assessing the importance of deterioration properties for the design decision-makings of low-rise asymmetric reinforced concrete buildings. *Structures*. https://www.sciencedirect.com/science/article/pii/S2352012421000345

Bajow, N., Alesa, S., Shaheen, F. A. Y., Almalki, A., Alshamrani, A., Alotaibi, R., Aloraifi, A., Montan, C., Lennquist, S., & Alotaibi, M. (2024). Assessment of the effectiveness of hospital external disaster functional drills on health care receivers' performance, using standardized patients and mass cards simulation: A pilot study from Saudi Arabia. *BMC Emergency Medicine*, *24*(1), 175. Advance online publication. DOI: 10.1186/s12873-024-01095-7 PMID: 39333890

Bao, Y., Chen, Z., Wei, S., Xu, Y., & Tang, Z. (2019). The state of the art of data science and engineering in structural health monitoring. *Engineering*. https://www.sciencedirect.com/science/article/pii/S2095809918308026

Basher, R. (2006). Global early warning systems for natural hazards: systematic and people-centred. *Philosophical transactions of the royal society a: mathematical, physical and engineering sciences, 364*(1845), 2167-2182.

Basher, R. (2006). Global early warning systems for natural hazards: systematic and people-centred. Philosophical Transactions. Series A, Mathematical, Physical, and Engineering Sciences, 364(1845), 2167–2182. DOI: 10.1098/rsta.2006.1819

Basher, R. (2023). Challenges in multi-hazard early warning systems: Last-mile connectivity. *International Journal of Disaster Risk Reduction*, •••, 92.

Bassetti, F. (2024, March 12). *The AI Revolution in Weather Forecasting: Reactions and Comments from the Mediasphere*. Retrieved from https://www.climateforesight.eu/articles/reacting-to-the-ai-revolution-in-weather-forecasting/

Behnia, A., & Chai, H. (2014). Advanced structural health monitoring of concrete structures with the aid of acoustic emission. *Construction and Building Materials*. https://www.sciencedirect.com/science/article/pii/S0950061814004292

Bell, L., & Hofmeyr, K. (2021). Enabling organisational ambidexterity: A leadership perspective. *South African Journal of Business Management*, *52*(1). Advance online publication. DOI: 10.4102/sajbm.v52i1.2268

Benzaouia, M., Hajji, B., Mellit, A., & Rabhi, A. (2023). Fuzzy-IoT smart irrigation system for precision scheduling and monitoring. *Computers and Electronics in Agriculture*, *215*, 108407. DOI: 10.1016/j.compag.2023.108407

Best, A., & Yalezo, B. (2022). Stakeholder engagement strategies assessment in expanded public works programme. *Journal of Governance and Regulation*, *11*(4), 46–58. DOI: 10.22495/jgrv11i4art5

Biglari, A. (2026). Evaluation effect of force equilibrium prediction on using secant stiffness in nonlinear displacement-based finite element method. *Journal of Engineering and Applied Science 2026 73:1, 73*(1), 42-. DOI: 10.1186/s44147-025-00840-4

Biglari, A., & Irani, F. (2006). The capability of Mutation and Convergent Functions in Nonlinear Analysis. *Amirkabir Journal of Science and Technology, 16.*

Biglari, A., & Nodeh, S. (2024). Evaluation Possibility of Concrete Frames Rehabilitation with Off-Grade Steel Bracing. *Modares Civil Engineering Journal, 24*(2), 31–44. https://doi.org/a-10-64884-2

Biglari, A. (2025). Innovative Modular House: Advancing Sustainability, Customization, and Community in Affordable Housing. *Sustainable Development Goals Series*, (Part F767), 343–411. DOI: 10.1007/978-981-96-5107-8_9

Biglari, A., & Aliarab, Z. (2022). Evaluation of Feasibility Study to Ensure Uniform Deformations Using Slender Steel Shearwalls with Partial Connections in the Rehabilitation Process. *Modares Civil Engineering Journal*, *22*(6), 165–178. DOI: 10.22034/22.6.165

Biglari, A., & Harrison, P. (2026). Proposing the enriched strain-based gradient as a general implicit method to solve time-independent constitutive equation. *Communications in Nonlinear Science and Numerical Simulation*, *157*, 109718. DOI: 10.1016/j.cnsns.2026.109718

Biglari, A., Harrison, P., & Bićanić, N. (2014). Quasi-hinge beam element implemented within the hybrid force-based method. *Computers & Structures*, *137*, 31–46. DOI: 10.1016/j.compstruc.2013.10.016

Bilal, M., Ji, Y., Wang, Y., & Akhter, M. (2022). An early warning system for earthquake prediction from seismic data using batch normalized graph convolutional neural network with attention mechanism. *Mdpi.Com*. https://www.mdpi.com/1424-8220/22/17/6482

Bodó, B., & Janssen, H. (2022). Maintaining trust in a technologized public sector. *Policy and Society*, *41*(3), 414–429. DOI: 10.1093/polsoc/puac019

Bofeng, Z., & Yue, L. (2005, November). Customized explanation in expert system for earthquake prediction. In *17th IEEE International Conference on Tools with Artificial Intelligence (ICTAI'05)* (pp. 5-pp). IEEE. DOI: 10.1109/ICTAI.2005.54

Bommer, J., & Abrahamson, N. A. (2006). Why do modern probabilistic seismic-hazard analyses often lead to increased hazard estimates? *Pubs.Geoscienceworld.Org*. https://pubs.geoscienceworld.org/ssa/bssa/article-abstract/96/6/1967/146678

Bompotas, A., Anagnostopoulos, C., Kalogeras, A., Kalogeras, G., Mylonas, G., Stefanidis, K., Alexakos, C., & Dandoulaki, M. (2022). A civil protection early warning system to improve the resilience of Adriatic-Ionian territories to natural and man-made risk. *arXiv*. https://arxiv.org/abs/2207.13941 DOI: 10.1109/ETFA52439.2022.9921697

Bondank, E. N., & Chester, M. V. (2020). Infrastructure Interdependency Failures from Extreme Weather Events as a Complex Process. *Frontiers in Water*, *2*, 21. Advance online publication. DOI: 10.3389/frwa.2020.00021

Bonev, B. K. (2025, July 16). *FourCastNet 3: A Geometric Approach to Probabilistic Machine-Learning Weather Forecasting at Scale*. Retrieved from NVIDIA Corporation: https://research.nvidia.com/publication/2025-07_fourcastnet-3-geometric-approach-probabilistic-machine-learning-weather

Botes, M. (2021). The use of geospatial surveillance data for public-health emergency and disaster management in South Africa: A review with legal recommendations. *Tydskrif vir die Suid-Afrikaanse Reg*, *3*(3), 474–503. DOI: 10.47348/TSAR/2021/i3a4

Botzen, W. J. W., Mol, J. M., Robinson, P. J., & Czajkowski, J. (2024). Drivers of natural disaster risk-reduction actions and their temporal dynamics: Insights from surveys during an imminent hurricane threat and its aftermath. *Risk Analysis*, *44*(10), 2448–2462. DOI: 10.1111/risa.14314 PMID: 38679462

Boublil, E. (2023). Vulnerability, Interdependence and the Care for the Living. *The Vulnerability of the Human World Springer*, *148*, 143–157. DOI: 10.1007/978-3-031-41824-2_10

Broekman, C., Coetzee, C., & Cilliers, D. (2025). Mapping pluralistic future flood risk scenarios for informal settlements in the eThekwini metropolitan municipality, South Africa. *Natural Hazards*, *121*(18), 20949–20983. DOI: 10.1007/s11069-025-07581-w

Bronfman, N. C., Cisternas, P. C., Repetto, P. B., & Castañeda, J. V. (2019). Natural disaster preparedness in a multi-hazard environment: Characterizing the sociodemographic profile of those better (worse) prepared. *PLoS One*, *14*(4), e0214249. DOI: 10.1371/journal.pone.0214249 PMID: 31017910

Brotzge, J., & Donner, W. (2013). The tornado warning process: A review of current research, challenges, and opportunities. *Bulletin of the American Meteorological Society*, *94*(11), 1715–1733. DOI: 10.1175/BAMS-D-12-00147.1

Browder, G., Sanchez, A. N., Jongman, B., Engle, N., Van Beek, E., Errea, M. C., & Hodgson, S. (2021). *An EPIC Response: Innovative Governance for Flood and Drought Risk Management—Executive Summary*. World Bank. DOI: 10.1596/35754

Bruinsma, W. S. (2024, June 3). *Introducing Aurora: The First Large-scale Foundation Model of the Atmosphere*. Retrieved from Microsoft Research Blog: https://www.microsoft.com/en-us/research/blog/introducing-aurora-the-first-large-scale-foundation-model-of-the-atmosphere/?msockid=370d1976c15d6cf10f670ca0c0226dad

Budimir, M., Šakić Trogrlić, R., Almeida, C., Arestegui, M., Chuquisengo Vásquez, O., Cisneros, A., Cuba Iriarte, M., Dia, A., Lizon, L., Madueño, G., Ndiaye, A., Ordoñez Caldas, M., Rahman, T., Rana Tharu, B., Sall, A., Uprety, D., Anderson, C., & McQuistan, C. (2025). Opportunities and challenges for people-centered multi-hazard early warning systems: Perspectives from the Global South. *iScience*, *28*(5), 112353. Advance online publication. DOI: 10.1016/j.isci.2025.112353 PMID: 40292326

Bülte, C. H. (2025). Uncertainty Quantification for Data-driven Weather Models. *Artificial Intelligence for the Earth Systems*, *39*, •••. https://journals.ametsoc.org/view/journals/aies/aop/AIES-D-24-0049.1/AIES-D-24-0049.1.xml

Burston, J., Ware, D., & Tomlinson, R. (2015). The real-time needs of emergency managers for tropical cyclone storm tide forecasting: Results of a participatory stakeholder engagement process. *Natural Hazards*, *78*(3), 1653–1668. DOI: 10.1007/s11069-015-1794-7

Bushnaq, O. M., Mishra, D., Natalizio, E., & Akyildiz, I. F. (2022). Unmanned aerial vehicles (UAVs) for disaster management. In *Nanotechnology-Based Smart Remote Sensing Networks for Disaster Prevention* (pp. 159–188). Elsevier., DOI: 10.1016/B978-0-323-91166-5.00013-6

Byaruhanga, N., Kibirige, D., & Mkhonta, G. (2025). Understanding South Africa's Flood Vulnerabilities and Resilience Pathways: A Comprehensive Overview. *Water (Basel)*, *17*(17), 2608. DOI: 10.3390/w17172608

Caballero Testón, J., Pérez-Mon, O., & R-Moreno, M. D. (2026). The role of automated planning in battle management systems for military tactics. *Expert Systems with Applications*, *297*, 129259. Advance online publication. DOI: 10.1016/j.eswa.2025.129259

Cai, S., Mao, Z., Wang, Z., Yin, M., & Karniadakis, G. E. (2021). Characterizing possible failure modes in physics-informed neural networks. *Proceedings.Neurips.Cc*, *37*(12), 1727–1738. DOI: 10.1007/s10409-021-01148-1

Cai, Z., Guldmann, J.-M., Tang, Y., & Han, G. (2022). Does city-water layout matter? Comparing the cooling effects of water bodies across 34 Chinese megacities. *Journal of Environmental Management, 324*, 116263. Advance online publication. DOI: 10.1016/j.jenvman.2022.116263 PMID: 36166866

Camps-Valls, G. F.-T. (2025). Artificial intelligence for Modeling and Understanding Extreme Weather and Climate Events. *Nature Communications,* https://www.nature.com/articles/s41467-025-56573-8

Camps-Valls, G., Fernández-Torres, M., Cohrs, K.-H., Höhl, A., Castelletti, A., Paçal, A., Robin, C. N., Martinuzzi, F., Papoutsis, I., Prapas, I., Pérez-Aracil, J., Weigel, K., Gonzalez-Calabuig, M., Reichstein, M., Rabel, M., Giuliani, M., Mahecha, M. D., Popescu, O.-I., Pellicer-Valero, Ó. J., . . . Williams, T. (2024). AI for Extreme Event Modeling and Understanding: Methodologies and Challenges. *arXiv (Cornell University).* https://doi.org//arxiv.2406.20080DOI: 10.48550

Cao, Y., Sheng, Q. Z., McAuley, J., & Yao, L. (2023). Reinforcement learning for generative AI: A survey. arXiv preprint arXiv:2308.14328.

Cao, B., Chen, Q., Du, M., Cheng, Q., Li, Y., & Li, R. (2022). Simulation Analysis of the Cooling Effect of Urban Water Bodies on the Local Thermal Environment. *Water (Basel), 14*(19), 3091. Advance online publication. DOI: 10.3390/w14193091

Cao, J., Zhou, T., Zhi, S., Lam, S., Ren, G., Zhang, Y., Wang, Y., Dong, Y., & Cai, J. (2024). Fuzzy inference system with interpretable fuzzy rules: Advancing explainable artificial intelligence for disease diagnosis—A comprehensive review. *Information Sciences, 662*, 120212. DOI: 10.1016/j.ins.2024.120212

Cataldi, D., Cataldi, G., & Straser, V. (2024). Earthquakes and AI: The Proposal of a New Research Model. New Concepts in Global Tectonics Journal, 12(1).

Cavadini, G. B., Cavadini, G. B., & Cavadini, G. B. (2024). Connecting blue-green infrastructure elements to reduce combined sewer overflows. *Journal of Environmental Management, 365*, 121465. Advance online publication. DOI: 10.1016/j.jenvman.2024.121465 PMID: 38901320

Chakrabarty, S., Deb, C. K., Marwaha, S., Haque, M. A., Kamil, D., Bheemanahalli, R., & Shashank, P. R. (2026). Application of artificial intelligence in insect pest identification - A review. *Artificial Intelligence in Agriculture, 16*(1), 44–61. DOI: 10.1016/j.aiia.2025.06.005

Chavula, P., Kayusi, F., Lungu, G., & Uwimbabazi, A. (2025). The Current Landscape of Early Warning Systems and Traditional Approaches to Disaster Detection. *LatIA, 3*, 77. DOI: 10.62486/latia202577

Cheng, F., Li, J., Zhou, L., & Lin, G. (2023). Fragility analysis of nuclear power plant structure under real and spectrum-compatible seismic waves considering soil-structure interaction effect. *Engineering Structures*, *280*, 115684. DOI: 10.1016/j.engstruct.2023.115684

Chen, J., Bach, P. M., Nice, K. A., & Leitão, J. P. (2024). Investigating the efficacy of a fast urban climate model for spatial planning of green and blue spaces for heat mitigation. *The Science of the Total Environment*, *955*, 176925. Advance online publication. DOI: 10.1016/j.scitotenv.2024.176925 PMID: 39454799

Chen, M., Chen, L., Zhou, Y., Hu, M., Jiang, Y., Huang, D., Gong, Y., & Xian, Y. (2023). Rising vulnerability of compound risk inequality to ageing and extreme heatwave exposure in global cities. *NPJ Urban Sustainability*, *3*(1), 38. Advance online publication. DOI: 10.1038/s42949-023-00118-9

Chen, W., Peng, J., Hong, H., Shahabi, H., Pradhan, B., Liu, J., Zhu, A.-X., Pei, X., & Duan, Z. (2018). Landslide susceptibility modelling using GIS-based machine learning techniques for Chongren County, Jiangxi Province, China. *The Science of the Total Environment*, *626*, 1121–1135. DOI: 10.1016/j.scitotenv.2018.01.124 PMID: 29898519

Chetty, K., Saal, P., Ntshayintshayi, N., Masuku, N., & Moosa, T. (2024). AI as a Catalyst for Good Governance: Transforming South Africa's Fight Against Corruption. *Development*, *67*(1-2), 50–60. DOI: 10.1057/s41301-024-00404-8

Chianumba, E. C., Forkuo, A. Y., Mustapha, A. Y., Osamika, D., & Komi, L. S. (2024). Advances in preventive care delivery through WhatsApp, SMS, and IVR messaging in high-need populations. *International Journal of Advanced Multidisciplinary Research and Studies*, *4*(6), 1967–1988. DOI: 10.62225/2583049X.2024.4.6.4248

Chipfupa, U., Tagwi, A., & Wale, E. (2021). Psychological capital and climate change adaptation: Empirical evidence from smallholder farmers in South Africa. *Jàambá*, *13*(1), 1–12. DOI: 10.4102/jamba.v13i1.1061 PMID: 34230848

Chowdhury, M. F., Biplob, M. R. A., & Uddin, J. (2018, June). Real time traffic density measurement using computer vision and dynamic traffic control. In 2018 Joint 7th International Conference on Informatics, Electronics & Vision (ICIEV) and 2018 2nd International Conference on Imaging, Vision & Pattern Recognition (icIVPR) (pp. 353-356). IEEE. DOI: 10.1109/ICIEV.2018.8641039

Cichowicz, R., & Bochenek, A. D. (2024). Assessing the effects of urban heat islands and air pollution on human quality of life. *Anthropocene*, *46*, 100433. Advance online publication. DOI: 10.1016/j.ancene.2024.100433

Cid, A., Siqueiros-García, J. M., Mazari-Hiriart, M., Guerra, A., & Lemer, A. M. (2024). Mobilizing institutional capacities to adapt to climate change: Local government collaboration networks for risk management in Mexico City. *npj. Npj Climate Action*, *3*(1), 20. DOI: 10.1038/s44168-024-00102-8

City of Cape Town. (2017). *Municipal Disaster Risk Management Plan*. City of Cape Town Disaster Risk Management Centre.

City of Johannesburg. (2021). *Climate Action Plan*. Johannesburg: Environment and Infrastructure Services Department (EISD). Available at: https://www.joburg.org.za/documents_/Pages/Key%20Documents/Climate%20Action%20Plan.aspx

City of Tshwane. (2024). *Disaster Risk Management 2024/2025 Annual Report*. City of Tshwane Metropolitan Municipality.

Clement, M. (2024). Case Studies: Successful Use of AI in Past Natural Disasters.

ClimaSens. (2017). *Transforming Climate Risk to Resilience*. Retrieved from ClimaSens: https://climasens.com/

Climavision. (2025). *AI in Weather Forecasting: The Future of Weather Prediction*. Retrieved from Climavision: https://climavision.com/resources/ai-weather-forecasting-guide/

Cloud, G. (2025, March 5). *Google Cloud Brings WeatherNext AI-Powered Weather Predictions to Enterprises*. Retrieved from Google Cloud: https://www.googlecloudpresscorner.com/2025-03-05-Google-Cloud-Brings-WeatherNext-AI-Powered-Weather-Predictions-to-Enterprises

Compendium of multi-hazard early warning cooperation. (n.d.). www.undrr.org/publication/global-status-multi-hazard-early-warning-systems-target-g

CONAGUA (Comisión Nacional del Agua). (2015). Actualización de la disponibilidad media anual de agua en el acuífero Chalco-Amecameca (1506), Estado de México: México, CONAGUA, Subdirección General Técnica Gerencia de Aguas Subterráneas, Mexico. https://sigagis.conagua.gob.mx/gas1/Edos_Acuiferos_18/edomex/DR_1506.pdf

CONAGUA. (2005). Sistema Cutzamala-Agua para millones de Mexicanos. Comision Nacional del Agua, SEMARNAT, Mexico. https://www.conagua.gob.mx/conagua07/publicaciones/publicaciones/sistema-cutzamala.pdf

Cooner, A. J., Shao, Y., & Campbell, J. B. (2016). Detection of urban damage using remote sensing and machine learning algorithms: Revisiting the 2010 Haiti earthquake. *Remote Sensing (Basel)*, *8*(10), 868. DOI: 10.3390/rs8100868

Cooper, A. C., Cooke, M. L., Takayama, K., Sumy, D. F., & Mcbride, S. (2024). From alert to action: Earthquake early warning and deaf communities. *Natural Hazards, 120*(14), 13573–13594. DOI: 10.1007/s11069-024-06719-6

Coppola, D. (2015). *Introduction to International Disaster Management. Butterworth-Heinemann*. Elsevier.

Costa, H., Floater, G., Hooyberghs, H., Verbeke, S., & Ridder, K. D. (2016). *Climate change, heat stress and labour productivity: A cost methodology for city economies*. Research Papers in Economics.

Creutin, J. D., Borga, M., Anquetin, S., Gaume, E., Gruntfest, E., & Lutoff, C. (2013). A space–time framework for analyzing human anticipation of flash floods. *Journal of Hydrology (Amsterdam), 482*, 14–24. DOI: 10.1016/j.jhydrol.2012.11.009

Csaszar, F. A., Ketkar, H. J., & Kim, H. (2024). Artificial Intelligence and Strategic Decision-Making: Evidence from Entrepreneurs and Investors. *Strategy Science, 9*(4), 322–345. DOI: 10.1287/stsc.2024.0190

Cui, J., Zhang, T., Shen, Y., Wang, X., Yang, L., Huang, X., Huang, Q., Yang, Y., Yang, W., & Li, Z. (2025). Probability-based early warning for seasonal influenza in China: Model development study. JMIR Medical Informatics, 13(v13i1e73631), e73631. DOI: 10.2196/73631

Dağdeviren, U., Demir, A. S., Erden, C., Kökçam, A. H., & Kurnaz, T. F. (2025). AI using ensemble machine learning with integrated SHapley additive explanations (SHAP)-Borda approach for estimation of the safety factor against soil liquefaction. *Environmental Earth Sciences, 84*(17), 507. Advance online publication. DOI: 10.1007/s12665-025-12466-z

Dahl, G. E., Yu, D., Deng, L., & Acero, A. (2011). Context-dependent pre-trained deep neural networks for large-vocabulary speech recognition. *IEEE Transactions on Audio, Speech, and Language Processing, 20*(1), 30–42. DOI: 10.1109/TASL.2011.2134090

Danziger, M. M., Shekhtman, L. M., Bashan, A., Berezin, Y., & Havlin, S. (2016). Vulnerability of interdependent networks and networks of networks. *Springer*, 79–99. DOI: 10.1007/978-3-319-23947-7_5

Das, T., & Choudhury, D. (2025). Bayesian network based probabilistic approach for uncertainty analysis of earthquake induced landslides. *Soils and Foundations*. https://www.sciencedirect.com/science/article/pii/S003808062500054X

Das, D. K., & Devadas, V. (2025). Smart and Resilient Infrastructure in the Wake of Climate Change. *Urban Planning*, *10*, 11373. Advance online publication. DOI: 10.17645/up.11373

Daud, S. M. S. M., Yusof, M. Y. P. M., Heo, C. C., Khoo, L. S., Singh, M. K. C., Mahmood, M. S., & Nawawi, H. (2022). Applications of drone in disaster management: A scoping review. *Science & Justice*, *62*(1), 30–42. DOI: 10.1016/j.scijus.2021.11.002 PMID: 35033326

De Cubber, G., Serrano, D., Berns, K., Chintamani, K., Sabino, R., Ourevitch, S., . . . Baudoin, Y. (2013, October). Search and rescue robots developed by the european icarus project. In 7th Int. Workshop on Robotics for Risky Environments. Citeseer.

De Lautour, O., & Omenzetter, P. (2009). Prediction of seismic-induced structural damage using artificial neural networks. *Elsevier*. https://www.sciencedirect.com/science/article/pii/S0141029608003805

Dedola, L., Ehrmann, M., Hoffmann, P., Lamo, A., Paz-Pardo, G., Slačálek, J., & Strasser, G. (2023). Digitalisation and the Economy. SSRN *Electronic Journal*. DOI: 10.2139/ssrn.4626463

DeepMind. G. (2025, March 5). *WeatherNext*. Retrieved from Google DeepMind: https://deepmind.google/science/weathernext

Deng, Z., Zhao, H., Li, L., Liu, G., Lin, H., & Adam, T. D. (2023). The climate adaptive characteristics of urban inside/outside water bodies based on their cooling effect in Poyang and Dongting lake regions, China. *Heliyon*, *9*(5), e15974. Advance online publication. DOI: 10.1016/j.heliyon.2023.e15974 PMID: 37215833

Department of Computer Science and Engineering Anantapur, S. R. I. of T. (2024). Adaptive artificial intelligence techniques for QoS and congestion management in 5G wireless networks: A literature. INTERANTIONAL JOURNAL OF SCIENTIFIC RESEARCH IN ENGINEERING AND MANAGEMENT, 08(06), 1–5. DOI: 10.55041/IJSREM36077

Department of Environment, Forestry and Fisheries. (2020). National Climate Risk and Vulnerability (CRV) Assessment Framework summary document. Pretoria, South Africa. DEFF.

Dhulipala, S. L. N. (2021). Gaussian Kernel Methods for Seismic Fragility and Risk Assessment of Mid-Rise Buildings. *Mdpi.Com*. https://www.mdpi.com/2071-1050/13/5/2973

Diekert, F., Heyen, D., Frikk, N., & Soheil, S. (2024). Balancing the Risk of Tipping: Early Warning Systems from Detection to Management, CESifo Working Paper, No. 10892, Center for Economic Studies and ifo Institute (CESifo), Munich, Germany. https://www.econstor.eu/bitstream/10419/295981/1/cesifo1_wp10892.pdf

Ding, W., Liu, M., Wu, Y., & Chen, H. (2024). How to expand the cooling capacity of blue and green spaces in peri-urban areas throughout the entire diurnal cycle: Evidence from an inland multilake city. *Journal of Cleaner Production*, *444*, 141165. Advance online publication. DOI: 10.1016/j.jclepro.2024.141165

Disaster Management Division. (n.d.). The Disaster Management Act, 2005. Ministry of Home Affairs. Government of India. https://ndmindia.mha.gov.in/images/The%20Disaster%20Management%20Act,%202005.pdf

Doğrusöz, L. A., & Yazıcı, S. (2025). Measuring organisational governance capacity in healthcare organisations: A scale development and validation study. *BMC Health Services Research*, *25*(1), 338. DOI: 10.1186/s12913-025-12442-0 PMID: 40045299

Domingos, B., & Nagamatsu, S. (2024). Why are community-based early warning systems inadequate? A case study of the Licungo River Basin in Mozambique. *Journal of Integrated Disaster Risk Management*, *14*(2), 95–126. DOI: 10.5595/001c.123357

Dore, P., Chakkor, S., El Oualkadi, A., & Baghouri, M. (2023). Real-time intelligent system for wind turbine monitoring using fuzzy system. *E-prime-Advances in Electrical Engineering. Electronics and Energy*, *3*, 100096. DOI: 10.1016/j.prime.2022.100096

Doshi, J., Basu, S., & Pang, G. (2018). From satellite imagery to disaster insights. arXiv preprint arXiv:1812.07033.

Došilović, F. K., Brčić, M., & Hlupić, N. (2018, May). Explainable artificial intelligence: A survey. In 2018 41st International convention on information and communication technology, electronics and microelectronics (MIPRO) (pp. 0210-0215). IEEE. DOI: 10.23919/MIPRO.2018.8400040

Du, H., & Zhou, F. (2022). Mitigating Extreme Summer Heat Waves with the Optimal Water-Cooling Island Effect Based on Remote Sensing Data from Shanghai, China. *International Journal of Environmental Research and Public Health*, *19*(15), 9149. Advance online publication. DOI: 10.3390/ijerph19159149 PMID: 35897518

Dunleavy, P., & Margetts, H. (2023). Data science, artificial intelligence and the third wave of digital era governance. *Public Policy and Administration*. Advance online publication. DOI: 10.1177/09520767231198737

Ebhuoma, E. E., Nene, N. J., & Leonard, L. (2024). Analysis of Urban Households' Preparedness and Municipal Interventions to Build Flood Resilience in Durban, South Africa: Implications for SDG 11. *Environmental and Sustainability Indicators, 23*, 100454. DOI: 10.1016/j.indic.2024.100454

ECMWF. (2025). *Unlocking the Black Box: The Potential of Explainable AI in Geoscience*. Retrieved from ECMWF: https://www.ecmwf.int/en/about/media-centre/focus/2025/unlocking-black-box-potential-explainable-ai-geoscience

Egusquiza, A., Cortese, M., & Perfido, D. (2019).Mapping of innovative governance models to overcome barriers for nature based urban regeneration. DOI: 10.1088/1755-1315/323/1/012081

Eisenstein, J. (2019). *Introduction to natural language processing*. MIT press.

Ejjami, R. (2024). The Adaptive Human-AI Synergy in logistics (AHASL) theory. Journal of Next-Generation Research 5.0. DOI: 10.70792/jngr5.0.v1i1.12

El Arab, R. A., Al Moosa, O. A., Albahrani, Z., Alkhalil, I., Somerville, J., & Abuadas, F. (2025). Integrating artificial intelligence into perinatal care pathways: A scoping review of reviews of applications, outcomes, and equity. *Nursing Reports, 15*(8), 281. DOI: 10.3390/nursrep15080281 PMID: 40863668

El Morr, C., Ozdemir, D., Asdaah, Y., Saab, A., El-Lahib, Y., & Sokhn, E. S. (2024). AI-based epidemic and pandemic early warning systems: A systematic scoping review. *Health Informatics Journal, 30*(3), 14604582241275844. DOI: 10.1177/14604582241275844 PMID: 39172555

Elhishi, S., & Elashry, A. (2023). Unboxing machine learning models for concrete strength prediction using XAI. *Nature.Com.* https://www.nature.com/articles/s41598-023-47169-7

Elmarakby, E., & Elkadi, H. (2024). Prioritising urban heat island mitigation intervention: Mapping a heat risk index. *The Science of the Total Environment, 948*, 174927. Advance online publication. DOI: 10.1016/j.scitotenv.2024.174927 PMID: 39038684

Eskom. (2023) Understanding load shedding. [Online] Available at: https://www.eskom.co.za/loadshedding

Fang, X., Li, J., & Ma, Q. (2023). Integrating green infrastructure, ecosystem services and nature-based solutions for urban sustainability: A comprehensive literature review. *Sustainable Cities and Society, 98*, 104843. Advance online publication. DOI: 10.1016/j.scs.2023.104843

Fanni, S. C., Febi, M., Aghakhanyan, G., & Neri, E. (2023). Natural language processing. In *Introduction to Artificial Intelligence* (pp. 87–99). Springer International Publishing. DOI: 10.1007/978-3-031-25928-9_5

Farrar, C. R., Park, G., Allen, D. W., & Todd, M. D. (2006). Sensor network paradigms for structural health monitoring. *Structural Control and Health Monitoring, 13*(1), 210–225. DOI: 10.1002/stc.125

Farrar, C., & Worden, K. (2012). *Structural health monitoring: a machine learning perspective*. John Wiley & Sons, Ltd. DOI: 10.1002/9781118443118

Farrokhzad, F., Choobbasti, A. J., Barari, A., & Ibsen, L. B. (2011). Assessing landslide hazard using artificial neural network: Case study of Mazandaran, Iran. *Carpathian Journal of Earth and Environmental Sciences, 6*(1).

Fatima, F., Hyatt, J. C., Rehman, S. U., Cruz, E. D. L., Nadella, G. S., & Meduri, K. (2024). Resilience and Risk Management in Cybersecurity: A Grounded Theory Study of Emotional, Psychological, and Organizational Dynamics. *Journal of Economy and Technology, 2*, 247–257. DOI: 10.1016/j.ject.2024.08.004

Fawad, M., Salamak, M., Poprawa, G., Koris, K., Jasinski, M., Lazinski, P., Piotrowski, D., Hasnain, M., & Gerges, M. (2023). Automation of structural health monitoring (SHM) system of a bridge using BIMification approach and BIM-based finite element model development. *Scientific Reports, 13*(1), 13215. DOI: 10.1038/s41598-023-40355-7 PMID: 37580350

Fazeli, S., Haghani, M., Mojtahedi, M., & Rashidi, T. H. (2024). The role of individual preparedness and behavioural training in natural hazards: A scoping review. *International Journal of Disaster Risk Reduction, 105*, 104379. DOI: 10.1016/j.ijdrr.2024.104379

Federici, F. M., & O'Brien, S. (Eds.). (2020). *Translation in Cascading Crises* (p. 246). Routledge.

Feizizadeh, B., Omarzadeh, D., Mohammadnejad, V., Khallaghi, H., Sharifi, A., & Karkarg, B. G. (2023). An integrated approach of artificial intelligence and geoinformation techniques applied to forest fire risk modeling in Gachsaran, Iran. *Journal of Environmental Planning and Management, 66*(6), 1369–1391. DOI: 10.1080/09640568.2022.2027747

Feng, R., & Balling, N. (2021). Bayesian convolutional neural networks for seismic facies classification. *Ieeexplore.Ieee. Org, 41*(10), 1809–1822. DOI: 10.1111/risa.13690

Ferguson, L., & Anna, M. (2024). The impact of heat adaptation on socioeconomically vulnerable populations. DOI: 10.1016/B978-0-443-13502-6.00010-5

Ferguson, M. (2025, May 26). *Microsoft's Aurora AI Revolutionizes Weather Forecasting*. Retrieved from OpenTools: https://opentools.ai/news/microsofts-aurora-ai-revolutionizes-weather-forecasting

Fernández, C. T., & Navarro, L M. (2024). *Analysis of the Impact of Urban Green Spaces on Mitigating the Urban Heat Island Effect*. Innovation Science and Technology., DOI: 10.56397/IST.2024.05.02

Fernández-Herrero, J. (2024). Evaluating recent advances in affective intelligent tutoring systems: A scoping review of educational impacts and future prospects. *Education Sciences*, *14*(8), 839. DOI: 10.3390/educsci14080839

Fernando, Y. S. (2023). TRANSFORMING EMERGENCY RESPONSE AND DISASTER MANAGEMENT WITH AI: DEVELOPING INTELLIGENT SYSTEMS FOR TRIAGE, RESOURCE ALLOCATION, AND SITUATIONAL AWARENESS IN HEALTHCARE CRISES. *International Journal of Applied Machine Learning and Computational Intelligence*, *13*(12), 11–20.

Fertier, A., Barthe-Delanoë, A. M., Montarnal, A., Truptil, S., & Bénaben, F. (2020). A new emergency decision support system: The automatic interpretation and contextualisation of events to model a crisis situation in real-time. *Decision Support Systems*, *133*, 113260. DOI: 10.1016/j.dss.2020.113260

Flint, M., Dhulipala, L., & Shahtaheri, Y. (2016). Developing a decision framework for multi-hazard design of resilient, sustainable buildings. *Researchgate. Net*, *142*(9). Advance online publication. DOI: 10.1061/(ASCE)ST.1943-541X.0001545

Florido, E., Martínez-Álvarez, F., Morales-Esteban, A., Reyes, J., & Aznarte-Mellado, J. L. (2015). Detecting precursory patterns to enhance earthquake prediction in Chile. *Computers & Geosciences*, *76*, 112–120. DOI: 10.1016/j.cageo.2014.12.002

Führer, K., Kwakkel, J. H., d, F. M., ', N. A., Hont, N. A., Rouwette, E. A. J. A., v, C. E., & Daalen, . (2025). Towards participatory decision-making under deep uncertainty: Benefits and challenges. *Inderscienceonline. Com*, *25*(2), 150–173. DOI: 10.1504/IJTPM.2025.147284

Fulton, R., Fulton, D., Hayes, N., & Kaplan, S. (2024). *The Transformation Risk-Benefit Model of Artificial Intelligence: Balancing Risks and Benefits Through Practical Solutions and Use Cases. arXiv*. Cornell University., DOI: 10.5121/ijaia.15201

Gallay, I., Olah, B., Murtinová, V., & Gallayová, Z. (2023). Quantification of the Cooling Effect and Cooling Distance of Urban Green Spaces Based on Their Vegetation Structure and Size as a Basis for Management Tools for Mitigating Urban Climate. *Sustainability (Basel), 15*(4), 3705. Advance online publication. DOI: 10.3390/su15043705

Gambhir, A., Albert, M., Doe, S. S. P., Donges, J. F., Farajalla, N., Giatti, L. L., Gundimeda, H., Hendel-Blackford, S., Homer-Dixon, T., Hoyer, D., Adán, S., Polit, D. J., Kemp, L., Korowicz, D., Kovacic, Z., Kwakkel, J., Laybourn-Langton, L., Lempert, R. J., Mahamoud, A., … Richardson, R. (2025). A systemic risk assessment methodological framework for the global polycrisis [Review of *A systemic risk assessment methodological framework for the global polycrisis*]. *Nature Communications, 16*(1), 7382. Nature Portfolio. DOI: 10.1038/s41467-025-62029-w

Gao, K., Haddad, S., Paolini, R., Altheeb, M., Mogirah, A. A., Moammar, A. B., & Santamouris, M. (2024). The use of green infrastructure and irrigation in the mitigation of urban heat in a desert city. *Building Simulation, 17*(5), 679–694. Advance online publication. DOI: 10.1007/s12273-024-1110-0

Gardezi, M., Joshi, B., Rizzo, D. M., Ryan, M., Prutzer, E., Brugler, S., & Dadkhah, A. (2024). Artificial intelligence in farming: Challenges and opportunities for building trust. *Agronomy Journal, 116*(3), 1217–1228. DOI: 10.1002/agj2.21353

Gcabashe, S. W. (2025). *An Evaluation of Disaster Risk Management in Local Government: The Case of the Nelson Mandela Bay Municipality*. Master's Dissertation. Gqeberha, South Africa: Nelson Mandela University.

Geman, B. (2025, July 17). *Nvidia says New AI Weather Forecasting Model boosts Speed, Accuracy*. Retrieved from Axios: Energy & Climate: https://www.axios.com/2025/07/17/nvidia-fourcastnet-3-ai-weather-forecasts

Geng, X., Zhang, D., Sun, W., Yuan, Y., Li, C., & Yu, Z. (2024). Heterogenous impact of background meteorological factors on cooling effect of two typical urban parks: Evidence from Shanghai, China. *Building and Environment, 265*, 112024. Advance online publication. DOI: 10.1016/j.buildenv.2024.112024

Ghani, S., Kumari, S., Jaiswal, S., & Sawant, V. A. (2022). Comparative and parametric study of AI-based models for risk assessment against soil liquefaction for high-intensity earthquakes. *Arabian Journal of Geosciences, 15*(14), 1262. DOI: 10.1007/s12517-022-10534-3

Giorgio, M., & Iervolino, I. (2016). On multisite probabilistic seismic hazard analysis. *Bulletin of the Seismological Society of America, 106*(3), 1223–1234. Advance online publication. DOI: 10.1785/0120150369

Givigi, S., & Jardine, P. T. (2018). Machine learning for data-driven control of robots. *IEEE Potentials*, *37*(4), 35–39. DOI: 10.1109/MPOT.2018.2824398

Gobatti, L., Bach, P. M., Scheidegger, A., & Leitão, J. P. (2023). Using satellite imagery to investigate Blue-Green Infrastructure establishment time for urban cooling. *Sustainable Cities and Society*, *97*, 104768. Advance online publication. DOI: 10.1016/j.scs.2023.104768

Govender, I. H., Reddy, M., & Pillay, R. P. (2025). A review of residual flood risks in South African-vulnerable coastal communities: Opportunities to influence policy. In *Climate Policies* (p. 143). IntechOpen., DOI: 10.5772/intechopen.1008977

Grabowski, Z. J., Teixeira, L. H., Fakirova, E., Meyer, S. T., Traidl-Hoffmann, C., Schloter, M., & Helmreich, B. (2023). Cosmopolitan conservation: The multi-scalar contributions of urban green infrastructure to biodiversity protection. *Biodiversity and Conservation*, *32*(11), 3595–3606. Advance online publication. DOI: 10.1007/s10531-023-02614-x

Graves, A., Mohamed, A. R., & Hinton, G. (2013, May). Speech recognition with deep recurrent neural networks. In 2013 IEEE international conference on acoustics, speech and signal processing (pp. 6645-6649). Ieee.

Greenwood, L., Hess, D., & Schneider, J. (2025). Building a Strategic System for Resilience in a Risky World. *Systems*, *13*(9), 805. DOI: 10.3390/systems13090805

Gregory, J., & Sovacool, B. K. (2019). Rethinking the governance of energy poverty in sub-Saharan Africa: Reviewing three academic perspectives on electricity infrastructure investment. *Renewable & Sustainable Energy Reviews*, *111*, 344–354. DOI: 10.1016/j.rser.2019.05.021

Grewal, J. S., Habahbeh, L., Acharyya, M., Aravind, R., Bhagaloo, S., Carey, M., Er, C., Farrugia, K., & Leung, K. (2022). COVID-19 and the effectiveness of ERM frameworks. *British Actuarial Journal*, *27*, e23. Advance online publication. DOI: 10.1017/S1357321722000174

Grossmann, T., Komorowska, U., & Latz, J. (2024). Can physics-informed neural networks beat the finite element method? *Academic.Oup.Com*. https://academic.oup.com/imamat/article-abstract/89/1/143/7680268

Guma, P. K. (2022). On tackling infrastructure: The need to learn from marginal cities and populations in the Global South. *Journal of the British Academy*, *10*, 29–37. DOI: 10.5871/jba/010.029

Gumiran, B. A., & Daag, A. (2021). Negotiated participatory action research for multi-stakeholder implementation of early warning systems for landslides. *International Journal of Disaster Risk Reduction*, *58*, 102184. DOI: 10.1016/j.ijdrr.2021.102184

Gunning, D., Stefik, M., Choi, J., Miller, T., Stumpf, S., & Yang, G. Z. (2019). XAI—Explainable artificial intelligence. *Science Robotics*, *4*(37), eaay7120. DOI: 10.1126/scirobotics.aay7120 PMID: 33137719

Guo, Y., Ren, Z., Dong, Y., Hu, N., Wang, C., Zhang, F., Jia, G., & He, X.-J. (2022). Strengthening of surface urban heat island effect driven primarily by urban size under rapid urbanization: National evidence from China. *GIScience & Remote Sensing*, *59*(1), 2127–2143. Advance online publication. DOI: 10.1080/15481603.2022.2147301

Gupta, A., & De, B. (2024).A systematic review on urban blue-green infrastructure in the south Asian region: recent advancements, applications, and challenges. DOI: 10.2166/wst.2024.014

Haase, A. (2024). Inequalities and Injustices of Urban Green Regeneration: Applying the Conflict Analysis Perspective. *Land (Basel)*, *13*(3), 296. Advance online publication. DOI: 10.3390/land13030296

Hadjimichael, A., & J Schlumberger. (2024). Data visualisation for decision making under deep uncertainty: current challenges and opportunities. *Iopscience.Iop.Org*. DOI: 10.1088/1748-9326/ad858b

Hadjinicolaou, P., Economou, T., Lazoglou, G., & Whitt, P. (2023).Quantification of the Urban Heat Island effect using paired station data in the Middle East and North Africa region. DOI: 10.5194/ems2023-13

Haffarian, S., Alidoost, F., Lagap, U., Chandramouli, P., Dzigan, Y., Grootes, M., Jalayer, F., & Kelman, I. (2024). Digital twins for early warning systems: Intricacies and solutions. DOI: 10.5194/egusphere-egu24-15738

Haigh, R. (2017). *Disaster Management Lifecycle*. Learning Package.

Hajikhodaverdikhan, P., Nazari, M., Mohsenizadeh, M., Shamshirband, S., & Chau, K. W. (2018). Earthquake prediction with meteorological data by particle filter-based support vector regression. *Engineering Applications of Computational Fluid Mechanics*, *12*(1), 679–688. DOI: 10.1080/19942060.2018.1512010

Han, H., Choi, C., Jung, J., & Kim, H. S. (2021). Deep Learning with Long Short Term Memory Based Sequence-to-Sequence Model for Rainfall-Runoff Simulation. *Water (Basel)*, *13*(4), 437. DOI: 10.3390/w13040437

Han, Z., & Wu, G. (2024). Why do people not prepare for disasters? A national survey from China. *Npj Natural Hazards*, *1*(1), 1. Advance online publication. DOI: 10.1038/s44304-024-00001-2

Haque, A. (2025). Responsible artificial intelligence (AI) in healthcare: A paradigm shift in leadership and strategic management. *Leadership in Health Services*, *38*(4), 644–656. Advance online publication. DOI: 10.1108/LHS-01-2025-0018 PMID: 40911384

Haque, U., Hashizume, M., Kolivras, K. N., Overgaard, H. J., Das, B., & Yamamoto, T. (2012). Reduced death rates from cyclones in Bangladesh: What more needs to be done? *Bulletin of the World Health Organization*, *90*(2), 150–156. DOI: 10.2471/BLT.11.088302 PMID: 22423166

Harper, S. (2020). Global Risks and the Resilience of Future Health Care Systems. *Journal of Population Ageing*, *13*(1), 1–3. DOI: 10.1007/s12062-020-09262-x PMID: 35368879

Hasan, M., Tsegaye, T., Shi, X., Schaefer, G., & Taylor, G. (2008). Model for predicting rainfall by fuzzy set theory using USDA scan data. *Agricultural Water Management*, *95*(12), 1350–1360. DOI: 10.1016/j.agwat.2008.07.015

Hassan, Z., Shahbaz, B., & Lopez, F. G. (2023).Enhancing Blue/Green Infrastructure for Resilient Urban Environments: Smart Solutions and Nature-Based Strategies. DOI: 10.61326/icelis.2023.18

Hay, J. C., Martin, F. C., & Wightman, C. W. (1960, January). The mark-1 perceptron-design and performance. *Proceedings of the Institute of Radio Engineers*, *48*(3), 398–399.

Hecker, N., & Domres, B. D. (2018). The German emergency and disaster medicine and management system—history and present. *Chinese Journal of Traumatology = Zhonghua Chuang Shang Za Zhi*, *21*(2), 64–72. DOI: 10.1016/j.cjtee.2017.09.003 PMID: 29622286

Helmrich, A., & Chester, M. (2022). Navigating Exploitative and Explorative Leadership in Support of Infrastructure Resilience. *Frontiers in Sustainable Cities*, *4*, 791474. Advance online publication. DOI: 10.3389/frsc.2022.791474

He, M., Yuan, C., Zhang, X., Wang, P., & Yao, C. (2023). Impacts of Blue-Green-grey infrastructures on high-density urban thermal environment at multiple spatial scales: A case study in Wuhan. *Urban Climate*, *52*, 101714. Advance online publication. DOI: 10.1016/j.uclim.2023.101714

Hendrickse, R. (2023). Ethnic leadership in the age of disruption: implications for South African municipalities. *International Journal of Research in Business and Social Science (2147-4478), 12*(7), 224. DOI: 10.20525/ijrbs.v12i7.2791

Henrico, S., & Els, S. (2025). Cyber Attacks in South Africa: Geopolitical and legal implications. *African Security Review, 1*(3), 312–336. Advance online publication. DOI: 10.1080/10246029.2025.2489352

Herald, L. I. V. E. (2024). *Heavy rains wreak havoc in Nelson Mandela Bay, causing death and destruction.* HeraldLIVE. https://www.theherald.co.za/news/2024-06-03-heavy-rains-wreak-havoc-in-nelson-mandela-bay-causing-death-and-destruction/

Hermans, T. D. G., Šakić Trogrlić, R., van den Homberg, M. J. C., Bailon, H., Sarku, R., & Mosurska, A. (2022). Exploring the integration of local and scientific knowledge in early warning systems for disaster risk reduction: a review. In *Natural Hazards* (Vol. 114, Issue 2, pp. 1125–1152). Springer Science and Business Media B.V. DOI: 10.1007/s11069-022-05468-8

He, X., Xu, P., An, B.-W., & Guo, Q. (2024). Urban green infrastructure: Bridging biodiversity conservation and sustainable urban development through adaptive management approach. *Frontiers in Ecology and Evolution, 12*, 1440477. Advance online publication. DOI: 10.3389/fevo.2024.1440477

Holling, C. S. (1973). Resilience and Stability of Ecological Systems. *Annual Review of Ecology and Systematics, 4*(1), 1–23. DOI: 10.1146/annurev.es.04.110173.000245

Hostettler, S., Besson, S. N., & Bolay, J. (2018). *Technologies for Development.* DOI: 10.1007/978-3-319-91068-0

Hou, R., Liu, Y., Zhang, H., & Li, J. (2024). The mediating role of satisfaction with warning services and community engagement in disaster response behaviour. *International Journal of Disaster Risk Reduction, 95*, 104187. DOI: 10.1016/j.ijdrr.2023.104187

Hsieh, C. H., Tai, H. H., & Lee, Y. N. (2014). Port vulnerability assessment from the perspective of critical infrastructure interdependency. *Taylor & Francis, 41*(6), 589–606. DOI: 10.1080/03088839.2013.856523

Huangfu, D., Wang, X., & Dai, H. (2025). Fire safety knowledge and awareness in high-rise residential buildings: An empirical study in China. *Scientific Reports, 15*(7), 41083. Advance online publication. DOI: 10.1038/s41598-025-24979-5 PMID: 41266510

Huang, M., Zhong, S., Mei, X., & He, J. (2024). Spatiotemporal Patterns in the Urban Heat Island Effect of Several Contemporary and Historical Chinese "Stove Cities.". *Sustainability (Basel), 16*(7), 3091. Advance online publication. DOI: 10.3390/su16073091

Huang, W. T. K., Masselot, P., Bou-Zeid, E., Fatichi, S., Paschalis, A., Sun, T., Gasparrini, A., & Manoli, G. (2023). Economic valuation of temperature-related mortality attributed to urban heat islands in European cities. *Nature Communications, 14*(1), 7438. Advance online publication. DOI: 10.1038/s41467-023-43135-z PMID: 37978178

Huang, X., & Jensen, J. R. (1997). A machine-learning approach to automated knowledge-base building for remote sensing image analysis with GIS data. *Photogrammetric Engineering and Remote Sensing, 63*(10), 1185–1193.

Huerta-Vergara, A. R., Arciniega-Esparza, S., Pedrozo-Acuña, A., Matus-Kramer, A., & Vega-López, E. (2022). Assessment of vulnerability to water shortage in the municipalities of Mexico City. *Boletín de la Sociedad Geológica Mexicana, 74*(1), A071021. DOI: 10.18268/BSGM2022v74n1a071021

Hu, N., Wang, G., Ma, Z., Ren, Z., Zhao, M., & Meng, J. (2023). The cooling effects of urban waterbodies and their driving forces in China. *Ecological Indicators, 156*, 111200. Advance online publication. DOI: 10.1016/j.ecolind.2023.111200

Hussain, A., Tahir, A., Hussain, Z., Sheikh, Z., Gogate, M., Dashtipour, K., Ali, A., & Sheikh, A. (2021). Artificial intelligence-enabled analysis of public attitudes on Facebook and twitter toward COVID-19 vaccines in the United Kingdom and the United States: Observational study. *Journal of Medical Internet Research, 23*(4), e26627. DOI: 10.2196/26627 PMID: 33724919

Ibrahim, R., Motshegwa, T., Benouar, D., Khodja, M., Ead, H. A., Zhang, L., Crawley, F. P., Mchunu, N. P., Hachigonta, S., Hodson, S., McBride, V. A., & Anshebo, T. Y. (2025). Sustainable North-South Africa Collaboration for Disaster and Crisis Management: A Strategic Capacity Development Framework using Open Science, Artificial Intelligence and Geoinformatics. *Data Science Journal, 24*, 11. Advance online publication. DOI: 10.5334/dsj-2025-011

Ifenthaler, D., Hofhues, S., Egloffstein, M., & Helbig, C. (2021). *Digital Transformation of Learning Organizations*. DOI: 10.1007/978-3-030-55878-9

Igini, M. (2025, February 26). *Europe's New AI Weather Forecasting Model Up to 20% More Accurate Than Conventional Methods*. Retrieved from Earth.Org: https://earth.org/europes-new-ai-weather-forecasting-model-up-to-20-more-accurate-than-conventional-methods/

Ikram, A., & Qamar, U. (2015). Developing an expert system based on association rules and predicate logic for earthquake prediction. *Knowledge-Based Systems*, *75*, 87–103. DOI: 10.1016/j.knosys.2014.11.024

Inamura, T., Yamada, H., Morinaga, K., Yamanobe, N., Hanai, R., & Domae, Y. (2025). Development and evaluation of a human-robot collaborative training system for retail stores using virtual reality and digital twin technologies. *Journal of Robotics and Mechatronics*, *37*(2), 478–487. DOI: 10.20965/jrm.2025.p0478

Independent Communications Authority of South Africa (ICASA). (2024). *ICASA conducts public hearings on the possible effects of load shedding on the electronic communications, broadcasting, and postal sectors*. https://www.icasa.org.za/news/2024/icasa-to-conducts-public-hearings-on-possible-effects-of-loadshedding-on-the-electronic-communications-broadcasting-and-postal-sectors

Independent Researcher Texas, U. S. (2025). Ethical challenges in data science: Navigating the complex landscape of responsibility and fairness. *International Journal of Current Science Research and Review*, *08*(03). Advance online publication. DOI: 10.47191/ijcsrr/V8-i3-09

INEGI. (2021). En México somos 126 014 024 habitantes: Censo de población y vivienda 2020. Comunicado de prensa Núm. 24/21, 25 de enero de 2021, 1-3. https://www.inegi.org.mx/contenidos/saladeprensa/boletines/2021/EstSociodemo/ResultCenso2020_Nal.pdf

International, H. (2024, March 26). *North.io and NVIDIA Advance Ocean Data Processing towards a Sustainable Future*. Retrieved from Hydro International: https://www.hydro-international.com/content/news/north-io-and-nvidia-advance-ocean-data-processing-towards-a-sustainable-future

IPCC. (2021). *Climate Change 2021: Synthesis report. Contribution of Working Group I to the Sixth Assessment Report of the Intergovernmental Panel on Climate Change* [Masson-Delmotte, V., P. Zhai, A. Pirani, S.L. Connors, C. Péan, S. Berger, N. Caud, Y. Chen, L. Goldfarb, M.I. Gomis, M. Huang, K. Leitzell, E. Lonnoy, J.B.R. Matthews, T.K. Maycock, T. Waterfield, O. Yelekçi, R. Yu, & B. Zhou (eds.)]. Cambridge University Press., DOI: 10.1017/9781009157896

Irwan, D., Ahmed, A., Ibrahim, S. L., Ibrahim, I., Mahmoud, M., Jacky, G., Nurhakim, A., Chah, M., Kumar, P., Sherif, M., & El-Shafie, A. (2025). Advancements in rainfall-runoff prediction: Exploring state-of-the-art neural computing modeling approaches. *Alexandria Engineering Journal*, *121*, 138–149. DOI: 10.1016/j.aej.2025.02.060

Islam, Md. M., Hasan, M., Mia, Md. S., Masud, A. A., & Islam, A. R. Md. T. (2025). Early Warning Systems in Climate Risk Management: Roles and Implementations in Eradicating Barriers and Overcoming Challenges. *Natural Hazards Research*. DOI: 10.1016/j.nhres.2025.01.007

Ivić, M. (2019). Artificial intelligence and geospatial analysis in disaster management. *The International Archives of the Photogrammetry, Remote Sensing and Spatial Information Sciences, 42*(W8), 161–166. DOI: 10.5194/isprs-archives-XLII-3-W8-161-2019

Iwuanyanwu, O., Gil-Ozoudeh, I., Okwandu, A. C., & Ike, C. S. (2024). *Cultural and social dimensions of green architecture: Designing for sustainability and community well-being*. International Journal of Applied Research in Social Sciences., DOI: 10.51594/ijarss.v6i8.1477

Jaadi, M. E., Haidi, T., & Bouabdallaoui, D. (2025). *IAES International Journal of Artificial Intelligence (IJ-AI)*. IAES International Journal of Artificial Intelligence., DOI: 10.11591/ijai

Jabbar, H. K., Hamoodi, M. N., & Al-Hameedawi, A. N. M. (2023). Urban heat islands: a review of contributing factors, effects and data.*IOP Conference Series: Earth and Environmental Science*. DOI: 10.1088/1755-1315/1129/1/012038

Janarthanan, R., Balamurali, R., Annapoorani, A., & Vimala, V. (2021). Prediction of rainfall using fuzzy logic. *Materials Today: Proceedings, 37*(2), 959–963. DOI: 10.1016/j.matpr.2020.06.179

Jandaghian, Z., & Colombo, A. (2024). The Role of Water Bodies in Climate Regulation: Insights from Recent Studies on Urban Heat Island Mitigation. *Buildings (Basel, Switzerland), 14*(9), 2945. Advance online publication. DOI: 10.3390/buildings14092945

Jang, S., Bae, J., & Kim, Y.-J. (2024). Street-level urban heat island mitigation: Assessing the cooling effect of green infrastructure using urban IoT sensor big data. *Sustainable Cities and Society, 100*, 105007. Advance online publication. DOI: 10.1016/j.scs.2023.105007

Jayasekara, R., Siriwardana, C., Amaratunga, D., & Haigh, R. (2022). Evaluating the network of stakeholders in Multi-Hazard Early Warning Systems for multiple hazards amidst biological outbreaks: Sri Lanka as a case in point. *Progress in Disaster Science, 14*, 100228. DOI: 10.1016/j.pdisas.2022.100228 PMID: 35402893

Jayasiri, G. P., Siriwardena, C. S. A., Hettiarachchi, S. S. L., Dissanayake, P. B. R., & Bandara, C. S. (2018). Evaluation of community resilience aspects of Sri Lankan coastal districts. *International Journal on Advanced Science, Engineering and Information Technology*, *8*(5), 2161–2167. DOI: 10.18517/ijaseit.8.5.7095

Jermutus, E., Kneale, D., Thomas, J., & Michie, S. (2022). Influences on user trust in healthcare artificial Intelligence: A systematic review. *Wellcome Open Research*, *7*, 65. DOI: 10.12688/wellcomeopenres.17550.1

Jia, S., & Wang, Y. (2023).Multiscale Evaluation of the Cooling Effect of Greenspace in Urban Environments. DOI: 10.1007/978-981-99-3626-7_75

Jiang, C., Wei, X., Cui, X., & You, D. (2009). Application of support vector machine to synthetic earthquake prediction. *Earth Science*, *22*(3), 315–320. DOI: 10.1007/s11589-009-0315-8

Jiang, J., El Naggar, H., Xu, C., Chen, G., & Du, X. (2023). Seismic fragility analysis for subway station considering varying ground motion ensembles. *Soil Dynamics and Earthquake Engineering*, *165*, 107705. DOI: 10.1016/j.soildyn.2022.107705

Jiang, Y., Huang, J., Shi, T., & Li, X. (2021). Cooling Island Effect of Blue-Green Corridors: Quantitative Comparison of Morphological Impacts. *International Journal of Environmental Research and Public Health*, *18*(22), 11917. Advance online publication. DOI: 10.3390/ijerph182211917 PMID: 34831672

Jiang, Y., Li, X., Luo, H., Yin, S., & Kaynak, O. (2022). Quo vadis artificial intelligence? *Discover Artificial Intelligence*, *2*(1), 4. DOI: 10.1007/s44163-022-00022-8

Ji, S., Zheng, Y., Wang, Z., & Li, T. (2019). A deep reinforcement learning-enabled dynamic redeployment system for mobile ambulances. *Proceedings of the ACM on Interactive, Mobile, Wearable and Ubiquitous Technologies*, *3*(1), 1–20. DOI: 10.1145/3314402 PMID: 40144218

Johnson, N., Zhao, G., Hunsader, E., Qi, H., Johnson, N., Meng, J., & Tivnan, B. (2013). Abrupt rise of new machine ecology beyond human response time. *Scientific Reports*, *3*(1), 2627. DOI: 10.1038/srep02627 PMID: 24022120

Jones, K. S. (1994). Natural language processing: a historical review. Current issues in computational linguistics: in honour of Don Walker, 3-16.

Jones, N. (2025, April 14). *A.I. Is Quietly Powering a Revolution in Weather Prediction.* Retrieved from E360: https://e360.yale.edu/features/artificial-intelligence-weather-forecasting

Jose Moises, D. R., Kgabi, N. A., & Kunguma, O. (2024). Policy implementation: Assessing institutional coordination and communication for flood warning in Namibia. *Jàambá, 16*(1), a1534. DOI: 10.4102/JAMBA.v16i1.1534 PMID: 39113930

Joshi, S. (2025). Leadership in the Age of AI: Review of Quantitative Models and Visualization for Managerial Decision-Making. *Preprints.Org*. DOI: 10.20944/preprints202504.1662.v1

Jung, D., Tran Tuan, V., Quoc Tran, D., Park, M., & Park, S. (2020). Conceptual framework of an intelligent decision support system for smart city disaster management. *Applied Sciences (Basel, Switzerland), 10*(2), 666. DOI: 10.3390/app10020666

Kalra, N., Hallegatte, S., Lempert, R., & Brown, C. (2014). Investment decision-making under deep uncertainty-application to climate change. *Enpc.Hal.Science*. https://enpc.hal.science/hal-00802049/document

Kangana, N., Kankanamge, N , De Silva, C., Mahamood, R., Ranasinghe, D., & Goonetilleke, A. (2025). Harnessing Mobile Technology for Flood Disaster Readiness and Response: A Comprehensive Review of Mobile Applications on the Google Play Store. *Urban Science (Basel, Switzerland), 9*(4), 106. DOI: 10.3390/urbansci9040106

Kar, B., & Meade, S. (2022). Urban Heat Island Impact on Energy Burden Among Socially Vulnerable Groups.*IEEE International Geoscience and Remote Sensing Symposium*. DOI: 10.1109/IGARSS46834.2022.9883806

Karimiziarani, M. (2023). Social Media Analytics in Disaster Response: A Comprehensive Review. arXiv preprint arXiv:2307.04046.

Kaynak, O. (2021). The Golden Age of Artificial Intelligence: Inaugural Editorial. *Discover Artificial Intelligence, 1*(1), 1–7. DOI: 10.1007/s44163-021-00009-x

Kebede, A., Warrach-Sagi, K., Schwitalla, T., Wulfmeyer, V., Abebe, T., & Tadesse, T. (2026). A multivariate approach to drought monitoring: Improving robustness and accuracy through a new drought index in regions with high climate variability, applied to the drought-prone region of Ethiopia. *Journal of Hydrology. Regional Studies, 64*, 103332. DOI: 10.1016/j.ejrh.2026.103332

Keinänen, J., Halonen, M., & Luoto, I. (2025). What can we learn about crisis management in the polycrisis era? Insights from Finland's international public policy network. *International Journal of Public Sector Management*. Advance online publication. DOI: 10.1108/IJPSM-06-2024-0191

Kelly, L., Hajistassi, M., & Ramasundram, S. (2024). Migrant and refugee communities strengthening disaster resilience. *Australian Journal of Emergency Management,* (No. 3), 49–58. https://doi.org/DOI: 10.47389/39

Kelman, I. (2020). *Disaster by choice: How our actions turn natural hazards into catastrophes*. Oxford University Press., https://global.oup.com/academic/product/disaster-by-choice-9780198841357

Kelman, I., & Fearnley, C. J. (2025). From multi-hazard early warning systems (MHEWS) to all-vulnerability warning systems (AVWS). *iScience*, *28*(7), 112977. Advance online publication. DOI: 10.1016/j.isci.2025.112977 PMID: 40687808

Khan, A., Hassan, B., Khan, S., Ahmed, R., & Abuassba, A. (2022). DeepFire: A Novel Dataset and Deep Transfer Learning Benchmark for Forest Fire Detection. *Mobile Information Systems*, *2022*, 1–14. DOI: 10.1155/2022/5358359

Kharazmi, E., Zhang, Z., & Karniadakis, G. E. (2019). *Variational Physics-Informed Neural Networks For Solving Partial Differential Equations*. https://arxiv.org/pdf/1912.00873

Khatoon, S., Alshamari, M. A., Asif, A., Hasan, M. M., Abdou, S., Elsayed, K. M., & Rashwan, M. (2021). Development of social media analytics system for emergency event detection and crisismanagement. *Computers, Materials & Continua*, *68*(3). Advance online publication. DOI: 10.32604/cmc.2021.017371

Kiani, B. H. (2022). *Towards Urban Sustainability: Impact of Blue and Green Infrastructure on Building Smart*. Climate Resilient and Livable Cities., DOI: 10.1201/9781003127819-7

Kibirige, D. (2025). I*ntegrated Flood Risk System for the Western Cape: Lessons from the September 2023 Floods*, EGU General Assembly 2025, Vienna, Austria, 27 April–2 May 2025, EGU25-9855. DOI: 10.5194/egusphere-egu25-9855

Kikasu, E. T.Eustache Tanzala Kikasu. (2025). Navigating through the Dynamic Landscape of Public Administration Governance through Innovative Adaptation and Resilience Practices and Strategies in Africa. *Journal of Information Systems Engineering & Management*, *10*(7s), 203–228. DOI: 10.52783/jisem.v10i7s.854

Kikasu, E. T., & Pillay, S. S. (2024). Forecasting an Inevitable Coalition Government at the National Level in South Africa: A New Path to Public Administration and Governance. *Open Journal of Political Science*, *14*(1), 28–51. DOI: 10.4236/ojps.2024.141003

Kim, J., Khouakhi, A., Corstanje, R., & Johnston, A. S. A. (2023). Greater local cooling effects of trees across globally distributed urban green spaces. *The Science of the Total Environment*. Advance online publication. DOI: 10.1016/j.scitotenv.2023.168494 PMID: 37979859

Kim, S., & Kim, D. (2023). Climate change and cooling equity: Spatial dynamics of vulnerable populations. *Growth and Change*. Advance online publication. DOI: 10.1111/grow.12701

Kiparsky, M., Sedlak, D. L., Thompson, B. H.Jr, & Truffer, B. (2013). The Innovation Deficit in Urban Water: The Need for an Integrated Perspective on Institutions, Organizations, and Technology. *Environmental Engineering Science*, *30*(8), 395–408. DOI: 10.1089/ees.2012.0427 PMID: 23983450

Kitazawa, K., & Hale, S. A. (2021). Social media and early warning systems for natural disasters: A case study of Typhoon Etau in Japan. *International Journal of Disaster Risk Reduction*, *52*, 101926. DOI: 10.1016/j.ijdrr.2020.101926

Koh, S. C. L., Suresh, K., Ralph, P. J., & Saccone, M. (2023). Quantifying organisational resilience: An integrated resource efficiency view. *International Journal of Production Research*, *62*(16), 5737–5756. DOI: 10.1080/00207543.2023.2296018

Kolivand, P., Azari, S., Bakhtiari, A., Namdar, P., Ayyoubzadeh, S. M., Rajaie, S., & Ramezani, M. (2025). AI applications in disaster governance with health approach: A scoping review [Review of *AI applications in disaster governance with health approach: A scoping review*]. *Archives of Public Health*, *83*(1). BioMed Central. DOI: 10.1186/s13690-025-01712-2

Konstantaras, A., Varley, M. R., Vallianatos, F., Makris, J. P., Collins, G., & Holifield, P. (2007). Detection of weak seismo-electric signals upon the recordings of the electrotelluric field by means of neuro-fuzzy technology. *IEEE Geoscience and Remote Sensing Letters*, *4*(1), 151–165. DOI: 10.1109/LGRS.2006.887068

Koshimura, S. Graduate School of Engineering Aramaki Aoba-ku Sendai 980-8579 Japan, T. U. A. 6-6-11-1104, Namegaya, Y., Yanagisawa, H., Geological Survey of Japan 1-1-1 Higashi Tsukuba 305-8567 Japan, N. I. of A. I. S. and T. C., & Tokyo Electric Power Services Co. Higashiueno 3-Chome Taito-ku Tokyo 110-0015 Japan, Ltd. 3-3. (2009). Tsunami fragility — A new measure to identify tsunami damage —. Journal of Disaster Research, 4(6), 479–488. DOI: 10.20965/jdr.2009.p0479

Krishna Pasupuleti, M. (2024a). AI and big data for climate resilience: Predictive analytics in environmental management. In AI and Big Data in Climate Change: Predictive Analytics for Environmental Management (pp. 255–280). National Education Services. DOI: 10.62311/nesx/46601

Krishna Pasupuleti, M. (2024b). AI-Powered Disaster Forecasting: Advanced tools for early warning and Risk Mitigation. In AI in Disaster Prediction: Tools for Early Warning Systems and Risk Mitigation (pp. 233–261). National Education Services. DOI: 10.62311/nesx/905759

Krizhevsky, A., Sutskever, I., & Hinton, G. E. (2017). ImageNet classification with deep convolutional neural networks. *Communications of the ACM, 60*(6), 84–90. DOI: 10.1145/3065386

Kumar, S. R. (2025). Weather Forecasting Using Artificial Intelligence and Machine Learning. *International Journal on Science and Technology (IJSAT), Volume 16, Issue 1.*

Kumar, A., & Arya, D. S. (2025). Spatiotemporal analysis and mechanisms of Drought flood abrupt alternation events in India. *Journal of Hydrology. Regional Studies, 62*, 102992. DOI: 10.1016/j.ejrh.2025.102992

Kumar, A., & Singh, R. (2024). Explainable AI for flood and air quality early warning systems. *Environmental Modelling & Software*, ●●●, 172.

Kumar, T. V., & Sud, K. (Eds.). (2020). *AI and robotics in disaster studies*. Palgrave Macmillan. DOI: 10.1007/978-981-15-4291-6

Kundu, S., Biswas, S., Tripathi, D., Karmakar, R., Majumdar, S., & Mandal, S. (2023). A review on rainfall forecasting using ensemble learning techniques. *E-prime-Advances in Electrical Engineering. Electronics and Energy, 6*, 100296. DOI: 10.1016/j.prime.2023.100296

Kunguma, O. (2020). *South African Disaster Management Framework: Assessing the Status and Dynamics of Establishing Information Management and Communication Systems in Provinces*. Doctoral dissertation. Bloemfontein, South Africa: University of the Free State. http://hdl.handle.net/11660/11002

Kunguma, O. (2022). A South African disaster legislative perspective of information management and communication systems. *South African Journal of Information Management, 24*(1), a1540. DOI: 10.4102/sajim.v24i1.1540

Kuriachan, A., Thomas, R. R., & Sukanya, R. (2024). *Place-Based Strategies, Multichannel Merger, and Context-Driven Alerts for Engagement With Mobile Marketing*. Igi Global., DOI: 10.4018/979-8-3693-1339-8.ch012

Kwag, S., & Hahm, D. (2020). Multi-objective-based seismic fragility relocation for a Korean nuclear power plant. *Springer, 103*(3), 3633–3659. DOI: 10.1007/s11069-020-04148-9

Kwon, M., Kwon, H.-H., & Han, D. (2020). A Hybrid Approach Combining Conceptual Hydrological Models, Support Vector Machines and Remote Sensing Data for Rainfall-Runoff Modeling. *Remote Sensing (Basel), 12*(11), 1801. DOI: 10.3390/rs12111801

Labib, N. S., Brust, M. R., Danoy, G., & Bouvry, P. (2021). The rise of drones in internet of things: A survey on the evolution, prospects and challenges of unmanned aerial vehicles. *IEEE Access : Practical Innovations, Open Solutions*, *9*, 115466–115487. DOI: 10.1109/ACCESS.2021.3104963

Lagmay, A. M. F., Bagtasa, G., Andal, D. F., Andal, F. D., Aldea, J., Bencito, D. C., Liporada, K., & Delmendo, P. (2024). An Impact-Based Flood Forecasting System for Citizen Empowerment. *Asian Journal of Agriculture and Development, 21*(SAI), 129–148. https://doi.org/DOI: 10.37801/ajad2024.21.20AI.8

Lakshmi, P. J., Apaza, R. A., Alkhayyat, A., Marhoon, H. A., & Alameri, A. A. (2022). Hybrid wavelet-gene expression programming and wavelet-support vector machine models for rainfall-runoff modelling. *Water Science and Technology*, *86*(12), 3205–3222. DOI: 10.2166/wst.2022.400

Lam, R. (2023, November 14). *GraphCast: AI model for Faster and More Accurate Global Weather Forecasting*. Retrieved from Google DeepMind: https://deepmind.google/discover/blog/graphcast-ai-model-for-faster-and-more-accurate-global-weather-forecasting/

Lamsal, Rabindra, and TV Vijay Kumar. (2020). "Artificial intelligence and early warning systems." AI and Robotics in Disaster Studies: 13-32.

Lanza, K., Jones, J., Acuña, F., Coudert, M., Bixler, R. P., Kamath, H. G., & Niyogi, D. (2023). Heat vulnerability of Latino and Black residents in a low-income community and their recommended adaptation strategies: A qualitative study. *Urban Climate*, *51*, 101656. Advance online publication. DOI: 10.1016/j.uclim.2023.101656

Lategan, L., Fisher-Holloway, B., Cilliers, E. J., & Cilliers, S. S. (2025). Moving on to Greener Pastures? A Review of South Africa's Housing Megaproject Literature [Review of *Moving on to Greener Pastures? A Review of South Africa's Housing Megaproject Literature*]. *Sustainability, 17*(4), 1677. Multidisciplinary Digital Publishing Institute. DOI: 10.3390/su17041677

Lawrence, M., Homer-Dixon, T., Janzwood, S., Rockstöm, J., Renn, O., & Donges, J. F. (2024). Global Polycrisis: The Causal Mechanisms of Crisis Entanglement. *Global Sustainability*, *7*, e6. Advance online publication. DOI: 10.1017/sus.2024.1

Leah. (2025, May 20). *How AI Is Revolutionizing Weather Forecasting: Models, Impacts, and the Future of Prediction*. Retrieved from BestAI: https://blog.bestai.com/how-ai-is-revolutionizing-weather-forecasting-models-impacts-and-the-future-of-prediction/

Leal, O. J. U., Fekete, A., Eudave, R. R., Matos, J. C., Sousa, H., & Teixeira, E. R. (2024). A systematic review of integrated frameworks for resilience and sustainability assessments for critical infrastructures. *Taylor & Francis, 34*(2), 266–280. DOI: 10.1080/10168664.2023.2265965

LeCun, Y., Bengio, Y., & Hinton, G. (2015). Deep learning. nature, 521(7553), 436-444.

Lee, S. (2025, June 13). *The Future of Weather Forecasting: AI-Driven Insights*. Retrieved from Number Analytics: https://www.numberanalytics.com/blog/future-of-weather-forecasting-ai-driven-insights#google_vignette

Lee, C., Wang, Z., Wang, D., Lyu, S., & Chen, C. (2025). Artificial-intelligence-driven governance: Addressing emerging risks with a comprehensive risk-prevention-centred model for public health crisis management. *Health Research Policy and Systems, 23*(1), 115. Advance online publication. DOI: 10.1186/s12961-025-01390-0 PMID: 41013547

Lefutso, D. (2025). Transformative policy and legislative strategies for flood disaster management in post-apartheid South Africa. *Journal of Disaster Science and Management, 1*(1), 15. DOI: 10.1007/s44367-025-00015-1

Lehmann, S. (2019). *Reconnecting Cities with Nature*. Building Resilience at the Urban Scale., DOI: 10.1007/978-3-030-04711-5_2

Lempert, R. J., Lawrence, J., Kopp, R. E., Haasnoot, M., Reisinger, A., Grubb, M., & Pasqualino, R. (2024). Reinforcement learning for decision-making under deep uncertainty. *Journal of Environmental Management, 6*. Advance online publication. DOI: 10.3389/fclim.2024.1380054

Lewartowska, E., Anguelovski, I., Oscilowicz, E., Triguero-Mas, M., Cole, H., Shokry, G., Perez-del-Pulgar, C., & Connolly, J. J. T. (2024). RACIAL INEQUITY IN GREEN INFRASTRUCTURE AND GENTRIFICATION: Challenging Compounded Environmental Racisms in the Green City. *International Journal of Urban and Regional Research, 48*(2), 294–322. Advance online publication. DOI: 10.1111/1468-2427.13232

Li, L. (2024, March 29). *Generative AI to quantify Uncertainty in Weather Forecasting*. Retrieved from Google Research: https://research.google/blog/generative-ai-to-quantify-uncertainty-in-weather-forecasting/

Li, A., & Kang, L. (2009, December). Knn-based modeling and its application in aftershock prediction. In *2009 International Asia Symposium on Intelligent Interaction and Affective Computing* (pp. 83-86). IEEE. DOI: 10.1109/ASIA.2009.21

Li, L., Wang, J., Wang, X., Peng, P., Shen, J., Zhu, H., & Zhang, Z. (2025). Big data and data science in global governance: Anticipating future needs and applications in the UN and beyond. *Frontiers in Political Science*, *7*, 1583772. Advance online publication. DOI: 10.3389/fpos.2025.1583772

Lin, T. Z. (2022). Suppressing Forest Fires in Global Climate Change Through Artificial Intelligence: A Case Study on British Columbia. Proceedings – 2022 International Conference on Big Data, Information and Computer Network, BDICN 2022. IEEE. DOI: 10.1109/BDICN55575.2022.00086

Li, P., Zhang, J., & Krebs, P. (2022). Prediction of Flow Based on a CNN-LSTM Combined Deep Learning Approach. *Water (Basel)*, *14*(6), 993. DOI: 10.3390/w14060993

Li, S., Liu, Y., Nguyen, A. H., Wu, Z., Al-Farsi, M. Z., Choi, T., Zhou, L., Her, Y., Li, F., Ren, D., & Romeiko, X. X. (2025). A framework for creating sustainable rainwater harvesting and reuse strategies for urban landscape irrigation in a changing climate. *Journal of Environmental Management*, *392*, 126852. DOI: 10.1016/j.jenvman.2025.126852

Lisdiono, P., Said, J., Yusoff, H., & Hermawan, A. A. (2022). Examining Leadership Capabilities, Risk Management Practices, and Organizational Resilience: The Case of State-Owned Enterprises in Indonesia. *Sustainability (Basel)*, *14*(10), 6268. DOI: 10.3390/su14106268

Liu, G., Tang, Z., Qin, H., Liu, S., Shen, Q., Qu, Y., & Zhou, J. (2022). Short-term runoff prediction using deep learning multi-dimensional ensemble method. *Journal of Hydrology (Amsterdam)*, *609*, 127762. DOI: 10.1016/j.jhydrol.2022.127762

Liu, L., Jin, H., Liu, Y., & Zhang, X. (2022). Intelligent Evacuation Route Planning Algorithm Based on Maximum Flow. *International Journal of Environmental Research and Public Health*, *19*(13), 7865. DOI: 10.3390/ijerph19137865 PMID: 35805524

Liu, M. (2024). *The role of urban green space in alleviating the urban heat island effect.* Environment, Resource and Ecology Journal., DOI: 10.23977/erej.2024.080102

Liu, Q., Dong, Q., Zhang, L., & Sun, C. (2024). Summer cooling island effects of blue-green spaces in severe cold regions: A case study of Harbin, China. *Building and Environment*, *257*, 111539. Advance online publication. DOI: 10.1016/j.buildenv.2024.111539

Liu, T., Zhang, D., Chu, Y., Ohashi, K., Hirai, Y., Ito, K., Ishibashi, K., Matsuo, Y., Yoshida, J., Ono, S., Kamiya, K., Li, H., & Yabu, H. (2026). Breaking the single-molecule paradigm: Multilayer cobalt phthalocyanine/carbon core-shell structure as the superior active unit for CO2-to-CO electroreduction. *Applied Catalysis B: Environmental, 381*, 125852. Advance online publication. DOI: 10.1016/j.apcatb.2025.125852

Liu, W., Li, A., Fang, W., Love, P. E. D., Hartmann, T., & Luo, H. (2023). A hybrid data-driven model for geotechnical reliability analysis. *Reliability Engineering & System Safety, 231*, 108985. Advance online publication. DOI: 10.1016/j.ress.2022.108985

Liu, Z. (2013). Seismic risk analysis using Bayesian belief networks. In *Handbook of Seismic Risk Analysis and Management of Civil Infrastructure Systems*. Woodhead Publishing Series in Civil and Structural Engineering., https://www.sciencedirect.com/science/article/pii/B9780857092687500074

Li, Y., Wu, B., Huang, Y., & Luan, S. (2024). Developing trustworthy artificial intelligence: Insights from research on interpersonal, human-automation, and human-AI trust. *Frontiers in Psychology, 15*, 1382693. DOI: 10.3389/fpsyg.2024.1382693 PMID: 38694439

Longato, D., Cortinovis, C., Balzan, M. V., & Geneletti, D. (2024). Identifying suitable policy instruments to promote nature-based solutions in urban plans. *Cities (London, England), 154*, 105348. Advance online publication. DOI: 10.1016/j.cities.2024.105348

Long, Y., Huang, Y., Ma, Q., Yu, W., & Tan, C. (2025). Future drought patterns in Southwest China: A CMIP6-based analysis using multiple drought indices. *Journal of Hydrology. Regional Studies, 61*, 102702. DOI: 10.1016/j.ejrh.2025.102702

Lumbroso, D., Vincent, K., Murambadoro, M., Steynor, A., Tsarouchi, G., & Nezi, M. (2024). Current uses and potential future needs for climate services in South Africa. *Climate Services, 36*, 100516. DOI: 10.1016/j.cliser.2024.100516

Lunga, W., Kaifa, J., Musarurwa, C., Malandela, G., Tshabalala, S., Baloyi, C., & Magampa, M. (2025). Preventing Disasters Before They Happen: Lessons from Successful Disaster Risk Reduction in Southern Africa. *Sustainability (Basel), 17*(20), 9131. DOI: 10.3390/su17209131

Lunga, W., Sithole, M., Musarurwa, C., Kombaté, E. N., & Marovah, T. (2025). Integrating Ubuntu: Embedding African philosophy into disaster risk reduction and management. *Frontiers in Sociology, 10*, 1637051. Advance online publication. DOI: 10.3389/fsoc.2025.1637051 PMID: 40969614

Lu, S., Christie, G. A., Nguyen, T. T., Freeman, J. D., & Hsu, E. B. (2022). Applications of artificial intelligence and machine learning in disasters and public health emergencies. *Disaster Medicine and Public Health Preparedness*, *16*(4), 1674–1681. DOI: 10.1017/dmp.2021.125 PMID: 34134815

Luther, J., Hainsworth, A., Tang, X., Harding, J., Torres, J., & Fanchiotti, M. (2017). World Meteorological Organization (WMO)—Concerted International Efforts for Advancing Multi-hazard Early Warning Systems. In *Advancing Culture of Living with Landslides* (pp. 129–141). Springer International Publishing. DOI: 10.1007/978-3-319-59469-9_9

Luxshi, K. (2024a). AI-driven Disaster Prediction and Early Warning Systems: A systematic literature review. *International Journal of Research in Computing*, *3*(1), 44. DOI: 10.64701/ijrc/345/8906

Lynch, M., Harris, F., Ierna, M., Mahomed, O., Henriquez-Mui, F., Gebreslasie, M., Ndzi, D., Viriri, S., Shakir, M. Z., Dickinson, N., Miller, C., Hursthouse, A., Nadesan-Reddy, N., Nkwanyana, F., Spencer, L. H., & Naidoo, S. (2025). Warning system for extreme weather events, awareness technology for healthcare, equitable delivery, and resilience (WEATHER) project: A mixed-methods research study protocol. *Climate (Basel)*, *13*(8), 170. Advance online publication. DOI: 10.3390/cli13080170

Ma, Y., Fu, Y., & Liu, C. (2024).Urban water bodies mitigate heat waves, the case of the Chinese city Kunming. DOI: 10.1007/s44327-024-00004-8

Macherera, M., & Chimbari, M. J. (2016). A review of studies on community based early warning systems. *Jàambá*, *8*(1), 206. DOI: 10.4102/jamba.v8i1.206 PMID: 29955285

Mada, G. S., Dethan, N. K. F., & Maharani, A. E. S. H. (2022). Defuzzification methods comparison of Mamdani fuzzy inference system in predicting Tofu production. *Journal Varian*, *5*(2), 137–148. DOI: 10.30812/varian.v5i2.1816

Magezi, C. (2023). *Uncertainty as certainty in a COVID-19 complex world: Reflections on leading congregations for effective missional praxis* (p. 119). DOI: 10.4102/aosis.2023.BK398.08

Makhura, N. K. (2025). Bridging the Governance Gap: Integrating Artificial Intelligence in South Africa's Public Policy Implementation. *The Asian Bulletin of Big Data Management*, *5*(2), 16–26. DOI: 10.62019/qq39na69

Makole, K. R., Ntshangase, B. A., & Adewumi, S. A. (2022). Coalition Governance: Unchartered Waters in South African Political Landscape. *Business Ethics and Leadership*, *6*(4), 23–37. DOI: 10.21272/bel.6(4).23-37.2022

Malik, F. S., & Terzidis, O. (2025). Thriving in turbulence: Resilience and strategic adaptation in global business. *Review of Managerial Science*. Advance online publication. DOI: 10.1007/s11846-025-00940-8

Manna, T., & Anitha, A. (2023). Precipitation prediction by integrating rough set on fuzzy approximation space with deep learning techniques. *Applied Soft Computing*, *139*, 110253. DOI: 10.1016/j.asoc.2023.110253

Manoharan, G., Razak, A., Rao, B. S., Singh, R., Ashtikar, S. P., & Nivedha, M. (2024). Navigating the Crescendo of Challenges in Harnessing Artificial Intelligence for Disaster Management. In Predicting Natural Disasters With AI and Machine Learning (pp. 64-94). IGI Global. DOI: 10.4018/979-8-3693-2280-2.ch003

Mao, Z., & Jagtap, A. (2020). Physics-informed neural networks for high-speed flows. *Computer Methods in Applied Mechanics and Engineering*. https://www.sciencedirect.com/science/article/pii/S0045782519306814

Maota, T. R. (2024). A conceptual framework for private higher educational institutions to respond to disruptions in South Africa. https://hdl.handle.net/10413/23345

Maota, T., Naido, V., & Darwish, S. (2024, April). Quantitative Assessment of Management Innovation in Navigating Disruptions Within Higher Education. In International Conference on Business and Technology (pp. 166-176). Cham: Springer Nature Switzerland. DOI: 10.1007/978-3-031-67444-0_16

Maota, T. (2026). Fortifying Municipal Governance: Leveraging IT Audits and Disruption Management for Sustainable Public Services. In Gupta, M., Walp, J., & Sharman, R. (Eds.), *Advancing IT Audits Through Integrative Approaches and Emerging Technologies* (pp. 339–368). IGI Global Scientific Publishing., DOI: 10.4018/979-8-3373-3078-5.ch011

Maota, T., & Naidoo, V. (2022). Response of private higher education leadership to pandemic crisis. *Research Highlights in Language. Literature and Education*, *2*(2), 127–148.

Marchau, V. A. W. J., Walker, W. E., & Bloemen, P. J. T. M. (2019). *Decision making under deep uncertainty: from theory to practice*. https://library.oapen.org/handle/20.500.12657/22900

Marengo, J. A., Alcântara, E. H., Moraes, O. L., Martinez, R., Seluchi, M., Alvalá, R. C., Dolif, G., & Gonçalves, D. (2025). Early warning services for disaster risk reduction in Brazil: The experience of CEMADEN during the floods of Rio Grande do Sul of May 2024. *International Journal of Disaster Risk Reduction, 126*, 105645. DOI: 10.1016/j.ijdrr.2025.105645

Marginean, I., Cuaresma, J. C., Hoffmann, R., Muttarak, R., Gao, J., & Daloz, A. S. (2024). High-Resolution Modeling and Projecting Local Dynamics of Differential Vulnerability to Urban Heat Stress. *Earth's Future, 12*(10), e2024EF004431. Advance online publication. DOI: 10.1029/2024EF004431

Maripe, K., Rankopo, M. J., Mwansa, L.-K., Coetzee, C., Khoza, S., Nemakonde, L. D., Shoroma, B. L., Wentink, G., Nyirenda, M., Chikuse, S., Kamanga, T., & Niekerk, D. V. (2022). Early Warning Systems in the Southern African Development Community: A Necessity. *Current Journal of Applied Science and Technology, 41*(48), 45–58. DOI: 10.9734/cjast/2022/v41i484033

Matikinca, P., Nyamakura, B., & Shackleton, S. (2024). Climate change adaptation and disaster risk reduction in South Africa's local municipal plans. *South African Journal of Science, 120*(7-8), 1–9. DOI: 10.17159/sajs.2024/15797

Matthews, S., & Morlidge, J. R. (2009). Performance based rehabilitation of reinforced concrete structuresNANo ranking found for "Concrete Repair, Rehabilitation and Retrofitting II." *Taylorfrancis. Com, 277–278*, 277–278. Advance online publication. DOI: 10.1201/9781439828403.ch100

Maulik, U., & Bandyopadhyay, S. (2002). Performance evaluation of some clustering algorithms and validity indices. *IEEE Transactions on Pattern Analysis and Machine Intelligence, 24*(12), 1650–1654. DOI: 10.1109/TPAMI.2002.1114856

Mbovane, T. (2024). Matanzima bridge in Kariega could take two years to repair. Available at: [Online] Available at: https://www.algoafm.co.za/local/matanzima-bridge-in-kariega-could-take-two-years-to-repair

Mbunge, E., Millham, R. C., Sibiya, M. N., Fashoto, S. G., Akinnuwesi, B., Simelane, S., & Ndumiso, N. (2021). Framework for ethical and acceptable use of social distancing tools and smart devices during COVID-19 pandemic in Zimbabwe. *Sustainable Operations and Computers, 2*, 190–199. DOI: 10.1016/j.susoc.2021.07.003

McCarthy, J., Minsky, M. L., Rochester, N., & Shannon, C. E. (2006). A proposal for the dartmouth summer research project on artificial intelligence, august 31, 1955. *AI Magazine, 27*(4), 12–12.

McClure, P. (2024, March 20). *Earth 2: Next-gen NVIDIA AI aims to solve weather and climate forecasting*. Retrieved from New Atlas: https://newatlas.com/technology/nvidia-ai-earth-2-weather-data-climate-action/

McCorduck, P., & Cfe, C. (2004). *Machines who think: A personal inquiry into the history and prospects of artificial intelligence*. AK Peters/CRC Press. DOI: 10.1201/9780429258985

Meemrot, R., Meena, R. L., Derwal, N., Meena, P., Balkh, S., & Jangid, J. B. (2024). Greening the Concrete Jungle: Enhancing Urban Resilience through Blue-Green Infrastructure. *International Journal for Research in Applied Science and Engineering Technology*, *12*(4), 2931–2935. Advance online publication. DOI: 10.22214/ijraset.2024.60578

Mehryar, S., Yazdanpanah, V., & Tong, J. (2024). AI and climate resilience governance. *iScience*, *27*(6), 109812. DOI: 10.1016/j.isci.2024.109812 PMID: 38784017

Mengesha, G. (2025). Integrating AI in Structural Health Monitoring (SHM): A Systematic Review on Advances, Challenges, and Future Directions. *University of Gondar, SSRN*. DOI: 10.2139/ssrn.5004977

Merchant, S. (2023, November 1). *MetNet-3: A state-of-the-art Neural Weather Model Available in Google Products*. Retrieved from Google Research: https://research.google/blog/metnet-3-a-state-of-the-art-neural-weather-model-available-in-google-products/

Merz, B., Kuhlicke, C., Kunz, M., Pittore, M., Babeyko, A., Bresch, D. N., Domeisen, D. I. V., Feser, F., Koszalka, I., Kreibich, H., Pantillon, F., Parolai, S., Pinto, J. G., Punge, H. J., Rivalta, E., Schröter, K., Strehlow, K., Weisse, R., & Wurpts, A. (2020). Impact Forecasting to Support Emergency Management of Natural Hazards. *Reviews of Geophysics*, *58*(4), e2020RG000704. Advance online publication. DOI: 10.1029/2020RG000704

Métivier, J., Bacchetta, M., Bekkers, E., & Koopman, R. (2023). International trade cooperation's impact on the world economy. *Journal of Policy Modeling*, *45*(4), 713–744. DOI: 10.1016/j.jpolmod.2023.06.007

Meuleman, L., & Fraser-Moleketi, G. (2022). Transforming Institutions and Governance to Build forward Better towards 2030. *Governance and Politics*, *1*(2), 7–29. DOI: 10.24833/2782-7062-2022-1-2-7-29

Microsoft. (n.d.). *Azure AI Foundry Labs*. Retrieved from Aurora: https://ai.azure.com/labs/projects/aurora

Milanović, S., Kaczmarowski, J., Ciesielski, M., Trailović, Z., Mielcarek, M., Szczygieł, R., Kwiatkowski, M., Bałazy, R., Zasada, M., & Milanović, S. D. (2023). Modeling and Mapping of Forest Fire Occurrence in the Lower Silesian Voivodeship of Poland Based on Machine Learning Methods. *Forests*, *14*(1), 46. DOI: 10.3390/f14010046

Mileti, D. S., & Sorensen, J. H. (1990). *Communication of emergency public warnings: A social science perspective and state-of-the-art assessment.* Oak Ridge National Laboratory., DOI: 10.2172/6137387

Mirrashid, M., Givehchi, M., Miri, M., & Madandoust, R. (2016). Performance investigation of neuro-fuzzy system for earthquake prediction.

Mishra, V. (2025). 'Very little time to react': Texas flash floods expose challenges in early warning. UN News Global perspective human stories. https://news.un.org/en/story/2025/07/1165355

Misyris, G., & Venzke, A. (2020). Physics-informed neural networks for power systems. *Ieeexplore.Ieee.Org*. https://ieeexplore.ieee.org/abstract/document/9282004/

Models, M. L. (n.d.). *Challenges and Solutions in Developing Weather AI Algorithms*. Retrieved from Machine Learning Models: https://machinelearningmodels.org/challenges-and-solutions-in-developing-weather-ai-algorithms/

Mohri, M., Rostamizadeh, A., & Talwalkar, A. (2018). *Foundations of machine learning*. MIT press.

Mojaddadi, H., Pradhan, B., Nampak, H., Ahmad, N., & Ghazali, A. H. B. (2017). Ensemble machine-learning-based geospatial approach for flood risk assessment using multi-sensor remote-sensing data and GIS. *Geomatics, Natural Hazards & Risk*, *8*(2), 1080–1102. DOI: 10.1080/19475705.2017.1294113

Mökander, J., & Axente, M. (2021). Ethics-based auditing of automated decision-making systems: Intervention points and policy implications. *AI & Society*, *38*(1), 153–171. DOI: 10.1007/s00146-021-01286-x

Mokhele, M. O., & Mvanyashe, A. (2025). The Role of Language as a Catalyst in Communicating Early Warning Messages: A Case Study in Walmer Airport Valley Informal Settlement, Gqeberha, South Africa. *International Journal of Disaster Risk Reduction*, *105573*, 105573. Advance online publication. DOI: 10.1016/j.ijdrr.2025.105573

Moravčík, M., Schmid, M., Burch, N., Lisý, V., Morrill, D., Bard, N., Davis, T., Waugh, K., Johanson, M., & Bowling, M. (2017). Deepstack: Expert-level artificial intelligence in heads-up no-limit poker. *Science*, *356*(6337), 508–513. DOI: 10.1126/science.aam6960 PMID: 28254783

Morrill, K. (2024). *AI Weather Forecasting: 10 Advances*. Retrieved from Yenra: https://yenra.com/ai-tech/weather-forecasting/

Mosavi, A., Ozturk, P., & Chau, K. W. (2018). Flood prediction using machine learning models: Literature review. *Water (Basel)*, *10*(11), 1536. DOI: 10.3390/w10111536

Moss, T., & Kar, B. (2020). Socio-economic vulnerability to urban heat in phoenix, arizona and dallas, texas during june 2020. *ISPRS Annals of the Photogrammetry, Remote Sensing and Spatial Information Sciences*, *VI-3/W1-2020*, 59–66. Advance online publication. DOI: 10.5194/isprs-annals-VI-3-W1-2020-59-2020

Motha, S. (2023). *East Coast Radio: Construction of flood-damaged KZN bridges to start in June*. https://www.ecr.co.za/news/news/construction-of-flood-damaged-kzn-bridges-to-start-in-june/

Motsumi, M. M., & Nemakonde, L. D. (2025). Indigenous early warning indicators for improving natural hazard predictions. *Jàambá*, *17*(1), 1–12. DOI: 10.4102/jamba.v17i1.1754 PMID: 40357010

Moudio, M. P. E., Bolin, R., Carpenter, A., Reese, S., Shehabi, A., & Rao, P. (2022). Characterizing manufacturing sector disruptions with targeted mitigation strategies. *Environmental Research Infrastructure and Sustainability*, *2*(4), 42001. DOI: 10.1088/2634-4505/ac9c8c

Mthambeka, N. H. (2022). *Urban Heat Islands (UHI) Assessment of Coastal Metropolitan Cities of South Africa*. Doctoral Thesis. Johannesburg, South Africa: University of Johannesburg. https://hdl.handle.net/10210/124987

Mtumane, G. (2024). *Nelson Mandela Bay Floods: six fatalities, hundreds displaced, and extensive damage reported.* News24. https://www.news24.com/nelson-mandela-bay-floods-six-fatalities-hundreds-displaced-and-extensive-damage-reported-20240603

Mudefi, E. (2023). Disaster management 'deeds' in the context of April 2022 KwaZulu-Natal floods: A scoping review. *International Journal of Disaster Risk Reduction*, *98*, 104122. DOI: 10.1016/j.ijdrr.2023.104122

Mugari, E., & Nethengwe, N. S. (2022). Mainstreaming ecosystem-based disaster risk reduction: Towards a sustainable and just transition in local development planning in rural South Africa. *Sustainability (Basel), 14*(19), 12368. DOI: 10.3390/su141912368

Muhtar, S. M., Amir, A. S., & Arya, N. (2024). Utilizing Social Media for Public Health Advocacy and Awareness in Digital Health Cummunication. *MSJ : Majority Science Journal, 2*(1), 270–278. DOI: 10.61942/msj.v2i1.96

Mukange, T., Mokoto, B., Moipolai, T. B., & Ndamase, Z. (2024). Accelerating the use of mobile phone capabilities to maximise the effectiveness of public emergency alerts in South Africa. *In Proceedings of the 17th International Conference on Theory and Practice of Electronic Governance* (pp. 75-80). DOI: 10.1145/3680127.3680144

Mulaudzi, P., Mokoto, B., Mothwa, L., Mukange, T., Koen, H., & Ndamase, Z. (2025). Evaluating the Impact and Challenges of Public Disaster Alert Systems in KwaZulu-Natal. *IST-Africa 2025 Conference Proceedings Miriam Cunningham and Paul Cunningham (Eds), IST-Africa Institute and IIMC, 2025.*

Municipal Infrastructure Support Agency. (2023). *Misa 2023/24 First Quarter Performance Report (01 April — 30 June 2023).* MISA.

Munsamy, L. (2019). Praxis of climate change adaptation in disaster risk reduction: A South African perspective *Transactions on Ecology and the Environment, 1*, 257. DOI: 10.2495/SC190231

Murphy, L. (2023). Disaster Relief. In The Role of Child Life Specialists in Community Settings (pp. 191-214). IGI Global.

Murphy, R. R., Tadokoro, S., & Kleiner, A. (2016). Disaster robotics. In *Springer handbook of robotics* (pp. 1577–1604). Springer. DOI: 10.1007/978-3-319-32552-1_60

Murray, V., Jonathan, A., Chadi, A., Kanza, A., Lucille, A., Djillali, B., Alonso, B. T., Hun, C., Simon, C., Jim, D., Lucy, F., Urbano, F. P., Qunli, H., John, H., Simon, H., Wirya, K., Lidia, M., Nick, M., Luiz Leal, M. O., & Wright, N. (2021). *Hazard Information Profiles: Supplement to UNDRR-ISC Hazard Definition & Classification Review: Technical Report: Geneva, Switzerland, United Nations Office for Disaster Risk Reduction.* International Science Council., DOI: 10.24948/2021.05

Murshed, S., Griffin, A. L., Islam, M. A., Wang, X. H., & Paull, D. J. (2023). Assessing sensitivity to climate-related disasters in the context of a developing country: Evidence from the coastal region of Bangladesh. *International Journal of Disaster Risk Reduction, 97*, 104023. DOI: 10.1016/j.ijdrr.2023.104023

Mutani, G., Scalise, A., Sufa, X., & Grasso, S. (2024).Synergizing Machine Learning and Remote Sensing for Urban Heat Island Dynamics: A Comprehensive Modelling Approach. DOI: 10.20944/preprints202409.2284.v1

Myeni, L., & Roffe, S. (2025). Towards effective weather and/or climate services in South Africa: Profiling sectoral needs and constraints. *Environmental Development*, *55*, 101240. DOI: 10.1016/j.envdev.2025.101240

Myeza, L., Ecim, D., & Maroun, W. (2023). The role of integrated thinking in corporate governance during the COVID-19 crisis: Perspectives from South Africa. *Journal of Public Budgeting, Accounting & Financial Management*, *35*(6), 52–77. DOI: 10.1108/JPBAFM-08-2022-0133

Nadkarni, P. M., Ohno-Machado, L., & Chapman, W. W. (2011). Natural language processing: An introduction. *Journal of the American Medical Informatics Association : JAMIA*, *18*(5), 544–551. DOI: 10.1136/amiajnl-2011-000464 PMID: 21846786

National Disaster Management Centre (NDMC). (2020). *Strategic Plan 2020–2025.* Pretoria: Department of Cooperative Governance and Traditional Affairs. Available at: https://www.gov.za/documents/national-disaster-management-centre-strategic-plan

National Institute of Standards and Technology (NIST). (2018). *Framework for Improving Critical Infrastructure Cybersecurity (Version 1.1).* Gaithersburg, MD. Available at: https://www.nist.gov/cyberframework

National Institute of Standards and Technology (NIST). (2023). *Artificial Intelligence Risk Management Framework (AI RMF 1.0).* Gaithersburg, MD: U.S. Department of Commerce. Available at: https://www.nist.gov/itl/ai-risk-management-framework

Nduli, S., Tekere, M., Masindi, V., & Foteinis, S. (2025). Assessment of the water quality and microbial regrowth in drinking water treatment plants and the distribution network. *Water Resources and Industry*, *33*, 100290. DOI: 10.1016/j.wri.2025.100290

Nemakonde, L. D., & Kunguma, O. (2024, October 31). Revealing the boon and bane of South Africa's disaster management legislation during COVID-19. *Jàambá*, *16*(1), 1689. DOI: 10.4102/JAMBA.v16i1.1689 PMID: 39507560

Neri, C., & Magaña, V. (2016). Estimation of Vulnerability and Risk to Meteorological Drought in Mexico. *Weather, Climate, and Society*, *8*(2), 95–110. DOI: 10.1175/WCAS-D-15-0005.1

Neyer, J. (2021). After global governance. *Zeitschrift für Politikwissenschaft*, *32*(2), 361–382. DOI: 10.1007/s41358-021-00290-3

Ngcamu, B. (2022). Climate Change and Disaster Preparedness Issues in the Eastern Cape and KwaZulu-Natal, South Africa. *Town and Regional Planning*, *81*, 53–66. DOI: 10.18820/2415-0495/trp81i1.5

Ngcamu, B. S., & Abrahams, G. (2024). The Preparedness Plans and Early Warning Systems During Flood Disasters in South Africa. In Jakaza, (Eds.), *The Palgrave Handbook of Language and Crisis Communication in Sub-Saharan Africa* (pp. 259–272)., DOI: 10.1007/978-3-031-43059-6_14

Nhamo, L., Magidi, J., Mpandeli, S., Mutanga, O., Liphadzi, S., & Mabhaudhi, T. (2025). Drought and flood risk mapping using a GIS-based multi-criteria decision method: A case of the Olifants Basin, South Africa. *Next Sustain*, *6*, 100100. DOI: 10.1016/j.nxsust.2025.100100

Nihad, B., & Nadjat, A. (2023). Drones Use in Disaster Management [Doctoral dissertation, university center of abdalhafid boussouf-MILA].

Nistor, M. S., Son, P. T., Pickl, S. W., Gaindric, C., & Cojocaru, S. (2020). A concise review of AI-based solutions for mass casualty management. *CEUR Workshop Proceedings*, *2805*, 222–232.

Nkombi, Z., & Wentink, G. J. (2022). The Role of Public Participation in Disaster Risk Reduction Initiatives: The Case of Katlehong Township. *Jàambá*, *14*(1), 1203. DOI: 10.4102/jamba.v14i1.1203 PMID: 35284043

Noll, S., & Bhar, T. (2023).The Five Pillars of Urban Environmental Justice: A Framework for Building Equitable Cities. DOI: 10.21827/potcj.1.1.41048

Norman, D. A. (2013). *The design of everyday things* (Revised and expanded ed.). Basic Books. https://www.hachettebookgroup.com/titles/don-norman/the-design-of-everyday-things/9780465050659/

Noy, S., & Zhang, W. (2023). Experimental evidence on the productivity effects of generative artificial intelligence. *Science*, *381*(6654), 187–192. DOI: 10.1126/science.adh2586

Ntali, M. A., Vubo, E. Y., & Ndonwi, A. S. (2023). Community preparedness to recurrent flood disasters in Babessi, rural Cameroon. *American Journal of Environment Studies*, *6*(2), 43–63. DOI: 10.47672/ajes.1608

Nunes, I. L., Calhamonas, G., Marques, M., & Teodoro, M. F. (2020). Building a decision support system to handle teams in disaster situations-a preliminary approach. In Hybrid Intelligent Systems: 18th International Conference on Hybrid Intelligent Systems (HIS 2018) Held in Porto, Portugal, December 13-15, 2018 18 (pp. 551-559). Springer International Publishing. DOI: 10.1007/978-3-030-14347-3_54

Nunes, A. R. (2025). Health and multihazard early warning systems in the media. *Perspectives in Public Health, 145*(6), 343–352. DOI: 10.1177/17579139251342156 PMID: 40838788

NVIDIA. (n.d.). *AI Weather Analytics with Earth-2*. Retrieved from NVIDIA Corporation: https://build.nvidia.com/nvidia/earth2-weather-analytics/blueprintcard

Nwachukwu, P. S., Chima, O. K., & Okolo, C. H.Priscilla Samuel NwachukwuOnyeka Kelvin ChimaChinelo Harriet Okolo. (2025). The artificial intelligence governance framework for finance: A control-by-design approach to algorithmic decision-making in accounting. *Finance & Accounting Research Journal, 7*(8), 350–379. DOI: 10.51594/farj.v7i8.2016

Nyamwanza, A. M., Johnson, K., Mthembu, A., Tshivhundo, Z. C., & Brown, N. (2026). Integrated Knowledge Systems Towards Flood Resilience and Sustainable Solid Waste Management in South African Urban Informal Settlements. *Sustainability (Basel), 18*(6), 2960. DOI: 10.3390/su18062960

Ocaña-Fernández, Y., & Fuster-Guillén, D. (2021). The bibliographical review as a research methodology. *Revista Tempos e Espaços em Educação, 14*(33), e15614–e15614. DOI: 10.20952/revtee.v14i33.15614

OECD. (2020). *Building Resilience to the COVID-19 Pandemic: The Role of Centres of Government.* Paris: OECD Publishing. Available at: https://www.oecd.org/coronavirus/policy-responses

OECD. (2024). Framework for Anticipatory Governance of Emerging Technologies. In *OECD science, technology and industry policy papers*. DOI: 10.1787/23074957

OECD. AI. (2025, May 19). *Artificial Intelligence Forecasting System (AIFS).* Retrieved from OECD.AI: https://oecd.ai/en/catalogue/tools/artificial-intelligence-forecasting-system-aifs

Ojo, O. S., & Ogunjo, S. T. (2022). Machine learning models for prediction of rainfall over Nigeria. *Scientific African, 16*, e01246. DOI: 10.1016/j.sciaf.2022.e01246

Oke, S. A. (2008). A literature review on artificial intelligence. International journal of information and management sciences, 19(4), 535-570.

Okunola, O. H. (2025). Beyond institutional silos: rethinking multilevel disaster risk governance in Africa a decade into the Sendai Framework implementation: OH Okunola. *International Journal of Disaster Risk Science, 16*(3), 321–332. DOI: 10.1007/s13753-025-00646-1

Olanrewaju, C., & Reddy, M. M. (2025). Assessment and Mitigation of Flood Risk in Urban Settlement: A Case of Ethekwini Metropolitan Area. *E-Journal of Humanities* [EHASS]. *Arts and Social Sciences*, *6*(2), 3340–3362. DOI: 10.38159/ehass.202561233

Olawuyi, D. A., Gbadegesin, A. S., Ajayi, D., Oyedele, P., Geiger, D., Seidemann, I., Geisemann, P., Sansone, S., Nasir, F., Antenyi, O. P., Salako, F., Agada, J., & Adaje, P. (2025). Anticipatory Action in River Flooding Risk Management in Nigeria: An Assessment of Community-Level Implementation. *Journal of Flood Risk Management*, *18*(4), e70117. Advance online publication. DOI: 10.1111/jfr3.70117

Omer, M. A., Abdi, A. M., & Mohamed, A. J. (2026). Assessing the effectiveness of early warning systems in Somaliland: Functionality, community response, and system gap. *European Journal of Sustainable Development Research*, *10*(1), em0344. Advance online publication. DOI: 10.29333/ejosdr/17281

Onduko, J. O., Kalombo, M. A., Kuol, M. D., Makale, B. G., & Saleh, M. A. (2024). AI-Driven Governance: Transforming Public and Addressing Legacy Issues in Post-Colonial Africa. *Proceedings of London International Conferences*, *11*(11), 52–63. DOI: 10.31039/plic.2024.11.243

Ongala, M., Kiraka, R., Choudrie, J., & Okello, J. (2025). Strategic integration of Artificial Intelligence in healthcare: Theoretical frameworks, adoption, enablers, and barriers — A scoping review. *Proceedings of the AAAI Symposium Series*, 6(1), 298–303. DOI: 10.1609/aaaiss.v6i1.36066

Organisation for Economic Co-operation and Development (OECD). (2014). *OECD Reviews of Risk Management Policies: Boosting Resilience through Innovative Risk Governance*. Paris: OECD Publishing. Available at: https://www.oecd.org/governance/risk/

Oscilowicz, E., Anguelovski, I., Triguero-Mas, M., García-Lamarca, M., Baró, F., & Cole, H. (2022).Green justice through policy and practice: a call for further research into tools that foster healthy green cities for all. DOI: 10.1080/23748834.2022.2072057

Ou, J., & Li, H. (2010). Structural health monitoring in mainland China: Review and future trends. *Structural Health Monitoring*, *9*(3), 219–231. DOI: 10.1177/1475921710365269

Ouyang, M., Hong, L., Mao, Z., & Yu, M. (2009). Vulnerability analysis of interdependent infrastructure systems: A methodological framework. *Physica A: Statistical Mechanics and Its Applications*. https://www.sciencedirect.com/science/article/pii/S0378437111009794

Ouyang, W., Morakinyo, T. E., & Ng, E. (2019).The Cooling Efficiency of Urban Greenery Coverage in a High-density City. DOI: 10.1088/1755-1315/329/1/012043

Pachar, S., Dudeja, D., Batra, N., Tomar, V., Bhimavarapu, J. P., & Singh, A. K. (2023). Recent Developments in Machine Learning Predictive Analytics for Disaster Resource Allocation. *Engineering Proceedings*, *59*(1), 19.

Padmapoorani, P., Senthilkumar, S., & Mohanraj, R. (2023). Machine learning techniques for structural health monitoring of concrete structures: A systematic review. *Civil Engineering (Shiraz)*, *47*(4), 1919–1931. DOI: 10.1007/s40996-023-01054-5

Page, M. J., McKenzie, J. E., Bossuyt, P. M., Boutron, I., Hoffmann, T. C., Mulrow, C. D., & Moher, D. (2021). The PRISMA 2020 statement: An updated guideline for reporting systematic reviews. *BMJ (Clinical Research Ed.)*, *372*, n71. Advance online publication. DOI: 10.1136/bmj.n71 PMID: 33782057

Pang, Y., Li, Y., Feng, Z., Feng, Z., Zhao, Z., Chen, S., & Zhang, H. (2022). Forest Fire Occurrence Prediction in China Based on Machine Learning Methods. *Remote Sensing (Basel)*, *14*(21), 5546. DOI: 10.3390/rs14215546

Partigöç, N. S. (2022). *The Role of Artificial Intelligence in Disaster Risk Management*. Bilişim Teknolojileri Dergisi.

Pashentsev, E., & Kolotaev, Y. Y. (2025). Information and communication technologies in political crisis management: Resilience, forecasting, and response. *Discover Global Society*, *3*(1), 84. Advance online publication. DOI: 10.1007/s44282-025-00216-2

Pasupuleti, M. K. (2024).Green Governance: Shaping Policies for a Sustainable Future. DOI: 10.62311/nesx/77622

Pasupuleti, M. K., & Services, N. E. (2025). Augmented Human Intelligence: Converging Generative AI, Quantum Computing, and XR for enhanced human-machine synergy. DOI: 10.62311/nesx/rrv525

Pasupuleti, M. K. (2024). *AI-driven leadership: Crisis management and virtual organizations in the digital age*. National Education Services., DOI: 10.62311/nesx/rb978-81-981179-1-5

Patil, D., Rane, N. L., & Rane, J. (2024). *Enhancing resilience in various business sectors with ChatGPT and generative artificial intelligence*. DOI: 10.70593/978-81-981367-8-7_4

Paton, D. (2019). Disaster risk reduction: Psychological perspectives on preparedness. *Australian Journal of Psychology*, *71*(4), 327–341. DOI: 10.1111/ajpy.12237

Patrucco, P. P., & Kähkönen, A. (2021). Agility, adaptability, and alignment: New capabilities for PSM in a post-pandemic world. *Journal of Purchasing and Supply Management, 27*(4), 100719. DOI: 10.1016/j.pursup.2021.100719

Paul, B. K. (2009). Why relatively fewer people died? The case of Bangladesh's Cyclone Sidr. *Natural Hazards, 50*(2), 289–304. DOI: 10.1007/s11069-008-9340-5

Paul, B. K., & Dutt, S. (2010). Hazard warnings and responses to evacuation orders: The case of Bangladesh's cyclone Sidr. *Geographical Review, 100*(3), 336–355. DOI: 10.1111/j.1931-0846.2010.00040.x PMID: 21061500

Pearson, E., Sharp, L., & Hampton, L. (2025). Resilience in context: A synthesis of theories and practices for educational psychologists. *Educational Psychology in Practice, 41*(4), 1–18. DOI: 10.1080/02667363.2025.2518990

Percival, V., Thoms, O. T., Oppenheim, B., Rowlands, D., Chisadza, C., Fewer, S., Yamey, G., Alexander, A. C., Allaham, C. L., Causevic, S., Daudelin, F., Gloppen, S., Guha-Sapir, D., Hadaf, M., Henderson, S., Hoffman, S. J., Langer, A., Lebbos, T. J., Leomil, L., … Friberg, P. (2023). The Lancet Commission on peaceful societies through health equity and gender equality [Review of *The Lancet Commission on peaceful societies through health equity and gender equality*]. *The Lancet, 402*(10413), 1661. Elsevier BV. DOI: 10.1016/S0140-6736(23)01348-X

Perera, D., Seidou, O., Agnihotri, J., Mehmood, H., & Rasmy, M. (2020). Challenges and technical advances in flood early warning systems (FEWSs). Flood impact mitigation and resilience enhancement.

Perrelet, K., Moretti, M., Dietzel, A., Altermatt, F., & Cook, L. M. (2024). Engineering blue-green infrastructure for and with biodiversity in cities. *NPJ Urban Sustainability, 4*(1), 27. Advance online publication. DOI: 10.1038/s42949-024-00163-y

Pescaroli, G., Dryhurst, S., & Karagiannis, G. M. (2025). Bridging gaps in research and practice for early warning systems: New datasets for public response. *Frontiers in Communication, 10*, 1451800. Advance online publication. DOI: 10.3389/fcomm.2025.1451800

Petrucci, A. L. (2023). *Environmental Retrofitting*. Fighting Urban Heat Island Toward NEZ Sustainable Smart Cities., DOI: 10.1007/978-3-031-35664-3_8

Pettorru, G., Fadda, M., Girau, R., Sole, M., Anedda, M., & Giusto, D. (2023). Using Artificial Intelligence and IoT Solution for Forest Fire Prevention. *2023 International Conference on Computing, Networking and Communications, ICNC 2023*. IEEE. DOI: 10.1109/ICNC57223.2023.10074289

Petty, H. (2023, July 5). *A Change in the Weather: AI, Accelerated Computing Promise Faster, More Efficient Predictions*. Retrieved from NVIDIA Corporation: https://blogs.nvidia.com/blog/ai-efficient-weather-predictions/

Pham, B. T., Le, L. M., Bui, K. T., Le, V. M., Ly, H. B., & Prakash, I. (2020). Development of advanced artificial intelligent models for daily rainfall prediction. *Atmospheric Research, 237*, 104845. DOI: 10.1016/j.atmosres.2020.104845

Pham, C. T., Nguyen, H. T., Le, H. H. T. C., Tran, N. Q. L., Do, K. Q., Bui, V., Phung, H., Phung, D., & Chu, C. (2025). Challenges and Strategies for the Development and Implementation of Climate-Informed Early Warning Systems for Vector-Borne Diseases: A Systematic Review. *Tropical Medicine & International Health, 31*(1), 10–21. Advance online publication. DOI: 10.1111/tmi.70045 PMID: 40999867

Pham, T. D. M., Shaw, R., & Takeuchi, Y. (2024). Community-based early warning systems in a changing climate: Enhancing resilience through participatory approaches. *Climate and Development, 16*(4), 389–402. DOI: 10.1080/17565529.2024.2307398

Phoon, K. K., & Zhang, W. (2023). Future of machine learning in geotechnics. In *Georisk* (Vol. 17, Number 1). DOI: 10.1080/17499518.2022.2087884

Platzer, E. K., & Knodt, M. (2023). Resilience beyond insurance: Coordination in crisis governance. *Environment Systems & Decisions, 43*(4), 569–576. DOI: 10.1007/s10669-023-09938-7

Plevris, V. (2024). AI-Driven Innovations in Earthquake Risk Mitigation: A Future-Focused Perspective. *Geosciences, 14*(9), 244. DOI: 10.3390/geosciences14090244

Pontius, J., & McIntosh, A. (2024).Urban Heat Islands. DOI: 10.1007/978-3-031-48762-0_10

Poole, L., Coughlan de Perez, E., & Easton-Calabria, E.. (2022). *Anticipatory action in motion: Recapping the most recent evidence and illuminating a pathway forward.* Feinstein International Center, Tufts University.

Poonia, V., & Tiwari, H. L. (2020). Rainfall-runoff modeling for the Hoshangabad Basin of Narmada River using artificial neural network. *Arabian Journal of Geosciences, 13*(18), 944. DOI: 10.1007/s12517-020-05930-6

Powers, C. J., Devaraj, A., Ashqeen, K., Dontula, A., Joshi, A., Shenoy, J., & Murthy, D. (2023). Using artificial intelligence to identify emergency messages on social media during a natural disaster: A deep learning approach. *International Journal of Information Management Data Insights, 3*(1), 100164. DOI: 10.1016/j.jjimei.2023.100164

Prayogi, S., & Verawati, N. N. S. P. (2024). Physics Learning Technology for Sustainable Development Goals (SDGs): A Literature Study. *International Journal of Ethnoscience and Technology in Education, 1*(2), 155. DOI: 10.33394/ijete.v1i2.12316

Presidential Climate Commission. (2024). *State of climate action in South Africa. Presidential Climate Commission.* https://pccommissionflo.imgix.net/uploads/images/24_REP_State-of-Climate-Action-in-South-Africa_v4.pdf

Pushparani, D. (2025). Review On: Collaboration Intelligence in Education. [IRJAEH]. *International Research Journal on Advanced Engineering Hub, 3*(09), 3606–3611. DOI: 10.47392/IRJAEH.2025.0526

Putu, I., Prasetia, W., Nyoman, I., & Kurniawan, H. (2021). Implementasi ETL (Extract, Transform, Load) pada Data warehouse Penjualan Menggunakan Tools Pentaho. *TIERS Information Technology Journal, 2*(1), 39–47. https://journal.undiknas.ac.id/index.php/tiers

Qiu, Y., & He, B. (2024).Review of urban water thermal environment cooling. DOI: 10.1088/1755-1315/1363/1/012073

Qi, W., Xu, C., & Xu, X. (2021). AutoGluon: A revolutionary framework for landslide hazard analysis. *Natural Hazards Review, 1*(3), 103–108. DOI: 10.1016/j.nhres.2021.07.002

Quinn, W., Angove, P., Buckley, J., Barrett, J., & Kelly, G. (2011). Design and performance analysis of an embedded wireless sensor for monitoring concrete curing and structural health. *Journal of Civil Structural Health Monitoring, 1*(1–2), 47–59. DOI: 10.1007/s13349-011-0005-9

Raftopoulos, M., Pertoldi-Bianchi, S., Eleftherios, P., & Kjelmann, K. G. (2024). Conceptualizing the Nexus between Disaster Risk Reduction and Climate Change Adaption and Mitigation in Governance. *Research Portal Denmark, 2024*(3), 172. https://local.forskningsportal.dk/local/dki-cgi/ws/cris-link?src=aau&id=aau-4383f7e9-8d76-4169-9202-c1936d86740c&ti=Conceptualizing%20the%20Nexus%20between%20Disaster%20Risk%20Reduction%20and%20Climate%20Change%20Adaption%20and%20Mitigation%20in%20Governance

Ragolane, M. (2024). Service Delivery Protests in the South African Government. *International Journal of Social Science Research and Review, 7*(4), 54–65. DOI: 10.47814/ijssrr.v7i4.1973

Ragus, E., Sutherland, C., Manderson, L., & Moyer, E. (2024). *Sculpting stories: methods to unsettle knowledge production in disasters.* Disaster Prevention and Management An International Journal., DOI: 10.1108/DPM-01-2024-0034

Rahman, M. A. (2020). Improvement of rainfall prediction model by using fuzzy logic. *American Journal of Climate Change*, *9*(4), 391–399. DOI: 10.4236/ajcc.2020.94024

Rahmat, B., Afiadi, F., & Joelianto, E. (2018, December). Earthquake prediction system using neuro-fuzzy and extreme learning machine. In *International Conference on Science and Technology (ICST 2018)* (pp. 452-458). Atlantis Press. DOI: 10.2991/icst-18.2018.95

Rahmatika, D. N., Firmansyah, F., & Wijaya, J. R. T. (2025). Audit Manajemen Perusahaan: Prosedur, Inovasi dan Implementasi. *Penerbiteureka.com*. https://repository.penerbiteureka.com/media/publications/603267-audit-manajemen-perusahaan-prosedur-inov-d2f86036.pdf

Rakkasagi, S., Goyal, M. K., & Jha, S. (2024). Evaluating the future risk of coastal Ramsar wetlands in India to extreme rainfalls using fuzzy logic. *Journal of Hydrology (Amsterdam)*, *632*, 130869. DOI: 10.1016/j.jhydrol.2024.130869

Ram, T., & Wang, G. (2013). Probabilistic seismic hazard analysis in Nepal. *Earthquake Engineering and Engineering Vibration*, *12*(4), 577–586. DOI: 10.1007/s11803-013-0191-z

Raphela, T. D. (2025). The ripple effect of flooding near one of the more prominent urban rivers in the Gauteng Province, South Africa. *Frontiers in Sustainable Cities*, *7*, 1656985. DOI: 10.3389/frsc.2025.1656985

Raphela, T. D., & Ekeke, N. (2025). Resilience and the dissemination of flood disaster early warning messages in a township in South Africa. *Frontiers in Communication*, *10*, 1503016. DOI: 10.3389/fcomm.2025.1503016

Rathnayake, N., Rathnayake, U., Chathuranika, I., Dang, T. L., & Hoshino, Y. (2023). Cascade-ANFIS to simulate nonlinear rainfall-runoff relationship. *Applied Soft Computing*, *147*, 110722. DOI: 10.1016/j.asoc.2023.110722

Ravikumar, R. N., & Aarthi, S. (2025a). Climate data without borders. In *Cross-Border Cooperation in the Age of Climate Change* (pp. 61–90). IGI Global Scientific Publishing., DOI: 10.4018/979-8-3373-3972-6.ch003

Razavi-Termeh, S. V., Sadeghi-Niaraki, A., Ali, F., Pradhan, B., & Choi, S. M. (2025). Optimizing ensemble learning for satellite-based multi-hazard monitoring and susceptibility assessment of landslides, land subsidence, floods, and wildfires. *Scientific Reports*, *15*(1), 30968. DOI: 10.1038/s41598-025-15381-2 PMID: 40846725

Reddy, M. M., Mbandlwa, Z., Pillay, S. S., Murwirapachena, G., & Ngema, S. S. (2024). Socio-economic Variables and Disaster Incidents in KwaZulu-Natal, South Africa. In Mbandlwa, S. (Ed.), *Challenges, Strategies, and Resiliency in Disaster and Risk Management* (pp. 23–55). IGI Global Scientific Publishing., DOI: 10.4018/979-8-3693-2721-0.ch002

Rehan, H. (2022). Enhancing Disaster Response Systems: Predicting and Mitigating the Impact of Natural Disasters Using AI. *Journal of Artificial Intelligence Research*, *2*(1), 501.

Reichstein, M., Benson, V., Blunk, J., Camps-Valls, G., Creutzig, F., Fearnley, C. J., Han, B., Kornhuber, K., Rahaman, N., Schölkopf, B., Tárraga, J. M., Vinuesa, R., Dall, K., Denzler, J., Frank, D., Martini, G., Nganga, N., Maddix, D. C., & Weldemariam, K. (2025). Early warning of complex climate risk with integrated artificial intelligence. *Nature Communications*, *16*(1), 2564. DOI: 10.1038/s41467-025-57640-w PMID: 40089483

Reichstein, M., Frank, D., Benson, V., Camps-Valls, G., Denzler, J., Kornhuber, K., Schoelkopf, B., Vinuesa, R., Han, B., Fearnley, C. J., Rahaman, N., Tárraga, J. M., Blunk, J., Dall, K., Martini, G., Nganga, N., & Robinson, D. E. (2024). Early warning of complex climate risk with integrated artificial intelligence. *Research Square (Research Square)*. DOI: 10.21203/rs.3.rs-4248340/v1

Renn, O. (2008). *Risk governance: Coping with uncertainty in a complex world*. Earthscan. https://www.taylorfrancis.com/books/mono/10.4324/9781849772440/risk-governance-ortwin-renn

Republic of South Africa. (2002). *Disaster Management Act 57 of 2002*. Pretoria: Government Printer. Available at: https://www.gov.za/documents/disaster-management-act-57-2002-0

Republic of South Africa. (2015). *Disaster Management Act - Amendment Bill 16 of 2015*. Government Printers.

Republic of South Africa. (2024). *Climate Change Act, 2024 (Act 22 of 2024)*. Government Printers.

Restas, A. (2017). Disaster management supported by unmanned aerial systems (UAS) focusing especially on natural disasters. Zeszyty Naukowe SGSP/Szkoła Główna Służby Pożarniczej, 61.

Reynoso Vanderhorst, H. D., Pathirage, C., & Proverbs, D. (2024). Navigating Flood Resilience: Challenges, Solutions, and Lessons Learnt from the Dominican Republic. *Water (Basel)*, *16*(3), 382. DOI: 10.3390/w16030382

Ribeiro, T. R. (2023). Justice in Urban Greening and Green Gentrification., DOI: 10.1007/978-3-031-22566-6_20

Rivera, L. M., & Faggian, R. (2024). Integrating Blue-Green Infrastructure strategies to enhance climate resilience in Colombia. *European Journal of Sustainable Development, 13*(4), 255. Advance online publication. DOI: 10.14207/ejsd.2024.v13n4p255

Rodina, L., Harris, L., Ziervogel, G., & Wilson, J. (2024). Resilience countercurrents: Water infrastructures, informality, and inequities in Cape Town, South Africa. *World Development, 180*, 106619. DOI: 10.1016/j.worlddev.2024.106619

Röhe, A. (2025). What Is the Meaning of Resilience? In *Organisational Resilience in Action: Strategies for Overcoming Future Crises* (pp. 5–13). Springer Berlin Heidelberg., DOI: 10.1007/978-3-662-70794-4_2

Rokhideh, M., Fearnley, C. J., & Budimir, M. (2025). Multi-hazard early warning systems in the Sendai framework for disaster risk reduction: Achievements, gaps, and future directions. *International Journal of Disaster Risk Science, 16*(1), 103–116. DOI: 10.1007/s13753-025-00622-9

Roxon, J., Ulm, F.-J., Pellenq, R. J.-M., & Pellenq, R. J.-M. (2020). Urban heat island impact on state residential energy cost and CO2 emissions in the United States. *Urban Climate, 31*, 100546. Advance online publication. DOI: 10.1016/j.uclim.2019.100546

Rumble, H., Angeoletto, F., Connop, S., Goddard, M. A., & Nash, C. (2019). Understanding and applying ecological principles in cities. DOI: 10.1007/978-3-030-01866-5_15

Russell, S. J., & Norvig, P. (2016). *Artificial intelligence: a modern approach*. Pearson.

Rust, B. W., & Rust, S. (2024). Place-based innovation for sustainable and resilient human systems. *Sustainable Earth Reviews, 7*(1), 19. Advance online publication. DOI: 10.1186/s42055-024-00089-4

Safar, N. Z. M., Ramli, A. A., Mahdin, H., Nzdi, D., & Khalif, K. M. N. (2019). Rain prediction using fuzzy rule based system in North-West Malaysia. *Indonesian Journal of Electrical Engineering and Computer Science, 14*(3), 1572–1581.

Sakalasuriya, M., Haigh, R., Hettige, S., Amaratunga, D., Basnayake, S., & Rahayu, H. P. (2020). Governance, Institutions and People within the Interface of a Tsunami Early Warning System. *Politics and Governance, 8*(4), 432–444. DOI: 10.17645/pag.v8i4.3159

Saka, T. N., Hormiga, E., & Valls-Pasola, J. (2024). Crisis response strategies: A digital reluctance perspective. *Review of Managerial Science*, *19*(8), 2569–2607. DOI: 10.1007/s11846-024-00822-5

Sakiyama, F. I. H., Lehmann, F., & Garrecht, H. (2021). Structural health monitoring of concrete structures using fibre-optic-based sensors: A review. *Magazine of Concrete Research*, *73*(4), 174–194. DOI: 10.1680/jmacr.19.00185

Salem, A., Azzam, S. M., Emam, O. E., & Abohany. A. A. (2024). Advancing cybersecurity: a comprehensive review of AI-driven detection techniques [Review of *Advancing cybersecurity: a comprehensive review of AI-driven detection techniques*]. *Journal Of Big Data*, *11*(1). Springer Science+Business Media. DOI: 10.1186/s40537-024-00957-y

SARACS. (2025). *Report on the drought safety management system for the case of Mexico City*. SEPI-ESIME, Instituto Politecnico Nacional.

Sargiacomo, M., Servalli, S., Potito, S., D'Andreamatteo, A., & Gitto, A. (2021). Accounting for natural disasters from a historical perspective: A literature review and research agenda. *Accounting History*, *26*(2), 179–204. DOI: 10.1177/10323732211003173

Sarli, P., Palar, P., Azhari, Y., & Setiawan, A. (2023). Gaussian process regression for seismic fragility assessment: Application to non-engineered residential buildings in Indonesia. *Mdpi.Com*. https://www.mdpi.com/2075-5309/13/1/59?utm_campaign=releaseissue_buildingsutm_medium=emailutm_source=releaseissueutm_term=titlelink16

Savoia, E., Piltch-Loeb, R., Stanton, E., & Koh, H. K. (2023). Learning from COVID-19: Government leaders' perspectives to improve emergency risk communication. *Globalization and Health*, *19*(1), 86. Advance online publication. DOI: 10.1186/s12992-023-00993-y PMID: 37968676

Savoie, É., Sapinski, J. P., & Laroche, A.-M. (2025). Key factors for revitalising heritage buildings through adaptive reuse. *Buildings &amp. Buildings & Cities*, *6*(1). Advance online publication. DOI: 10.5334/bc.495

Schiavi, G. S., Behr, A., & Marcolin, C. B. (2023). Institutional theory in accounting information systems research: Shedding light on digital transformation and institutional change. *International Journal of Accounting Information Systems*, *52*, 100662. DOI: 10.1016/j.accinf.2023.100662

Schlösser, P., Rockström, J., Edwards, C., Mirazo, P., Heilemann, A., Kitzmann, N. H., & Krobjinski, S. L. (2023). Accelerating transformations for a just, sustainable future: 10 'Must Haves.'. *Global Sustainability*, *6*, e17. Advance online publication. DOI: 10.1017/sus.2023.14

Schoenherr, J. R., Abbas, R., Michael, K., Rivas, P., & Anderson, T. D. (2023). Designing AI using a human-centered approach: Explainability and accuracy toward trustworthiness. *IEEE Transactions on Technology and Society*, *4*(1), 9–23. DOI: 10.1109/TTS.2023.3257627

Seddighi, H., Lopez, M. L., Zwitter, A., Muldoon, M. L., Sajjadi, H., & Yousefzadeh, S. (2023). Non-formal disaster education programs for school students in Iran: A qualitative study of the challenges experienced by stakeholders. *International Journal of Disaster Risk Reduction*, *86*, 103531. DOI: 10.1016/j.ijdrr.2023.103531

Sena, M. (2024). IoT-based early warning systems for natural disaster management. *International Journal of Innovation Studies*.

Seneviratne, S. X.-S. (2021). Weather and Climate Extreme Events in a Changing Climate. In V. P. Masson-Delmotte, *In Climate Change 2021: The Physical Science Basis. Contribution of Working Group I to the Sixth Assessment Report of the Intergovernmental Panel on Climate Change,* https://www.ipcc.ch/report/ar6/wg1/chapter/chapter-11/ (pp. pp. 1513–1766). Cambridge University Press. Retrieved from https://www.ipcc.ch/report/ar6/wg1/chapter/chapter-11/

Sevinç, V. (2023). Mapping the forest fire risk zones using artificial intelligence with risk factors data. *Environmental Science and Pollution Research International*, *30*(2), 4721–4732. DOI: 10.1007/s11356-022-22515-w PMID: 35974271

Seyama, E. S., Maocha, M., & Dube, T. (2019). Evaluation of TAMSAT satellite rainfall estimates for southern Africa: A comparative approach. *Physics and Chemistry of the Earth Parts A/B/C*, *112*, 141–153. DOI: 10.1016/j.pce.2019.02.008

Shaamala, A., Yigitcanlar, T., Nili, A., & Nyandega, D. (2024). Algorithmic Green Infrastructure Optimisation: Review of Artificial Intelligence Driven Approaches for Tackling Climate Change. *Sustainable Cities and Society*, *101*, 105182. Advance online publication. DOI: 10.1016/j.scs.2024.105182

Shafik, W. (2024). Artificial Intelligence Models to Prevent Forest Fires. In AI and IoT for Proactive Disaster Management (pp. 78-106). IGI Global. DOI: 10.4018/979-8-3693-3896-4.ch005

Shah, A. M., Liu, G., Nawab, A., Li, H., Xu, D., Yeboah, F. K., Yang, Q., & Zhang, L. (2024). *Sustainability and resilience interface at typical urban green and blue infrastructures: costs, benefits and impacts assessment*. Frontiers in Sustainable Cities., DOI: 10.3389/frsc.2024.1453829

Shahbazi, Z., & Byun, Y. C. (2021). Blockchain-based event detection and trust verification using natural language processing and machine learning. *IEEE Access : Practical Innovations, Open Solutions*, *10*, 5790–5800. DOI: 10.1109/ACCESS.2021.3139586

Shahfahad, N., Naikoo, M. W., Talukdar, S., Das, T., & Rahman, A. (2022). Identification of homogenous rainfall regions with trend analysis using fuzzy logic and clustering approach coupled with advanced trend analysis techniques in Mumbai city. *Urban Climate*, *46*, 101306. DOI: 10.1016/j.uclim.2022.101306

Shah, R., Flemming, J. A., Chinnery, J., & Heaner, G. (2024). Building back better? The role of education sector responses in strengthening or eroding societal resilience during the COVID-19 pandemic. *Frontiers in Education*, *9*, 1089422. Advance online publication. DOI: 10.3389/feduc.2024.1089422

Shaw, J. (2019, January–February). Harvard Magazine. Retrieved May 18, 2024, from Harvard Magazine: https://www.harvardmagazine.com/2019/01/artificial-intelligence-limitations#

Shekar, P. R., & Mathew, A. (2023). Assessing groundwater potential zones and artificial recharge sites in the monsoon-fed Merredu river basin, India: An integrated approach using GIS, AHP, and Fuzzy-AHP. *Groundwater for Sustainable Development*, *23*, 100994. DOI: 10.1016/j.gsd.2023.100994

Sheng, S., & Wang, Y. (2024). Configuration characteristics of Blue-Green spaces for efficient cooling in urban environments. *Sustainable Cities and Society*, *100*, 105040. Advance online publication. DOI: 10.1016/j.scs.2023.105040

Shi, W., Zeng, S., & Aihemaiti, N. A. M. A. I. T. I.SHI. (2024, August 1). W., ZENG, S., & Aihemaiti, N. (2024). Research on the Spatial Heterogeneity of the Impact of Blue-Green Space Within Urban Block on Urban Thermal Environment. *Feng Jing Yuan Lin = Landscape Architecture*, *31*(10), 98–105. Advance online publication. DOI: 10.3724/j.fjyl.202312070549

Shlomi, J., & Battaglia, P. (2020). Graph neural networks in particle physics. *Iopscience.Iop.Org*. DOI: 10.1088/2632-2153/abbf9a

Shortridge, J., Aven, T., & Guikema, S. (2017). Risk assessment under deep uncertainty: A methodological comparison. *Elsevier*. https://www.sciencedirect.com/science/article/pii/S095183201630713X

Siahkoohi, A., Rizzuti, G., & Herrmann, F. J. (2022). Deep Bayesian inference for seismic imaging with tasks. *Library.Seg. Geophysics, 87*(5), 1–109. DOI: 10.1190/geo2021-0666.1

Siddique, A.B., Rahman, M.A., Gazi, E., Khatun, M.N., Khan, A.S., & Al Helal, A.S. (2026). Groundwater Contamination and Human Health Risk Assessment of Drinking Water in Kaliganj Municipality, Bangladesh. H2Open Journal. https://doi.org/DOI: 10.1016/j.htopen.2026.100007

Sidiqui, P., Tariq, M., & Ng, A. W. M. (2022). An Investigation to Identify the Effectiveness of Socioeconomic, Demographic, and Buildings' Characteristics on Surface Urban Heat Island Patterns. *Sustainability (Basel), 14*(5), 2777. Advance online publication. DOI: 10.3390/su14052777

Sigfrids, A., Leikas, J., Salo-Pöntinen, H., & Koskimies, E. (2023). Human-centricity in AI governance: A systemic approach. *Frontiers in Artificial Intelligence, 6*, 976887. Advance online publication. DOI: 10.3389/frai.2023.976887 PMID: 36872934

Simon, D., Goodness, J., Lwasa, S., Oliveira, J. A. P. de, Macedo, L. S. V. de, Kavonic, J., Török, E. H., & Elmqvist, T. (2021).Urban Governance of and for Urban Green and Blue Infrastructure. DOI: 10.1007/978-3-030-67650-6_16

Simonyan, K., & Zisserman, A. (2014). Very deep convolutional networks for large-scale image recognition. arXiv preprint arXiv:1409.1556.

Singh, C. R., & Manoharan, G. (2024). Strengthening Resilience: AI and Machine Learning in Emergency Decision-Making for Natural Disasters. In Internet of Things and AI for Natural Disaster Management and Prediction (pp. 249-278). IGI Global.

Singh, D. (2025, May 26). *Microsoft's Aurora AI Sets New Standard for Environmental Forecasting*. Retrieved from Aadhunik AI: https://aadhunik.ai/blog/ai-weather-forecasting-microsoft-aurora/

Singh, R. (2023). Turkey Earthquake 2023: Reducing Risks in the Indian Context. MP-IDSA. https://idsa.in/idsacomments/turkey-earthquake-2023-rsingh-120423

Singh, A. K., Kumar, P., Ali, R., Al-Ansari, N., Vishwakarma, D. K., Kushwaha, K. S., Panda, K. C., Sagar, A., Mirzania, E., Elbeltagi, A., Kuriqi, A., & Heddam, S. (2022). An Integrated Statistical-Machine Learning Approach for Runoff Prediction. *Sustainability (Basel), 14*(13), 8209. DOI: 10.3390/su14138209

Singh, P. (2024). *Generative AI through the Lens of Institutional Theory*. DOI: 10.2139/ssrn.4953180

Singh, R., Manoharan, G., & Singh, S. (2024). Strategic Deployment of AI and Drones Enhancing Disaster Management in Natural Disasters. In Ouaissa, M., Ouaissa, M., Boulouard, Z., Iwendi, C., & Krichen, M. (Eds.), *AI and IoT for Proactive Disaster Management* (pp. 172–203). IGI Global., DOI: 10.4018/979-8-3693-3896-4.ch009

Singla, M. K., Kar, H. D., & Nijhawan, P. (2019). Rain prediction using fuzzy logic. *International Journal of Engineering and Advanced Technology*, *9*(1), 2796–2799.

Sitar, N., MacLaughlin, M. M., & Doolin, D. M. (2005). Influence of Kinematics on Landslide Mobility and Failure Mode. *Journal of Geotechnical and Geoenvironmental Engineering*, *131*(6), 716–728. DOI: 10.1061/(ASCE)1090-0241(2005)131:6(716)

Sithole, B. (2014). *Municipal Disaster Management in South Africa: Intergovernmental Relations as a Planning Instrument*. Doctoral Thesis. Bloemfontein, South Africa: Central University of Technology, Free State.

Sitnicka, S., Mursalov, M., Mammadov, H., Babaiev, D., & Zhou, Y. (2025). Ethics, Institutions, Infrastructure, and Governance in AI National-Level Readiness: A Hidden Driver of Banking Transformation. *Business Ethics and Leadership*, *9*(3), 290–304. DOI: 10.61093/bel.9(3).290-304.2025

Siyongwana, P. Q., Heijne, D., & Tele, A. (2015). The Vulnerability of Low-Income Communities to Flood Hazards in Missionvale, South Africa. *Journal of Human Ecology (Delhi, India)*, *52*(1-2), 104–115. DOI: 10.1080/09709274.2015.11906935

Socher, R., Lin, C. C., Manning, C., & Ng, A. Y. (2011). Parsing natural scenes and natural language with recursive neural networks. In *Proceedings of the 28th international conference on machine learning (ICML-11)* (pp. 129-136).

Sohn, H., Farrar, C., & Hemez, F. (1999). A review of structural health monitoring literature: 1996–2001NANo ranking found for "IEE Colloquium on Condition Monitoring Machinery, External Structures and Health." *Academia.Edu*, (34), 3–6. DOI: 10.1049/ic:19990185

Soundrapandiyan, R., Manickam, A., Akhloufi, M., Murthy, Y. V. S., Sundaram, R. D. M., & Thirugnanasambandam, S. (2023). An efficient COVID-19 mortality risk prediction model using deep synthetic minority oversampling technique and convolution neural networks. *BioMedInformatics*, *3*(2), 339–368. DOI: 10.3390/biomedinformatics3020023

South African History Online. (2014). The Group Areas Act of 1950. South African History. https://sahistory.org.za/article/group-areas-act-1950

South African Weather Service (SAWS). (2023). *How will the climate change by 2050 in South Africa?* https://www.weathersa.co.za/

South African Weather Services [SAWS]. (2018). *A WeatherSmart Nation Annual Report 2018/2019*. Pretoria, South Africa: SAWS.

Southern African Development Community (SADC). (2016). *SADC Disaster Risk Reduction Strategy and Plan of Action 2016–2030*. Gaborone: SADC Secretariat. Available at: https://www.sadc.int/document/sadc-disaster-risk-reduction-strategy-and-plan-action-2016-2030

Stander, J., Theron, A., & Du Plessis, J. (2011). Storm Surge Early Warning System Guidelines for South Africa. Pretoria: South African Weather Service (SAWS) and Disaster Mitigation for Sustainable Livelihoods Programme (DMISA), University of Cape Town.

Stefanidou, S., & Kappos, A. J. (2021). Fragility-informed selection of bridge retrofit scheme based on performance criteria. *Engineering Structures*. https://www.sciencedirect.com/science/article/pii/S0141029621001267

Stenvall, J., Pinheiro, R., Johanson, J., Pekkola, E., Trondal, J., & Rannisto, P.-H. (2025). Governing uncertainty by anticipating risk: assessing crisis management during COVID-19. In *Edward Elgar Publishing eBooks* (p. 137). Edward Elgar Publishing., DOI: 10.4337/9781035318674.00019

Stephen, F., Yairen, J. C., Martin, L. T., & Barry, O. (2021). *Creating Resilient Futures*. DOI: 10.1007/978-3-030-80791-7

Subandi, A., Alim, S., Haryanti, F., & Prabandari, Y. S. (2019). Training on modified model of programme for enhancement of emergency response flood preparedness based on the local wisdom of Jambi community. *Jàambá, 11*(1), 801. DOI: 10.4102/jamba.v11i1.801 PMID: 31534642

Sufi, F., & Alsulami, M. (2025). AI-Driven Global Disaster Intelligence from News Media. *Mathematics, 13*(7), 1083. DOI: 10.3390/math13071083

Sugiyanto, S., Muryani, C., & Ni'Matussyahara, D. (2024). Strengthening student empathy in GeoCapabilities: Digital learning innovations and pedagogical strategies for disaster mitigation. *Contemporary Educational Technology, 16*(3), ep521. DOI: 10.30935/cedtech/14913

Sulaiman, N. A. F., Shaharudin, S. M., Ismail, S., Zainuddin, N. H., Tan, M. L., & Abd Jalil, Y. (2022). Predictive Modelling of Statistical Downscaling Based on Hybrid Machine Learning Model for Daily Rainfall in East-Coast Peninsular Malaysia. *Symmetry, 14*(5), 927. DOI: 10.3390/sym14050927

Suljic, V. (2025). Strategic Leadership in AI-Driven Digital Transformation: Ethical Governance, Innovation Management, and Sustainable Practices for Global Enterprises and SMEs. *Strategic Leadership in AI-Driven Digital Transformation: Ethical Governance, Innovation Management, and Sustainable Practices for Global Enterprises and SMEs.*, *13*(1), 17. DOI: 10.70301/JOUR/SBS-JABR/2025/13/1/2

Sun, L., Li, H., Nagel, J., & Yang, S. (2024). Convergence of AI and urban emergency responses: Emerging pathway toward resilient and equitable communities. *Applied Sciences (Basel, Switzerland)*, *14*(17), 7949. DOI: 10.3390/app14177949

Sun, W., Bocchini, P., & Davison, B. D. (2020). Applications of artificial intelligence for disaster management. *Natural Hazards*, *103*(3), 2631–2689. DOI: 10.1007/s11069-020-04124-3

Szabo, L. (2024, March 22). *Understanding Nvidia's Earth-2: A New Age of Climate Forecasting*. Retrieved from NowadAIs: https://www.nowadais.com/nvidias-earth-2-new-age-of-climate-forecasting/

Tadokoro, S. (2020). Disaster Response Robot. In Baillieul, J., & Samad, T. (Eds.), *Encyclopedia of Systems and Control.* Springer., DOI: 10.1007/978-1-4471-5102-9_181-2

Tan, J., Kuang, W., Yu, D., Zhou, W., Li, X., Liu, Q., & He, L. (2024). Investigating the cooling effects of land cover and landscape patterns surrounding rivers: Insights from the subtropical city of Changsha, China. *Urban Climate*, *55*, 101975. Advance online publication. DOI: 10.1016/j.uclim.2024.101975

Tanzi, T. J., Chandra, M., Isnard, J., Camara, D., Sebastien, O., & Harivelo, F. (2016). Towards" drone-borne" disaster management: future application scenarios. In XXIII ISPRS Congress, Commission VIII (Volume III-8) (Vol. 3, pp. 181-189). Copernicus GmbH.

Tariq, M. U. (2024). AI and IoT in Flood Forecasting and Mitigation: A Comprehensive Approach. In AI and IoT for Proactive Disaster Management (pp. 26-60). IGI Global. DOI: 10.4018/979-8-3693-3896-4.ch003

Taylor, A. (2025). *South Africa needs R250 billion just to meet basic climate adaptation needs over the next decade.* Conversations., DOI: 10.64628/AAJ.h5tngphmn

Team, C. C. (2025, July 5). *How Artificial Intelligence Is Transforming Weather Forecasting*. Retrieved from Climate Cosmos: https://climatecosmos.com/climate-science/how-artificial-intelligence-is-transforming-weather-forecasting/

Team, G. (2025). *AI-Enhanced Weather Forecasting: Advanced Prediction Models*. Retrieved from GPAI: https://blog.gpai.app/solver/ai-enhanced-weather-forecasting-advanced-prediction#:~:text=AI%20algorithms%20can%20analyze%20meteorological%20data%20with%20unprecedented,leading%20to%20more%20reliable%20and%20timely%20weather%20predictions

Terblanche, T., De Sousa, L. O., & Van Niekerk, D. (2022). Disaster Resilience Framework Indicators for a City's Disaster Resilience Planning Strategy. *Jàambá*, *14*(1), 1264. DOI: 10.4102/jamba.v14i1.1264 PMID: 35747366

Thebe, T. P. (2016). Community participation: Reality and rhetoric in the development and implementation of the Integrated Development Plan (IDP) within municipalities of South Africa. *Journal of Public Administration*, *51*(4), 712–723. https://hdl.handle.net/10520/EJC-9903b8a5b

Tiggeloven, T., Pfeiffer, S., Matanó, A., van den Homberg, M., Thalheimer, L., Reichstein, M., & Torresan, S. (2025). The role of artificial intelligence for early warning systems: Status, applicability, guardrails, and ways forward. *iScience*, *28*(11), 113689. Advance online publication. DOI: 10.1016/j.isci.2025.113689 PMID: 41280665

Tim, Y., & Leidner, D. E. (2023). Digital Resilience: A Conceptual Framework for Information Systems Research. *Journal of the Association for Information Systems*, *24*(5), 1184–1198. DOI: 10.17705/1jais.00842

Tirivangasi, H. M. (2018). Regional disaster risk management strategies for food security: Probing Southern African Development Community channels for influencing national policy [Review of *Regional disaster risk management strategies for food security: Probing Southern African Development Community channels for influencing national policy*]. *Jàmbá Journal of Disaster Risk Studies*, *10*(1). AOSIS. DOI: 10.4102/jamba.v10i1.468

Tiwari, M., Lodorfos, G., McClelland, Z., & Nair, S. (2025). Unraveling the complexities associated with leadership during times of supply chain crisis: A study on the healthcare sector. *International Studies of Management & Organization*, *1*(4), 442–475. Advance online publication. DOI: 10.1080/00208825.2025.2465002

Tokognon, C., & Gao, B. (2017). Structural health monitoring framework based on Internet of Things: A survey. *IEEE Internet of Things Journal*. https://ieeexplore.ieee.org/abstract/document/7842584/

Tora, T. T., Andarge, L. T., Abebe, A. T., Utallo, A. U., Meshesha, D. T., & Wubie, A. F. (2025). An expert-based assessment of early warning systems effectiveness in South Ethiopia Regional State. *Discover Sustainability*, *6*(1), 188. Advance online publication. DOI: 10.1007/s43621-025-00943-1

Tounsi, A., & Temimi, M. (2023). A systematic review of natural language processing applications for hydrometeorological hazards assessment. *Natural Hazards*, *116*(3), 2819–2870. DOI: 10.1007/s11069-023-05842-0 PMID: 36776702

Trist, E. L., & Bamforth, K. W. (1951). Some social and psychological consequences of the longwall method of coal-getting. *Human Relations*, *4*(1), 3–38. DOI: 10.1177/001872675100400101

Trogrlić, R. Š., van den Homberg, M., Budimir, M., McQuistan, C., Sneddon, A., & Golding, B. (2022). *Early Warning Systems and Their Role in Disaster Risk Reduction.*, DOI: 10.1007/978-3-030-98989-7_2

Tsheola, J., Sebola, M. P., & Mamabolo, M. A. (2023). South Africa's "Triple Crisis of Governance" and Societal Leadership Vacuum. *Insights into Regional Development*, *5*(4), 23–35. DOI: 10.9770/IRD.2023.5.4(2)

UN. (2025). The world's cities in 2025. Department of Economic and Social Affairs. https://www.un.org/development/desa/pd/sites/www.un.org.development.desa.pd/files/undesa_pd_2025_data-booklet_world_cities_in_2025.pdf

UNDRR & WMO. (2025). Global status of multi-hazard early warning systems. Geneva, Switzerland. https://www.undrr.org/publication/documents-and-publications/global-status-multi-hazard-early-warning-systems-2025

UNISDR (UN-International Strategy for Disaster Reduction). (2009). *UNISDR terminology on disaster risk reduction.* UNISDR.

United Nations Office for Disaster Risk Reduction (UNDRR). (2015). *Sendai Framework for Disaster Risk Reduction 2015–2030.* Geneva: United Nations. Available at: https://www.undrr.org/publication/sendai-framework-disaster-risk-reduction-2015-2030

United Nations Office for Disaster Risk Reduction (UNDRR). (2018). *Multi-Hazard Early Warning Systems: A Checklist.* Geneva: UNDRR. Available at: https://www.undrr.org/publication/multi-hazard-early-warning-systems-checklist

United Nations Office for Disaster Risk Reduction. (2015). *Sendai Framework for Disaster Risk Reduction 2015–2030.* UNDRR. https://www.undrr.org/publication/sendai-framework-disaster-risk-reduction-2015-2030

United Nations Office for Disaster Risk Reduction. (2015). *Sendai framework for disaster risk reduction 2015–2030*. United Nations. https://www.undrr.org/publication/sendai-framework-disaster-risk-reduction-2015-2030

United Nations Office for Disaster Risk Reduction. (2022). *Global assessment report on disaster risk reduction 2022: Our world at risk*. United Nations. https://www.undrr.org/gar

United Nations Office for Disaster Risk Reduction. (2023). *Early warnings for all: Executive action plan 2023–2027*. United Nations. https://www.preventionweb.net/publication/early-warnings-all-executive-action-plan-2023-2027

United Nations Office for Disaster Risk Reduction. (n,d,). Disaster. https://www.undrr.org/terminology/disaster

United Nations. (1999). Activities of the International Decade for Natural Disaster Reduction; Report of the General Secretary. UN. https://www.un.org/esa/documents/ecosoc/docs/1999/e1999-80.htm

United Nations. (2022). *Early Warnings for All: The UN Global Initiative for Early Warning Systems*. New York: United Nations. Available at: https://www.un.org/en/climatechange/early-warnings-for-all

Urban heat island. (2022). DOI: 10.1016/B978-0-323-85539-6.00009-3

Valls, G. T. (2025, February 24). Artificial Intelligence for Modeling and Understanding Extreme Weather and Climate Events. *Nature Communications*, *16*(1), 1919. Advance online publication. https://www.nature.com/articles/s41467-025-56573-8. DOI: 10.1038/s41467-025-56573-8 PMID: 39994190

Van Niekerk, D. (2014). A critical analysis of the South African Disaster Management Act and Policy Framework. *Disasters*, *38*(4), 858–877. DOI: 10.1111/disa.12081 PMID: 25196341

Van Wyk De Vries, M. (2025). All hazards are multihazards, few of them are natural. *Npj Natural Hazards*, *2*(1), 18. Advance online publication. DOI: 10.1038/s44304-025-00071-w

Van, S. P., Le, H. M., Thanh, D. V., Dang, T. D., Loc, H. H., & Anh, D. T. (2020). Deep learning convolutional neural network in rainfall–runoff modelling. *Journal of Hydroinformatics*, *22*(3), 541–561. DOI: 10.2166/hydro.2020.095

Vermiglio, C., Noto, G., Rodríguez Bolívar, M. P., & Zarone, V. (2022). Disaster management and emerging technologies: A performance-based perspective. *Meditari Accountancy Research*, *30*(4), 1093–1117. DOI: 10.1108/MEDAR-02-2021-1206

Verschuur, J., Murgatroyd, A., Vittis, Y., Mosnier, A., Obersteiner, M., Godfray, H. C. J., & Hall, J. W. (2024). The impacts of polycrises on global grain availability and prices. *Research Square (Research Square)*. DOI: 10.21203/rs.3.rs-3969801/v1

Vujovic, S., Haddad, B., Karaky, H., Sebaibi, N., & Boutouil, M. (2021). *Urban Heat Island: Causes*. Consequences, and Mitigation Measures with Emphasis on Reflective and Permeable Pavements., DOI: 10.3390/civileng2020026

Wachinger, G., Renn, O., Begg, C., & Kuhlicke, C. (2013). The risk perception paradox—Implications for governance and communication of natural hazards. *Risk Analysis*, *33*(6), 1049–1065. DOI: 10.1111/j.1539-6924.2012.01942.x PMID: 23278120

Wagner, J., Rodríguez-Mega, E., & Sengupta, S. (2024, May 18). La crisis del agua empeora en Ciudad de México. New York times. https://www.nytimes.com/es/2024/05/18/espanol/mexico-agua-crisis.html

Walker, T. (2024, March 19). *Climate Pioneers: 3 Startups Harnessing NVIDIA's AI and Earth-2 Platforms*. Retrieved from NVIDIA Corporation: https://blogs.nvidia.com/blog/climate-startups-ai-earth-2/

Walters, L., Hall, L. A., & Grigsby, J. D. (2023). Ready or Not, It Will Come: Preparing Accountancy for the Next Disruption. *Journal of Applied Business and Economics*, *25*(3). Advance online publication. DOI: 10.33423/jabe.v25i3.6212

Wang, S., & Sankaran, S. (2023). Self-adaptive physics-informed neural networks. *Journal of Computational Physics*. https://www.sciencedirect.com/science/article/pii/S0021999122007859

Wang, X., Zhu, Y., Liang, L., Chao, Y., Li, J., & Wang, Y. (2026). Study on the propagation characteristics and triggering thresholds of meteorological, hydrological and agricultural drought in the Mu Us sandy land. *Climate Services*, *42*, 100648. DOI: 10.1016/j.cliser.2026.100648

Wang, Y., Song, H., Zhu, W., & Wang, Y. (2023). The Cooling Effects of Landscape Configurations of Green–Blue Spaces in Urban Waterfront Community. *Atmosphere (Basel)*, *14*(5), 833. Advance online publication. DOI: 10.3390/atmos14050833

Wang, Z., & Ye, X. (2018). Social media analytics for natural disaster management. *International Journal of Geographical Information Science*, *32*(1), 49–72. DOI: 10.1080/13658816.2017.1367003

Waqas, M. H. (2025). Natural Hazards Research, Volume 5, Issue 2. *Artificial Intelligence and Numerical Weather Prediction Models: A Technical Survey*, 306-320.

Webber, M. K., & Samaras, C. (2022). A review of decision making under deep uncertainty applications using green infrastructure for flood management. *Earth's Future, 10*(7), e2021EF002322. Advance online publication. DOI: 10.1029/2021EF002322

Weber, N., & Bieroza, M. (2026). The impact of data quality and outlier detection in high-frequency water quality data on water management and process understanding. *Journal of Hydrology (Amsterdam), 670*, 135171. DOI: 10.1016/j.jhydrol.2026.135171

Wen-Bing, J. (2024). Implementing advanced techniques for urban mountain torrent surveillance and early warning using rainfall predictive analysis. *Urban Climate, 53*, 101782. DOI: 10.1016/j.uclim.2023.101782

Wen, H., Yan, F., Huang, J., & Li, Y. (2025). Interpretable machine learning models and decision-making mechanisms for landslide hazard assessment under different rainfall conditions. *Expert Systems with Applications, 270*, 126582. DOI: 10.1016/j.eswa.2025.126582

West, D. M. (2018). *The future of work: Robots, AI, and automation*. Brookings Institution Press. DOI: 10.5040/9780815751878

Wibowo, A., Amri, I., Surahmat, A., & Rusdah, R. (2025). Leveraging artificial intelligence in disaster management: A comprehensive bibliometric review. *Jàambá, 17*(1), 1776. DOI: 10.4102/jamba.v17i1.1776 PMID: 40357012

Wibowo, Y. A., Septiningrum, U. A. D., Dewi, R. P., & Ronggowulan, L. (2024). Disaster risk reduction in elementary schools in flood-prone areas (Case: Disaster risk reduction in Grogol flood-prone areas, Indonesia). *IOP Conference Series. Earth and Environmental Science, 1314*(1), 012056. DOI: 10.1088/1755-1315/1314/1/012056

Wilhite, D. A. (2000). Chapter 35 Preparing for Drought: A Methodology" (2000). Drought Mitigation Center Faculty Publications. 72. https://digitalcommons.unl.edu/droughtfacpub/72

Wilhite, D. A., Botterill, L., & Monnik, K. (2005a). National Drought Policy: Lessons Learned from Australia, South Africa, and the Unit ed States. In: Wilhite, D. (ed.). *Drought and water crises: science, technology, and management issues*, CRC Press, 137-172.

Wilhite, D. A., Hayes, M. J., & Knutson, C. L. (2005b). Drought preparedness planning: building institutional capacity. In: Wilhite, D. (ed.), 2005. *Drought and water crises: science, technology, and management issues*, CRC Press, 93-121.

Wisner, B., Blaikie, P., Cannon, T., & Davis, I. (2004). *At risk: Natural hazards, people's vulnerability, and disasters* (2nd ed.). Routledge.

Witesman, E. (2021). *Public Value Governance: A Framework.*, DOI: 10.1007/978-3-030-29980-4_84

WMO. (2025, August 06). Devastating floods highlight need and challenges for warnings. https://wmo.int/media/news/devastating-floods-highlight-need-and-challenges-warnings?utm_source=miragenews&utm_medium=miragenews&utm_campaign=news

Wong, N. H., Tan, C. L., Kolokotsa, D., & Takebayashi, H. (2021).Greenery as a mitigation and adaptation strategy to urban heat. DOI: 10.1038/s43017-020-00129-5

World Meteorological Organization & United Nations Office for Disaster Risk Reduction. (2023). *Early warnings for all: Executive action plan 2023–2027*. Geneva: United Nations. https://digitallibrary.un.org/record/4001006

World Meteorological Organization (WMO). (2018). *Guidelines on Multi-Hazard Impact-Based Forecast and Warning Services.* Geneva: WMO. Available at: https://library.wmo.int/idurl/4/55733

World Meteorological Organization. (2020). *Multi-hazard early warning systems: A checklist.* WMO. https://community.wmo.int/sites/default/files/EWS_Checklist_0.pdf

World Meteorological Organization. (2023). *Global status of multi-hazard early warning systems: Target G of the Sendai Framework for Disaster Risk Reduction.* World Meteorological Organization. https://library.wmo.int/idurl/4/68696

Wtodarczyk-Makuta, M., Tchorzewska-Cieslak, B., & Rak, J. R. (2025). Risk assessment method for pharmaceutics removal in treated wastewater-challenges of new Directive (EU) on urban wastewater treatment. *The Science of the Total Environment, 996*, 180141. DOI: 10.1016/j.scitotenv.2025.180141

Wu, Q., Huang, Y., Irga, P. J., Kumar, P., Li, W., Wei, S., Shon, H. K., Lei, C., & Zhou, J. L. (2024). Synergistic control of urban heat island and urban pollution island effects using green infrastructure. *Journal of Environmental Management, 370*, 122985. Advance online publication. DOI: 10.1016/j.jenvman.2024.122985 PMID: 39461153

Wu, T., Wang, S., & Xu, Q. (2024). Spatio-Temporal Heterogeneity of the Urban Heat Effect and Its Socio-Ecological Drivers in Yangzhou City, China. *Land (Basel), 13*(9), 1470. Advance online publication. DOI: 10.3390/land13091470

Xie, Q., Ren, L., & Yang, C. (2023). Regulation of water bodies to urban thermal environment: Evidence from Wuhan, China. *Frontiers in Ecology and Evolution, 11*, 983567. Advance online publication. DOI: 10.3389/fevo.2023.983567

Xiong, Y., Shi, Y., Pu, Q., & Liu, N. (2024). More trust or more risk? User acceptance of artificial intelligence virtual assistant. *Human Factors and Ergonomics in Manufacturing*, *34*(3), 190–205. DOI: 10.1002/hfm.21020

Xi, Y., Zhou, L., & Wang, H. (2025). Ambidextrous leadership and innovation: A process perspective and an experimental approach. *Humanities & Social Sciences Communications*, *12*(1), 612. Advance online publication. DOI: 10.1057/s41599-025-04922-9

Xu, C., Huang, Q., Haase, D., Dong, Q., Teng, Y., Su, M., & Yang, Z. (2024). Cooling Effect of Green Spaces on Urban Heat Island in a Chinese Megacity: Increasing Coverage versus Optimizing Spatial Distribution. *Environmental Science & Technology*, *58*(13), 5811–5820. Advance online publication. DOI: 10.1021/acs.est.3c11048 PMID: 38502088

Xu, H., Lu, Y., Chen, Y., & Fan, Y. (2026). A broadly applicable warning response index: Cross-hazard validation for enhanced early warning communication. *Risk Analysis*, *46*(2), e70189. Advance online publication. DOI: 10.1111/risa.70189 PMID: 41657267

Xu, R., Liu, L., Liu, Y., Yi, X., & Qiu, H. (2025). Multi-scenario modelling of urban spatial growth under water resources and aquatic ecological environmental constraints. *Ecological Indicators*, *177*, 113803. DOI: 10.1016/j.ecolind.2025.113803

Yan, L., Mahmud, S., Shen, H., Foutz, N. Z., & Anton, J. (2020, November). MobiRescue: Reinforcement learning based rescue team dispatching in a flooding disaster. In 2020 IEEE 40th International Conference on Distributed Computing Systems (ICDCS) (pp. 111-121). IEEE.

Yang, R. H. (2024). Interpretable Machine Learning for Weather and Climate Prediction: A Review. *Atmospheric Environment, Volume 338, https://doi.org/,* https://www.sciencedirect.com/science/article/abs/pii/S1352231024004722#:~:text=Advanced%20machine%20learning%20models%20have%20recently%20achieved%20high,impede%20user%2.DOI: 10.1016/j.atmosenv.2024.120797

Yang, F., Yousefpour, R., Zhang, Y., & Wang, H. (2023). The assessment of cooling capacity of blue-green spaces in rapidly developing cities: A case study of Tianjin's central urban area. *Sustainable Cities and Society*, *99*, 104918. Advance online publication. DOI: 10.1016/j.scs.2023.104918

Yang, M., Wang, H., & Cao, S.-J. (2022). A global challenge of accurately predicting building energy consumption under urban heat island effect. *Indoor and Built Environment*. Advance online publication. DOI: 10.1177/1420326X221123222

Yang, Y., Li, X., Wang, D., & Liu, Z. (2024). *Optimizing urban ecosystems through green and blue infrastructure: Strategies for sustainability, resilience, and justice*. Applied and Computational Engineering., DOI: 10.54254/2755-2721/72/20240985

Yang, Z., Nguyen, L., Zhu, J., Pan, Z., Li, J., & Jin, F. (2020, December). Coordinating disaster emergency response with heuristic reinforcement learning. In *2020 IEEE/ACM International Conference on Advances in Social Networks Analysis and Mining (ASONAM)* (pp. 565-572). IEEE. DOI: 10.1109/ASONAM49781.2020.9381416

Yang, Z., Xu, C., & Li, L. (2022). Landslide detection based on ResU-Net with transformer and CBAM embedded: Two examples with geologically different environments. *Remote Sensing (Basel)*, *14*(12), 2885. DOI: 10.3390/rs14122885

Yaseen, Z. M., El-Shafie, A., Jaafar, O., Afan, H. A., & Sayl, K. N. (2015). Artificial intelligence based models for stream-flow forecasting: 2000–2015. *Journal of Hydrology (Amsterdam)*, *530*, 829–844. DOI: 10.1016/j.jhydrol.2015.10.038

Yende, N. E. (2025). The Ripple Effect of Violent Service Delivery Protests in South Africa: Causes and Implications. In *IntechOpen eBooks*. IntechOpen. DOI: 10.5772/intechopen.1008853

Yhee, Y., & Koo, C. (2026). Seeing AI as human or machine? Effects of transparency, valence, and readability on review summary helpfulness. *Tourism Management*, *113*, 105305. DOI: 10.1016/j.tourman.2025.105305

Yin, S., & Kaynak, O. (2015). Big data for modern industry: Challenges and trends [point of view]. *Proceedings of the IEEE*, *103*(2), 143–146. DOI: 10.1109/JPROC.2015.2388958

Yin, Y., Li, S. H., Xing, X., Zhou, X., Kang, Y., Hu, Q., & Li, Y. (2024). Cooling Benefits of Urban Tree Canopy: A Systematic Review. *Sustainability (Basel)*, *16*(12), 4955. Advance online publication. DOI: 10.3390/su16124955

Yuan, L., Jiang, H., Shen, H., Shi, L., & Cheng, N. (2023). Sustainable Development of Information Dissemination: A Review of Current Fake News Detection Research and Practice. *Systems*, *11*(9), 458. DOI: 10.3390/systems11090458

Yuan, Q., Yao, G., Liu, W., Weragoda, S. K., Weerasooriya, R., Meng, Y., & Luan, F. (2025). Assessment of harvested rainwater quality and surrogate parameter development for drinking water provision in rural Sri Lanka. *Journal of Environmental Chemical Engineering*, *13*(5), 117632. DOI: 10.1016/j.jece.2025.117632

Yu, B., Xu, C., Chen, F., Wang, N., & Wang, L. (2022). HADeenNet: A hierarchical-attention multi-scale deconvolution network for landslide detection. *International Journal of Applied Earth Observation and Geoinformation, 111*, 102853. DOI: 10.1016/j.jag.2022.102853

Yu, G., & Qi, X. (2004). *Disruption management: framework, models, and applications.* World Scientific. DOI: 10.1142/5632

Yu, Z., Chen, J., Chen, J., Zhan, W., Wang, C., Zhou, S., & Zhu, K. (2024). Enhanced observations from an optimized soil-canopy-photosynthesis and energy flux model revealed evapotranspiration-shading cooling dynamics of urban vegetation during extreme heat. *Remote Sensing of Environment, 305*, 114098. Advance online publication. DOI: 10.1016/j.rse.2024.114098

Yu, Z., Uchiyama, Y., & Sato, M. (2024). *Combined effects of urban blue–green spaces on the thermal environment: a case study of Kobe*. International Journal of Economic Policy Studies., DOI: 10.1007/s42495-024-00140-4

Zadeh, L. A. (1965). Fuzzy sets. *Information and Control, 8*(3), 338–353. DOI: 10.1016/S0019-9958(65)90241-X

Zafar, U., Shah, M. A., Wahid, A., Akhunzada, A., & Arif, S. (2019). Exploring IoT applications for disaster management: identifying key factors and proposing future directions. Recent trends and advances in wireless and IoT-enabled networks, 291-309.

Zahran, B., Ayyoub, B., Abu-Ain, W., Hadi, W., & Al-Hawary, S. (2023). A fuzzy based model for rainfall prediction. *International Journal of Data and Network Science, 7*(1), 97–106. DOI: 10.5267/j.ijdns.2022.12.001

Zenda, M. (2025). Climate change adaptation and mitigation in different livestock production systems and agro-ecological zones in South Africa: A systematic review [Review of *Climate change adaptation and mitigation in different livestock production systems and agro-ecological zones in South Africa: A systematic review*]. *Tropical Animal Health and Production, 57*(8). Springer Science+Business Media. DOI: 10.1007/s11250-025-04660-9

Zepp, H., Gessner, M., Gruenhagen, L., & Bührs, M. (2023).Modeling the cooling effect of urban green spaces: the neglected control variables of 'soil moisture' and 'biotope types.'DOI: 10.1016/j.ufug.2023.128137

Zhang, B., Ren, Y., He, S., Gao, Z., & Li, B. (2025). A review of methods and applications in structural health monitoring (SHM) for bridgesQ1Measurement. *Measurement.* https://www.sciencedirect.com/science/article/pii/S0263224124024606

Zhang, H., Luo, M., Pei, T., Liu, X.-P., Wang, L., Zhang, W., Lin, L., Ge, E., Liu, Z., & Liao, W. (2023). Unequal urban heat burdens impede climate justice and equity goals. DOI: 10.1016/j.xinn.2023.100488

Zhang, D., Han, X., & Deng, C. (2018). Review on the research and practice of deep learning and reinforcement learning in smart grids. *CSEE Journal of Power and Energy Systems*, *4*(3), 362–370. DOI: 10.17775/CSEEJPES.2018.00520

Zhang, H., Dolan, C., Jing, S. M., Uyimleshi, J., & Dodd, P. (2019). Bounce forward: Economic recovery in post-disaster Fukushima. *Sustainability (Basel)*, *11*(23), 6736. DOI: 10.3390/su11236736

Zhang, J.-Y. (2023). *Impact of Urban Heat Island: A Local-Level Urban Climate Phenomenon on Urban Ecology and Human Health*. Advances in Geographical and Environmental Sciences., DOI: 10.1007/978-981-99-3006-7_5

Zhang, W., Zhang, Y., Gu, X., Wu, C., & Han, L. (2022). Application of Soft Computing, Machine Learning, Deep Learning and Optimizations in Geo engineering and Geoscience. In *Application of Soft Computing, Machine Learning, Deep Learning and Optimizations in Geo engineering and Geoscience*. Springer Nature Link., DOI: 10.1007/978-981-16-6835-7

Zhang, Y., Geng, P., Sivaparthipan, C. B., & Muthu, B. A. (2021). Big data and artificial intelligence based early risk warning system of fire hazard for smart cities. *Sustainable Energy Technologies and Assessments*, *45*, 100986. DOI: 10.1016/j.seta.2020.100986

Zhao, X., Hu, S., & Cho, J. (2019). Uncertainty-based decision making using deep reinforcement learning. *Ieeexplore.Ieee.Org*. https://ieeexplore.ieee.org/abstract/document/9011218/

Zhao, L., & Chen, X. (2021). Artificial intelligence applications in disaster prediction and response. *International Journal of Disaster Risk Science*, *12*(3), 421–437.

Zhao, M., Lee, J. K. W., Kjellstrom, T., & Cai, W. (2021). Assessment of the economic impact of heat-related labor productivity loss: A systematic review. *Climatic Change*, *167*(1-2), 22. Advance online publication. DOI: 10.1007/s10584-021-03160-7

Zheng, B., & Chu, V. W. (2025). Multi-hazard early warning systems for agriculture with featural-temporal explanations. *arXiv*. https://arxiv.org/abs/2507.22962

Zheng, X., Li, H., & Shi, Z.-Q. (2023). Hybrid AI-Bayesian-based demand models and fragility estimates for tall buildings against multi-hazard of earthquakes and winds. *Thin-walled Structures*, *52*(1), 248–267. DOI: 10.1016/j.tws.2023.110749

Zhong, M., & Zhang, Q. (2010). Prediction of reservoir-induced earthquake based on fuzzy theory. *Proc. ISNNS*, 101-104.

Zhou, L., Wu, X., Xu, Z., & Fujita, H. (2018). Emergency decision making for natural disasters: An overview. *International Journal of Disaster Risk Reduction*, *27*, 567–576. DOI: 10.1016/j.ijdrr.2017.09.037

Zhou, T., Jia, W., Yan, L., Hong, B., & Wang, K. (2024). Urban park's vertical canopy structure and its varied cooling effect under continuous warming climate. *Urban Climate*, *53*, 101819. Advance online publication. DOI: 10.1016/j.uclim.2024.101819

Zhyvko, M. (2024). Fragmentation of Globalized Economy and Macroeconomic Cost of the War in Ukraine. *Problèmes Économiques*, *3*(61), 15–30. DOI: 10.32983/2222-0712-2024-3-15-30

Zia, A., & Wagner, C. H. (2015). Mainstreaming Early Warning Systems in Development and Planning Processes: Multilevel Implementation of Sendai Framework in Indus and Sahel. *International Journal of Disaster Risk Science*, *6*(2), 189–199. DOI: 10.1007/s13753-015-0048-3

Ziervogel, G., Lennard, C., Midgley, G., New, M., Simpson, N. P., Trisos, C. H., & Zvobgo, L. (2022). Climate change in South Africa: Risks and opportunities for climate-resilient development in the IPCC Sixth Assessment WGII Report. *South African Journal of Science*, *118*(9-10), 1–5. DOI: 10.17159/sajs.2022/14492

Ziervogel, G., & Taylor, A. (2023). A co-produced national climate change risk and vulnerability assessment framework for South Africa. *Frontiers in Climate*, *5*, 1197167. Advance online publication. DOI: 10.3389/fclim.2023.1197167

Zub, C.-I., Stratan, A., Dogariu, A., & Vulcu, C. (2018). Development of Two Types of Buckling Restrained Braces Using Finite Element Modelling. *Seismic Hazard and Risk Assessment*, 373–387. DOI: 10.1007/978-3-319-74724-8_25

About the Contributors

Jaime Santos-Reyes is a lecturer at the National Polytechnic Institute (IPN), Mexico. He obtained a PhD from Heriot-Watt University, Edinburgh, Scotland, UK, in 2001. His main research interests are modelling disaster/safety management systems, early warning systems, diagnosing systems failure, technological and natural hazards risk perceptions, accident and risk analysis, past disaster analysis, critical infrastructure interdependency modelling, human reliability analysis, human stampede disasters, futures scenarios. He edited the following book "Earthquakes: Impact, community vulnerability and resilience" (2009). IntechOpen, Croatia. ISBN: 978-1-78984-667-6. Over 40 peer review papers published in the Web of Science/Clarivate Journals. Invited Editor (2020-2021) – Sustainability journal. Invited Editor (2024, handling a paper on risk communication) – Frontiers in Public Health (WOS/Clarivate), Invited Editor (2025, handling a paper on human stampede disasters) – Frontiers in Public Health (WOS/Clarivate). Over 300 articles have been reviewed of over 40 WOS/Clarivate Journals.

***

Eduard Babulak is accomplished international scholar, researcher, consultant, educator, professional engineer and polyglot, with more than thirty years of experience. He served as Editor, reviewer, Conference Chair, has successfully published. He served as Chair of the IEEE Vancouver Ethics, Professional and Conference Committee. He was Invited Plenary & Keynote Speaker in Tokyo (2023), Oxford (2020 & 2023), University of Cambridge (2005 & 2010), MIT (2005 & 2010), Yokohama National University & University of Electro Communications in Tokyo (2009 – 2010), Kyushu Institute of Technology (2016), Japan, Shanghai Jiao Tong University (2013), Sungkyunkwan University in Korea (2012 – 2014), University of Surrey, Purdue University, Penn State, Graz University of Technology, Austria, Czech Technical University, and other prestigious academic institutions worldwide. His academic and engineering work was recognized internationally by the Engineering Council in UK, the European Federation of Engineers and credited by the Ontario Society of Professional Engineers and APEG in British Columbia in Canada. He was awarded higher postdoctoral degree DOCENT - Doctor of Science (D.Sc.) in the Czech Republic, Ph.D., M.Sc., and High National Certificate (HNC) diplomas in the United Kingdom, as well as, the Diploma Engineer (MSc and B.Sc.) in Electrical Engineering in Slovakia. He is Fellow of the Royal Society RSA, London, UK; Chartered Fellow. Mentor and Member of the ELITE Group of the British Computer Society, London, UK; 2013-2014 Invited Panel Member for National Science Foundation, USA; Expert Consultant for HORIZON 2020 & CORDIS FP6 - FP7 European Commission, Brussellex, Belgium; Mentor and Senior Member of the IEEE and ACM, USA; Nominated Fellow of the Institution of Engineering and Technology, London, UK and Distinguished Member of the ACM, USA; Chartered Member of the IET, London, UK. He serves as the Editor-in-Chief, Associate Editor-in-Chief, Co-Editor, and Guest-Editor. He speaks 16 languages and his biography was cited in the Cambridge Blue Book, Cambridge Index of Biographies, Stanford Who's Who, and number of issues of Who's Who in the World and America.

Ali Biglarifadafan is a Senior Lecturer in Civil Engineering within the Faculty of Engineering at the University of Golestan. He/She earned a Ph.D. in Civil Engineering (Structural Engineering) from the University of Glasgow in 2014, supported by an elite scholarship from the MSRT. His/Her doctoral research was conducted under the supervision of Professors P. Harrison, N. Bicanic, C. Pearce, and Z. Guo, contributing to the field of computational mechanics. This was followed by a postdoctoral research fellowship in Geotechnical Engineering at the University of Warwick. Appointed as an Assistant Professor at Golestan University in 2015 and subsequently promoted to Senior Lecturer, his/her primary research and teaching agenda focuses on sustainable and resilient design methodologies. Dr. Biglari currently leads a research team of doctoral and postgraduate students investigating a Resilient Seismic Design framework for structures and geo-structures, employing integrated traditional and artificial intelligence (AI) techniques. In addition to research and teaching, Dr. Biglari has held significant administrative and leadership roles, including Course Coordinator, Head of Subject, Head of the In

Mary Subaja Christo is associated with the Department of Networking and Communications, School of Computing, SRM Institute of Science and Technology, Kattankulathur, India. Her academic and research work focuses on emerging technologies in the domains of computer networks, distributed systems, and intelligent computing. She has contributed to research in areas such as networking technologies, wireless communication, Internet of Things (IoT), and data-driven computing systems. Dr. Christo is actively involved in teaching, mentoring students, and guiding research initiatives at the undergraduate and postgraduate levels. She has participated in various academic collaborations and research projects that aim to address contemporary challenges in networked systems and digital infrastructure. Her scholarly contributions include research publications in conferences and journals, along with involvement in academic activities such as curriculum development, technical reviews, and collaborative research initiatives. Through her academic and research engagements, she continues to contribute to advancements in networking technologies and computing education.

Sai Puneeth Damarla is an Architect and urban planner currently serving as Assistant Professor at Woxsen University's School of Architecture and Planning, Hyderabad. Holds an MCP from IIT Kharagpur; B.Arch. from NIT Calicut, and Prior research experience as a Project Associate at IIT Kharagpur. Research focuses on urban heat-island mitigation and the performance of blue-green infrastructure in Indian cities. Published in Scopus-indexed proceedings (IOP, Springer, IGI Global) with first-authored work on multi-temporal remote sensing analysis of Land Surface Temperature and Green Infrastructure. Registered Architect (CoA) and licensed planner (AITP). Maintains an active freelance practice in architecture and urban design.

Sreenivasarao Devavarapu is an Assistant Professor in Computer Science and Enginerring at Sreenidhi Insitute of Science and Technology, Hderabad. He is an academician and researcher. He has published various number of research papers and book chapters, books etc. His research area includes Digital Image Processing, Machine Learning, and Deep Learning, Alogortithms and Analysis.

Nayla Fadiya is an information technology professional with research interests in the Internet of Things (IoT) and Early Warning Systems. Her work focuses on the design and implementation of sensor-based systems, real-time data acquisition, and intelligent monitoring solutions to support risk mitigation and decision-making processes. She has been involved in projects that integrate hardware, software, and network technologies to improve system reliability and responsiveness. With a strong commitment to applied research and technological innovation, Nayla continues to contribute to the development of smart systems for environmental monitoring and safety applications.

Amina Jakoet-Salie is the Head of Department and a Senior Lecturer in the Department of Public Management and Leadership in the School of Governmental Studies at the Nelson Mandela University. She completed a Bachelor of Social

Science and Social Science Honours at the University of Cape Town and furthered her studies and received her Masters (Cum Laude) and a PhD in Public Administration at the Nelson Mandela University. She received the National Institute of Humanities and Social Sciences (NIHSS) Doctoral Fellowship in 2016.She has published manuscripts in local and international peer-reviewed academic journals. She has supervised several Honours and Masters students to completion as well as the lead supervisor for Doctoral Studies. Her research niche areas include co-production, good governance and ethical leadership, disaster management, digital governance and curriculum development.

Ismatulla Khayrullayev is affiliated with the Department of Information Technology and Exact Sciences at Termez University of Economics and Service, Termez, Uzbekistan. His academic interests encompass information technology, applied computing, and quantitative analytical methods, with an emphasis on data-driven problem solving and the application of exact sciences in interdisciplinary research. He is actively involved in teaching and scholarly activities, contributing to the advancement of technical education and research within the university.

Geetha Manoharan is currently working in Telangana as an assistant professor at SR University. She is the university-level PhD program coordinator and has also been given the additional responsibility of In Charge Director of Publications and Patents under the Research Division at SR University. Under her tutelage, students are inspired to reach their full potential in all areas of their education and beyond through experiential learning. It creates an atmosphere conducive to the growth of students into independent thinkers and avid readers. She has more than ten years of experience across the board in the business world, academia, and the academy. She has a keen interest in the study of organizational behavior and management. More than forty articles and books have been published in scholarly venues such as UGC-refereed, SCOPUS, Web of Science, and Springer. Over the past six-plus years, she has participated in varied research and student exchange programs at both the national and international levels. A total of five of her collaborative innovations in this area have already been published and patented. Emotional intelligence, self-efficacy, and work-life balance are among her specialties. She organizes programs for academic organizations. She belongs to several professional organizations, including the CMA and the CPC. The TIPSGLOBAL Institute of Coimbatore has recognized her twice (in 2017 and 2018) for her outstanding academic performance.

Tshepo Maota is a leading expert in IT auditing, disruption management, and governance frameworks, combining his PhD research, industry expertise, and professional certifications (CISA, COBIT) to drive innovative solutions in

IT governance and risk management. His contributions continue to shape audit methodologies, cybersecurity frameworks, and IT governance strategies that enhance resilience, compliance, and organizational performance. His commitment to academic research and industry best practices positions him as a key thought leader in technology-driven audit methodologies and digital transformation strategies. With a vision for enhancing IT governance, integrating AI into risk assessments, and fortifying cybersecurity frameworks, Dr. Maota remains dedicated to advancing the body of knowledge in IT auditing and governance while providing strategic leadership in the evolving digital landscape.

Daniel Martinez is a distinguished academic with a Ph.D. in Advanced Technology from CICATA-Leg at the National Polytechnic Institute (Mexico), as well as a postdoctoral fellowship in Applied Mathematics from the Faculty of Physical and Mathematical Sciences at BUAP (Mexico). He has enriched scientific knowledge with numerous published articles and book chapters. Currently, he serves as a full-time professor-researcher in the Industrial Engineering and Administration program at UANL, contributing to the advancement of higher education.

Barno Matchanova is affiliated with the Department of National Idea and Philosophy at Urgench State Pedagogical Institute, Urgench, Uzbekistan. Her academic interests focus on philosophy, national ideology, ethics, and socio-cultural studies, with particular emphasis on value education and the philosophical foundations of social development. She is actively involved in teaching and academic research, contributing to the advancement of philosophical inquiry and civic education in higher education.

Mosekama Osia Mokhele is an emerging scholar and lecturer within the Department of Public Management and Leadership at Nelson Mandela University. Furthermore, he serves as a Research Associate at the Raymond Mhlaba Centre for Governance and Leadership, also housed at NMU. His robust academic foundation is evident in his dual master's qualifications: an MA in Public Management and Policy Analysis, earned from the International University of Japan, and a master's degree in governance and political Transformation, obtained from the University of the Free State. The scholar is currently pursuing a joint doctoral degree in Disaster Management, jointly supervised by the University of the Free State in South Africa and Vrije Universiteit Amsterdam in the Netherlands. Mokhele's central research agenda is sharply focused on critical areas within contemporary governance and crisis response. Specifically, his work focuses on decentralisation in disaster management. A second key area is the analysis of legislation and policies governing disaster response, often assessing gaps between policy intent and practical implementation, particularly in terms of resource allocation and rapid deployment protocols. Perhaps the most impactful dimension of his research is its focus on disaster inclusivity and participation, which cuts across essential operational dynamics. This focus encompasses the complexities of central-local coordination, robust stakeholder collaboration, and subjective disaster preparedness among marginalised populations, which transcend quantitative metrics to capture the lived experiences of risk. He has published articles in reputable national and international journals and presented papers at national and international conferences.

Yanelisa Ndatshe is a postgraduate student in the Department of Public Management and Leadership at Nelson Mandela University (NMU), pursuing a Master of Arts in Public Administration. Building on a solid foundation, Yanelisa holds both a Bachelor of Arts and a BA Honours in Public Administration from NMU. His scholarly focus centres on three interrelated domains: disaster management, infrastructure resilience, and institutional knowledge. Beyond academic accolades, Yanelisa has actively contributed to community development and educational initiatives. Additionally, Yanelisa spearheaded the "Sustainable Futures" programme. As an emerging scholar, Yanelisa has already disseminated their research on institutional knowledge loss in a peer-reviewed international journal, "International Journal of Research in Business and Social Science." The article, titled "The effects of employee turnover on the loss of organisational knowledge in South African municipalities: Balancing rhetoric with actual practice," has been cited 5 times. Looking ahead, Yanelisa plans to expand his work by developing a framework for municipal disaster centres in South Africa, integrating lessons learned from his recent work on the Multi-Hazard Early Warning System with best-practice guidelines from the United Nations Office for Disaster Risk Reduction (UNDRR). By bridging theory and practice, he aims to strengthen the capacity and resilience of the forgotten assets: critical infrastructure and vulnerable communities across South Africa to withstand disasters that are likely to occur in the near future.

Marcos Angel Olvera received the B.Sc. (2002) and M.Sc. (2005) degrees in Electrical Engineering, both with honors, and the Ph.D. degree in Engineering (2009), all from Universidad Nacional Autónoma de México (UNAM). He is currently a Professor at Universidad Autónoma de la Ciudad de México (UACM). His research interests include nonlinear control, fractional-order controllers, optimization, and dynamic modeling. He has been a Board member of the Association of Automatic Control of Mexico (AMCA).

Ravikumar R N is an experienced academician with over 12+ years of teaching and research expertise across institutions in India, the UAE, and the Maldives. He currently serves as an Assistant Professor at Marwadi University, Rajkot, specializing in Machine Learning, Recommender Systems, and AI-driven applications in healthcare and security. He holds a Ph.D. in Machine Learning and Recommender Systems from Amity University, Jaipur, a Master of Engineering in CSE from VMRF University, Salem, and a Bachelor of Engineering in CSE from Anna University. He also has a MiniMBA from the Swiss eLearning Institute. Dr. Ravikumar has authored more than 8+ multiple SCIE/Scopus-indexed journal articles, 25+ IEEE/ Springer conference papers, and co-authored a book on Artificial Intelligence. He has also published 9 patents and contributed 35+ book chapters in emerging domains such as Federated Learning, Blockchain, and Digital Twins for IGI Global, Wiley and Springer publisher. He actively mentors students through initiatives like the Research Activities Club (RAC) and Student Research Internship Training (SRIT), promoting innovation and research culture.

V. Muthu Ruben is a distinguished Associate Professor at the School of Law, Christ University, situated in Bengaluru, Karnataka, India. Driven by an extensive scholarly foundation and a steadfast dedication to legal education, Dr. Ruben makes substantial contributions to both the academic community and the domain of law. His proficiency transcends numerous spheres within the realm of law, and he significantly influences the educational journeys of students enrolled at Christ University. Motivated by an ardent interest in legal research, he actively participates in scholarly endeavors, thereby making significant contributions to the ongoing legal dialogue. In addition to his scholarly responsibilities, Dr. V. Muthu Ruben is renowned for his commitment to cultivating an atmosphere that encourages intellectual engagement. His mentorship and counsel motivate pupils to explore into the complexities of the law, fostering the development of critical thinking skills and a profound comprehension of legal principles. In his role as an Associate Professor, Dr. Ruben provides the School of Law with an abundance of expertise, experience, and a forward-thinking perspective, thereby enhancing the institution's dedication to providing superior legal education.

Aarthi S is a dedicated Assistant Professor in Computer Science with over six years of teaching experience. She is currently pursuing her Ph.D. at Marwadi University, Gujarat, and holds an M.Phil. and MCA from Bharathiar University, Tamil Nadu. Throughout her academic career, Aarthi has specialized in programming languages such as C++, Java, and C, and contributed to research in recommender systems and AI-related data privacy. She has published journal articles, contributed to book chapters, and actively participated in technical workshops. Known for her student mentorship and engagement in academic events, Aarthi remains committed to fostering innovation and excellence in computer science education.

Binastya Anggara Sekti has been a lecturer at the Faculty of Computer Science, Esa Unggul University, Jakarta, since 2016. He is actively involved in the Artificial Intelligence and Information Security Study Group. He holds a Bachelor's degree in Informatics Engineering from the Bandung Institute of Technology and a Master's degree in Information Systems Management from Trisakti University, Jakarta. Prior to entering academia, he accumulated over 25 years of professional experience as an IT practitioner in various companies across Jakarta. His work focused on developing data centers for internet service providers and delivering IT solutions for the manufacturing and oil & gas sectors. His research interests include artificial intelligence, cybersecurity, and the development of IT infrastructure.

Ravinder Singh, boasting 35 years of unwavering service in the Indian Army, has extensive military expertise across diverse terrains. Beyond his military acumen, he is a scholar and practitioner in Disaster Risk Management, holding a postgraduate diploma in the field alongside certifications from the esteemed Gujarat Institute of Disaster Management. His academic portfolio further includes an MBA specializing in HR and a postgraduate diploma in Mass Communication and Journalism. Particularly noteworthy are his seminal contributions to the discourse on disaster management, focusing notably on the integration of artificial intelligence within the field. His commitment to education and training shines through his development of a comprehensive training module and numerous trainings for the Gujarat Institute of Disaster Management. As a prolific writer, his scholarly articles delve into critical topics such as Ladakh's susceptibility to flash floods and strategies for earthquake risk mitigation. Notably, his impactful training sessions on disaster risk management, underscore his efforts in imparting invaluable insights and skills to a diverse audience, which includes professors under the aegis of UGC, government officials, biological incident first responders, officials and professionals from PSU, banks, companies, and industries, students of schools and colleges (Engineering, Business Administration, Mass Communication, and Journalism), NCC, and NSS, further cementing his commitment in the realm of disaster management. His adeptness in curating courses is evident in his conceptualizing and redesign of the existing disaster management courses into a 'Disaster Risk Management' training curriculum for B. Tech. and MBA students. Through meticulous identification of deficiencies and gaps in the existing structure, he has formulated a blended approach aimed at achieving the course objectives effectively.

Samrath Singh is a student currently pursuing a Bachelor's degree in Computer Science and Engineering. With a fervent passion for innovation and technology, Samrath has actively participated in several competitions to showcase his talents and contribute to societal betterment. One of his notable achievements is his participation in the esteemed IEEE Yesist Junior track, an international event that celebrates innovation and creativity. Here, he demonstrated his compassion and technical acumen by developing Ultrasonic Obstacle Detecting Goggles tailored specifically for the visually impaired. His groundbreaking project not only showcased his technical skills but also highlighted his commitment to making a positive impact on the lives of others. For this innovative solution, Samrath received honorable recognition, earning acclaim for his dedication and ingenuity. In addition to his international endeavors, Samrath has also made significant contributions at the local level. He participated in the Inter APS Science Competition, where his innovative spirit shone once again. His project, a wireless firecracker igniter, captured the attention of the judges and earned him the prestigious first prize. His ability to combine technical expertise with creative problem-solving underscores his potential to effect meaningful change in the world. Through his participation in various competitions and projects, Samrath Singh exemplifies the qualities of a forward-thinking innovator, driven by a desire to make a positive difference in society. As he continues his academic journey in Computer Science and Engineering, Samrath remains committed to leveraging technology for the betterment of humanity.

Ranjit Singha is a Doctorate Research Fellow at Christ (Deemed to be University) and holds a prestigious membership in the American Psychological Association (APA). With a strong background in Research and Development, he has significantly contributed to various fields such as Mindfulness, Addiction Psychology, Women Empowerment, UN Sustainable Development Goals, and Data Science. With over 15 years of experience in Administration, Teaching, and Research, both in Industry and Higher Education Institutions (HEI), Mr Ranjit has established himself as a seasoned professional. Mr Ranjit is dedicated to research and teaching endeavours, primarily focusing on mindfulness and compassion-based interventions. His work in these areas aims to promote well-being and foster positive change in individuals and communities.

Surjit Singha is an Associate Professor, PhD Research Guide, and researcher in Commerce and Management at Kristu Jayanti University. His expertise spans CSR, sustainability, entrepreneurship, policy studies, and emerging interdisciplinary domains. As of 2 January 2026, he has published 121 Scopus-indexed research papers, ranking him No. 1 across Kristu Jayanti University based on document count. He serves as an editor and reviewer for international journals and actively collaborates across disciplines. His research emphasizes evidence-based insights with strong practical and societal impact, and he is open to international research collaboration and academic mobility. With an impressive academic portfolio, Dr. Singha has published over 160+ articles, book chapters, and textbooks, including

100+ SCOPUS-indexed publications. He also holds 2+ US Copyrights and two Patent, demonstrating his dedication to innovation and originality. Additionally, he has successfully completed two fully funded minor research projects from Kristu Jayanti (Deemed to be University). As a research mentor, Dr. Singha has guided numerous students and young scholars toward academic excellence. His academic qualifications include a PhD, M.Phil, and MBA (Distinction) from Christ University, M.Com, BBM from Kristu Jayanti College (Bangalore University), and PGDHE (Higher Education) from IGNOU (Central University). He is also a UGC-NET (Commerce) qualifier and serves as an Examiner for International Baccalaureate (IB).

Anusha Thakur is an "INDEPENDENT RESEARCHER" with nearly "50 SOLO RESEARCH PAPERS" in the World's reputed books and journals. Anusha's exemplary contributions without the conventional Ph.D. degree, and in a span of nearly three years, have earned her the title of the "BEST WOMAN RESEARCHER OF THE YEAR" - (Category - Below 35 years) during the 4th International Inspirational Women Awards - (IIWA), 2023, held in New Delhi, in June 2023. Anusha has been conferred the prestigious "Shiksha Bharati Award", in October 2023 by the "Indian Achievers Forum" in recognition of her extraordinary professional achievements and contribution towards nation-building. Anusha has also been awarded the "YOUNG TURKS AWARD - 2024" at the 3AI GCC-X Summit in July 2024. One of her published Chapter titled "Impact of Mood of the Millennial Customers on Purchase of Apparels Online", published by IGI Global, USA was chosen to be included as a "Reprinted Chapter" in 2021 for the book "Research Anthology on E-Commerce Adoption, Models, and Applications for Modern Business (3 Volumes)" by the "Information Resources Management Association (USA). One of her published Chapters include "Market for Plant-Based Meat Alternatives" for the Book titled "Environmental, Health, and Business Opportunities in the New Meat Alternatives Market". The Book won the #BEST BOOK IN THE WORLD in the Prestigious 25th GOURMAND AWARDS, in 2020. GOURMAND AWARDS is considered the OSCARS OF FOOD AND WINE PUBLICATIONS in the world. Also, this book won "NATIONAL WINNER IN AUSTRALIA" in 2019, in the category of "Vegetarian Writing". Anusha serves as an "EDITORIAL REVIEW BOARD MEMBER" for Journals "IJEIS, IJAMTR, IJSMI, and IJAIBM (IGI Global, USA)". She is also a Peer Reviewer for several publications with renowned publishers such as IGI Global, USA, and Common Ground Research Networks, USA. Anusha has pursued B.Tech in "Electronics and Instrumentation Technology" from Tumkur, Karnataka, and MBA in "The Department of Economics and International Business" from the University of Petroleum & Energy Studies, (UPES) Dehradun. She has work experience of approximately 4+ years in the

Market Research Industry as a Senior Research Analyst with extensive knowledge of market research in diversified industries.

Khayrulla Urozboev is associated with the Department of International Scientific Journals and Ratings at Alfraganus University, Tashkent, Uzbekistan. His academic and professional work focuses on scholarly publishing, research evaluation, and the advancement of international scientific visibility for academic institutions. He is actively involved in activities related to journal indexing, academic rankings, and the promotion of high-quality research publications in global databases. His interests include research management, bibliometric analysis, and strategies for improving the impact and quality of scientific journals. Through his work, he contributes to strengthening research standards and enhancing the international recognition of academic research produced by universities and scholars.

Index

A

B

C

D

E

F

G

H

I

L

www.ingramcontent.com/pod-product-compliance
Lightning Source LLC
LaVergne TN
LVHW062339300926
849161LV00007B/1748